CAMBRIDGE MONOGRAPHS ON
MECHANICS AND APPLIED MATHEMATICS

General Editors

G. K. BATCHELOR, F. R. S.
Department of Applied Mathematics, University of Cambridge

C. WUNSCH
Department of Earth, Atmospheric, and Planetary Sciences,
Massachusetts Institute of Technology

J. RICE
Division of Applied Sciences, Harvard University

DYNAMIC FRACTURE MECHANICS

Dynamic Fracture Mechanics

L. B. FREUND

Brown University

CAMBRIDGE UNIVERSITY PRESS

PUBLISHED BY THE PRESS SYNDICATE OF THE UNIVERSITY OF CAMBRIDGE
The Pitt Building, Trumpington Street, Cambridge CB2 1RP

CAMBRIDGE UNIVERSITY PRESS
The Edinburgh Building, Cambridge CB2 2RU, United Kingdom
40 West 20th Street, New York, NY 10011-4211, USA
10 Stamford Road, Oakleigh, Melbourne 3166, Australia

First published 1990
First paperback edition 1998

Library of Congress Cataloging-in-Publication Data is available.

A catalog record for this book is available from the British Library.

ISBN   0-521-30330-3 hardback
ISBN   0-521-62922-5 paperback

Transferred to digital printing 2004

CONTENTS

PREFACE

This book is an outgrowth of my involvement in the field of dynamic fracture mechanics over a period of nearly twenty years. This subbranch of fracture mechanics has been wonderfully rich in scope and diversity, attracting the attention of both researchers and practitioners with backgrounds in the mechanics of solids, applied mathematics, structural engineering, materials science, and earth science. A wide range of analytical, experimental, and computational methods have been brought to bear on the area. Overall, the field of dynamic fracture is highly interdisciplinary, it provides a wealth of challenging fundamental issues for study, and new results have the potential for immediate practical application. In my view, this combination of characteristics accounts for its continued vitality.

I have written this book in an effort to summarize the current state of the mechanics of dynamic fracture. The emphasis is on fundamental concepts, the development of mathematical models of phenomena which are dominated by mechanical features, and the analysis of these models. Mathematical problems which are representative of the problem classes that comprise the area are stated formally, and they are also described in common language in an effort to make their features clear. These problems are solved using mathematical methods that are developed to the degree required to make the presentation more or less self-contained. Experimental and computational approaches have been of central importance in this field, and relevant results are cited in the course of discussion. The extraordinary contributions of the few individuals who pioneered the area of dynamic fracture mechanics occupy prominent positions in this discussion. One hope in preparing this book is that people with new perspectives will be attracted to the field, which continues to provide fascinating and technically important challenges. Perhaps the book can serve as a guide to further development of the area.

The reader is assumed to be familiar with concepts of continuum mechanics and methods of applied mathematics to the level normally provided through the first year of study in a graduate program in solid mechanics in the United States. A brief summary of relevant results is included in the first chapter in order to establish notation and to provide a common source for reference in later chapters. Some background in equilibrium fracture mechanics would be helpful, of course, but none is presumed. In terms of graduate instruction in fracture mechanics, the book could serve as a text for a course devoted to dynamic fracture mechanics or, for a more general course, as a supplement to other books which provide broader coverage of the whole of fracture mechanics. The overall organization of the book is evident from the chapter titles. The brief overview included in Chapter 1 can serve as a guide to those readers interested in using the book as a source of reference for specific results.

The bibliography is an important part of the book. In view of contemporary publication practices, the compilation of an all-inclusive bibliography in any technical area is an impossible task. Nonetheless, the bibliography is intended to be comprehensive in the sense that it includes entries which describe research results on essentially all aspects of dynamic fracture. With only a few exceptions, the entries are either articles published in the open literature or relevant textbooks and monographs. Thus, most of the references should be available in a reasonably complete technical library. All references cited in the text are included in the bibliography, and many additional references are included as well. Some judgment was required in the selection of references for citation in the text, and I have done my best to accurately identify sources for key steps in the evolution of ideas.

I am indebted to a number of colleagues who read drafts of various chapters of the book. Those who offered suggestions and encouragement in this way are John Hutchinson, Fred Nilsson, Ares Rosakis, and John Willis. A special thanks goes to Jim Rice, Editor of this series, who generously read a draft of the entire manuscript. It has been among my greatest fortunes to be a member of the Solid Mechanics Group at Brown University, which has provided an intellectually stimulating and most congenial environment over the years. I am especially grateful to my colleagues Rod Clifton, Alan Needleman, Michael Ortiz, Fong Shih, and Jerry Weiner for their willingness to read and discuss some of the material that is included here. My own views on the mechanics of dynamic fracture and its fundamental precepts have been formed over a long period of time through interactions with people far too numerous to mention individually. This group includes the many colleagues and students with whom I have collaborated and written joint papers; they are identified in the bibliography. It also includes those who, through questions and discussions, showed me where my own understanding of certain points had been incomplete.

I also thank Peter-John Leone, Earth Sciences Editor, Rhona Johnson, Manuscript Development Editor, and Louise Calabro Gruendel, Production Editor, of Cambridge University Press in New York for the efficient way in which they have managed the preparation of the book and for their sensitivity in dealing with my concerns on the matter. It has been a pleasure to be involved in long-term programs of research on fracture mechanics funded by the Office of Naval Research and by the National Science Foundation. These programs, and the collaborations that they have fostered, have been invaluable.

Finally, I thank my wife Colleen and our sons, Jon, Jeff, and Steve, who enthusiastically adopted the mission of writing this book as their own. Their interest and unwavering devotion have lightened the task immensely.

L. B. Freund

LIST OF SYMBOLS

Mathematical symbols and functions are defined the first time that they are used in the book. Brief definitions of the most frequently used symbols and functions are listed below. Some symbols necessarily have different definitions in different sections. In such cases, definitions are stated locally and are used consistently within sections.

a	Inverse dilatational wave speed c_d^{-1}; half length of a Griffith crack
$A_I(v)$	Universal function of crack tip speed for mode I deformation (similar for modes II and III)
b	Inverse shear wave speed c_s^{-1}; a material parameter
c	Inverse Rayleigh wave speed c_R^{-1}; an elastic wave speed
c_d	Elastic dilatational wave speed
c_o	Speed of longitudinal waves in an elastic bar
c_R	Elastic Rayleigh surface wave speed
c_s	Elastic shear wave speed
C_I	Dimensionless factor for the mode I asymptotic stress field (similar for modes II and III)
C_{ijkl}	Components of the elastic stiffness tensor
d	Inverse crack tip speed v^{-1}
d_{ij}	Components of the symmetric part of the velocity gradient tensor
$D(v)$	The quantity $4\alpha_d\alpha_s - (1+\alpha_s^2)^2$; a function of crack tip speed
E	Young's elastic modulus
E_R	Far-field radiated energy
$F(\Gamma)$	Energy flux through contour Γ
G	Energy release rate; dynamic energy release rate

G_a — Crack arrest fracture energy

G_c — Critical value of energy release rate

h — Thickness of a beam; width of a strip

$h_i(x,t)$ — Components of the dynamic weight function

$H(t)$ — Unit step function

$J_N(\Gamma, s)$ — Path independent integral of Laplace transformed fields

$k(v)$ — Universal function of crack tip speed for elastic crack growth in mode I

$k_{II}(v)$ — Universal function of crack tip speed for elastic crack growth in mode II (similar for mode III)

K_I — Elastic stress intensity factor for mode I (similar for modes II and III)

K_{Ia} — Value of stress intensity factor at crack arrest; crack arrest toughness (similar for modes II and III)

K_{Iappl} — Remotely applied stress intensity factor

K_{Ic} — Value of stress intensity factor at fracture initiation; fracture toughness (similar for modes II and III)

K_{Id} — Value of stress intensity factor during crack growth; dynamic fracture toughness (similar for modes II and III)

$l, l(t)$ — Crack length; amount of crack growth

$\dot{l}$ — Instantaneous crack tip speed

$l^{\pm}$ — Limit as position $x = l$ is approached through values of x greater than l $(+)$ or less than l $(-)$

l_c — Critical crack length

l_0 — Initial crack length

m — Normalized crack tip speed v/c_s or v/c_R

M_{ijkl} — Components of the elastic compliance tensor

n — Crack tip bluntness parameter; a material parameter

n_i — Components of unit vector; normal to surface or curve

$o(f(x))$ — Asymptotically dominated by f as $x \to$ a limit point

$O(f(x))$ — Asymptotically proportional to f as $x \to$ a limit point

$p(x)$ — Pressure distribution

p^* — Magnitude of a concentrated normal force

$P(\zeta)$ — Amplitude of Φ

P_c — Dimensionless combination of material parameters for a strain-rate-dependent elastic-plastic solid

q^* — Magnitude of a concentrated shear force

$Q(\zeta)$ — Amplitude of Ψ

r	Polar coordinate
$r_d \exp(i\theta_d)$	Polar form of the complex variable $x + i\alpha_d y$
r_p	Plastic zone size
$r_s \exp(i\theta_s)$	Polar form of the complex variable $x + i\alpha_s y$
R	Region in space; region in a plane
$R(\zeta)$	Rayleigh wave function
s	Laplace transform parameter; a real variable
s_{ij}	Components of the deviatoric stress tensor
$S_\pm(\zeta)$	Factors of the Rayleigh wave function that are nonzero and analytic in overlapping half planes
t	Time coordinate
T	Kinetic energy density
T_{tot}	Total kinetic energy of a body
$\mathbf{T}, \mathcal{T}_i$	Traction vector; components of traction vector
u_i	Components of particle displacement vector ($i = 1, 2, 3$ or $i = x, y, z$)
$\dot{u}_i$	Components of particle velocity vector
$\ddot{u}_i$	Components of particle acceleration vector
$u_y^L(x,t)$	Normal surface displacement for Lamb's problem
$u_-(x,t)$	Displacement distribution for $x < 0$
U	Stress work density; elastic strain energy density
U_{tot}	Total stress work; total strain energy
v	Crack tip speed
w	Displacement for antiplane shear deformation
x_i	Rectangular coordinates ($i = 1, 2, 3$)
x, y, z	Rectangular coordinates
$\alpha(\zeta)$	The function $(a^2 - \zeta^2)^{1/2}$
α_d, α_s	The quantities $\sqrt{1 - v^2/c_d^2}$ and $\sqrt{1 - v^2/c_s^2}$
$\beta(\zeta)$	The function $(b^2 - \zeta^2)^{1/2}$
γ	A material parameter
$\dot{\gamma}_0$	Viscosity parameter of a rate dependent plastic material
Γ	Crack tip contour; specific fracture energy
$\Gamma(\cdot)$	Gamma (factorial) function
$\Gamma_c, \Gamma_m, \Gamma_o$	Constant values of specific fracture energy
δ_t	Crack tip opening displacement
$\delta(t)$	Dirac delta function
Δ	Amplitude of an elastic dislocation
ϵ	A small real parameter
ϵ_{ij}	Components of the small strain tensor

ϵ_{ij}^{e}	Components of the elastic strain tensor
ϵ_{ij}^{p}	Components of the plastic strain tensor
ζ	Complex variable; Laplace transform parameter
η	Rectangular coordinate
θ	Polar coordinate
κ	Parameter determined by the behavior of the Rayleigh wave function as $|\zeta| \to \infty$
λ	Lamé elastic constant
Λ	Length of cohesive zone
μ	Lamé elastic constant; elastic shear modulus
ν	Poisson's ratio
$\boldsymbol{\nu}, \nu_i$	Unit vector normal to a surface or curve
ξ	Rectangular coordinate
ρ	Material mass density
σ	Mean stress; effective stress
$\sigma(x)$	Normal traction within a cohesive zone
σ_{ij}	Components of the stress tensor
σ_∞	Amplitude of remotely applied tension
σ_0	Tensile flow stress of an ideally plastic material
$\sigma_+(x,t)$	Normal traction distribution for $x > 0$
σ^*	Magnitude of applied normal traction
Σ_{ij}^{I}	Angular variation of asymptotic crack tip stress field for mode I (similar for modes II and III)
τ_∞	Amplitude of remotely applied shear traction
τ_0	Shear flow stress of an ideally plastic material
$\tau_+(x,t)$	Shear traction distribution on $x > 0$
τ^*	Magnitude of applied shear traction
ϕ	Lamé scalar displacement potential function
Φ	Double Laplace transform of ϕ
$\boldsymbol{\psi}, \psi_i$	Lamé vector displacement potential function
ψ	Magnitude of $\boldsymbol{\psi}$ for plane deformation; local shear angle in plastically deforming region
Ψ	Double Laplace transform of ψ
Ω	Potential energy
Ω_S	Potential energy increase due to creation of free surface
$(cc.ss.nn)$	Equation number nn in Section ss of Chapter cc
$(c.s.n)_m$	The mth equation in a group of equations identified by the single number $(c.s.n)$

1

BACKGROUND AND
OVERVIEW

1.1 Introduction

The field of fracture mechanics is concerned with the quantitative description of the mechanical state of a deformable body containing a crack or cracks, with a view toward characterizing and measuring the resistance of materials to crack growth. The process of describing the mechanical state of a particular system is tantamount to devising a mathematical model of it, and then drawing inferences from the model by applying methods of mathematical or numerical analysis. The mathematical model typically consists of an idealized description of the geometrical configuration of the deformable body, an empirical relationship between internal stress and deformation, and the pertinent balance laws of physics dealing with mechanical quantities. For a given physical system, modeling can usually be done at different levels of sophistication and detail. For example, a particular material may be idealized as being elastic for some purposes but elastic-plastic for other purposes, or a particular body may be idealized as a one-dimensional structure in one case but as a three-dimensional structure in another case. It should be noted that the results of most significance for the field have not always been derived from the most sophisticated and detailed models.

A question of central importance in the development of a fracture mechanics theory is the following. Is there any particular feature of the mechanical state of a cracked solid that can be interpreted as a "driving force" acting on the crack, that is, an effect that is correlated with a tendency for the crack to extend? A viewpoint that underlies this important concept is that of the crack as an entity which itself behaves according to a "law" of mechanics expressed in terms of a relationship between driving force and motion. The modeling phase provides the language for considering the strength of real materials, and it concludes with hypotheses on the behavior of materials. It is only through observation of the fracturing of materials that the hypotheses can be verified or refuted. This synergistic process has led to standard methods of practice whereby fracture mechanics is used routinely in materials selection and engineering design.

1.1.1 Inertial effects in fracture mechanics

Dynamic fracture mechanics is the subfield of fracture mechanics concerned with fracture phenomena for which the role of material inertia becomes significant. Phenomena for which strain rate dependent material properties have a significant effect are also typically included. Inertial effects can arise either from rapidly applied loading on a cracked solid or from rapid crack propagation. In the case of rapid loading, the influence of the loads is transferred to the crack by means of stress waves through the material. To determine whether or not a crack will advance due to the stress wave loading, it is necessary to determine the transient driving force acting on the crack. In the case of rapid crack propagation, material particles on opposite crack faces displace with respect to each other once the crack edge has passed. The inertial resistance to this motion can also influence the driving force, and it must be taken into account in a complete description of the process. There is also a connection between the details of rapid crack motion and the stress wave field radiated from a moving crack that is important in seismology, as well as in some material testing techniques.

Progress toward understanding dynamic fracture phenomena has been impeded by several complicating features. The inherent time dependence of a dynamic fracture process results in mathematical models that are more complex than the equivalent equilibrium models for the same configuration and material class. From the experimental point of view, the time dependence requires that many accurate sequential measurements of quantities of interest must be made in an extremely short time period in a way that does not interfere with the process being observed. In the case of crack growth, the regions of the boundary of the body over which certain conditions must be enforced in a mathematical model change with time. From the experimental point of view, this feature implies that the place where quantities of interest must be measured varies, usually in a nonuniform way, during the process.

The question of whether or not inertial effects are significant in any given fracture situation depends on the loading conditions, the material characteristics, and the geometrical configuration of the body. Circumstances under which they are indeed significant cannot be specified unambiguously, but some guidelines are evident. For example, suppose that there is a characteristic time associated with the applied loading, say a load maximum divided by the rate of load increase. Unless this time is large compared to the time required for a stress wave to travel at a characteristic wave speed of the material over a representative length of the body, say the crack length or the distance from the crack edge to the loaded boundary, it can be expected that inertial effects will be significant. In the case of crack growth, inertial effects will probably be important if the speed of the crack tip or edge is a significant fraction of the lowest characteristic wave speed of the material. If the boundaries of the body must be taken into account, then the potential for the body to function as a waveguide must be considered. The group velocity of guided waves is frequently much less than the characteristic bulk wave speeds of the material. For such cases, inertial effects could be important for crack speeds that are much less than the bulk wave speeds but comparable to relevant group velocities.

1.1.2 Historical origins

Empirical studies concerned with the bursting of military cannon or with the impact loading of industrial machines were carried out in the 19th century. For example, an interesting early series of observations on "new" industrial iron-based materials is reported by Kirkaldy (1863). However, no single scientific discovery or other event can be identified as the stimulus for launching dynamic fracture mechanics as an area of research. Instead, the area as it is summarized in this book is the conjunction of numerous paths of investigation that were motivated by pressing practical needs in engineering design and material selection, by scientific curiosity about earthquakes and other natural phenomena, by challenges of classes of mathematical problems, and by advances in experimental and observational techniques.

The area has been under continuous development as a subfield of the engineering science of fracture mechanics since the 1940s and as a topical area in geophysics since the late 1960s. The view in the early days of fracture mechanics was that once a crack in a structure began to grow the structure had failed (Tipper 1962). The rapid crack growth phase of failure was viewed as interesting but not particularly important. Beginning about 1970, however, the importance of understanding crack propagation and arrest was recognized in both engineering and earth sciences, and significant progress has been made since then.

A few of the early contributions to the area that provided ideas of lasting significance are cited here. In some cases, these ideas were developed in connection with fracture under equilibrium conditions, but they have been important in the area of dynamic fracture as well. Among these is the work of Griffith (1920), which is commonly acknowledged as representing the start of equilibrium fracture mechanics as a quantitative science of material behavior. In considering an ideally brittle elastic body containing a crack, he recognized that the macroscopic potential energy of the system, consisting of the internal stored elastic energy and the external potential energy of the applied loads, varied with the size of the crack. He also recognized that extension of the crack resulted in the creation of new crack surface, and he postulated that a certain amount of work per unit area of crack surface must be expended at a microscopic level to create that area. In this context, the term "microscopic" implies that this work is not included in a continuum description of the process. It is common to characterize this work per unit area by assuming that a certain force-displacement relationship governs the reversible interaction of atoms or molecules across the fracture plane. The area under the force-displacement relationship from equilibrium to full separation, averaged over the fracture surface, is assumed to represent this work of separation. This is one way to define the *surface energy* of the material. Griffith simply included this work as an additional potential energy of the system, and then invoked the equilibrium principle of minimum potential energy for conservative systems. That is, he considered the system to be in equilibrium with a particular fixed loading and a particular crack length. He postulated that the crack was at a *critical state of incipient growth* if the reduction in macroscopic potential energy associated with a small virtual crack advance from that state was equal to the microscopic

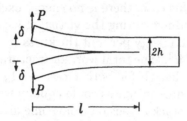

Figure 1.1. A crack of length l in a rectangular strip. Forces P act to separate the crack faces and the displacement of each load point is δ.

work of separation for the crack surface area created by the virtual crack advance. A particular attraction of Griffith's energy fracture condition is that it obviates the need to examine the actual fracture process at the crack tip in detail.

Griffith's original work was based on the plane elasticity solution for a crack of finite length in a body subjected to a uniform remote tension in a direction normal to the crack plane. The concept is illustrated here by means of a simple example based on elementary elastic beam theory. Consider the split rectangular strip shown in Figure 1.1 which is of unit thickness in the direction perpendicular to the plane of the figure. Suppose that the slit or crack of length l is opened symmetrically by imposing either forces or displacements at the ends of the arms as shown. In either case, the end force per unit thickness is denoted by P and the displacement of the load point of each arm from the undeflected equilibrium position is denoted by δ. Suppose that the isotropic material is elastic with Young's modulus E and that the deformation can be approximated by assuming that each arm deforms as a Bernoulli-Euler beam of length l cantilevered at the crack tip end. The relationship between the end force per unit thickness and the end displacement is then

$$P = \frac{Eh^3}{4l^3}\delta \qquad (1.1.1)$$

and the total stored elastic energy per unit thickness in both arms is

$$\delta P = \frac{Eh^3}{4l^3}\delta^2 = \frac{4l^3}{Eh^3}P^2 . \qquad (1.1.2)$$

The work of separation per unit area of surface created is denoted by γ.

Consider first the case of imposed end displacement δ^*. What is the critical crack size for this fixed end displacement according to the Griffith theory? In this case, there is no energy exchange between the body and its surroundings during the virtual crack extension, and the macroscopic potential energy per unit thickness Ω is the stored elastic energy $\Omega = Eh^3\delta^{*2}/4l^3$. The total work-of-fracture per unit thickness is γ times the crack length for both faces, or $\Omega_S = 2\gamma l$. The total work-of-fracture per unit thickness could equally well be written as 2γ times the amount of crack extension from some arbitrary initial crack length, a work measure that would differ from Ω_S by a constant. This constant is arbitrary, and it has no significance in considering crack advance. The total potential energy is $\Omega + \Omega_S$. According to the Griffith postulate, the critical crack length for incipient growth is determined by the condition

$$\frac{\partial}{\partial l}(\Omega + \Omega_S) = 0 \quad \Rightarrow \quad l_c = \left[\frac{3Eh^3\delta^{*2}}{8\gamma}\right]^{1/4}. \tag{1.1.3}$$

The individual potential energies vary with crack length as shown schematically in Figure 1.2. Note that

$$\frac{\partial^2}{\partial l^2}(\Omega + \Omega_S) > 0 \tag{1.1.4}$$

at the critical crack length. This means that the system is stable in the sense of mechanical equilibrium. If the crack length is at the critical value and δ^* is slowly increased to a larger value, then the crack will slowly advance under equilibrium conditions to a new length that is related to the increased δ^* through $(1.1.3)_2$. A useful interpretation of $(1.1.3)_1$ is that $-\partial\Omega/\partial l$ is the feature of the mechanical state that can be identified as a crack driving force. The driving force is work-conjugate to crack position l, and the fracture mechanics balance law governing crack behavior is $-\partial\Omega/\partial l = 2\gamma$.

Consider next the case of imposed end force P^*. What is the critical crack size for this end condition? The stored elastic energy is again given by (1.1.2). The external potential energy of the loading on both arms is $-2\delta P^* = -8l^3 P^{*2}/Eh^3$, so that the macroscopic

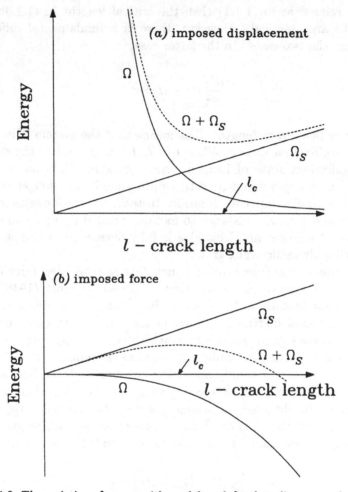

Figure 1.2. The variation of energy with crack length for the split rectangular strip in Figure 1.1 for (a) imposed displacement or (b) imposed force. The variation of the continuum potential energy plus the surface energy is shown by the dashed line.

potential energy is $\Omega = -4l^3 P^{*2}/Eh^3$. Imposition of the Griffith condition in this case yields

$$\frac{\partial}{\partial l}(\Omega + \Omega_S) = 0 \quad \Rightarrow \quad l_c = \left[\frac{Eh^3\gamma}{6P^{*2}}\right]^{1/2} . \qquad (1.1.5)$$

Again, the way in which the individual potential energies vary with crack size is illustrated schematically in Figure 1.2. If δ^*, P^*, and

l_c are related as in (1.1.1) then the critical lengths in (1.1.3b) and (1.1.5b) are identical. However, there is a fundamental difference between the two cases. In the latter case,

$$\frac{\partial^2}{\partial l^2}\left(\Omega + \Omega_S\right) < 0 \qquad (1.1.6)$$

at the critical crack length. This means that the system is unstable in an equilibrium sense. When $l = l_c$ for a given P^*, the state is an equilibrium state of incipient crack growth. However, if P^* is increased by any amount from this state, then it is no longer possible to find an equilibrium configuration. Instead, the crack begins to grow rapidly and inertial resistance to motion is called into play to ensure balance of momentum. This idea is fully developed in Chapter 7 in discussing dynamic crack growth.

A theoretical framework for including inertial effects during the rapid crack growth phase was first proposed by Mott (1948), who adopted the features of Griffith's analysis as a point of departure. He recognized that inertial resistance of the material to crack opening could become significant at high crack speeds. To estimate the crack speed for a particular loading system, he assumed that the crack growth process was steady state, that is, time independent as seen by an observer moving at speed v with the crack tip. Under these conditions, he obtained an estimate for the total kinetic energy of the system T_{tot} in the form of v^2 times a function of crack length l. He then argued that the total energy of the system $\Omega + \Omega_S + T$ is constant, so that

$$\frac{\partial}{\partial l}\left(\Omega + T_{tot} + \Omega_S\right) = 0\,. \qquad (1.1.7)$$

This condition provides a relationship involving the loading parameters, the crack length l, and the crack speed v. In implementing this condition for the Griffith plane elasticity crack model, Mott assumed that Ω and T could be calculated on the basis of the equilibrium field.

The idea underlying (1.1.7) is correct if the terms are strictly interpreted. In fact, it is a special case for steady state crack growth of the general energy-rate balance: Rate of work of applied loads = rate of increase of stored elastic energy + rate of increase of kinetic energy − rate of energy loss at the crack tip. However, later work has shown that the assumptions made by Mott in his original analysis of the dynamic Griffith crack are not valid. Consequently, the conclusions inferred from this model are generally not valid. For example, this analysis has often been cited as establishing that the maximum crack speed in an ideal brittle material is some fixed fraction, typically given as about one-half, of the shear wave speed of the material. If the model is analyzed more thoroughly, it is found that this is not the case. Indeed, the result that the theoretical limiting speed of a tensile crack must be the Rayleigh wave speed, as established in Chapter 7, was anticipated by Stroh (1957) on the basis of an intuitive argument. Nonetheless, Mott's basic idea of energy balance during rapid crack growth was very important to the development of the subject. A number of energy concepts pertinent to dynamic fracture are discussed in Chapter 5, and the crack tip equation of motion of the type originally sought by Mott is presented in Chapter 7.

Another important theoretical idea for applying work and energy methods to fracture dynamics was provided by Irwin (1948). He was concerned with cleavage fracture of structural steels, a cleavage process that is invariably accompanied by some amount of plastic flow of the material adjacent to the fracture path. Irwin adopted Mott's postulate of stationary total energy, including a work of fracture, denoted above by Ω_S. However, Irwin proposed that this term Ω_S may be approximately represented as the sum of two terms, one proportional to the area of fracture surface and the other proportional to the volume of material affected by plastic deformation. Within the context of the Griffith theory outlined above, this assumption is invoked by writing $\Omega_S = 2(\gamma_s + \gamma_p)l$ when the thickness of the plastically deformed layer adjacent to the crack faces is small. Thus, γ_s represents the surface energy of the material associated with cleavage fracture and γ_p represents the plastic work dissipated in the surrounding material per unit crack surface area created. Irwin's idea therefore extended the range of applicability of the energy balance fracture theory to include materials that undergo some plastic deformation during crack growth. This particular extension of Griffith's theory was also proposed for fracture initiation and crack propagation in steels by Orowan (1955). Early experiments aimed at measuring γ_p for cleavage of metal single crystals and polycrystals were carried out by Hall (1953). He showed that $\gamma_p \gg \gamma_s$ for metals, particularly for polycrystals. Furthermore, both Hall and Irwin predicted strong connections between the fracture behavior of polycrystalline metals and their microstructures. An extensive survey of early work on the connection between metallurgical properties and brittle fracture of welded steel plates was given by Wells (1961) and Tipper (1962).

The continuum field approach to fracture of solids was launched with the introduction of the *elastic stress intensity factor*, usually denoted by K, as a crack tip field characterizing parameter by Irwin (1957). This idea provided an alternate framework for discussing the strength of cracked solids of nominally elastic material. Irwin proposed that a crack will begin to grow in a cracked body with limited plastic deformation when K is increased to a value called the *fracture toughness* of the material. The equivalence of the Irwin stress intensity factor criterion and the Griffith energy criterion for onset of growth of a tensile crack in a two-dimensional body of nominally elastic material under plane stress conditions was demonstrated by Irwin (1957, 1960), who showed that

$$-\frac{\partial \Omega}{\partial l} = \frac{K^2}{E},\tag{1.1.8}$$

where E is Young's modulus of the material. The quantity on the left side of (1.1.8) is usually denoted by G and it is called the *energy release rate*. Much of the analysis of this book is devoted to determining the stress intensity factor or the energy release rate for various conditions. These concepts are developed more thoroughly in a way relevant to dynamic fracture mechanics in the chapters to follow.

Another idea that has been important in the evolution of fracture mechanics concerns the size scale over which different phenomena dominate a fracture process. The idea is implicit in Irwin's stress intensity factor concept and it is a central feature of the crack tip cohesive zone model introduced by Barenblatt (1959a). Consider a planar crack in a body subjected to tensile loading normal to the plane of the crack. Suppose that the material response is linearly elastic, except for a region adjacent to the crack edge where the response departs from linearity. The source of nonlinearity can be plastic deformation, diffuse microcracking, nonlinear interatomic forces, or some other physical mechanism. The crack tip region is said to be *autonomous* at fracture initiation or during crack growth if the following two conditions are met: (i) the extent of the region of nonlinearity from the crack edge is very small compared to all other length dimensions of the body and loading system, and (ii) the mechanical state within this end region at incipient growth or during growth is independent of loading and geometrical configuration. For the particular case of an elastic-plastic material, the property of autonomy implies that the crack tip plastic zone is completely surrounded by an elastic stress intensity factor field and that the state within the plastic zone is determined by the level of stress intensity of the surrounding field. This situation, which was termed *small-scale yielding* by Rice (1968), is the situation in which the stress intensity factor is a useful fracture characterizing parameter. Furthermore, this close connection between effects at adjacent size scales provided the basis for development of an asymptotic formulation of nonlinear crack tip field problems in general. In its simplest form, the asymptotic formulation can be summarized in the following way. Suppose that, at one scale of observation, the mechanical state near a crack edge is dominated by an asymptotic *near crack tip solution* with some characterizing parameter. Then, to study the phenomenon at a physical size scale that is an order of magnitude smaller, a boundary value problem is formulated with a condition that its solution must match the aforementioned asymptotic solution *in the far-field*. It is essentially this asymptotic formulation that underlies the focus on semi-infinite cracks and near tip fields in this monograph, as well as in fundamental fracture mechanics studies in general. The central idea of autonomy has been extended to more general material classes than those mentioned in these introductory remarks.

A good deal of very important experimental research on dynamic fracture of materials was carried out in parallel with the development of dynamic fracture mechanics concepts for analyzing cracked solids. These experiments were often done on specimens of glass or other very brittle materials. The results of these experiments, in the form of both qualitative observations and quantitative data, have been extremely important in the development of dynamic fracture mechanics. Indeed, a number of specific observations on the behavior of growing cracks in glass were made in the 1940s and 1950s for which generally accepted explanations based on physical principles are still incomplete. For example, quantitative modeling as a basis for understanding terminal crack speed in glass or dynamic crack bifurcation under symmetric tensile loading is still in a preliminary stage. In addition to the importance of the experimental results obtained in the early stages of dynamic fracture research, the significance of the experimental techniques developed in the course of this work must also be recognized. The full-field optical methods of observation and special techniques

of high-speed photography have been particularly notable in dynamic
fracture research.

The research group led by Schardin played a major role in the
development of optical methods and high-speed photography. Much
of this work is reviewed by Schardin (1959). For example, in 1937
it was found that running cracks in glass seem to have a maximum
speed that depends on the material, but not on loading or geometrical
configuration. This phenomenon was also observed for fracture of
glass by Edgerton and Barstow (1941). The multiple spark camera,
now known as the Cranz-Schardin camera in its most common configuration, was developed to obtain a sequence of photographs of a
fracturing specimen in a very short time interval. It was also shown
that stress waves obliquely incident on the crack path could cause
the path to gradually curve toward the local direction of maximum
tensile stress. It was observed that there was no delay in the change in
crack growth direction due to incident stress waves, suggesting that
the crack tip itself does not have an effective inertia (cf. Section
7.4). This phenomenon was exploited in the 1950s by Kerkhof, who
superimposed ultrasonic waves on a crack growing rapidly in a brittle
body. Because the frequency of the ultrasonic waves is known, the
small undulations produced on the fracture surface due to the interaction of the crack with the ultrasonic waves provided a continuous
record of crack tip speed (Kerkhof 1970, 1973).

The optical shadow spot method was also introduced through the
work of this group (Manogg 1966). This technique made it possible
to infer directly a value of the instantaneous crack tip stress intensity
factor and the crack tip position. When used in conjunction with the
multiple spark camera, a record of the stress intensity factor and crack
motion could be obtained for a crack growth process. A multiple spark
arrangement was also used by Wells and Post (1958) to determine the
isochromatic fringe pattern representing the near crack tip stress field
during rapid crack growth in a photoelastic brittle plastic. A review
of early work on the formation of fractures in brittle materials by
means of stress waves is given by Kolsky (1953). Modern surveys of
experimental results in dynamic fracture of brittle materials are given
by Dally (1987) for photoelastic methods and by Kalthoff (1987) for
shadow spot methods.

1.2 Continuum mechanics

In this section, some notational conventions are established and basic equations of continuum mechanics are presented in order to provide a framework for discussing dynamic fracture mechanics. Field equations and integral relations are stated without derivation; the underlying developments can be found in the references cited. In presenting these results, it is tacitly assumed that the mathematical operations implied by the expressions can indeed be carried out. Whether or not this is the case in the study of any particular model must be determined in the course of analysis.

1.2.1 Notation

Consider a three-dimensional Euclidean point space in which the behavior of a deformable body is to be described. A rectangular coordinate frame is introduced, consisting of a set of mutually orthogonal unit base vectors $\mathbf{e}_1$, $\mathbf{e}_2$, $\mathbf{e}_3$. The position vector $\mathbf{p}$ of a point in the space is

$$\mathbf{p} = x_1\mathbf{e}_1 + x_2\mathbf{e}_2 + x_3\mathbf{e}_3 = \sum_{i=1}^{3} x_i\mathbf{e}_i = x_i\mathbf{e}_i \,, \qquad (1.2.1)$$

where the summation over the repeated index implied by the last entry in the continued equality is adopted as a convention. The quantities x_1, x_2, x_3 are the rectangular coordinates of the point. Vector and tensor valued quantities are represented by boldface type $\mathbf{v}$ throughout the book, and components are represented by means of subscripts.

Suppose that $\mathbf{v}$ and $\mathbf{T}$ are a vector field and a second-order tensor field, respectively, over the space. At each point, the vector is an element of a vector space associated with the point space, and the second-order tensor is a linear transformation of elements of the vector space into possibly another vector space or, as in this case, the same one. The rectangular components of these fields at any point are

$$v_i = \mathbf{v} \cdot \mathbf{e}_i \quad \text{and} \quad T_{ij} = \mathbf{e}_i \cdot \mathbf{T} \cdot \mathbf{e}_j \,, \qquad (1.2.2)$$

where the dot denotes the binary *inner product* operator. This notation is commonly called the Gibbs dyadic notation (Malvern 1969). The inner product of two vectors $\mathbf{v}$ and $\mathbf{u}$ is

$$\mathbf{u} \cdot \mathbf{v} = \mathbf{v} \cdot \mathbf{u} = u_i v_i \,. \qquad (1.2.3)$$

Likewise, the inner product from the left (right) of vector $\mathbf{v}$ with the tensor $\mathbf{T}$ is $\mathbf{v} \cdot \mathbf{T} = v_i T_{ij} \mathbf{e}_j$ ($\mathbf{T} \cdot \mathbf{v} = \mathbf{e}_i T_{ij} v_j$). Inner products involving tensors of second- or higher order are not commutative unless the tensors possess certain symmetry properties. The cross product of two vectors $\mathbf{u}$ and $\mathbf{v}$ is a vector with components

$$(\mathbf{u} \times \mathbf{v})_i = -(\mathbf{v} \times \mathbf{u})_i = e_{ijk} u_j v_k \,, \qquad (1.2.4)$$

where e_{ijk} is the alternating symbol with values

$$e_{ijk} = \begin{cases} +1 & \text{if } ijk \text{ is an even permutation of 123,} \\ -1 & \text{if } ijk \text{ is an odd permutation of 123,} \\ 0 & \text{otherwise.} \end{cases} \qquad (1.2.5)$$

The array of components of the second-order unit tensor $\mathbf{I}$ in the rectangular frame is the Kronecker delta

$$\delta_{ij} = \begin{cases} 1 & \text{for } i = j, \\ 0 & \text{for } i \neq j. \end{cases} \qquad (1.2.6)$$

Thus, for any vector $\mathbf{v}$, $\mathbf{v} = \mathbf{I} \cdot \mathbf{v}$ or $v_i = \delta_{ij} v_j$. The trace of a second-order tensor $\mathbf{T}$ is

$$\text{tr}(\mathbf{T}) = \mathbf{e}_i \cdot \mathbf{T} \cdot \mathbf{e}_i = T_{ii} \,. \qquad (1.2.7)$$

Let p represent the set of coordinates x_1, x_2, x_3. The gradient of a scalar field $f(p)$, denoted by ∇f, is the vector with components

$$(\nabla f)_i = \frac{\partial f}{\partial x_i} = f_{,i} \,, \qquad (1.2.8)$$

where the last entry of the continued equality introduces the convention whereby a subscript comma is used to denote partial differentiation. The divergence of a vector field $\mathbf{v}(p)$, denoted by $\nabla \cdot \mathbf{v}$, is the scalar

$$\nabla \cdot \mathbf{v} = \frac{\partial v_i}{\partial x_i} = v_{i,i} \,. \qquad (1.2.9)$$

The curl of a vector field $\mathbf{v}(p)$, denoted by $\nabla \times \mathbf{v}$, is a vector with components

$$(\nabla \times \mathbf{v})_i = e_{ijk} v_{k,j} \,. \qquad (1.2.10)$$

The Laplacian operator acting on a scalar field f, defined by $\nabla \cdot \nabla f$ and denoted by $\nabla^2 f$, is

$$\nabla^2 f = f_{,ii} \tag{1.2.11}$$

in indicial notation. If the gradient operator

$$\nabla = \mathbf{e}_i \frac{\partial}{\partial x_i} \tag{1.2.12}$$

is viewed as a vector, then useful generalizations of the foregoing results become obvious. For example, the gradient of a vector from the left or the right is the dyad $\nabla \mathbf{v}$ or $\mathbf{v} \nabla$ which is a second-order tensor with rectangular components

$$(\nabla \mathbf{v})_{ij} = v_{j,i} \qquad \text{or} \qquad (\mathbf{v} \nabla)_{ij} = v_{i,j} \,. \tag{1.2.13}$$

Likewise, the divergence of a second-order tensor from the left $\nabla \cdot \mathbf{T}$ or from the right $\mathbf{T} \cdot \nabla$ is a vector with rectangular components

$$(\nabla \cdot \mathbf{T})_j = T_{ij,i} \qquad \text{or} \qquad (\mathbf{T} \cdot \nabla)_j = T_{ji,i} \,, \tag{1.2.14}$$

and so on.

Gauss's divergence theorem on integrals of vector fields plays a key role in continuum mechanics. To state the theorem in its simplest form, consider a vector field $\mathbf{v} = v_i \mathbf{e}_i$ over a region R of the point space. Let S be the closed boundary surface of R and let n_i be the outward unit normal to S. Then the divergence theorem states that

$$\int_R \nabla \cdot \mathbf{v} \, dR = \int_S \mathbf{n} \cdot \mathbf{v} \, dS \qquad \text{or} \qquad \int_R v_{i,i} \, dR = \int_S n_j v_j \, dS \,. \tag{1.2.15}$$

Again, the result is readily generalized to tensor fields of second- and higher order if the divergence and inner product operators are properly interpreted. For the integral of the divergence from the left of the second-rank tensor field $\mathbf{T}$, for example, the divergence theorem states that

$$\int_R \nabla \cdot \mathbf{T} \, dR = \int_S \mathbf{n} \cdot \mathbf{T} \, dS \qquad \text{or} \qquad \int_R T_{ij,i} \, dR = \int_S T_{ij} n_i \, dS \,. \tag{1.2.16}$$

Most of the fields to be introduced to represent physical quantities depend on time t as well as on position in a three-dimensional space. In a few situations, it is advantageous to treat the time coordinate in the same way as a spatial coordinate in a four-dimensional point space with coordinates (p, t), as in Section 5.2. For the most part, however, the time coordinate is treated separately from spatial coordinates in applying the above conventions. Partial differentiation of a time dependent scalar field $f(p, t)$ is denoted by

$$\frac{\partial f}{\partial t} = f_{,t} . \qquad (1.2.17)$$

1.2.2 Balance equations

Consider a deformable body occupying the region R of the three-dimensional point space at time t. The closed boundary surface of R is S with outward unit normal vector n_i. The components of the second-order Cauchy stress tensor field with respect to the rectangular reference frame are denoted by $\sigma_{ij}(p, t)$. Consider any surface in R, possibly the boundary surface, and let ν_i be a unit normal vector at a particular point on the surface. According to the Cauchy stress principle, the vector representing local traction, or force per unit area, as a function of position on the surface is

$$T_i = \sigma_{ji}\nu_j . \qquad (1.2.18)$$

This is the traction exerted on the material on the side of the surface in the direction of $-\nu_i$. If S is the boundary surface, this traction may arise from external sources. For a surface interior to R, the relationship describes the action of the material on one side of the surface on the material on the other side. The validity of this relationship between the stress tensor and the traction vector does not require that the fields be equilibrium fields. No couple stress effects are included in the discussion of fracture mechanics in this book.

Suppose that a particular fixed configuration of the body is identified as a reference configuration. Material particles are labeled by associating each with its position in the reference configuration, say the point with coordinates x_1^o, x_2^o, x_3^o, represented collectively by p^o. The deformation is specified by a nonsingular mapping of points $x_i^o \mathbf{e}_i$

in the reference configuration to points $x_i e_i$ at time t in the "current" configuration

$$x_i = \chi_i(x_1^o, x_2^o, x_3^o, t) \qquad (1.2.19)$$

with inverse

$$x_i^o = \chi_i^o(x_1, x_2, x_3, t). \qquad (1.2.20)$$

The particle displacement $\mathbf{u} = u_i e_i$ is given by

$$u_i = \chi_i(x_1^o, x_2^o, x_3^o, t) - x_i^o \quad \text{or} \quad u_i = x_i - \chi_i^o(x_1, x_2, x_3, t) \quad (1.2.21)$$

in a material description or a spatial description , respectively, of the deformation. The particle velocity field $\mathbf{v} = v_i e_i$ is given by

$$v_i = \frac{\partial \chi_i}{\partial t} = \left. \frac{\partial u_i}{\partial t} \right|_{p^o}, \qquad (1.2.22)$$

which is naturally a function of (p^o, t) but which can be viewed as a function of (p, t) through (1.2.20). The particle acceleration $\mathbf{a} = a_i e_i$ is given by either

$$a_i = \left. \frac{\partial v_i}{\partial t} \right|_{p^o} = \left. \frac{\partial^2 u_i}{\partial t^2} \right|_{p^o} \qquad (1.2.23)$$

or

$$a_i = \left. \frac{\partial v_i}{\partial t} \right|_p + v_j \frac{\partial v_i}{\partial x_j}. \qquad (1.2.24)$$

The two expressions are both forms of the material time derivative of particle velocity, the first in the material description of motion and the second in the spatial description. The material time derivative will be denoted by a superposed dot, that is, $\mathbf{a} = \dot{\mathbf{v}}$, which implies that $a_i = \dot{v}_i$ in the rectangular frame.

Suppose that the mass density of the material occupying R is $\rho(p, t)$. Conservation of mass during a deformation with particle velocity field v_i requires that

$$\dot{\rho} + \rho v_{i,i} = 0. \qquad (1.2.25)$$

If the material is subjected to a body force with rectangular components $f_i(p, t)$ per unit mass, the balance of linear momentum for an arbitrary *material* volume requires that the fields must satisfy

$$\frac{\partial \sigma_{ji}}{\partial x_j} + \rho f_i = \rho a_i. \qquad (1.2.26)$$

The balance of angular momentum requires that the stress tensor is symmetric, or

$$\sigma_{ij} = \sigma_{ji}.$$
(1.2.27)

If traction and particle velocity are discontinuous across a surface moving through the material, then the momentum balance equation (1.2.26) is replaced by

$$[\![\sigma_{ij}\nu_i + \rho V v_j]\!] = 0,$$
(1.2.28)

where V is the local speed of the surface in the direction of the unit normal ν_i with respect to material particles instantaneously on the surface. The double square bracket notation is adopted for the jump in the value of a function f that is discontinuous across a surface whereby

$$[\![f]\!] = f^+ - f^-,$$
(1.2.29)

where $f^\pm$ is the limiting value of f on the surface as it is approached in the direction of $\mp\nu_i$.

The instantaneous rate of work of a traction distribution on a surface is defined in the following way. Suppose that S is a *material* surface in the body, that is, material particles on the surface at any one time are on the surface at all times. If T_i is the traction distribution on the surface and v_i is the velocity of material particles on the surface then the instantaneous rate of work being done on the material on the $-\nu_i$ side of S by the traction is

$$\int_S T_i v_i \, dS = \int_S \sigma_{ji} \nu_j v_i \, dS.$$
(1.2.30)

The surface S may be an interior surface or any part of the boundary surface of the body.

The kinetic energy per unit volume is defined as

$$T = \tfrac{1}{2}\rho \mathbf{v} \cdot \mathbf{v} = \tfrac{1}{2}\rho v_i v_i,$$
(1.2.31)

where ρ is the material mass density. Total kinetic energy of a material volume, say T_{tot}, is the integral of T over the volume. The stress power per unit volume is defined as

$$P = \sigma_{ij} \frac{\partial v_i}{\partial x_j}.$$
(1.2.32)

Suppose that a certain volume of material instantaneously occupies a region of space R, and that the boundary surface of the region is S with outward normal n_i. Then the rate of work being done on the body by the traction on S and by the body force f_i throughout R is equal to the net stress power plus the rate of increase of kinetic energy throughout R, or

$$\int_S \sigma_{ij} n_i v_j \, dS + \int_R \rho f_j v_j \, dR = \int_R (P + \dot{T}) \, dR. \qquad (1.2.33)$$

This relation can be established by applying the divergence theorem (1.2.16) to the first term on the left side, and by incorporating the momentum balance equations (1.2.26), conservation of mass (1.2.25), and the kinematic definitions.

The field equations and integral relations introduced above with respect to the current configuration of the deformable body can be transformed into equivalent expressions with respect to the reference configuration. Suppose that the body occupies the region R_o of space with boundary surface S_o and outward normal n_i^o in its reference configuration. The mass density in the reference configuration is ρ_o. The nonsymmetric nominal stress tensor field (or the transpose of the first Piola-Kirchhoff stress tensor) is related to the Cauchy stress by

$$\sigma_{ij}^o = \frac{\rho_o}{\rho} \frac{\partial \chi_i^o}{\partial x_k} \sigma_{kj}. \qquad (1.2.34)$$

If f_i^o is the body force per unit mass in the reference configuration, then the balance of linear momentum takes the form

$$\frac{\partial \sigma_{ij}^o}{\partial x_i^o} + \rho_o f_j^o = \rho_o \frac{\partial^2 \chi_j}{\partial t^2}. \qquad (1.2.35)$$

Similarly, the work rate balance equation takes the form

$$\int_{S_o} \sigma_{ij}^o n_i^o \frac{\partial \chi_j}{\partial t} \, dS_o + \int_{R_o} \rho_o f_j^o \frac{\partial \chi_j^o}{\partial t} \, dR_o =$$

$$\int_{R_o} \sigma_{ij}^o \frac{\partial^2 \chi_j}{\partial x_i^o \partial t} \, dR_o + \frac{\partial}{\partial t} \int_{R_o} \tfrac{1}{2} \rho_o \frac{\partial \chi_i}{\partial t} \frac{\partial \chi_i}{\partial t} \, dR_o, \qquad (1.2.36)$$

where $T_o = \frac{1}{2}\rho_o \chi_{i,t}\chi_{i,t}$ is the kinetic energy per unit volume in the reference configuration. Because the reference configuration is fixed, differentiation with respect to time commutes with integration over R_o. Thus, the first term on the right side of (1.2.36) can be rewritten as

$$\int_{R_o} \sigma_{ij}^o \frac{\partial^2 \chi_j}{\partial x_i \partial t}\, dR_o = \frac{\partial}{\partial t}\int_{R_o} U_o\, dR_o\,, \qquad (1.2.37)$$

where U_o is the accumulated stress work per unit volume, or the stress work density

$$U_o = \int_{-\infty}^t \sigma_{ij}^o \frac{\partial^2 \chi_j}{\partial x_i^o \partial \tau}\, d\tau \qquad (1.2.38)$$

at each material point in the reference configuration.

Suppose that $\omega(p,t)$ is a time-dependent field which represents the density per unit mass of some physical quantity throughout a deformable body. Then the net measure of the quantity over a volume of material occupying a time-dependent region of space $R(t)$ is

$$\Omega(t) = \int_{R(t)} \rho(p,t)\omega(p,t)\, dR\,. \qquad (1.2.39)$$

Suppose that $S(t)$ is the time dependent boundary surface of $R(t)$ with outward normal n_i, and that the velocity of any point on S is v_i^S. The time rate of change of the net measure of Ω for the material instantaneously in $R(t)$ is

$$\dot{\Omega} = \int_{R(t)} \frac{\partial(\rho\omega)}{\partial t}\, dR + \int_{S(t)} \rho\omega V_n\, dS\,, \qquad (1.2.40)$$

where $V_n = v_i^S n_i$ is the local normal velocity of the surface S (Truesdell and Toupin 1960). The result states that the value of Ω for the material in the region $R(t)$ changes for two reasons. The first term on the right side of (1.2.40) is a contribution due to the pointwise rate of change of density ω throughout R. The second term on the right side of (1.2.40) accounts for the fact that, as the surface S moves, there is a local flux of ω through the boundary. The net value of this flux also contributes to the rate of change of Ω. In the special case when V_n is the component of velocity of the material particles on $S(t)$ in the direction of n_i, or $V_n = v_i n_i$, the surface S is a material surface and the result (1.2.40) is an expression for the rate of change of Ω for a fixed material volume. In the latter case, (1.2.40) is called the Reynold's transport theorem. For the general case when the normal speed V_n has no connection to the motion of the material particles currently on the surface, the result will be called the *generalized transport theorem*. This theorem plays a central role in computing energy flux through a tubular surface surrounding a crack edge and moving with it, for example.

If the same volume of material is considered in the reference configuration of the body, then the statement of the generalized transport theorem has the form

$$\dot{\Omega}(t) = \int_{R_o} \rho_o \dot{\omega}_o \, dR_o + \int_{S_o} \rho_o \omega_o V_n^o \, dS_o \,, \tag{1.2.41}$$

where $V_n^o = (v_i^S - v_i) n_k^o \chi_{k,i}^o$ is the normal speed of the surface S mapped back to the reference configuration. The normal speed was introduced as an absolute speed of the surface, so in the reference configuration it assumes the form of a speed with respect to the speed of the particles currently on S_o.

Finally, some specific deformation measures are adopted. The rate of deformation tensor $\mathbf{d}$ is the symmetric part of the particle velocity gradient in the current configuration, and its rectangular components are

$$d_{ij} = \frac{1}{2} \left(\frac{\partial v_i}{\partial x_j} + \frac{\partial v_j}{\partial x_i} \right) . \tag{1.2.42}$$

The strain tensor $\mathbf{E}$ in terms of displacement (1.2.21) as a function of coordinates in the reference configuration has components

$$E_{ij} = \frac{1}{2} \left(\frac{\partial u_i}{\partial x_j^o} + \frac{\partial u_j}{\partial x_i^o} + \frac{\partial u_k}{\partial x_i^o} \frac{\partial u_k}{\partial x_j^o} \right) . \tag{1.2.43}$$

Likewise, the strain tensor $\mathbf{E}^*$ in terms of displacement components as a function of coordinates in the current configuration has components

$$E_{ij}^* = \frac{1}{2} \left(\frac{\partial u_i}{\partial x_j} + \frac{\partial u_j}{\partial x_i} + \frac{\partial u_k}{\partial x_i} \frac{\partial u_k}{\partial x_j} \right) . \tag{1.2.44}$$

When all components of displacement gradients in either description are small, the terms involving products of gradient components in

(1.2.43) and (1.2.44) will be small compared to the terms linear in
gradient components. The two strain expressions then have the same
form, except that the gradients are taken with respect to material
coordinates in the former case and spatial coordinates in the latter
case. When the displacements and the displacement gradients are
sufficiently small, the distinction between the two descriptions is ignored and a single set of coordinates x_1, x_2, x_3 is used. In this case,
the strain tensor components are

$$\epsilon_{ij} = \frac{1}{2}\left(\frac{\partial u_i}{\partial x_j} + \frac{\partial u_j}{\partial x_i}\right). \qquad (1.2.45)$$

A description of deformation based on (1.2.45) is referred to as a small
strain formulation or an infinitesimal strain formulation. Although
the small strain formulation can be viewed as an approximation to a
theory of deformable bodies undergoing arbitrary deformations, the
process of extracting the small strain formulation from the general
theory of deformation as an asymptotic result is somewhat ambigu-
ous (Gurtin 1972; Malvern 1969). An interesting perspective on the
conditions necessary for the small strain conditions to be met for
elastic-plastic materials is given by Rice et al. (1979). Formulations
based on small strain concepts are themselves mathematically complete, however, and they provide a consistent framework for solving
boundary value problems in mechanics. Most of the results in this
book are based on small strain concepts. The few exceptions are
clearly identified.

1.2.3 Linear elastodynamics

The fundamental set of field equations governing the motion of a
homogeneous and isotropic elastic body consists of the strain displacement relation for small strain (1.2.45)

$$\epsilon = \tfrac{1}{2}\left(\nabla \mathbf{u} + \mathbf{u}\nabla\right) \quad \text{or} \quad \epsilon_{ij} = \tfrac{1}{2}\left(u_{j,i} + u_{i,j}\right), \qquad (1.2.46)$$

the momentum balance equations (1.2.26) and (1.2.27)

$$\nabla \cdot \boldsymbol{\sigma} + \rho\mathbf{f} = \rho\ddot{\mathbf{u}} \quad \text{or} \quad \sigma_{ij,i} + \rho f_j = \rho\ddot{u}_j, \qquad (1.2.47)$$

and the linear stress–strain relation

$$\boldsymbol{\sigma} = \lambda\mathbf{I}\,\mathrm{tr}(\boldsymbol{\epsilon}) + 2\mu\boldsymbol{\epsilon} \quad \text{or} \quad \sigma_{ij} = \lambda\delta_{ij}\epsilon_{kk} + 2\mu\epsilon_{ij}, \qquad (1.2.48)$$

where λ and μ are the positive Lamé elastic constants. The field equations are to be satisfied throughout a region of space occupied by the elastic solid in its initial, undeformed configuration. No distinction is made between the deformed and undeformed configurations in the linear theory. Thus, fields representing the same physical quantities in the current "deformed" configuration (without superscript or subscript "o") and in the reference "undeformed" configuration in Section 1.2.2 are viewed as being indistinguishable, except that a displacement vector is associated with each material particle. The superscript or subscript "o" designation is dropped in the linear formulation. Thus, for example, the rate of work balance (1.2.33) or (1.2.36) takes the form

$$\int_S \sigma_{ij} n_i \frac{\partial u_j}{\partial t}\, dS + \int_R \rho f_j \frac{\partial u_j}{\partial t}\, dR = \frac{\partial}{\partial t} \int_R (U+T)\, dR, \qquad (1.2.49)$$

where the body occupies region R with boundary S, T is the kinetic energy density, and U is the stress work density for general material response. In the case of linear elasticity, U is the elastic strain energy density $\frac{1}{2}\sigma_{ij}\epsilon_{ij}$.

If the momentum equation is written in terms of displacement $\mathbf{u}$ by substituting first from (1.2.48) and then from (1.2.46), Navier's equation of motion

$$(\lambda + \mu)\nabla(\nabla \cdot \mathbf{u}) + \mu\nabla^2 \mathbf{u} + \rho\mathbf{f} = \rho\ddot{\mathbf{u}} \qquad (1.2.50)$$

is obtained. In light of the vector identity $\nabla \times (\nabla \times \mathbf{u}) = \nabla(\nabla \cdot \mathbf{u}) - \nabla^2 \mathbf{u}$, (1.2.50) can be rewritten as

$$c_d^2 \nabla(\nabla \cdot \mathbf{u}) - c_s^2 \nabla \times (\nabla \times \mathbf{u}) + \mathbf{f} = \ddot{\mathbf{u}}, \qquad (1.2.51)$$

where

$$c_d = \sqrt{\frac{\lambda + 2\mu}{\rho}} \qquad \text{and} \qquad c_s = \sqrt{\frac{\mu}{\rho}}. \qquad (1.2.52)$$

Let the body force $\mathbf{f}$ be zero for the time being. The result of operating on each term of (1.2.51) with the divergence operator $\nabla\cdot$ is

$$c_d^2 \nabla^2 (\nabla \cdot \mathbf{u}) = (\nabla \cdot \mathbf{u})_{,tt}. \qquad (1.2.53)$$

Therefore, the dilatation $\nabla \cdot \mathbf{u}$ satisfies the elementary wave equation
with characteristic wave speed c_d, called the dilatational wave speed.
Similarly, the result of operating on each term of (1.2.51) with the
curl operator $\nabla \times$ is

$$c_s^2 \nabla^2 (\nabla \times \mathbf{u}) = (\nabla \times \mathbf{u})_{,tt} \qquad (1.2.54)$$

so that the rotation vector $\frac{1}{2} \nabla \times \mathbf{u}$ satisfies the elementary wave equation with characteristic wave speed c_s, called the shear wave speed or
the rotational wave speed. In an unbounded isotropic elastic solid,
waves of pure dilatation (rotation) propagate at normal speed c_d (c_s).
Furthermore, the two types of wave are independent in the interior of
a solid. At boundaries, such as the faces of a crack, for example, the
two types of wave interact. This is the feature of elastodynamic fields
that accounts for most of the complication in solving initial-boundary
value problems. A complete discussion of plane wave propagation in
unbounded bodies and plane wave reflection from boundaries is given
by Achenbach (1973). He also presents the basic singular solutions
of elastodynamics, principally, Stokes's solution for a time-dependent
concentrated force in an unbounded solid and Love's solution for a
center of dilatation of time-dependent strength in an unbounded solid.

The dilatational and shear wave speeds for an isotropic elastic
material are defined in (1.2.52) in terms of mass density and the Lamé
elastic constants λ and μ. It is sometimes more convenient to express
the wave speeds in terms of Young's modulus E and Poisson's ratio
ν. The relationships among these elastic constants are

$$E = \frac{\mu(3\lambda + 2\mu)}{\lambda + \mu}, \qquad \nu = \frac{\lambda}{2(\lambda + \mu)} \qquad (1.2.55)$$

which can be inverted to yield

$$\lambda = \frac{E\nu}{(1 + \nu)(1 - 2\nu)}, \qquad \mu = \frac{E}{2(1 + \nu)}. \qquad (1.2.56)$$

In terms of E and ν, the wave speeds are

$$c_d = \sqrt{\frac{E(1 - \nu)}{\rho(1 + \nu)(1 - 2\nu)}}, \qquad c_s = \sqrt{\frac{E}{2\rho(1 + \nu)}}. \qquad (1.2.57)$$

Another useful relationship follows from recognizing that the ratio c_s/c_d has a form depending only on Poisson's ratio,

$$\frac{c_s^2}{c_d^2} = \frac{1 - 2\nu}{2(1 - \nu)}. \qquad (1.2.58)$$

To formulate an initial-boundary value problem in elastodynamics, an appropriate set of initial and boundary conditions is required. In view of the Navier equation, values of $\mathbf{u}$ and $\dot{\mathbf{u}}$ must be specified throughout the body initially, or some conditions equivalent to these must be specified. Boundary conditions must be specified on the boundary of the body for all time beyond the initial time. Conditions on the boundary surface may be:

    i. the traction is specified,

    ii. the displacement is specified,

    iii. the component of traction in a certain direction is specified and the displacement in the plane perpendicular to this direction is also specified, or

    iv. the component of displacement in a certain direction is specified and the traction in the plane perpendicular to this direction is also specified.

The boundary may be divided into any number of parts, but one and only one of the conditions (i) through (iv) must be specified over each part. For an appropriate set of initial and boundary conditions, Neumann's uniqueness theorem ensures that the initial-boundary value problem has at most one solution, provided that energy is added to the body or taken from it only through the loaded part of the boundary (Gurtin 1972; Achenbach 1973). This property of uniqueness of solutions has enormous practical value, for it means that an elastodynamic field obtained by any means whatsoever which satisfies the governing equations, initial conditions, and boundary conditions is the only solution of the problem. In obtaining solutions to elastic crack problems, conditions are often assumed in constructing a candidate solution that cannot be verified a priori. Nonetheless, if such a candidate solution can be shown to satisfy the governing equations and other conditions, then it is certain that it represents the unique solution. For dynamically growing cracks, the statement of the uniqueness theorem requires some modification due to the possibility of energy flow out of the body through the crack tip, as discussed in Section 7.7.

The elastodynamic reciprocal theorem provides a useful relationship between solutions of two different boundary value problems for the same elastic body. Suppose that solutions of two different boundary value problems for a given body exist, say $\mathbf{u}$ and $\mathbf{u}^*$. Furthermore, suppose that the displacement and particle velocity are zero for both solutions at the initial time, say $t = 0$. In addition, suppose that no body forces are present in either boundary value problem. Then, if $\mathbf{T}$ and $\mathbf{T}^*$ are the tractions on the boundary surface, say S, corresponding to the two solutions, then the solutions are related by

$$\int_S \int_0^t \mathbf{T}(p_S, t - \tau) \cdot \mathbf{u}^*(p_S, \tau) \, d\tau \, dS$$

$$= \int_S \int_0^t \mathbf{T}^*(p_S, t - \tau) \cdot \mathbf{u}(p_S, \tau) \, d\tau \, dS. \quad (1.2.59)$$

The generalization of Betti's reciprocal theorem to elastodynamics, including body forces and general initial conditions, is discussed in detail by Gurtin (1972) and Achenbach (1973). The theorem was extended to the case of unbounded domains by Wheeler and Sternberg (1968). There is no difficulty in extending the theorem to apply for general initial conditions and body forces.

The practical matter of integrating (1.2.51) to determine particular solutions often hinges on a representation of $\mathbf{u}$ in terms of other fields which satisfy the elementary wave equations with characteristic speed c_d or c_s. The advantage gained through such a representation is that various means are available for solving the wave equation. A difficulty with this approach concerns the completeness of the representation. Does every solution of the Navier equation have a representation of the assumed form? This matter is discussed by Sternberg (1960) for the representations to be introduced here.

A complete representation of a solution of the wave equation is obtained by means of the Helmholtz additive decomposition of the displacement vector into the gradient of a scalar field ϕ and the curl of a divergence free vector field $\boldsymbol{\psi}$,

$$\mathbf{u} = \nabla\phi + \nabla \times \boldsymbol{\psi} \quad \text{or} \quad u_i = \phi_{,i} + e_{ijk}\psi_{k,j}, \quad (1.2.60)$$

where the dilatational displacement potential ϕ and the shear displacement potential $\boldsymbol{\psi}$ are wave functions satisfying

$$c_d^2 \nabla^2 \phi - \ddot{\phi} = 0, \quad c_s^2 \nabla^2 \boldsymbol{\psi} - \ddot{\boldsymbol{\psi}} = 0, \quad \nabla \cdot \boldsymbol{\psi} = 0. \quad (1.2.61)$$

In terms of the displacement potential functions ϕ and ψ, the dilatation of the deformation is $\nabla \cdot \mathbf{u} = \nabla^2 \phi$ and the rotation vector of the deformation is $\frac{1}{2}\nabla \times \mathbf{u} = -\frac{1}{2}\nabla^2 \psi$. The relationship (1.2.60) is called the Lamé representation of the displacement vector.

Sternberg (1960) also discussed other representations. For example, the Poisson representation is

$$\mathbf{u} = \mathbf{u}^d + \mathbf{u}^s, \qquad (1.2.62)$$

where

$$\nabla \times \mathbf{u}^d = 0, \qquad \nabla \cdot \mathbf{u}^s = 0 \qquad (1.2.63)$$

and

$$c_d^2 \nabla^2 \mathbf{u}^d - \ddot{\mathbf{u}}^d = 0, \qquad c_s^2 \nabla^2 \mathbf{u}^s - \ddot{\mathbf{u}}^s = 0. \qquad (1.2.64)$$

This representation has been applied effectively in two-dimensional cases when the elastodynamic fields are steady with respect to a moving observer or are self-similar, so that the wave equations reduce to Laplace's equation for which a general solution is available. For the special case of antiplane shear deformation, $u_1 = u_2 = 0$ and $u_3 = w(x_1, x_2, t)$. The out-of-plane displacement satisfies the elementary wave equation

$$c_s^2 \nabla^2 w - \ddot{w} = 0. \qquad (1.2.65)$$

Yet another representation that is shown by Sternberg (1960) to be equivalent to the Lamé representation is

$$\mathbf{u} = 2(1 - \nu)\left[c_d^2 \nabla^2 \mathbf{G} - \ddot{\mathbf{G}}\right] - \nabla(\nabla \cdot \mathbf{G}), \qquad (1.2.66)$$

where the vector fields $\mathbf{G}$ and $\mathbf{H}$ satisfy

$$c_s^2 \nabla^2 \mathbf{G} - \ddot{\mathbf{G}} = \mathbf{H}, \qquad c_d^2 \nabla^2 \mathbf{H} - \ddot{\mathbf{H}} = 0. \qquad (1.2.67)$$

The Lamé representation is used extensively in this book, and expressions for rectangular components of displacement and stress in terms of ϕ and components of ψ are recorded here for later reference. From (1.2.60), the components of displacement in terms of the displacement potentials are

$$u_1 = \frac{\partial \phi}{\partial x_1} + \frac{\partial \psi_3}{\partial x_2} - \frac{\partial \psi_2}{\partial x_3},$$

$$u_2 = \frac{\partial \phi}{\partial x_2} + \frac{\partial \psi_1}{\partial x_3} - \frac{\partial \psi_3}{\partial x_1}, \tag{1.2.68}$$

$$u_3 = \frac{\partial \phi}{\partial x_3} + \frac{\partial \psi_2}{\partial x_1} - \frac{\partial \psi_1}{\partial x_2}.$$

If these expressions are substituted into (1.2.48), then the components of stress in terms of the displacement potentials are

$$\sigma_{11} = \mu \left[\frac{c_d^2}{c_s^2} \nabla^2 \phi - 2\frac{\partial^2 \phi}{\partial x_2^2} - 2\frac{\partial^2 \phi}{\partial x_3^2} + 2\frac{\partial^2 \psi_3}{\partial x_1 \partial x_2} - 2\frac{\partial^2 \psi_2}{\partial x_3 \partial x_1} \right],$$

$$\sigma_{22} = \mu \left[\frac{c_d^2}{c_s^2} \nabla^2 \phi - 2\frac{\partial^2 \phi}{\partial x_3^2} - 2\frac{\partial^2 \phi}{\partial x_1^2} + 2\frac{\partial^2 \psi_1}{\partial x_2 \partial x_3} - 2\frac{\partial^2 \psi_3}{\partial x_1 \partial x_2} \right],$$

$$\sigma_{33} = \mu \left[\frac{c_d^2}{c_s^2} \nabla^2 \phi - 2\frac{\partial^2 \phi}{\partial x_1^2} - 2\frac{\partial^2 \phi}{\partial x_2^2} + 2\frac{\partial^2 \psi_2}{\partial x_3 \partial x_1} - 2\frac{\partial^2 \psi_1}{\partial x_2 \partial x_3} \right],$$

$$\sigma_{12} = \mu \left[2\frac{\partial^2 \phi}{\partial x_1 \partial x_2} + \frac{\partial^2 \psi_3}{\partial x_2^2} - \frac{\partial^2 \psi_2}{\partial x_2 \partial x_3} - \frac{\partial^2 \psi_3}{\partial x_1^2} + \frac{\partial^2 \psi_1}{\partial x_3 \partial x_1} \right],$$

$$\sigma_{23} = \mu \left[2\frac{\partial^2 \phi}{\partial x_2 \partial x_3} + \frac{\partial^2 \psi_1}{\partial x_3^2} - \frac{\partial^2 \psi_3}{\partial x_3 \partial x_1} - \frac{\partial^2 \psi_1}{\partial x_2^2} + \frac{\partial^2 \psi_2}{\partial x_1 \partial x_2} \right],$$

$$\sigma_{31} = \mu \left[2\frac{\partial^2 \phi}{\partial x_3 \partial x_1} + \frac{\partial^2 \psi_2}{\partial x_1^2} - \frac{\partial^2 \psi_1}{\partial x_1 \partial x_2} - \frac{\partial^2 \psi_2}{\partial x_3^2} + \frac{\partial^2 \psi_3}{\partial x_2 \partial x_3} \right].$$

$$\tag{1.2.69}$$

Although no specific solutions are presented for anisotropic elastic material response, some of the general developments are carried out for an elastic material of general symmetry. The most general stress–strain relation for a material possessing a strain energy function is

$$\sigma_{ij} = C_{ijkl} \epsilon_{kl}. \tag{1.2.70}$$

The array of elastic constants satisfies the symmetry conditions

$$C_{ijkl} = C_{klij} = C_{jikl} = C_{ijlk} \qquad (1.2.71)$$

so that there are 21 elastic constants at most. In terms of the stiffness array C_{ijkl}, the strain energy density of the material is

$$U = \tfrac{1}{2}C_{ijkl}\epsilon_{ij}\epsilon_{kl} = \tfrac{1}{2}C_{ijkl}u_{i,j}\,u_{k,l}\ . \qquad (1.2.72)$$

It is assumed that the relationship is invertible, so that there exists an array M_{ijkl} such that

$$\epsilon_{ij} = M_{ijkl}\sigma_{kl}\ . \qquad (1.2.73)$$

The array M_{ijkl} of elastic compliances possesses the same symmetries as C_{ijkl} does.

1.2.4 Inelastic materials

The limited results available on dynamic fracture of inelastic materials are also discussed in the chapters to follow. General analytical procedures do not exist for solving boundary value problems involving inelastic material response, except for the case of linear viscoelasticity, and results are obtained on the basis of ad hoc procedures. The relevant field equations for formulating such problems, aside from the constitutive equations, are included in Section 1.2.2. Field equations describing material response are presented along with the specific problems studied. Among these are the determination of asymptotic crack tip fields for crack growth in a rate independent elastic-plastic material in Sections 4.4 and 4.5. Crack tip fields for rapid crack growth in an elastic-viscous material are considered in Section 4.6, and in an elastic-viscoplastic material in Section 4.7.

Chapter 8 is devoted to the study of plasticity and strain-rate effects in dynamic fracture. Here, a few results of fairly broad applicability are obtained for rapid crack growth in elastic-plastic and elastic-viscoplastic materials. This chapter also includes brief discussions of ductile void growth in plastic materials and the apparent rate sensitivity exhibited by brittle materials undergoing rapid microcracking. In all cases, constitutive models are developed as part of the problem formulation.

1.3 Analytic functions and Laplace transforms

Some basic mathematical results to be used subsequently are summarized in this section. These results concern the general properties of an analytic function of a complex variable and the properties of Laplace transforms, both the one-sided and the two-sided types. Theorems are stated without proof, and only to a degree of generality important to the applications to follow.

The theory of analytic functions of a complex variable is fully developed by Hille (1959) and, within the context of applications, by Carrier, Krook, and Pearson (1966). A useful summary of main results is given by Seebass (1983). Integral transforms are developed by Carrier et al. (1966), and aspects relevant to the Wiener-Hopf method are summarized by Noble (1958).

1.3.1 Analytic functions of a complex variable

Consider the function $f(\zeta)$ of the complex variable $\zeta = x + iy = re^{i\theta}$ defined in a neighborhood of a particular point. If $f(\zeta)$ is differentiable with respect to ζ at that point, then it is an analytic function at that point. If $f(\zeta)$ is analytic at every point in a region, it is said to be analytic in the region. Analyticity of $f(\zeta)$, in turn, implies that derivatives of all orders exist.

If $f(\zeta)$ is analytic then, with $f(\zeta) = u(x,y) + iv(x,y)$, it follows that

$$\frac{\partial u}{\partial x} = \frac{\partial v}{\partial y}, \qquad \frac{\partial u}{\partial y} = -\frac{\partial v}{\partial x}. \qquad (1.3.1)$$

These relations are called the Cauchy-Riemann equations. If $f(\zeta)$ is analytic in a region, then the Cauchy-Riemann relations imply that

$$\nabla^2 u = 0, \qquad \nabla^2 v = 0 \qquad (1.3.2)$$

in that region. Real functions that satisfy Laplace's equation are said to be harmonic functions. A general solution of the partial differential equation $\nabla^2 u = 0$ over some region is

$$u = \mathrm{Re}\{F(\zeta)\} \quad \text{or} \quad u = \mathrm{Im}\{G(\zeta)\}, \qquad (1.3.3)$$

where $F(\zeta)$ or $G(\zeta)$ is an analytic function in the region that must be determined from the boundary conditions.

Among the most useful results on analytic functions for purposes of applied analysis are Cauchy's integral theorem and its consequences. If $f(\zeta)$ is analytic on and inside a simple closed curve C in the complex ζ-plane, then

$$\oint_C f(\zeta)\, d\zeta = 0.$$ (1.3.4)

Similarly, if $f(\zeta)$ is analytic on and outside of a simple closed curve, including at infinity, and if $\left|f(re^{i\theta})\right| = o(1/r)$ as $r \to \infty$ then (1.3.4) also applies. Either form of the theorem can be extended to apply for multiple simple closed curves by interconnecting the simple curves with paths through regions of analyticity. If these paths are traversed in both directions, then the multiple curves and interconnections can be viewed as a simple closed curve.

A consequence of Cauchy's theorem is Cauchy's integral formula. Suppose that $f(\zeta)$ is analytic on and inside a simple curve C enclosing a region of the ζ-plane. If ζ_0 is a point inside C then

$$f(\zeta_0) = \frac{1}{2\pi i} \oint_C \frac{f(\zeta)}{\zeta - \zeta_0}\, d\zeta,$$ (1.3.5)

where the path is traversed in the counterclockwise direction. Similarly, suppose that $f(\zeta)$ is analytic on and outside of a simple closed curve C in the ζ-plane, including at infinity. Then, if $\left|f(re^{i\theta})\right| = o(1)$ as $r \to \infty$ and if ζ_0 is outside of C then

$$f(\zeta_0) = -\frac{1}{2\pi i} \oint_C \frac{f(\zeta)}{\zeta - \zeta_0}\, d\zeta,$$ (1.3.6)

where the curve C is counterclockwise oriented.

Various other extensions of the integral formula (1.3.5) are useful in different circumstances. For example, if $f(\zeta)$ is analytic on and inside the simple closed curve C and ζ_0 is inside C then

$$f^{(n)}(\zeta_0) = \frac{n!}{2\pi i} \oint_C \frac{f(\zeta)}{(\zeta - \zeta_0)^{n+1}}\, d\zeta,$$

where the left side denotes the n-th derivative of $f(\zeta)$. Cauchy's integral formula can also be used to establish other useful properties of

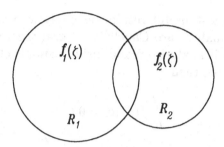

Figure 1.3. Overlapping regions in the complex ζ-plane used to illustrate the role of the identity theorem in analytic continuation.

analytic functions. For example, according to the maximum modulus theorem, if $f(\zeta)$ is analytic on and inside a simple closed curve C, then either the maximum value of $|f(\zeta)|$ is attained only on C or $f(\zeta)$ is constant throughout the region. Another very powerful result is Liouville's theorem, which states that a bounded entire function of a complex variable is necessarily a constant. A generalization of Liouville's theorem is stated in Section 2.5.2 in connection with the Wiener-Hopf technique. Yet another result, called the zero counting formula, is used to great advantage in Section 2.5.

A theorem which provides the basis for techniques of *analytic continuation* is the identity theorem or uniqueness theorem of analytic function theory. Suppose that $f(\zeta)$ is analytic in a region of the complex ζ-plane and that $f(\zeta) = 0$ on any dense subset of that region, for example, along a segment of a curve or over a small area. Under these conditions, $f(\zeta)$ is identically zero throughout the region. This very important theorem plays a central role in most techniques of applied mathematics based on analytic function theory and on integral transforms. It follows immediately from the identity theorem that if $f_1(\zeta)$ and $f_2(\zeta)$ are both analytic throughout a region of the ζ-plane and if $f_1(\zeta) = f_2(\zeta)$ on a dense subset of the region, then the two functions are equal to each other and, indeed, each represents precisely the same function over the whole region.

To see how the identity theorem provides the basis for analytic continuation, consider functions $f_1(\zeta)$ and $f_2(\zeta)$ that are analytic in regions R_1 and R_2, respectively, of the complex plane. Furthermore, suppose that the two regions have some overlap, as shown schematically in Figure 1.3. Then, if $f_1(\zeta) = f_2(\zeta)$ within the common region of overlap, $f_1(\zeta)$ or $f_2(\zeta)$ represents the analytic continuation of

$f_2(\zeta)$ or $f_1(\zeta)$ into R_1 or R_2, respectively, and the functions together represent a single analytic function throughout the combined region $R_1 + R_2$. This analytic continuation idea is the basis for an important step in the Wiener-Hopf technique.

1.3.2 Laplace transforms

Laplace transforms are used in a rather formal way to obtain solutions to boundary value problems in the chapters to follow. The transforms are defined here, and certain properties which are exploited later are identified. The equations to be solved are typically partial differential equations involving functions of time and one or more spatial variables. For this general discussion, it is sufficient to consider a function of time t and a single spatial variable, say x.

Consider a function $w(x,t)$ defined for $t > 0$ and $-\infty < x < \infty$. The Laplace transform on time of w is denoted by a superposed hat on the function and is defined by

$$\widehat{w}(x,s) = \int_0^\infty w(x,t)\, e^{-st}\, dt\,, \qquad (1.3.7)$$

where s is the transform parameter. Suppose that $w(x,t)$ is integrable over any finite interval of t and $|w(x,t)| = O(e^{\sigma_+ t})$ as $t \to \infty$, where σ_+ is a real constant. Then the integral in (1.3.7) converges for $\mathrm{Re}(s) > \sigma_+$ and it defines some function of s in that half plane. Furthermore, the integral can be differentiated with respect to s, and the derivative converges for $\mathrm{Re}(s) > \sigma_+$. Thus, $\widehat{w}(x,s)$ is an analytic function of the complex variable s in $\mathrm{Re}(s) > \sigma_+$ for fixed x. In view of the identity theorem for analytic functions, it is sufficient to view $\widehat{w}(x,s)$ as a function of a real variable s over some segment of the real axis in the half plane of analyticity. Once $\widehat{w}(x,s)$ is determined as an explicit function of s in the course of solving the transformed differential equations, the definition of $\widehat{w}(x,s)$ can be extended to the entire complex s-plane, except for isolated singular points, by analytic continuation.

The function $w(x,t)$ is obtained by means of the transform inversion integral

$$w(x,t) = \frac{1}{2\pi i} \int_{\sigma - i\infty}^{\sigma + i\infty} \widehat{w}(x,s)\, e^{st}\, ds\,, \qquad (1.3.8)$$

where σ is any real parameter greater than σ_+. The integration path is a line parallel to the imaginary axis in the s-plane and within the region of analyticity of $\widehat{w}(x, s)$.

In the particular case when a transform is factorable into a product of transforms, say

$$\widehat{w}(x, s) = \widehat{w}_1(x, s)\widehat{w}_2(x, s), \qquad (1.3.9)$$

the function $w(x, t)$ itself is a convolution of $w_1(x, t)$ and $w_2(x, t)$, that is,

$$w(x, t) = \int_0^t w_1(x, t - \tau)w_2(x, \tau)\, d\tau. \qquad (1.3.10)$$

Transform inversion integrals of the form (1.3.8) are often evaluated by changing the path of integration from the infinite line parallel to the imaginary axis to a path in a left half plane that encloses all of the singular points of the integrand in the complex s-plane. The change in path is accomplished by forming a closed path made up of the inversion path and the path enclosing the singularities plus circular arcs of indefinitely large radius, and then invoking Cauchy's theorem (1.3.4). For this process to be successful, it is essential that the integrals taken along the circular arcs of large radius vanish as the radius becomes indefinitely large. This will be the case provided that $\widehat{w}(x, s) \to 0$ uniformly as $|s| \to \infty$. This result follows from Jordan's lemma as applied to integrals of the form (1.3.8) (Seebass 1983).

The transform of the time derivative $\partial w(x, t)/\partial t$ is

$$\int_{0^+}^{\infty} \frac{\partial w}{\partial t} e^{-st}\, dt = s\widehat{w}(x, s) - w(x, 0^+). \qquad (1.3.11)$$

If w is identically zero for $t < 0$ and if it increases discontinuously at $t = 0$ to $w(x, 0^+)$ then, in the sense of distributions (Stakgold 1968),

$$\int_{0^-}^{\infty} \frac{\partial w}{\partial t} e^{-st}\, dt = s\widehat{w}(x, s). \qquad (1.3.12)$$

The utility of the Laplace transform methods in solving linear differential equations is represented by (1.3.11). Multiplication of each term in the differential equation by e^{-st} and integration over t from 0^+ to ∞, in effect, replaces time differentiation with multiplication by the

parameter s. The parameter s can be manipulated algebraically. Once
a solution of the transformed equations is found, the solution in the
time domain is obtained by superposition over the continuous range
of s represented by (1.3.8). The completeness of the representation
of functions in terms of integral transforms is established by means
of Fourier's integral theorem (Carrier et al. 1966). In the present
case, completeness implies that every function (within an appropriate
class of integrable functions) has the representation (1.3.8), where the
density function $\widehat{w}(x, s)$ is given by (1.3.7).

Consider now the function $\widehat{w}(x, s)$ for any fixed real s in the region
of convergence of $\widehat{w}(x, s)$ and for $-\infty < x < \infty$. The two-sided or
bilateral Laplace transform of $\widehat{w}(x, s)$ is denoted by a corresponding
uppercase letter and it is defined by the integral

$$
\begin{aligned}
W(\xi, s) &= \int_{-\infty}^{\infty} \widehat{w}(x, s)\, e^{-\xi x}\, dx \\
&= \int_{-\infty}^{0} \widehat{w}(x, s)\, e^{-\xi x}\, dx + \int_{0}^{\infty} \widehat{w}(x, s)\, e^{-\xi x}\, dx\,,
\end{aligned}
\tag{1.3.13}
$$

where ξ is the complex transform parameter. Suppose that $\widehat{w}(x, s)$ is
integrable over any finite interval of x and that

$$
|\widehat{w}(x, s)| = \begin{cases} O\left(e^{\tau_+ + x}\right) & \text{as } x \to +\infty, \\ O\left(e^{-\tau_- - x}\right) & \text{as } x \to -\infty, \end{cases}
\tag{1.3.14}
$$

where τ_+ and τ_- are real constants. The integral over positive values
of x in (1.3.13) then converges and defines an analytic function of ξ in
the half plane $\text{Re}(\xi) > \tau_+$, whereas the integral over negative values of
x converges and defines an analytic function in the half plane $\text{Re}(\xi) <
\tau_-$. Thus, $W(\xi, s)$ exists and represents an analytic function in the
strip $\tau_+ < \text{Re}(\xi) < \tau_-$. Once sufficient information is available about
the dependence of $W(\xi, s)$ on ξ from the boundary value problem,
the definition can be extended to the entire complex plane, except for
isolated singular points, by means of analytic continuation. If $\tau_+ = \tau_-$
the strip of analyticity degenerates to a line, and if $\tau_+ > \tau_-$ then the
two-sided transform does not exist.

The inversion integral for the two-sided Laplace transform is

$$
\widehat{w}(x, s) = \frac{1}{2\pi i} \int_{\tau - i\infty}^{\tau + i\infty} W(\xi, s)\, e^{\xi x}\, d\xi\,,
\tag{1.3.15}
$$

where τ is a real parameter in the interval $\tau_+ < \tau < \tau_-$. The integration path is a line parallel to the imaginary axis in the ξ-plane within the strip of analyticity of $W(\xi, s)$.

Consider the special case of a function of x and t that is zero for all x and $t < 0$ and that vanishes for all $x \leq 0$ and $t > 0$, say $w_+(x, t)$. The Laplace transform on time $\widehat{w}_+(x, s)$ then also vanishes for $x \leq 0$ and s in a suitable interval of the real axis. The "two-sided" Laplace transform of $\widehat{w}_+(x, t)$ is then no different from a one-sided transform, that is,

$$W_+(\xi, s) = \int_0^\infty \widehat{w}_+(x, s) \, e^{-\xi x} \, dx. \qquad (1.3.16)$$

Evidently, the value of $W_+(\xi, s)$ at any particular value of ξ depends on $\widehat{w}_+$ over the entire range of x. Likewise, from the inversion integral, the value of $\widehat{w}_+(x, s)$ at any x depends on W_+ over the full range of ξ. However, transform pairs of the kind defined by (1.3.16) have remarkable asymptotic properties relating limiting values of a function and its transform. According to van der Pol and Bremmer (1955), theorems that yield an asymptotic property of a transform from a known asymptotic property of the function are called Abel theorems. On the other hand, theorems that provide an asymptotic property of a function from a known asymptotic property of its transform are called Tauber theorems.

The basic Abel theorem has been useful in elastic fracture mechanics studies. In this case, direct asymptotic analysis of the differential equations often leads to the conclusion that a particular field depends on a spatial coordinate near the crack tip in a certain way. Through an Abel theorem, it is then possible to make definite statements about the behavior of the transform at remote points in the transform parameter plane.

For example, suppose that $\widehat{w}_+(x, s)$ in (1.3.16) has the behavior

$$\widehat{w}_+(x, s) \sim x^\alpha \qquad (1.3.17)$$

near $x = 0$, where α is a real constant with $\alpha > -1$. The Abel theorem then states that

$$\lim_{x \to 0+} \frac{\Gamma(1 + \alpha)}{x^\alpha} \widehat{w}_+(x, s) = \lim_{\xi \to +\infty} \xi^{1+\alpha} W_+(\xi, s), \qquad (1.3.18)$$

where $\Gamma(\cdot)$ is the Gamma or factorial function. If the region of convergence of the integral defining W_+ includes the imaginary axis then it can also be established that

$$\lim_{x \to +\infty} \frac{\Gamma(1+\beta)}{x^\beta} \widehat{w}_+(x,s) = \lim_{\xi \to 0^+} \xi^{1+\beta} W_+(\xi,s) \qquad (1.3.19)$$

for an appropriate value of $\beta > -1$.

1.4 Overview of dynamic fracture mechanics

Analytical results of a permanent nature in the area of dynamic fracture mechanics are gathered together and summarized in Chapters 2 through 8. For the most part, the results are obtained within the small strain formulation of continuum mechanics as described in Section 1.2.2. Cracks are assumed to be planar and the opposite faces are assumed to occupy the same plane in the undeformed configuration. The geometrical configurations considered are also very simple, in general. For example, numerous developments are based on the assumption of a half plane crack in an otherwise unbounded body under conditions that result in a two-dimensional state of deformation. Though this assumption provides the basis for the formulation of a well-posed boundary value problem, it does have practical limitations. The results can only be applied in cases when the material points of interest are much closer to the crack edge than to other boundaries of the body and when the local deformation field is indeed essentially two-dimensional. Other configurations considered include crack growth in a long uniform strip and idealizations of bodies as structural elements, typically beams or strings.

The purpose here is to provide a very brief overview of the material in the chapters to follow, including a summary of the main results and comments on directions for further study. A number of books and monographs are available to provide background on fracture mechanics from various perspectives. For example, the books by Kanninen and Popelar (1985), Cherepanov (1979), and Hellan (1984) develop the topic from the mechanics perspective. The books by Knott (1979) and Lawn and Wilshaw (1975) introduce the subject from a materials science perspective, whereas Barsom and Rolfe (1987) discuss fracture within the context of materials selection and structural applications.

A number of interesting but specialized issues in dynamic fracture are discussed only briefly in this monograph. Among these topics are: the existence of an apparent terminal crack speed well below

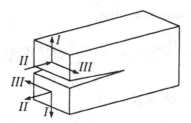

Figure 1.4. The classification of crack opening modes I, II, and III based on the components of the relative displacement of the crack faces near the crack edge.

the Rayleigh wave speed in glass and some other very brittle materials; the contact interaction of crack faces during dynamic fracture events in materials; the symmetric bifurcation of a straight, rapidly running crack in a tensile field into two branches; the influence of heat generated during rapid fracture on the separation process or on the stress distribution; atomistic modeling of dynamic cleavage; and the erosion of materials by impact of solid particles or liquid droplets. Some experimental research and analytical modeling have been done on each of these problems using the approaches described in the following chapters, but the phenomena are not yet completely understood. Available references are included in the Bibliography. Other topical areas that are not yet fully developed are identified in the overview to follow or in the later chapters.

1.4.1 Basic elastodynamic solutions for a stationary crack

In Chapter 2, the fundamental stationary crack problems of linear elastodynamics are analyzed. The two-dimensional configuration of a semi-infinite crack in an otherwise unbounded body is considered. The body is initially stress free and at rest. At time $t = 0$, equal and opposite spatially uniform tractions begin to act on the crack faces. Formulated in this way, the problem has neither a characteristic length nor a characteristic time. The in-plane opening mode of crack deformation (mode I), the in-plane shearing mode (mode II), and the antiplane shearing mode (mode III) are treated separately. The direction of the relative crack opening displacement for each mode is indicated in Figure 1.4, and the classification is further discussed in Section 2.1.

Two mathematical methods for solving initial-boundary value problems are introduced, namely, Green's method based on superposition over a fundamental singular solution of the governing differential equation and the Wiener-Hopf method based on Laplace transform techniques. These methods are presented in a formal way, and they are used in later chapters as well. The main results of the analysis are very simple. Indeed, aside from a scalar multiplier of order unity, the stress distribution near the crack edge can be obtained for each mode through dimensional analysis. With reference to the coordinate system shown in Figure 1.4, if the crack face traction has magnitude τ^* per unit area and if it acts in a direction parallel to the crack edge, then the state of deformation is antiplane shear and the crack opens in mode III. The shear stress on the plane $y = 0$ directly ahead of the crack tip, and close to it in some sense, is

$$\sigma_{zy}(x,0,t) \sim \frac{K_{III}(t)}{\sqrt{2\pi x}}, \qquad K_{III}(t) = 2\tau^*\sqrt{\frac{2c_s t}{\pi}},$$

where c_s is the elastic shear wave speed. This result is derived in Section 2.3 where it appears as (2.3.18). Likewise, if a normal pressure of magnitude σ^* begins to act on the crack faces at time $t = 0$ then the state of deformation is plane strain and the crack opens in mode I. The normal stress on the plane $y = 0$ near the crack tip in this case is

$$\sigma_{yy}(x,0,t) \sim \frac{K_I(t)}{\sqrt{2\pi x}}, \qquad K_I(t) = 2\sigma^*\frac{\sqrt{c_d t(1-2\nu)/\pi}}{(1-\nu)},$$

where c_d is the dilatational wave speed of the material and ν is Poisson's ratio. This result is derived in Section 2.5.3 where it appears as (2.5.44). Lastly, if a shear traction of magnitude τ^* begins to act on the crack faces at time $t = 0$ in a direction perpendicular to the crack edge, then the state of deformation is plane strain and the crack opens in mode II. The shear stress on the plane directly ahead of the crack tip in this case is

$$\sigma_{xy}(x,0,t) \sim \frac{K_{II}(t)}{\sqrt{2\pi x}}, \qquad K_{II}(t) = 2\tau^*\sqrt{\frac{2c_s t}{\pi(1-\nu)}}.$$

This result is given in Section 2.6 as equation (2.6.9). The solutions obtained in Chapter 2 are readily generalized to the case of applied crack face loading with arbitrary time variation. Furthermore, the solution for mode I, II, or III is identical to that for a plane tensile, transverse shear, or horizontal shear wave carrying a simple jump in stress that is normally incident on the crack plane, except for an elementary one-dimensional plane pulse.

1.4.2 Further results for a stationary crack

The problems analyzed in detail in Chapter 2 have been extremely important in establishing the connection between transient loading applied to a cracked elastic body and the time dependence of the resulting stress intensity factor. The fact that the fields are strictly two-dimensional and that the loading/configuration systems do not involve a characteristic length limits the applicability of the results in analyzing certain phenomena. Thus, Chapter 3 is devoted to extensions that overcome some of these limitations.

In Section 3.2, the mode I problem of a pair of concentrated normal forces suddenly applied to the faces of a semi-infinite crack at a fixed distance from the crack tip is considered. The fixed distance from the load point to the crack tip is a characteristic length in this case. The loading on the crack faces generates a displacement of the surface, and a moving dislocation problem is formulated and solved to make it possible to cancel this displacement for points beyond the crack tip. The exact stress intensity factor history is obtained by constructing a solution as a superposition over a fundamental moving dislocation field for the configuration. For loads that tend to open the crack, the stress intensity factor is initially negative, as could be anticipated on the basis of known elastodynamic wave fields. The surprising result is that the transient stress intensity factor takes on its long-time equilibrium limiting value *instantaneously* at a time corresponding to the arrival time at the crack tip of a Rayleigh wave generated at the load point at $t = 0$, rather than asymptotically as $t \to \infty$. The result is given as equation (3.2.14). With this result in hand, the stress intensity factor for general time variation of the forces can be obtained as in (3.2.15) or for general spatial distribution of traction as in (3.2.16).

The mode I situation of suddenly applied pressure on the faces of a crack of *finite* length is considered in Section 3.3. The characteristic length in this case is clearly the length of the crack. It is recognized in this case that the complete solution of the problem can be viewed as a series of solutions of half plane crack problems of ever increasing complexity. The first problem in the series is that solved in Section 2.5 and the second is formulated and solved in Section 3.3. These two solutions together give a complete picture of the influence of crack length through first-order wave scattering. This picture reveals that the stress intensity factor at either crack tip increases to a level about 30 percent *greater* than the long-time equilibrium stress intensity factor before unloading waves from the far end of the crack cause it to again decrease. This phenomenon, which is illustrated in Figure 3.5, is called *dynamic overshoot* of the stress intensity factor. Subsequent wave interactions lead to an oscillation of the stress intensity factor that decays over time to the long-time limit, which is the equilibrium result for the specified loading.

The problems considered in Sections 3.4 and 3.5 are again concerned with the half plane crack configuration. However, in the cases considered, the loading is nonuniform in the direction *parallel* to the crack edge. Consequently, all field variables, including the stress intensity factor, vary with position along the crack edge. In Section 3.4, a plane tensile pulse is obliquely incident on the edge of the crack. If it is noted that the resulting transient deformation field is time independent as seen by an observer traveling along the crack edge at a certain speed, then the problem can be reduced to two-dimensional problems of the kind solved exactly by the methods introduced in Chapter 2. For the particular problem analyzed, the crack opening occurs as a combination of mode I and mode III, and the transient stress intensity factors are given in (3.4.10) and (3.4.14), respectively. For an incident pulse carrying a step increase in tensile stress and for any given point on the crack edge, the stress intensity factor increases in proportion to the square root of the time elapsed after the wavefront reaches that point. The coefficient depends on the angle of incidence, of course.

In Section 3.5, an exact analysis of a truly three-dimensional elastic crack problem is presented, although for a case without a characteristic length. A traction is suddenly applied to the crack faces in the form of opposed line loads oriented in a direction *perpendicular* to the crack edge, with the loads acting in a direction normal to the crack faces. The transient stress intensity factor at any point along the crack edge is obtained by a modification of the transform methods introduced in Chapter 2. An expression for the stress intensity factor is given in (3.5.19) and a graph is shown in Figure 3.6. The stress intensity factor is zero until the arrival of the first dilatational wave. It is then negative until just after the arrival of the Rayleigh surface wave traveling on the crack faces. Thereafter, the stress intensity factor is positive, and it approaches the equilibrium stress intensity

factor as a long-time limit. Another problem in this same class is that
for an opposed pair of point forces suddenly applied to the crack faces
at a fixed distance behind the crack edge. The availability of this solution could have major impact on further study of three-dimensional
elastodynamic crack fields. However, this problem, which is the three-
dimensional equivalent of the two-dimensional case studied in Section
3.2, has a characteristic length and a solution is not yet available.

Chapter 3 concludes with a discussion of the Irwin stress intensity
factor criterion for fracture initiation from a preexisting crack under
stress wave loading. According to this criterion, a crack will begin
to extend when the crack tip stress intensity factor is increased to
a level characteristic of the material, called the fracture toughness.
The discussion is based on the elastodynamic solutions in Chapters 2
and 3. The important role of stress wave effects in fracture initiation
is demonstrated. The duration of an incident pulse, as well as its
amplitude, is important in considering fracture initiation under stress
wave loading. Some examples of experimental data showing the strong
dependence of fracture toughness on the rate of loading are included
in the discussion.

1.4.3 Asymptotic fields near a moving crack tip

Chapter 4 is concerned with the description of stress and deformation fields for points very close to the crack tip compared to any
other length scale in the configuration. Strictly speaking, the results
are valid only asymptotically as the observation point approaches the
crack tip. However, it is expected that in some cases the asymptotic
field provides a reasonably accurate description of the mechanical
state near the crack tip over a region large enough to be of some
practical significance for real materials. For example, the size of the
region of validity of an asymptotic field should be large enough to
completely include within it any finite strain zone or region dominated
by effects beyond the realm of the continuum mechanical models used.

The asymptotic crack tip fields for linear elastodynamic crack
growth are extracted in Section 4.2 for the antiplane shear mode and
in Section 4.3 for the plane strain modes, all for general motion of
the crack tip in a plane. The analysis is based on standard methods
of asymptotic analysis, applied in a systematic way. As in the case
of equilibrium crack tip fields, the asymptotic solution is found to
have universal spatial dependence. All information about loading and
configuration are embedded in a scalar multiplier called the dynamic

stress intensity factor. The leading term in the asymptotic expansion
in each case, which is the most singular term in the local stress or
strain distribution corresponding to bounded total mechanical energy,
is found to vary with radial distance from the crack tip r as $r^{-1/2}$
and to have a characteristic angular variation. Thus, for any stress
component,

$$\sigma_{ij} \sim \frac{K_I}{\sqrt{2\pi r}}\Sigma_{ij}^I(\theta, v) + \frac{K_{II}}{\sqrt{2\pi r}}\Sigma_{ij}^{II}(\theta, v) + \frac{K_{III}}{\sqrt{2\pi r}}\Sigma_{ij}^{III}(\theta, v)$$

as $r \to 0$, where r, θ are polar coordinates in the plane of deformation for each mode. The functions prescribing angular variation are
given explicitly in (4.2.17) for the case of mode III crack growth, in
(4.3.11) for mode I crack growth, and in (4.3.24) for mode II. The
angular variation of stress components is found to differ little from
the corresponding equilibrium results for crack speed less than about
40 percent of the shear wave speed c_s of the material, but to show
significant variation with crack speed for higher speeds. Some steps
are taken toward determining the terms in an asymptotic expansion
beyond the first for nonplanar crack growth and for nonsteady growth,
but this is a line of investigation that is not yet fully developed.

The case of steady crack growth in the antiplane shear mode
through an elastic-ideally plastic material is considered in Section 4.4.
The material is assumed to be governed by an associated flow rule. In
this case, the stress components are bounded at the crack tip but the
total strain components are logarithmically singular. In particular,
the total shear strain on the crack line directly ahead of the moving
crack tip is found to vary as $\ln r$, where r is the radial distance from
the crack tip. This is in contrast to the corresponding result for steady
crack growth under equilibrium conditions for which this shear strain
component varies as $\ln^2 r$. This fundamental inconsistency cannot
be resolved through asymptotic analysis alone because the range of
validity of asymptotic fields cannot be established without the benefit
of a more complete solution. For this particular case, the inconsistency
is eventually resolved in Section 8.3.

The case of steady crack growth under plane strain conditions
through an elastic-ideally plastic material is considered in Section 4.5.
Again, the material is assumed to respond according to an associated
flow rule. Furthermore, it is assumed that the material is elastically
incompressible, as well as plastically incompressible. A complete

asymptotic solution is constructed which satisfies all of the imposed conditions. It has the feature that all angular sectors around the crack tip are either regions of active plastic flow or regions of incipient flow in which the stress state satisfies the yield condition. In addition, this asymptotic solution has the feature that the plastic strain components are *bounded* at the crack tip. One way to admit more general asymptotic solutions in this case is to permit discontinuities in the angular variation of stress and particle velocity. However, it is shown that this possibility is ruled out if the material is required to satisfy the principle of maximum plastic work through the discontinuities. The possibility of elastic unloading sectors is also ruled out on the basis of similar arguments. Thus, the fully plastic asymptotic solution with bounded plastic strain components appears to be the only asymptotic solution, at least for elastic incompressibility. No information on the range of validity of this solution is available, and asymptotic fields of this kind require further study.

Chapter 4 concludes with brief sections on the asymptotic fields for dynamic crack growth in elastic-viscous and elastic-viscoplastic materials of certain special kinds. It is shown that if the rate sensitivity of the material is very strong then the behavior of the material very near to the crack tip is dominated by the instantaneous elastic properties, so the stress is found to be square root singular at the crack tip. For weaker rate sensitivity, on the other hand, the elastic and plastic strain rates are of the same order in magnitude and the stress is found to have a weaker singularity. The material models in these two sections differ from each other only in the fact that viscous effects are present at all levels of shear stress in the case of elastic-viscous response whereas they are present only if the stress state satisfies some yield condition in the case of elastic-viscoplastic response. The crack tip response of such materials is not yet fully understood, and only ad hoc estimates of the range of validity of the available asymptotic results exist.

1.4.4 Energy concepts in dynamic fracture

Analytical results concerned with mechanical energy or energy-like quantities are considered in Chapter 5. Following a brief review of Irwin's calculation leading to the relationship between energy release rate and stress intensity factor for elastic cracks, suggested by (1.1.8) above and stated for mode I cracks in (5.3.9), the matter of a general crack tip energy flux integral is addressed. An expression for energy flux F through a contour Γ surrounding the tip of a crack moving nonuniformly through a material with arbitrary mechanical response is

$$F(\Gamma) = \int_{\Gamma} \left[\sigma_{ij} n_j \frac{\partial u_i}{\partial t} + (U + T) v n_1 \right] ds,$$

where the crack grows at speed v in the x_1-direction. This result is derived as (5.2.7). The energy flux includes both recoverable and irrecoverable work done on the material. The development is based on mechanical quantities alone. However, the restrictions required for a thermodynamic interpretation of the result are outlined. Furthermore, it is noted that the energy flux integral applies for deformations with finite strain with a proper interpretation of the mechanical field quantities involved in the integrand. The crack tip integral is path independent only in the case of steady state crack growth. It is also noted that its limiting value as Γ is shrunk onto the crack tip is independent of the shape of Γ for all known asymptotic fields.

The energy flux integral is first applied to the case of elastodynamic crack growth. The limiting value of the crack tip energy flux as the integration contour is shrunk onto the crack tip is identified as the dynamic energy release rate G. By means of a direct calculation based on the asymptotic fields established in Chapter 4, a generalization of the Irwin relationship between stress intensity factor and energy release rate is established for the case of dynamic crack growth. The result has the form

$$G = \frac{1 - \nu^2}{E} A_I(v) K_I^2$$

for mode I crack growth at speed v, where E and ν are the elastic moduli and $A_I(v)$ is a universal function of instantaneous crack tip speed. The general result for all three modes of crack advance is given in (5.3.10).

The property of path independence of the crack tip energy integral for steady crack growth is then exploited to establish several results. First, for a situation with small-scale yielding, the connection between some properties of a crack tip cohesive zone and the dynamic stress intensity factor of the elastic field surrounding this nonlinear zone is determined. For example, if the cohesive stress in the zone has the uniform value σ_o, the crack tip opening displacement is δ_t, and K_{Iappl} is the remote applied stress intensity factor, then

$$A_I(v) K_{Iappl}^2 = E \sigma_o \delta_t$$

under plane stress mode I conditions. The case of a slip weakening cohesive zone in mode II is also considered. Next, the case of steady crack growth along an infinite strip with uniform edge conditions is recognized as an ideal case for application of the path independent energy integral, and several specific examples are described in Section 5.4.

In Section 5.5, some crack growth models based on simple structural elements, such as strings or beams, are considered. The central idea is to determine the relationship between the crack tip energy release rate and crack tip field parameters. Due to the kinematic constraints on the deformation imposed by the structural model, a stress intensity factor field will not develop, in general. Nonetheless, the dynamic energy release rate can be expressed in terms of other quantities. For example, for a string model G is expressed in terms of the slope of the string at the crack tip in (5.5.5), or for a beam model it is expressed in terms of the bending moment at the crack tip in (5.5.14). The value of such results is that, even though a stress intensity factor field does not develop in the model at the level of detail employed, a value of stress intensity factor can be inferred nonetheless by calculating G and then invoking the general relationship between the energy release rate and the stress intensity factor.

For transient elastodynamics, there are no path independent integrals that involve actual work and energy quantities of the system. However, there are path integrals of convolutions of stress and strain, or force and particle velocity, that are path independent. Because these integrals do not represent quantities of physical interest, they are not useful in establishing field characterizing parameters. Instead, their value must be sought in their usefulness in extracting information from boundary value problems. One such path independent integral is introduced in Section 5.6 in the Laplace transform domain. The connection between this integral and the Laplace transform of the stress intensity factor of a stationary crack is established in (5.6.22). The integral is used to determine the stress intensity for the problem considered in Section 2.5, but without actually solving the boundary value problem in this case.

For elastic crack problems in general, it is reasonable to expect that the stress intensity factor is a linear functional of the load distribution. This idea is developed in Section 5.7 for a body containing a stationary crack and subjected to transient loading. The kernel of the linear functional is called the *weight function* for the configuration, and two means of determining this function for mode I cracks are outlined. In the first method, the weight function is defined in terms of the solution of a particular boundary value problem, whereas in the second method a boundary value problem is formulated for the weight function itself. Use of the second method is illustrated by determining the stress intensity factor for the mode I point load problem considered in Section 3.2.

The last section of Chapter 5 is concerned with the far field energy radiation from an expanding crack. The radiated energy is defined as the net work done on a remote material surface, and various expressions for this energy flux in terms of crack plane features are obtained. This issue is of central importance in interpreting seismic crustal faulting data in the earth, and the connection between the radiated energy and certain average fault parameters is discussed.

1.4.5 Elastic crack growth at constant speed

Chapter 6 is concerned with the solution of particular crack growth problems under the restriction that the crack speed is constant once growth begins. The problems selected are representative of the classes of crack growth problems that are amenable to analysis. The first class considered is characterized by the feature that all fields are time independent as seen by an observer moving with the crack tip. Historically, the most important problem in this class is the Yoffe problem, consisting of a mode I crack of fixed length traveling at constant speed under the action of uniform remote tensile loading. The direction of loading is perpendicular to the plane of the crack. For steady crack growth at speeds less than the characteristic wave speeds, the governing equations are elliptic and the problem can be reduced to a boundary value problem for two analytic functions. Analysis leads to the conclusion that the stress intensity factor, which is given in (6.2.20), is identical to that for the same configuration with a stationary crack under equilibrium conditions. However, the angular variation of the near tip stress field has the general properties established in Section 4.3 for asymptotic mode I cracks. A case of steady crack growth in mode II is also analyzed, and the solution is used to construct solutions for steady crack growth accompanied by a crack tip cohesive zone. The length of the cohesive zone for uniform cohesive traction is given in (6.2.33).

The next problem class addressed is that of self-similar crack growth. Problems in this class are characterized by the property that fields at any one time are obtained from those at any other time simply by rescaling, or by the property that the functions representing mechanical fields are homogeneous functions of spatial coordinates and time. By assuming that the fields are self-similar, the governing partial differential equations can be recast into a form that is elliptic behind the wavefronts but hyperbolic ahead of the wavefronts. Thus, initial data on characteristics are carried to the wavefronts, jump conditions provide boundary conditions on the elliptic region, and methods of analytic function theory yield a solution in the elliptic region. The method is presented in detail for the Broberg problem. The essential features of this problem are that a mode I crack grows symmetrically at constant speed from zero initial length under the action of remote tensile loading in a direction perpendicular to the plane of the crack. If crack growth begins at time $t = 0$, it is found that the stress intensity factor increases in proportion to $\sqrt{t}$. The full expression is given in (6.3.48). The higher order terms in the asymptotic field and the amount of work done on the body that goes into fracture are also considered in Section 6.3. The corresponding solutions for a mode II shear crack and for nonsymmetric crack growth are outlined as well.

The matter of extension of a preexisting half plane crack in an unbounded body subjected to time independent loading is addressed in Section 6.4. The development proceeds by first finding the full mode I deformation field for constant speed crack growth when an opposed pair of concentrated forces on the crack faces is left behind as the crack tip begins to move at constant speed v. The analysis is based on Laplace transform methods and the Wiener-Hopf technique, essentially as developed in Chapter 2. With the fundamental solution in hand, the stress intensity factor field for crack advance which negates any time independent load can be constructed by linear superposition. If $p(x)$ is a tensile traction distribution on the crack plane ahead of the crack tip that is to be negated by crack advance, then it is shown in (6.4.31) that the dynamic stress intensity factor at any time after the onset of crack growth is

$$K_I(vt, v) = k(v)\sqrt{\frac{2}{\pi}} \int_0^{vt} \frac{p(x)}{\sqrt{vt - x}}\, dx \,,$$

where $k(v)$ is a universal function of crack speed given in (6.4.26). This function decreases monotonically, and roughly linearly, from $k = 1$

at speed zero to $k = 0$ at the Rayleigh wave speed c_R. Thus, the dynamic stress intensity factor has the form of a universal function of crack tip speed multiplied by the corresponding equilibrium stress intensity factor for the same loading conditions and the instantaneous crack length. The corresponding results for the other modes of crack advance, which are summarized in the same section, have this same general feature.

The chapter concludes with an analysis of the similar situation of growth of a preexisting crack at constant speed under the action of transient loading. The approach is illustrated by considering again the mode I problem analyzed in Section 2.5. Thus, at time $t = 0$ a pressure begins to act on the crack faces and a transient field is radiated from the loaded crack faces. In the present case, however, the crack begins to grow at constant speed at some arbitrary time interval, called the delay time, after the loading is applied. For the case of a normally incident plane stress pulse that carries a jump in stress of magnitude σ^*, the stress intensity factor is shown in (6.5.16) to be

$$K_I(t, v) = 2\sigma^* k(v) \frac{\sqrt{c_d t (1 - 2\nu)/\pi}}{(1 - \nu)},$$

where $k(v)$ is the same universal function that appeared for the case of time independent loading. Note that the resulting dynamic stress intensity factor is independent of the arbitrary delay time, and that it has the form of a universal function of crack tip speed times the stress intensity factor that would exist if the crack tip has been stationary and at its instantaneous position for all time. The result is readily generalized to arbitrary time variation of the incident wave loading.

1.4.6 Elastic crack growth at nonuniform speed

An objective of dynamic fracture mechanics is to predict the motion of a crack tip under given conditions of loading and geometrical configuration. Certainly, if the motion is specified a priori then there is no hope of predicting motion in any literal sense. In order to establish an equation of motion for the position of a crack tip under certain conditions, it is necessary to know the mechanical conditions that prevail for *all possible motions.* Then, a crack growth criterion can be imposed to select the "actual" motion according to that criterion from among all possible motions. From the analytical point of view, this means that a solution of the relevant boundary value problem valid for an arbitrary, nonuniform motion of the crack tip must be known. The few results of this kind that are available are summarized and applied in Chapter 7.

The case of nonsteady growth of a semi-infinite antiplane shear crack in an otherwise unbounded body is considered first. This problem can be solved directly for quite general loading conditions by means of Green's method as presented in Section 2.3. Some of the details are presented in Section 7.2 for the case of time independent loading, and the results for general loading are also summarized. The solution is found to have several features that are very unusual among two-dimensional wave propagation problems. For example, it is found that the stress intensity factor during nonuniform growth is a universal function of crack speed times the equilibrium stress intensity factor for the specified loading and the instantaneous crack tip position. Furthermore, it is found that if the crack speed is abruptly reduced to zero, then the stress intensity factor changes discontinuously to the appropriate equilibrium value for the given loading and crack position. In fact, it is found that the fully established equilibrium field radiates out behind the shear wavefront emitted when the crack tip suddenly stops.

It is demonstrated in Section 7.3, by means of an inverse argument, that these same remarkable features are present for the case of mode I crack growth under general time independent loading, except that the equilibrium field is not fully established in the region behind the shear wavefront. Instead, it is shown that the equilibrium traction distribution is radiated ahead of the crack tip behind the shear wavefront and the equilibrium displacement distribution is radiated on the crack faces behind a point moving with the Rayleigh wave speed. These features are exploited to show that the dynamic stress intensity factor for arbitrary motion of the crack tip, described by the time dependent amount of crack advance $l(t)$, say, has the form

$$K_I(l,\dot{l}) = k(\dot{l})K_I(l,0),$$

as given in (7.3.19). That is, the dynamic stress intensity factor has the form of a universal function of crack tip speed times the corresponding equilibrium stress intensity factor for the given loading and instantaneous crack tip position. Thus, the dynamic stress intensity factor depends on instantaneous crack tip position and crack speed, but otherwise not on the history of crack motion. The corresponding results for the other modes of crack advance have the same general features.

With the dynamic stress intensity factor available for nonuniform crack growth, the dynamic energy release rate is also available through the generalized Irwin relationship. Then, if the Griffith energy balance crack growth criterion is applied in the form $G = 2\gamma$ as discussed in Section 1.1, then an *equation of motion* for the crack tip is obtained in the form

$$\frac{2E\gamma}{(1 - \nu^2)K_I(l, 0)^2} = A_I(\dot{l})k(\dot{l})^2 \approx 1 - \dot{l}/c_R.$$

A slightly more general form of this equation of motion appears as equation (7.4.4). If the circumstances of Mott's original analysis are assumed, as outlined in Section 1.1, then this equation of motion implies that the maximum speed of crack growth is the Rayleigh wave speed. In addition, it is seen that if the work of separation γ is a constant then the equation of motion is a first order differential equation. Thus, if an analogy with the motion of a mass particle is made, the equation of motion implies that the crack tip has no effective inertia. It is also noted, on the basis of the equation of motion, that if an interface with no cohesive strength is suddenly opened then the opening boundary will travel with the Rayleigh wave speed c_R.

Some illustrations of the equation of motion based on an energy balance crack growth criterion are discussed in Section 7.4. Included is the situation of crack growth through a material with periodic fracture resistance for which the interpretation of fracture energy is different at different size scales of observation. Also included is the consideration of mode II crack growth in a situation involving spatially nonuniform applied stress or fracture resistance. This analysis illustrates the important role of nonuniform fields in controlling crack propagation and arrest. Finally, the crack tip equation of motion for some structural configurations is analyzed. Because these configurations are relatively simple, the implications of the equation of motion can be presented in greater detail than is possible in general continuum analysis. This feature makes it possible to examine the role of reflected waves on crack growth and arrest in bounded bodies, for example. The idea of a crack tip equation of motion is extended to the case of transient loading in Section 7.5.

Chapter 7 concludes with a brief discussion concerning uniqueness of elastodynamic solutions involving dynamic crack growth. The
proof of the standard uniqueness theorem of elastodynamics is based
on a demonstration of the fact that the energy of a difference solution
for a particular boundary value problem is always zero. Because a
moving crack tip acts as a sink or source of energy, the conditions
under which a solution can be expected to be unique must be reconsidered. It is shown that the uniqueness theorem applies as long as
the crack speed is less than the Rayleigh wave speed of the material.

1.4.7 Plasticity and rate effects during crack growth

The concluding chapter is a collection of essays on various topics
concerned with the dynamic fracture of inelastic materials that are
representative of the most recent developments in the area. The
treatment of crack growth in linear viscoelastic materials is very brief.
It is noted that for the case of the growth of a sharp crack through a
viscoelastic material, the near tip field is determined by the instantaneous elastic moduli and it is independent of the relaxation properties
of the material. This paradox is resolved by introducing a failure zone
or cohesive zone of finite length. The crack growth process is then
"fast" or "slow" depending on the time required to advance one zone
length compared to a typical relaxation time of the material.

The matter of crack growth in an elastic-plastic material is considered next. The particular problem analyzed in some detail is the
steady growth of an antiplane shear crack through a rate independent
elastic-ideally plastic material under small-scale yielding conditions.
This is the only case of dynamic elastic-plastic crack growth for which
an exact analytical result is available, and this result is limited to
the distribution of strain on the crack line within the active plastic
zone. Nonetheless, this result has been very important in resolving
an inconsistency between equilibrium and dynamic asymptotic fields
for the problem, and in guiding the development of computational
approaches to the various problems in the same general class. The
distribution of shear strain on the crack line within the active plastic
zone is given parametrically in (8.3.20), and a plot of this strain
distribution is shown in Figure 8.4. An important conclusion is that
inertial resistance has a strong effect in suppressing the development of
plastic strain ahead of the crack tip. The growth criterion adopted in
order to extract a theoretical dynamic fracture toughness versus crack
speed relationship for the material from the model is a requirement

that the crack must grow in such a way that the strain at a fixed distance ahead of the crack tip has a certain fixed value. The influence of inertial resistance on a small scale within the active plastic zone has a strong effect on the applied driving force required to sustain crack growth according to the assumed criterion.

The next topic addressed is high strain-rate crack growth in a plastic solid. In this case, it is assumed that the material is viscoplastic, but that crack growth occurs rapidly enough so that there is little time for inelastic strain relaxation near the crack tip. In other words, the crack is running fast enough through an elastic-viscoplastic material to outrun any substantial accumulation of plastic strain. This is viewed as a model of cleavage crack propagation in a material that can undergo a transition in fracture mode from brittle to ductile, with the transition being influenced by the rate of deformation in the crack tip region. The main point of the analysis is to establish minimum conditions for the sustained growth of a sharp crack in this material in terms of the crack driving force, the crack speed, and the material properties. Temperature has an important effect through the temperature dependence of material properties. The minimum crack driving force required to sustain crack growth at a given temperature (or for certain values of material parameters) is interpreted as the crack arrest toughness of the material at that temperature.

The study of the influence of material rate effects in the dynamic fracture of materials is carried over into Section 8.5. Here, steady growth of a mode I crack accompanied by a crack tip cohesive zone is considered. The material response within the cohesive zone is time dependent, in the sense that the cohesive stress depends on the local rate of opening of the cohesive zone. The point of view is adopted that, for a given remote applied stress intensity factor, the crack will advance according to either a brittle fracture criterion or a ductile fracture criterion, depending on whichever is satisfied at the lower level of applied load. Interpretation of the results obtained by analysis of this model leads to the conclusion that, for a given level of strain rate sensitivity of the material in the cohesive zone, the crack will accelerate from zero speed under rising applied load as a ductile fracture. However, once the speed reaches a certain level the stress in the cohesive zone is elevated through rate effects to a sufficiently high level to activate a brittle fracture mechanism. The crack then accelerates very rapidly to a significantly higher speed as a brittle fracture. Subsequent behavior depends on the details of loading and other features of the system. In any case, the model permits a demonstration of a rate induced mode transition in dynamic fracture response.

The last two sections of the chapter are concerned with mechanisms of material separation that operate on a microscopic scale in certain materials. The first mechanism considered is the ductile growth of voids in a plastic material. It is demonstrated, on the basis of very simple analytical models, that the influence of the inertial resistance of the material is to elevate the apparent resistance to deformation at high rates of straining. For very small voids, however, is seems that material viscosity has a stronger influence on resistance than does inertia for many materials. In view of the number of competing effects and the geometrical complexities of large deformations and of interacting and coalescing voids, this is an area that requires significant further analytical modeling and experimental work.

The last topic considered is the apparent rate sensitivity exhibited by a brittle material experiencing distributed microcracking while undergoing a high average rate of straining. The origin of the rate dependence appears to be associated with the nucleation and coalescence of the many microcracks required to form a complete separation of the solid. Sample calculations based on crude models of the process are described to illustrate the qualitative features of this time dependent strength characteristic. This, too, is a problem area that requires further work.

2

BASIC ELASTODYNAMIC SOLUTIONS
FOR A STATIONARY CRACK

2.1 Introduction

Consider a body of nominally elastic material that contains a crack. For the time being, the idealized crack is assumed to have no thickness, that is, in the absence of applied loads the two faces of the crack coincide with the same surface in space. The edge of the crack is a smooth simple space curve, either a closed curve for an internal crack or an open curve intersecting the boundary of the body at two points for an edge or surface crack.

Under the action of applied loads on the boundary of the body or on the crack faces, the crack edge is a potential site for stress concentration. If the rate at which loads are applied is sufficiently small, in some sense, then the internal stress field is essentially an equilibrium field. The body of knowledge that has been developed for describing the relationships between crack tip fields and the loads applied to a solid of specified configuration is Linear Elastic Fracture Mechanics (LEFM). This a well-developed branch of engineering science which forms the basis for results to be discussed in this chapter and the next.

If loads are rapidly applied to a cracked solid, on the other hand, the internal stress field is not, in general, an equilibrium field and inertial effects must be taken into account. There is no unambiguous criterion for deciding whether or not loads are "rapidly" applied in a particular situation. Instead, the decision must normally be based on qualitative reasoning and experience with known solutions to specific problems. For example, if the applied loading has some characteristic time associated with it, say the time for a load to increase from zero to its final value or the period of a cyclic load, then this time may be compared to the transit time for an elastic wave over a representative dimension of the cracked solid. If the ratio of the former to the latter is much greater than unity, then inertial effects are expected to be minimal. If the ratio is much less than unity, on the other hand, then wave effects are expected to be significant.

It is worthwhile to think about the phenomenon in general terms at this point in order to appreciate the way in which inertial or stress wave effects can influence the relationship between crack edge stress concentrations and applied loads. In the case of slowly applied loads, the influence of each load point is felt "simultaneously" throughout the body. The body resists the applied loading by means of its stiffness; that is, the body develops a compatible deformation field which, through the elastic stress–strain response, corresponds to an equilibrium stress field to balance the loads. In the case of rapidly applied loads, on the other hand, resistance of the body to the applied loads derives not only from material stiffness, but also from inertial resistance of the material to acceleration. Consequently, the influence of each load point is radiated into the body as a mechanical wave. Different load points first affect the crack tip field at different times, resulting in a transient crack tip field.

But this relationship has another, more subtle, feature which may be appreciated by considering some particular situations. For example, consider the case of a suddenly applied pressure on the surface of a spherical cavity in an otherwise unbounded elastic solid (Achenbach 1973). The resulting wave field is spherically symmetric. The equilibrium stress field for this case varies with radial distance r from the center of the sphere as r^{-3}. If attention is focused on the wavefronts of the wave field resulting from the suddenly applied pressure, then it is found that the strength of the stress field carried on the wavefront varies as r^{-1}. This is an illustration of a general, fundamental difference between equilibrium and wave fields. The difference arises from the fact that the governing equations are elliptic in the former case, whereas they are hyperbolic in the latter case. In a sense, hyperbolic systems are more effective than elliptic systems in transmitting detailed information about the load points throughout the body. After the initial transients have passed a field point, the difference diminishes there. In the case of dynamically loaded cracked solids, the effect is manifested in the phenomenon of *dynamic overshoot*. In some cases, the intensity of the crack tip field resulting from a suddenly applied load reaches levels *greater* than the equilibrium intensity level corresponding to the slow application of the same loads. That is, in the early stages of the process, the level overshoots its long-time limiting value before eventually decaying to the limiting value. An example of this phenomenon is discussed in Section 3.3 where the situation of suddenly applied pressure on the faces of a crack of finite length is analyzed.

For points close to the crack edge compared to the principal radius of curvature of the edge and to the distance to the nearest wavefront or boundary, the local fields are essentially two-dimensional. This can be seen by considering the gradient of the stress field at points very close to the crack line. In qualitative terms, the gradient in a direction parallel to the crack edge tangent is not sensitive to a crack edge singularity, whereas the gradient in a normal direction is more singular than the field itself. Consequently, if the governing partial differential equations are expressed with reference to coordinate directions locally normal and tangential to the crack edge tangent, then the local *singular* field is determined by the terms in the governing equations with the highest order derivatives in the direction normal to the crack edge. Furthermore, the structure of the asymptotic field does not vary along the crack edge, although the intensity may vary. This argument is based on the tacit assumption of a crack edge without corners or cusps.

It is convenient to follow Irwin (1960) in resolving these local fields into three distinct two-dimensional fields. The basis for classification is the way in which the components of relative displacement of initially contiguous particles on opposite faces of the crack contribute to crack face separation very close to the crack edge: Mode I is the in-plane opening mode due to normal separation of the crack faces, mode II is the in-plane shearing mode due to relative sliding of the crack faces in a direction locally perpendicular to the crack edge, and mode III is the antiplane (out-of-plane) shearing mode due to relative sliding of the crack faces in a direction tangent to the crack edge (see Figure 1.4). The components of the near tip stress field with reference to a rectangular coordinate system are expressed for each of these mode contributions as

$$\sigma_{ij} = \frac{K(t)}{\sqrt{2\pi r}}\Sigma_{ij}(\theta) + \sigma_{ij}^{(1)} + o(1) \qquad (2.1.1)$$

as $r \to 0$, where r, θ are polar coordinates in a plane perpendicular to the crack edge. The point $r = 0$ coincides with the crack edge, and the line $\theta = 0$ is tangent to the crack surface at the crack edge and in the direction ahead of the crack. The dimensionless function $\Sigma_{ij}(\theta)$ represents the angular variation of each crack tip stress component. It is a universal function, independent of the configuration of the body, the details of the applied loads, and the elastic constants. This function, which is given explicitly below for each stress component for each mode, is normalized to satisfy $\Sigma_{ij}(0) = 1$. The multiplier $K(t)$, which has physical dimensions of force/length$^{3/2}$, is the time-dependent *elastic stress intensity factor*. The spatial variation of the dominant term in the asymptotic crack tip field is universal, and it is only the factor $K(t)$ which reflects the influence of the geometrical configuration of the body and the details of the loading in a specific problem. The determination of this influence is a central problem in fracture mechanics.

The term $\sigma_{ij}^{(1)}$ in (2.1.1) represents a contribution to the crack tip field of order unity. More precisely, in each mode,

$$\sigma_{ij}^{(1)} = \lim_{r \to 0} \left\{ \sigma_{ij} - \frac{K(t)}{\sqrt{2\pi r}} \Sigma_{ij}(\theta) \right\} . \qquad (2.1.2)$$

Information on this contribution of order unity will be included in the discussion of problems to be considered in this chapter.

Consider a right-handed rectangular local coordinate system oriented so that the x_2-axis is normal to the crack surface at a point on the crack edge (that is, in the direction $\theta = \pi/2$), the x_1-axis is tangent to the crack surface and in the direction of prospective crack advance ($\theta = 0$), and the x_3-axis is tangent to the crack edge. The functions $\Sigma_{ij}(\theta)$ of the angular coordinate θ for the three modes of crack opening with respect to this local coordinate system are given below. For mode I, $K(t) = K_I(t)$ and

$$\Sigma_{11}(\theta) = \cos \tfrac{1}{2}\theta \{ 1 - \sin \tfrac{1}{2}\theta \sin \tfrac{3}{2}\theta \} ,$$

$$\Sigma_{12}(\theta) = \cos \tfrac{1}{2}\theta \sin \tfrac{1}{2}\theta \cos \tfrac{3}{2}\theta , \qquad (2.1.3)$$

$$\Sigma_{22}(\theta) = \cos \tfrac{1}{2}\theta \{ 1 + \sin \tfrac{1}{2}\theta \sin \tfrac{3}{2}\theta \} .$$

For mode II, $K(t) = K_{II}(t)$ and

$$\Sigma_{11}(\theta) = -\sin\tfrac{1}{2}\theta\{2 + \cos\tfrac{1}{2}\theta\cos\tfrac{3}{2}\theta\},$$

$$\Sigma_{12}(\theta) = \cos\tfrac{1}{2}\theta\{1 - \sin\tfrac{1}{2}\theta\sin\tfrac{3}{2}\theta\}, \qquad (2.1.4)$$

$$\Sigma_{22}(\theta) = \sin\tfrac{1}{2}\theta\cos\tfrac{1}{2}\theta\cos\tfrac{3}{2}\theta.$$

For mode III, $K(t) = K_{III}(t)$ and

$$\Sigma_{31}(\theta) = -\sin\tfrac{1}{2}\theta \qquad \Sigma_{32}(\theta) = \cos\tfrac{1}{2}\theta. \qquad (2.1.5)$$

Graphs of $\Sigma_{ij}(\theta)$ for modes I and II of crack opening are included with graphs of the corresponding functions for dynamic crack growth in Figures 4.3–4.7. Concerning the terms of order unity in the crack tip stress field (2.1.2) for the case of traction-free crack faces, the boundary conditions require that $\sigma_{i2}^{(1)} = 0$ for $i = 1, 2, 3$. The terms of order unity that are not determined by the crack face boundary conditions can only be found from a solution for the complete field.

Clearly, a stress distribution which is singular at the crack tip is a mathematical idealization. No real material can actually support such a stress distribution. The rationalization for admitting the singular stress distribution in the study of the fracture of real materials is based on the *universal* spatial dependence of the crack tip stress field and on the concept of *small-scale yielding*. The small-scale yielding situation arises when the potentially large stresses in the vicinity of the crack edge are relieved through plastic flow, or some other inelastic process, throughout a region which has lateral dimensions that are small compared to the crack length and other body dimensions. Under these conditions, the stress distribution in the elastic material surrounding the inelastic crack tip zone is adequately described by the dominant singular term in the elasticity solution based on a sharp elastic crack. This surrounding field is completely determined by the stress intensity factor and, consequently, the stress intensity factor provides a one-parameter characterization of the load level applied to the material in the inelastic fracture zone. The stress intensity factor itself does not provide information on the way in which the material in this zone responds to the applied loading. The main implication of this viewpoint is the following. Consider two bodies of

the same material, but having different shapes and/or having cracks
of different size. Suppose that the two bodies are loaded to result
in the same mode of crack tip deformation. If the loading results in
the same stress intensity factor in the two cases, then the material
in the crack tip region is assumed to respond in the same way in the
two cases. This idea is exploited in engineering practice by measuring
the stress intensity factor at which a crack will begin to advance in
a well-characterized laboratory specimen, and then assuming that a
cracked structure will experience crack growth at the same level of
stress intensity. This condition for fracture is commonly known as
the Irwin criterion for fracture initiation, after G. R. Irwin, and the
critical value of the stress intensity factor is denoted by a subscript c,
for example, K_{Ic} in mode I under plane strain conditions.

Several dynamic fracture problems are examined in the following
sections within the general framework introduced. Attention is limited
to the fundamental dynamic fracture problem for each mode of crack
tip deformation separately. In each case, the system studied is a
half plane crack in an otherwise unbounded body of homogeneous
and isotropic elastic material. Loading is spatially uniform and suddenly applied over the crack faces, resulting in two-dimensional fields.
Transient stress intensity factor histories are derived in each case.
Extensions of the basic results to other situations of dynamically
loaded stationary cracks are described in Chapter 3.

2.2 Suddenly applied antiplane shear loading

Consider a body of elastic material that contains a half plane crack,
but that is otherwise unbounded. Introduce a right-handed rectangular coordinate system in the body so that the z-axis (or x_3-axis) lies
along the crack edge, and the y-axis (or x_2-axis) is normal to the plane
of the crack. The crack occupies the half plane $y = 0$, $-\infty < x \le 0$;
see Figure 2.1.

The opposite faces of the crack are subjected to opposite, suddenly applied uniform traction in the z-direction, say $T_z(x,z,t) =
\pm\tau^* H(t)$ on $y = \pm 0$, where τ^* is a constant traction magnitude and
$H(\cdot)$ is the unit step function. The remaining traction components
are $T_x(x,z,t) = T_y(x,z,t) = 0$ on $y = \pm 0$. Thus, the crack faces are
traction-free for time $t < 0$, and are subjected to uniformly distributed
shear traction for $t \ge 0$. The material is stress free and at rest
everywhere for $t < 0$.

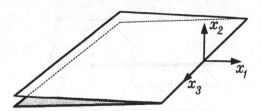

Figure 2.1. Configuration for analysis of the antiplane shear and plane strain deformations considered in Chapters 2 and 3. The coordinate axes are alternately labeled x, y, z or x_1, x_2, x_3, and the crack is in the plane $y = 0$ or $x_2 = 0$.

The components of displacement in the x, y, z-coordinate directions are u_x, u_y, u_z, respectively. Because the complete system is invariant under translation in the z direction, the displacement is independent of z. Furthermore, the symmetries of the system can be used to show that u_x and u_y are zero for all time if they are zero initially. To illustrate this point, consider the outcome of reversing the sense of the applied loading in two different ways. Suppose that the displacement in the y-direction for any point (x, y, z) is u_y. If the direction of the traction vector on the crack faces is reversed, then, because the response of the system in linear, the displacement at the same point will also be reversed to become $-u_y$. The sense of the traction can also be reversed by reflecting the system in the plane $z = 0$. Because u_y is independent of z, its sense is not changed in this case, so that the displacement at the point of interest will remain u_y. Linearity and reflective symmetry can be satisfied at an arbitrary point only if $u_y = 0$. A similar argument leads to $u_x = 0$. Consequently, the only nonzero component of displacement is $u_z(x, y, t) = w(x, y, t)$, the displacement in the z-direction. This field is two-dimensional, and the state of deformation at any point is antiplane shear deformation.

If $u_z = w$ is the displacement at any point (x, y), then reversal of the sense of the applied loads followed by reflection of the system in the plane $y = 0$ leads to the symmetry property that $w(x, -y, t) = -w(x, y, t)$; in particular, $w(x, 0, t) = 0$ for $x > 0$. The field over the entire x, y-plane is determined by solving the field equations for $y > 0$ only, subject to this symmetry condition.

Before actually developing a solution to the field equations, it is worthwhile to consider the physical features of the phenomenon to be analyzed. Prior to application of the loads, the material is stress free

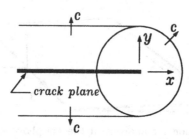

Figure 2.2. Diagram of the wavefronts generated by sudden application of crack face traction acting in the z-direction. The state of deformation is antiplane shear.

and at rest everywhere. The loading is suddenly applied to the crack faces at time $t = 0$ and, according to Huygens's principle, the influence of each loaded boundary point spreads out from that point behind an elementary wavefront traveling with the characteristic wave speed of the material $c = c_s$. The envelopes of the elementary wavefronts are the actual wavefronts of the transient field, and these wavefronts are shown in a plane $z =$ constant in Figure 2.2. For points near a crack face compared to the distance to the crack edge, the transient field consists only of a plane wave parallel to the crack face and traveling away from it at speed c. As this wavefront passes a material point, the component of stress $\sigma_{zy} = \tau_y$ (or σ_{32}) changes discontinuously from zero to $-\tau^*$ and the particle velocity $\partial w/\partial t$ changes discontinuously from zero to $\pm\tau^*/\rho c$ for $\pm y > 0$.

Near the crack edge, on the other hand, the field is more complex and some nonuniform field exists behind a cylindrical wavefront (circular in two dimensions) of radius ct that is centered on the crack edge (tip in two dimensions). This is the region in which the stress concentration develops.

It is noteworthy that the process being described involves neither a characteristic length nor a characteristic time with respect to which the independent variables may be scaled. That is, the sketch in Figure 2.2 represents the wave propagation process not only at a particular time, but at any time, provided only that the length scale in the coordinate directions is adjusted to make the radius of the cylindrical wavefront equal to ct. Suppose that $\tau_\beta(x, y, t)$ is a stress component $\sigma_{\beta z}$ at point (x, y) in the plane at time t. If the same stress component is considered at another point $(\alpha x, \alpha y)$, where α is any positive real number, then the original value will be recovered if it is considered at time αt. That is,

$$\tau_\beta(\alpha x, \alpha y, \alpha t) = \tau_\beta(x, y, t) \qquad (2.2.1)$$

for any positive real number α. The observation (2.2.1) shows that stress is a homogeneous function of its arguments of degree zero. By making a particular choice for α in (2.2.1), say $\alpha = t^{-1}$, it becomes clear that stress can be expressed as a function of two variables, rather than the apparent three. It follows that particle velocity is also a homogeneous function of degree zero, displacement is a homogeneous function of degree one, and so on. It is clearly important in deducing (2.2.1) that the nonzero boundary data are all expressible in terms of a fixed level of stress. If the nonzero boundary data were given in terms of a fixed level of stress *rate*, for example, then stress rate would necessarily be homogeneous of degree zero, stress would be homogeneous of degree one, and so on. The property of homogeneity can be exploited in developing solution techniques for this problem class, and an illustration is given in Section 6.3.

For points close to the crack tip, the stress field will be dominated by the square root singular stress intensity factor field given in (2.1.1). Indeed, the observations that the system is linear, that the fields are homogeneous, and that the stress varies inversely with the square root of distance from the crack tip lead directly to the statement that

$$\tau_y(x, 0, t) \sim C_{III}\tau^* \sqrt{\frac{ct}{x}} \qquad (2.2.2)$$

as $x/ct \to 0$, where C_{III} is an undetermined dimensionless real constant. From the definition of stress intensity factor in (2.1.1),

$$K_{III}(t) = \lim_{x \to 0^+} \sqrt{2\pi x}\, \tau_y(x, 0, t) = C_{III}\tau^* \sqrt{2\pi ct}. \qquad (2.2.3)$$

Thus, the stress intensity factor is known up to a dimensionless multiplier C_{III}, and a main purpose in solving the field equations from the fracture mechanics point of view is to determine the value of C_{III}. The fact that the stress intensity factor increases indefinitely with time is coupled to the fact that the static elasticity problem equivalent to the transient problem being discussed here does not have a solution.

Finally, it is noted that (2.2.2) is only an asymptotic result. It is reasonable to expect, however, that (2.2.2) provides a good approximation to the stress component for values of x small compared to the distance from the crack tip to the cylindrical wavefront ct, the only length available for comparison with distance from the crack tip. This idea cannot be made more precise until a complete solution to the problem is available.

The mathematical problem that leads to the solution is stated in the following way. A function $w(x, y, t)$ is sought that satisfies the wave equation in two space dimensions and time,

$$\frac{\partial^2 w}{\partial x^2} + \frac{\partial^2 w}{\partial y^2} - \frac{1}{c^2}\frac{\partial^2 w}{\partial t^2} = 0 \qquad (2.2.4)$$

in the half plane $-\infty < x < \infty$, $0 < y < \infty$ for time in the range $0 < t < \infty$. The solution of the wave equation (2.2.4) is subject to the boundary conditions that

$$\tau_y(x, 0^+, t) = \mu\frac{\partial w}{\partial y}(x, 0^+, t) = -\tau^* H(t), \qquad -\infty < x < 0,$$
$$\qquad (2.2.5)$$
$$w(x, 0^+, t) = 0, \qquad 0 < x < \infty,$$

for all time, where μ is the shear modulus, and the initial conditions that

$$w(x, y, 0) = 0, \qquad \frac{\partial w}{\partial t}(x, y, 0) = 0 \qquad (2.2.6)$$

for all points in the half plane. Only solutions that result in stress components that vary inversely with the square root of distance from the crack tip are admitted; stresses that have singularities stronger than an inverse square root are rejected on physical grounds because they correspond to deformation states with unbounded elastic energy. The mathematical statement of this requirement, which is already implicit in (2.2.2), is

$$\lim_{x \to 0^+} \sqrt{x}\left|\frac{\partial w}{\partial y}(x, 0^+, t)\right| < \infty. \qquad (2.2.7)$$

A solution of this boundary value problem may be obtained through application of any of several techniques of analysis. Solution by Green's method and by the Wiener-Hopf method are illustrated in

Sections 2.3 and 2.5, respectively. It is observed in Section 2.3 that the solution is equivalent to the solution for diffraction of a plane stress pulse by a crack with traction-free faces. It may be observed that, once a solution is available for boundary loading with step function time dependence, extension to boundary loading with other time dependence is quite simple.

2.3 Green's method of solution

The initial-boundary value problem outlined in Section 2.2 and stated in (2.2.4)–(2.2.7) is now solved by means of Green's method. This approach was originally developed for solving problems in potential theory, but it has been applied with great success to virtually all categories of linear boundary value problems in mathematical physics. Suppose that in addition to displacement $w(x, y, t)$, a second wave field $g(x, y, t)$ is defined over the half plane $-\infty < x < \infty$, $0 < y < \infty$ for $-\infty < t < \infty$. Then Green's second formula for the wave operator in two space dimensions is

$$\int_{t_1}^{t_2} \int_A \left\{ \left(\nabla^2 w - \frac{1}{c^2} \frac{\partial^2 w}{\partial t^2} \right) g - \left(\nabla^2 g - \frac{1}{c^2} \frac{\partial^2 g}{\partial t^2} \right) w \right\} dA\, dt$$

$$= \frac{1}{c^2} \int_A \left[w \frac{\partial g}{\partial t} - g \frac{\partial w}{\partial t} \right]_{t_1}^{t_2} dA + \frac{1}{\mu} \int_{t_1}^{t_2} \int_{-\infty}^{\infty} \left[g \frac{\partial w}{\partial y} - w \frac{\partial g}{\partial y} \right]_{y=0+} dx\, dt$$

$$(2.3.1)$$

for arbitrary times $t_2 > t_1$. The area A is the entire area of the half plane, and ∇^2 denotes the Laplacian operator in two space dimensions. The functions w and g are assumed to have properties such that the integrals in (2.3.1) exist. Integrals taken along paths in remote parts of the plane have been assumed to vanish in writing (2.3.1) due to the wave propagation character of the fields.

The great utility of (2.3.1) is demonstrated when g is selected to be a fundamental singular solution, or Green's function, for the problem. The fundamental singular solution may be defined in a number of ways, depending on the details of the problem. In the present case, g is required to satisfy

$$\frac{\partial^2 g}{\partial x^2} + \frac{\partial^2 g}{\partial y^2} - \frac{1}{c^2} \frac{\partial^2 g}{\partial t^2} = -\frac{P}{\mu} \delta(x - x_o) \delta(y - y_o) \delta(t - t_o), \quad (2.3.2)$$

where $\delta(\cdot)$ is the Dirac delta function, (x_0, y_0) is an arbitrary point in the unbounded x, y-plane, t_0 is an arbitrary time, and P is an amplitude that has physical dimensions of force$\times$time/length if g is to have a physical dimension of length. The solution of (2.3.2), which is written as $g_0(x, y, t)$ to make explicit the special significance of the point (x_0, y_0) and the time t_0, is subject to no boundary conditions, that is, it is the *free space* Green's function. To render (2.3.1) particularly useful, however, it is required that

$$g_0 = 0, \qquad \frac{\partial g_0}{\partial t} = 0 \qquad \text{for} \qquad t \geq t_0 \qquad (2.3.3)$$

for all points (x, y).

A function that satisfies (2.3.2) and (2.3.3) is

$$g_0(x, y, t) = \frac{P}{2\pi\mu} \begin{cases} \dfrac{1}{\sqrt{(t_0 - t)^2 - R^2/c^2}} & \text{if } t < t_0 - R/c, \\ 0 & \text{if } t > t_0 - R/c, \end{cases} \qquad (2.3.4)$$

where $R = \sqrt{(x_0 - x)^2 + (y_0 - y)^2} \geq 0$. In the three-dimensional x, y, t-space, the function g_0 represents a wave field that is nonzero inside the cone $t_0 - t > R/c$ with apex at (x_0, y_0, z_0), and that is identically zero elsewhere. This is the displacement that would result if a line of body forces acting in the z-direction and uniformly distributed along $x = x_0$, $y = y_0$, $-\infty < z < \infty$ were applied impulsively at time $t = t_0$, provided that the sense of time was reversed. For example, this is the interpretation if $(t_0 - t)$ is the measure of elapsed time.

If g in (2.3.1) is now understood to be the free space Green's function g_0, and the choices $t_1 = 0^+$, $t_2 = t_0^+$ are made, then

$$Pw(x_0, y_0, t_0) = \int_{B_0} \left\{ g_0(x, 0, t) \frac{\partial w}{\partial y}(x, 0^+, t) \right.$$

$$\left. - w(x, 0^+, t) \frac{\partial g_0}{\partial y}(x, 0, t) \right\} dx \, dt, \qquad (2.3.5)$$

where B_0 is the portion of the boundary surface $y = 0$ that falls inside the cone $(t_0 - t) = \sqrt{(x_0 - x)^2 + y_0^2}$ for time in the range

$0 < t < t_o$. The first term on the right side of (2.3.1) vanishes because both w and g_o have initial values of zero. The result (2.3.5) is Green's representation formula for the solution of the wave equation (2.2.4). It states that the solution at any interior point in the half space is determined by boundary values of the wave function and its derivative (and by the initial conditions if nonzero initial conditions are specified). Furthermore, it is only those boundary data at points that fall inside the cone $(t_o - t) = R/c$ with apex at the observation point (x_o, y_o, z_o) that influence the field at the observation point. This backward running cone defines the domain of dependence for the observation point. Both the boundary displacement on the crack faces $w(x, 0^+, t)$ for $x < 0$ and the boundary traction $\tau_y(x, 0, t) = \mu \partial w(x, 0^+, t)/\partial y$ for $x > 0$ are unknown at this point.

If the observation point in (2.3.5) approaches a boundary point ahead of the crack tip, that is, $y_o \to 0^+$ for $x_o > 0$, $t_o > 0$, then (2.3.5) becomes

$$\int_0^{t_o} \int_{x_o - c(t_o - t)}^{x_o + c(t_o - t)} \frac{\tau_y(x, 0, t)\, dx\, dt}{\sqrt{c^2(t_o - t)^2 - (x_o - x)^2}} = 0, \quad x_o > 0. \quad (2.3.6)$$

The condition that the displacement vanishes on the crack plane ahead of the tip [equation $(2.2.5)_2$] has been used in writing (2.3.6), and the observation that $\partial g_o/\partial y = 0$ for all x when $y = y_o = 0$ has been incorporated as well. The unknown stress distribution on the boundary ahead of the crack tip may be determined from (2.3.6). With this result in hand, the observation point in the representation formula (2.3.5) can be taken to be on the crack face, which yields an expression for the displacement distribution on the crack face. Then, having determined the boundary values appearing in the representation formula (2.3.5), this formula provides a complete solution. For present purposes, it is information about the stress ahead of the crack tip that is of primary interest, and such information is now extracted from (2.3.6).

The region of the x, t-plane that is of interest in considering (2.3.6) is shown in Figure 2.3, where the time coordinate has been scaled by the wave speed c for convenience. With reference to Figure 2.3, the area of integration in (2.3.6) is the triangular region OBD. First, it is observed that if the observation point was anywhere in the sector DCE, then the integration area in (2.3.6) would not include any part of the loaded crack face $x < 0$. Consequently, the equation

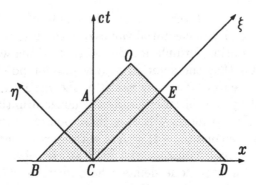

Figure 2.3. The x, t-plane showing the range of integration for the integral equation (2.3.6). The time axis has been rescaled by the wave speed c for convenience. The observation point O has coordinates x_0, ct_0.

for τ_y is homogeneous under these conditions and a partial solution of (2.3.6) is that

$$\tau_y(x, 0, t) = 0, \qquad x > ct. \tag{2.3.7}$$

This result is obvious from the discussion in the preceding section concerning the wave propagation character of the transient field. For points that are more distant than ct from the nearest loaded boundary point, the initial rest state (2.2.6) persists and the stress is zero. With reference to Figure 2.2, the partial solution (2.3.7) applies for points on $y = 0$ ahead of the advancing cylindrical wavefront.

For an observation point in sector ECA in Figure 2.3, on the other hand, the region of integration includes the part of the loaded crack face in the triangular region ABC in Figure 2.3 and, consequently, (2.3.6) is not a homogeneous equation. For this situation, solution of (2.3.6) is facilitated by a change of integration variables to

$$\xi = ct + x, \qquad \eta = ct - x. \tag{2.3.8}$$

The inverse transformation is

$$ct = \frac{1}{2}(\xi + \eta), \qquad x = \frac{1}{2}(\xi - \eta). \tag{2.3.9}$$

The transformation (2.3.8) carries the coordinates of the observation point (x_0, t_0) into (ξ_0, η_0), where $\xi_0 = ct_0 + x_0$, $\eta_0 = ct_0 - x_0$. In view of the partial solution (2.3.7), introduction of the transformation (2.3.8) into (2.3.6) yields

$$\int_0^{\eta_o} \frac{1}{\sqrt{\eta_o - \eta}} \int_{-\eta}^{\xi_o} \frac{\tau_y(\xi, 0, \eta)}{\sqrt{\xi_o - \xi}} \, d\xi \, d\eta = 0 \,, \qquad \xi_o > \eta_o > 0 \,. \quad (2.3.10)$$

The name of the variable representing the y-component of stress has been preserved as τ_y in writing (2.3.10) even though it is obviously a different function of its arguments than it was in the original equation (2.3.6).

The transformed integral equation (2.3.10) will be satisfied if the inner integral vanishes for all η in the range $0 < \eta < \eta_o$. This condition can be written as a Volterra integral equation of the first kind

$$\int_{\eta_o}^{\xi_o} \frac{\tau_y(\xi, 0, \eta_o)}{\sqrt{\xi_o - \xi}} \, d\xi = \int_{-\eta_o}^{\eta_o} \frac{\tau^*}{\sqrt{\xi_o - \xi}} \, d\xi \,, \qquad \xi_o > \eta_o > 0 \,, \quad (2.3.11)$$

where the traction boundary condition on the crack face [equation $(2.2.5)_1$] for $x < 0$ has been made explicit. If the range of integration on the left side of (2.3.11) is shifted to make the lower limit of integration equal to zero, then (2.3.11) takes the standard form of an Abel integral equation with a square root singular kernel

$$\int_0^{\zeta_o} \frac{\tau(\zeta)}{\sqrt{\zeta_o - \zeta}} \, d\zeta = \sigma(\zeta_o) \,, \qquad \zeta_o > 0 \,, \quad (2.3.12)$$

where the function σ is given. A solution $\tau(\zeta)$ of (2.3.12) is readily obtained by application of Laplace transform methods. Denoting the Laplace transform of a function by a superposed hat [equation (1.3.7)], the application of a transform to the terms in (2.3.12) yields

$$\left(\frac{\pi}{s}\right)^{1/2} \widehat{\tau}(s) = \widehat{\sigma}(s) \,, \quad (2.3.13)$$

where s is the transform parameter. This equation is then solved for $\widehat{\tau}(s)$ in the form

$$\widehat{\tau}(s) = \left(\frac{s}{\pi}\right)^{1/2} \widehat{\sigma}(s) = \frac{s\widehat{\sigma}(s) - \sigma(0)}{(\pi s)^{1/2}} + \frac{\sigma(0)}{(\pi s)^{1/2}} \,. \quad (2.3.14)$$

Application of the inverse transform provides the solution of (2.3.12)
as

$$\tau(\zeta_0) = \frac{1}{\pi} \int_0^{\zeta_0} \frac{\sigma'(\zeta)}{\sqrt{\zeta_0 - \zeta}} \, d\zeta + \frac{\sigma(0)}{\pi\sqrt{\zeta_0}}, \qquad \zeta_0 > 0, \qquad (2.3.15)$$

where the prime denotes differentiation.

For the problem at hand, the general solution (2.3.15) implies
that the solution to (2.3.11) is

$$\tau_y(x, 0, t) = \frac{2\tau^*}{\pi} \left\{ \sqrt{\frac{(ct - x)}{x}} - \tan^{-1} \sqrt{\frac{(ct - x)}{x}} \right\} \qquad (2.3.16)$$

for $0 < x < ct$, where the rotated coordinates ξ, η have been replaced
by the physical coordinates x, t.

Some general features of the shear stress distribution on the plane
$y = 0$ are evident from (2.3.16). First, $\tau_y(x, 0, t)$ is continuous across
the wavefront, as could have been anticipated on the basis of the
continuity of displacement and the fact that the discontinuity of a
tangential derivative is the tangential derivative of a discontinuity,
as deduced from Hadamard's lemma (Truesdell and Toupin 1960).
The magnitude of τ_y increases monotonically with distance from the
wavefront toward the crack tip at $x = 0$. For points very close to the
tip,

$$\tau_y(x, 0, t) \sim \frac{2\tau^*}{\pi} \sqrt{\frac{ct}{x}}. \qquad (2.3.17)$$

The form of this result was anticipated in (2.2.2). With the solution
now available, it is clear that the undetermined constant in (2.2.2) has
the value $C_{III} = 2/\pi$. Thus, from (2.2.3), the transient stress intensity
factor for suddenly applied antiplane shear loading on the faces of a
half plane crack is

$$K_{III}(t) = 2\tau^* \sqrt{2ct/\pi}. \qquad (2.3.18)$$

This is the main result of this section.

Although the stress intensity factor (2.3.18) was obtained for the
case of crack face loading, it can be given another interpretation. Once
again, consider the unbounded body containing a half plane crack
depicted in Figure 2.1. Suppose that the crack faces are traction-free,
but that a plane horizontal shear pulse described by

$$w(x, y, t) = \frac{\tau^*}{\mu}(y - ct)H(ct - y) \qquad (2.3.19)$$

is incident on the crack plane. The front of the pulse is parallel to the crack plane, and it propagates toward the crack plane at speed c, carrying a jump in stress τ_y from zero to τ^*. At time $t = 0$, the incident pulse strikes the crack. For points far to the left of the crack tip ($x \ll -ct$), the pulse reflects back onto itself as a plane wave with a change in sign of the stress. For points far to the right ($x \gg ct$), the pulse continues to propagate as a plane wave without being influenced by the crack. The incident pulse is diffracted at the edge of the crack, and the diffracted wave radiates out from the tip behind a cylindrical wavefront of radius ct.

The complete deformation field of this stress pulse diffraction problem differs from the deformation field of the problem described in Section 2.2 only by the plane pulse propagating in an *uncracked* solid. Because the governing equations are linear, and because the plane pulse, by itself, results in no stress singularity at the crack tip, it follows immediately that (2.3.18) is also the stress intensity factor history for the phenomenon of stress pulse diffraction. Furthermore, the stress distribution on the crack plane ahead of the tip for the diffraction problem is given by (2.3.16) plus the constant τ^*, that is,

$$\tau_y(x,0,t) = \frac{2\tau^*}{\pi} \left\{ \sqrt{\frac{(ct-x)}{x}} + \frac{\pi}{2} - \tan^{-1}\sqrt{\frac{(ct-x)}{x}} \right\} . \qquad (2.3.20)$$

This distribution, because it is exact, provides the opportunity to gain some understanding of the size of the region near the crack over which the stress intensity factor field (2.3.17) dominates the complete stress distribution in a situation that involves no characteristic length. If the stress distribution (2.3.20) is expanded in powers of the small quantity x/ct, then

$$\tau_y \sim \frac{2\tau^*}{\pi}\sqrt{\frac{ct}{x}}\left\{1 + \frac{x}{2ct}\right\}, \qquad \frac{x}{ct} \ll 1. \qquad (2.3.21)$$

Consequently, the stress intensity factor field describes the full stress field to within 5 percent of the exact result for points closer to the tip than 10 percent of the distance from the tip to the cylindrical wavefront along $y = 0$.

This completes the discussion of the basic problem of determining transient stress intensity factors associated with antiplane shear deformation fields in cracked solids. The discussion is resumed in Section 7.2 where nonuniform crack growth is analyzed by means of the same general approach. This basic problem is a useful vehicle for introducing many ideas of dynamic fracture analysis, both physical and mathematical, that have much broader applicability. The problem is relatively simple, and it can be analyzed completely by direct application of classical methods of analysis. The plane strain counterpart of this basic problem, which is more involved but more significant from the practical point of view, is considered next.

2.4 Suddenly applied crack face pressure

Consider once again the body of elastic material that contains a half plane crack but that is otherwise unbounded (see Figure 2.1). In the present instance, the crack faces are subjected to a suddenly applied, spatially uniform pressure of magnitude σ^*. In terms of boundary traction, $T_x(x, z, t) = T_z(x, z, t) = 0$ and $T_y(x, z, t) = \pm\sigma^*H(t)$ on $y = \pm 0$, where $\sigma^* > 0$ and $H(\cdot)$ is the unit step function. Thus, the crack faces are traction-free for time $t < 0$, and they are subjected to uniform normal pressure for $t \geq 0$. The material is stress free and at rest everywhere for $t < 0$.

The components of displacement in the x, y, z-coordinate directions are u_x, u_y, u_z, respectively. Arguments similar to those used in Section 2.2 can be applied here to conclude that $u_z = 0$ for all time if it is zero initially, and that u_x and u_y are independent of the z-coordinate, so the state of deformation is plane strain. Furthermore, the nonzero components of displacement satisfy the symmetry conditions $u_x(x, -y, t) = u_x(x, y, t)$ and $u_y(x, -y, t) = -u_y(x, y, t)$. In particular, these conditions imply that both $u_y(x, 0, t) = 0$ and $\sigma_{xy}(x, 0, t) = 0$ for $x > 0$, due to symmetry.

As in the case of the antiplane shear loading considered in Section 2.2, it is helpful at this point to survey the features of the phenomenon to be analyzed. Before application of the loads, the material is stress free and at rest everywhere. The pressure is suddenly applied to the crack faces at time $t = 0$, and the influence of each loaded boundary point spreads out from that point behind elementary wavefronts traveling with the characteristic speeds of the material, the speed of dilatational waves c_d and the speed of shear waves c_s, in this case. The envelopes of the elementary wavefronts are the anticipated wavefronts of the transient field, and these wavefronts are shown in a

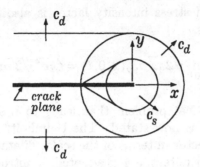

Figure 2.4. The wavefronts generated by sudden application of a spatially uniform pressure on the faces of the crack. The state of deformation is plane strain.

plane $z = $ constant in Figure 2.4. For points near to the crack face compared to the distance to the crack edge, the transient field consists only of a plane wave parallel to the crack face and traveling away from it at speed c_d. The uniform pressure loading does not induce a corresponding plane shear wave. As the plane dilatational wavefront passes a material point in its path, the component of stress σ_{yy} (or σ_{22}) changes discontinuously from zero to $-\sigma^*$ and the particle velocity $\partial u_y/\partial t$ (or $\partial u_2/\partial t$) changes discontinuously from zero to $\pm\sigma^*/\rho c_d$ for $\pm y > 0$.

Near the crack edge, on the other hand, the deformation field is more complex. A nonuniform scattered field radiates out from the crack edge behind a cylindrical wavefront of radius $c_d t$ that is centered on the crack edge. Due to the coupling of dilatational and shear waves at a boundary, this scattered field also includes a cylindrical shear wavefront of radius $c_s t$ that is centered on the crack edge plus the associated plane-fronted headwaves traveling at speed c_s.

As was the case for antiplane shear loading discussed in Section 2.2, the process being described here involves neither a characteristic length nor a characteristic time. The arguments presented in connection with (2.2.1) can be applied here as well to conclude that the components of stress and particle velocity are homogeneous functions x, y, t of degree zero. An immediate consequence is that, for points very close to the crack edge,

$$\sigma_{yy} \sim C_I \sigma^* \sqrt{\frac{c_d t}{x}}, \qquad (2.4.1)$$

where C_I is an undetermined dimensionless constant. In view of

(2.4.1), the mode I stress intensity factor is also known up to the constant C_I, that is,

$$K_I(t) = \lim_{x \to 0^+} \sqrt{2\pi x}\, \sigma_{yy}(x,0,t) = C_I \sigma^* \sqrt{2\pi c_d t}. \qquad (2.4.2)$$

The mathematical problem that leads to a solution with the properties outlined is now stated. The Helmholtz representation of the displacement vector in terms of the scalar dilatational potential ϕ and the vector shear potential ψ is adopted, as introduced in (1.2.60). For plane strain deformation that is independent of z, the vector potential ψ has only one nonzero component; this component is in the z direction and it will be denoted by ψ. Thus, two functions $\phi(x,y,t)$ and $\psi(x,y,t)$ are sought that satisfy the wave equations in two space dimensions and time

$$\frac{\partial^2 \phi}{\partial x^2} + \frac{\partial^2 \phi}{\partial y^2} - \frac{1}{c_d^2}\frac{\partial^2 \phi}{\partial t^2} = 0\,, \qquad \frac{\partial^2 \psi}{\partial x^2} + \frac{\partial^2 \psi}{\partial y^2} - \frac{1}{c_s^2}\frac{\partial^2 \psi}{\partial t^2} = 0 \quad (2.4.3)$$

in the half plane $-\infty < x < \infty$, $0 < y < \infty$ for time in the range $0 < t < \infty$. The wave functions satisfying (2.4.3) are subject to the boundary conditions that

$$\sigma_{yy}(x,0,t) = -\sigma^* H(t)\,, \qquad -\infty < x < 0\,,$$

$$\sigma_{xy}(x,0,t) = 0\,, \qquad -\infty < x < \infty\,, \qquad (2.4.4)$$

$$u_y(x,0,t) = 0\,, \qquad 0 < x < \infty\,,$$

for all time, where stress and displacement components are interpreted as their representations in terms of displacement potentials ϕ and ψ. The initial conditions that

$$\phi(x,y,0) = \frac{\partial \phi}{\partial t}(x,y,0) = \psi(x,y,0) = \frac{\partial \psi}{\partial t}(x,y,0) = 0 \qquad (2.4.5)$$

for all points in the half plane ensure that the body is stress free and at rest until the crack face pressure is applied. Finally, only solutions that result in stress components that vary inversely with the square root of distance from the crack tip are admitted. As before, stresses that have singularities stronger than inverse square root are rejected on the physical grounds that they correspond to states with unbounded total energy. The total energy density is proportional to $\sigma_{ij}\epsilon_{ij} + \rho\dot{u}_i\dot{u}_i$. The integral of this density over any bounded portion of the body, even a portion including the crack tip, must have a finite value.

The solution method to be developed in the next few sections is based on the application of integral transforms. Solution of partial differential equations by means of integral transforms is only a formal procedure under the best of circumstances. That is, convergence of the improper integrals of unknown functions must be assumed *a priori*, and it is only after a solution has been obtained that the validity of the assumptions can be checked. Nonetheless, knowledge of certain qualitative features of the transforms can be used to great advantage in guiding progress toward a solution. This is the case in the Wiener-Hopf technique, which is to be outlined in the following section. Indeed, some knowledge of the domains of convergence of certain transforms is *essential* in applying the Wiener-Hopf technique. In the present context, such qualitative features are suggested by the wave propagation character of the fields being analyzed.

The one-sided Laplace transform will be used to suppress dependence on time. According to the convention established in (1.3.7), the Laplace transform of $\phi(x, y, t)$ is

$$\widehat{\phi}(x, y, s) = \int_0^\infty \phi(x, y, t)\, e^{-st} dt, \qquad (2.4.6)$$

where, for the time being, the transform parameter s is a real, positive variable. Some understanding of the behavior of $\widehat{\phi}(x, y, s)$, as well as other transformed fields, as the magnitude of x or y becomes very large is important in the Wiener-Hopf technique.

Consider any fixed point (x, y) in the half plane. Let $t^*(x, y)$ denote the time at which the first wavefront arrives at (x, y) to disturb the initial rest state. Under these circumstances, the lower limit of integration in (2.4.6) can be replaced by t^*. If the change of variable $t = \tau + t^*$ is then made, (2.4.6) takes the form

$$\widehat{\phi}(x, y, s) = e^{-st^*(x,y)} \int_0^\infty \phi(x, y, \tau + t^*)\, e^{-s\tau} d\tau. \qquad (2.4.7)$$

Consequently, if the integral has the form of an algebraic function of x and y, then the behavior of the transform for large x or y is determined by the arrival time t^* of the first motion. For example, for any fixed value of x in the problem under consideration, the dilatational wave function ϕ appears to be dominated in $y > 0$ by the plane wave traveling at speed c_d away from the crack face. In this case, $t^*(x, y) = y/c_d$ so that

$$\widehat{\phi}(x, y, s) = o\left(e^{-sy(1-\epsilon)/c_d}\right) \tag{2.4.8}$$

for any fixed x as $y \to \infty$, where ϵ is a small positive real number. The plane wave itself is described by $\phi = -\left(\sigma^*/2\rho c_d^2\right)(c_d t - y)^2 H(c_d t - y)$ for $x < 0$. A straightforward calculation based on (2.4.7) then yields $\widehat{\phi} = -\left(\sigma^*/\rho c_d^2 s^3\right) e^{-sy/c_d}$. As a second illustration involving a cylindrical wave, consider the wave function $\phi(x, y, t)$ describing the wave response to an impulsive source of strength $2\pi c_d^2$ per unit mass concentrated at the origin of coordinates in an unbounded plane. Though the relevance of this problem to mode III crack growth could be established, only the mathematical features are of interest here. The wave function satisfies

$$\nabla^2\phi - c_d^{-2}\phi_{,tt} = -\delta(x)\delta(y)\delta(t) \tag{2.4.9}$$

where $\delta(\cdot)$ is the Dirac delta function. If ϕ and $\phi_{,t}$ are assumed to be zero at time $t = 0^-$, then the solution of (2.4.9) is

$$\phi(x, y, t) = \frac{H(t - r/c_d)}{\sqrt{t^2 - r^2/c_d^2}}, \tag{2.4.10}$$

where $H(\cdot)$ is the unit step function and $r = \sqrt{x^2 + y^2}$. Note that $t^*(x, y) = r/c_d$ in this case. The Laplace transform on time of (2.4.10) is

$$\widehat{\phi}(x, y, s) = \int_0^\infty \phi(x, y, t)\, e^{-st}\, dt = K_0(sr/c_d), \tag{2.4.11}$$

where $K_0(\cdot)$ is the modified Bessel function of the second kind. From the asymptotic property that

$$K_0(sr/c_d) \sim \sqrt{\frac{\pi c_d}{2rs}}\, e^{-sr/c_d} \quad \text{as} \quad r \to \infty \tag{2.4.12}$$

it is evident that the asymptotic behavior of $\widehat{\phi}(x, y, s)$ at remote points in the plane is determined by $t^*(x, y)$.

It is noted again that solution of differential equations by means of integral transforms is a formal procedure and (2.4.7) has no fundamental significance. However, it does provide guidance at a key point in applying the Wiener-Hopf technique, as will be demonstrated in the next section. Unlike the analysis of harmonic or steady wave phenomena, these transient wave propagation problems do not involve so-called radiation conditions. In a sense, the qualitative conditions such as (2.4.11)–(2.4.12), which follow from the wave propagation nature of the mechanical fields, play a role much like the role played by the radiation conditions in steady wave motion (Courant and Hilbert 1962; Kupradze 1963).

2.5 The Wiener-Hopf technique

The mathematical method known as the Wiener-Hopf technique was developed originally to solve a particular linear integral equation of the first kind which arose in the analysis of electromagnetic wave phenomena. Two distinguishing features of the integral equation are (i) it is enforced on a semi-infinite interval and (ii) the kernel depends only on the difference between the independent variable and the integration variable. The evolution of the method and a number of extensions are described in the book by Noble (1958). The application of the technique to transient problems in elastodynamics was pioneered by de Hoop (1958) in a study of several half plane diffraction problems, including the one being considered here. He observed that, if time dependence was suppressed through application of a Laplace transform, then an integral equation amenable to solution by the Wiener-Hopf technique could be obtained. Once a solution in the transformed domain is in hand, the task of inverting the transforms to obtain a solution in the physical domain remains. To execute this step, de Hoop (1961) extended an idea introduced earlier by Cagniard (1962). The result, which is now known as the Cagniard-de Hoop technique, is a powerful and elegantly simple method for inverting transforms in a wide range of elastodynamic wave propagation problems.

Although each of these techniques has taken on an existence separate from its foundations, each is perhaps best viewed in the context of its origins. Each technique is based on the judicious application of certain powerful theorems from the theory of analytic functions of a complex variable, Cauchy's integral formulas and the property of uniqueness of an analytic function in its domain of analyticity. The

plane strain problem of a crack with suddenly applied pressure loading
on the faces is now used as a vehicle for introducing these techniques.
Additional illustrations will be included in subsequent chapters. The
approach to be used, in fact, will not rely on reformulation of the problem in terms of a standard Wiener-Hopf integral equation. Instead,
the governing differential equations and boundary conditions will be
solved directly. This direct approach was apparently introduced by
Jones (1952) for application to scalar wave problems.

2.5.1 Application of integral transforms

Solution of the problem proceeds by application of a one-sided Laplace
transform on time, and then a two-sided Laplace transform on x, to
the governing differential equations and boundary conditions. The
use of the exponential Fourier transform with a complex transform
parameter is completely equivalent to the use of the two-sided Laplace
transform; the choice of one over the other is a matter of taste. It is
noted that the boundary conditions (2.4.4) are defined only on half
of the range of x. Consequently, the two-sided Laplace transform
cannot be applied to these boundary conditions as they stand. To
remedy the situation, the boundary conditions must be extended to
apply on the full range of x. To this end, two unknown functions
$u_-(x,t)$ and $\sigma_+(x,t)$ are introduced. The function u_- is *defined* to be
the (at present unknown) displacement of the crack face $y = 0^+$ in
the y direction for $-\infty < x < 0$, $0 < t < \infty$, and to be identically
zero for $0 < x < \infty$, $0 < t < \infty$. Likewise, σ_+ is defined to be the
tensile stress in the y-direction on the plane $y = 0$ for $0 < x < \infty$,
$0 < t < \infty$. With these definitions, the boundary conditions can be
rewritten as

$$\sigma_{yy}(x, 0^+, t) = \sigma_+(x,t) - \sigma^* H(t) H(-x),$$

$$\sigma_{xy}(x, 0^+, t) = 0, \tag{2.5.1}$$

$$u_y(x, 0^+, t) = u_-(x,t)$$

for the full range $-\infty < x < \infty$, $0 < t < \infty$. Obviously, the
subscripts $+$ and $-$ are used at this point to indicate on which half of
the x-axis a subscripted function is nonzero. The notation is carried
over into the transformed domain, where it is found that the same

subscript symbols are useful for designating a particular half plane of analyticity.

Next, the one-sided Laplace transform (2.4.6) is applied to the wave equations (2.4.3), in light of the initial conditions (2.4.5), and to the boundary conditions (2.5.1). Then, the two-sided Laplace transform defined by

$$\Phi(\zeta, y, s) = \int_{-\infty}^{\infty} \widehat{\phi}(x, y, s)\, e^{-s\zeta x} dx$$

$$= \int_{-\infty}^{0} \widehat{\phi}(x, y, s)\, e^{-s\zeta x} dx + \int_{0}^{\infty} \widehat{\phi}(x, y, s)\, e^{-s\zeta x} dx \qquad (2.5.2)$$

is applied. The notation in (2.5.2), whereby an upper case symbol is used to denote the double transform of the function represented by the corresponding lower case symbol, is adopted as a convention for the remainder of this chapter. The transform variable ζ is understood to be a complex variable, and the transform plane is scaled by the real parameter s simply for convenience. The product $s\zeta$ could be viewed as some other transform parameter, say ζ', at this point, but eventually the change of variable $\zeta' \rightarrow s\zeta$ would be found to be indispensable. The transform Φ in (2.5.2) is written as the sum of two semi-infinite integrals in order to consider the convergence property discussed in Section 1.3.2. In general, the second semi-infinite integral will converge for values of ζ in some *right* half plane. That is, there exists a real number ξ_+ such that the integral converges for $\mathrm{Re}(\zeta) > \xi_+$, and, consequently, this integral defines an analytic function in that half plane. Likewise, the first semi-infinite integral will converge for values of ζ in some *left* half plane, say $\mathrm{Re}(\zeta) < \xi_-$, and it defines an analytic function there. It is evident that the two-sided transform exists only if $\xi_+ \leq \xi_-$. The case $\xi_+ = \xi_-$ requires special consideration, and it can be avoided in most cases through the introduction of vanishingly small dissipation in mechanical problems. If $\xi_+ < \xi_-$, then the doubly infinite integral in (2.5.2) converges and defines an analytic function of ζ in the strip $\xi_+ < \mathrm{Re}(\zeta) < \xi_-$ of the complex ζ-plane.

Returning to the idea concerning asymptotic behavior of transforms embodied in (2.4.7), a strip of analyticity of $\Phi(\zeta, y, s)$ for fixed $y \geq 0$ and s may now be anticipated. For any $x > 0$, ϕ will be zero at least until time $t^* = r/c_d \geq x/c_d$. For any $x < 0$, on the other hand, ϕ takes on nonzero values upon passage of the plane wave at time $t^* = y/c_d$. Then, from (2.4.7), it is anticipated that

$$\widehat{\phi} = \begin{cases} o\left(e^{-sx(1-\epsilon)/c_d}\right) & \text{as } x \to +\infty, \, y > 0, \\ o\left(e^{-sx\epsilon/c_d}\right) & \text{as } x \to -\infty, \, y > 0, \end{cases} \tag{2.5.3}$$

for arbitrarily small real positive values of ϵ. If the two semi-infinite integrals in (2.5.2) are now considered in light of (2.5.3), it is expected that the first semi-infinite integral will converge for $\mathrm{Re}(\zeta) < 0$ for any values of y and s, and that the second will converge for $\mathrm{Re}(\zeta) > -1/c_d$. These domains of analyticity will be illustrated in Figure 2.5 once all relevant transforms have been defined. In any case, it is expected that $\Phi(\zeta, y, s)$ is an analytic function of ζ in the strip $-1/c_d < \mathrm{Re}(\zeta) < 0$. Similar reasoning applied in the case of other functions leads to the conclusion that this is a domain of analyticity common to all doubly transformed functions in the problem.

To develop further insight into the properties of the transforms, consider once again the wave motion due to an impulsive source as introduced in Section 2.4. If the two-sided Laplace transform is applied to $\widehat{\phi}(x, y, s)$ in (2.4.11), then (Watson 1966)

$$\Phi(\zeta, y, s) = \int_{-\infty}^{\infty} K_0(sr/c_d) \, e^{-s\zeta x} \, dx$$

$$= \frac{\pi}{s(c_d^{-2} - \zeta^2)^{1/2}} \, e^{-sy(c_d^{-2} - \zeta^2)^{1/2}} \tag{2.5.4}$$

for $y \geq 0$. It was shown in Section 2.4 that Φ is expected to decay exponentially as $y \to \infty$ if ϕ is a plane wave propagating in the y direction. The result (2.5.4) suggests that the same qualitative behavior should be expected for a cylindrical wave if ζ is in the strip of convergence of the transform integral defining Φ.

If the one-sided Laplace transform is applied to the wave equations (2.4.3), which are subject to the initial conditions (2.4.5), then

$$\nabla^2 \widehat{\phi} - a^2 s^2 \widehat{\phi} = 0, \qquad \nabla^2 \widehat{\psi} - b^2 s^2 \widehat{\psi} = 0, \tag{2.5.5}$$

where s is the real positive transform parameter. In (2.5.5), and for the remainder of the chapter, the inverse wave speeds are denoted by

$a = 1/c_d$ and $b = 1/c_s$. The transformed boundary conditions are

$$\hat{\sigma}_{yy}(x, 0^+, s) = \hat{\sigma}_+(x, s) - \sigma^* H(-x)/s \,,$$

$$\hat{\sigma}_{xy}(x, 0^+, s) = 0 \,,$$

$$\hat{u}_y(x, 0^+, s) = \hat{u}_-(x, s) \,, \tag{2.5.6}$$

which are valid for $-\infty < x < \infty$. Based on the foregoing observations on the wave propagation nature of the fields, it is anticipated that $\hat{\sigma}_+(x, s)$ is bounded by an algebraic function of x times e^{-asx} as $x \to +\infty$ for any positive real value of s. Likewise, $\hat{u}_-(x, s)$ is anticipated to have algebraic behavior as $x \to -\infty$ for any real positive value of s. In fact, the displacement of the crack faces is dominated by the plane waves propagating away from the faces for large negative values, so that the dominant part of $\hat{u}_-(x, s)$ will be independent of x as $x \to -\infty$.

Next, the two-sided Laplace transform (2.5.2) is applied to (2.5.5) and (2.5.6). The boundary conditions are considered first, and the representation of the left side of each boundary condition in (2.5.6) in terms of the transformed displacement potentials is introduced at this point. Thus,

$$\mu \left[(b^2 - 2a^2)s^2 \Phi + 2\frac{d^2\Phi}{dy^2} - 2s\zeta\frac{d\Psi}{dy} \right]_{y=0^+} = \frac{\Sigma_+(\zeta)}{s^2} + \frac{\sigma^*}{s^2\zeta} \,,$$

$$\mu \left[2s\zeta\frac{d\Phi}{dy} + \frac{d^2\Psi}{dy^2} - s^2\zeta^2\Psi \right]_{y=0^+} = 0 \,, \tag{2.5.7}$$

$$\left[\frac{d\Phi}{dy} - s\zeta\Psi \right]_{y=0^+} = \frac{U_-(\zeta)}{s^3} \,.$$

The special form of the terms on the right side of (2.5.7) has been selected for convenience, and the selection is based on hindsight, to a large extent. At this point, it is still *possible* that

$$\Sigma_+(\zeta) = s^2 \int_0^\infty \hat{\sigma}_+(x, s)\, e^{-s\zeta x} dx \,,$$

$$U_-(\zeta) = s^3 \int_{-\infty}^0 \hat{u}_-(x, s)\, e^{-s\zeta x} dx \tag{2.5.8}$$

will depend on s as well as on ζ. As suggested by the special form of (2.5.7), however, these functions will be found to depend only on ζ. From the foregoing observations on the rate of decay of transformed fields, Σ_+ is expected to be analytic in $\mathrm{Re}(\zeta) > -a$ and U_- is expected to be analytic in the overlapping half plane $\mathrm{Re}(\zeta) < 0$. In particular, both functions are expected to be analytic in the strip $-a < \mathrm{Re}(\zeta) < 0$.

Application of the two-sided Laplace transform (2.5.2) to the differential equations in (2.5.5) yields linear second order differential equations in y for $\Phi(\zeta, y, s)$ and $\Psi(\zeta, y, s)$, involving ζ and s as parameters. For values of ζ in the common strip of analyticity, each equation has two independent solutions, one growing exponentially and one decaying exponentially as $y \to \infty$. The wave propagation nature of the fields precludes the possibility of exponential growth as y becomes large with $-a < \mathrm{Re}(\zeta) < 0$. Consequently, only the solutions decaying as $y \to \infty$ are admitted, so that

$$\Phi(\zeta, y, s) = \frac{1}{s^4} P(\zeta) e^{-s\alpha y}, \qquad \Psi(\zeta, y, s) = \frac{1}{s^4} Q(\zeta) e^{-s\beta y}, \quad (2.5.9)$$

where

$$\alpha = \alpha(\zeta) = \left(a^2 - \zeta^2\right)^{1/2}, \qquad \beta = \beta(\zeta) = \left(b^2 - \zeta^2\right)^{1/2}. \quad (2.5.10)$$

Again, the special form selected for the undetermined coefficients in (2.5.9) is based on the expectation that P and Q will not depend on s. The functions α and β in (2.5.10) are multiple valued functions of ζ in the complex ζ-plane with branch points at $\zeta = \pm a$ and $\zeta = \pm b$, respectively. The branch of the square root in α is chosen as the one yielding values of α with a positive real part at each interior point of the common strip. For future reference, the definition of α is extended to the entire ζ-plane. Branch cuts are introduced along $a \le |\mathrm{Re}(\zeta)| < \infty$, $\mathrm{Im}(\zeta) = 0$ and the branch of α which is chosen is the one which has a positive real part everywhere in the cut plane except on the branch cuts. If the point ζ approaches the cut in the right (left) half plane with $\mathrm{Im}(\zeta) \to 0^+$ then the limiting value of α is a negative (positive) imaginary value. The limiting value on the opposite side of the branch cut has the same magnitude but the opposite sign. Likewise, $\beta(\zeta)$ is defined for the entire ζ-plane cut along $b \le |\mathrm{Re}(\zeta)| < \infty$, $\mathrm{Im}(\zeta) = 0$ and the branch with nonnegative real part is understood in (2.5.10).

Next, the transformed potentials (2.5.9) are substituted into the transformed boundary conditions (2.5.7). The result is a system of three linear algebraic equations for the four unknown functions $P(\zeta)$, $Q(\zeta)$, $\Sigma_+(\zeta)$, and $U_-(\zeta)$. The system of equations is valid for all values of ζ in the common strip. As was anticipated, the system of equations does *not* involve the parameter s. If two of the equations are used to eliminate P and Q in favor of Σ_+ and U_-, then the remaining equation is

$$\frac{\sigma^*}{\zeta} + \Sigma_+(\zeta) = -\frac{\mu}{b^2} \frac{R(\zeta)}{\alpha(\zeta)} U_-(\zeta), \qquad (2.5.11)$$

which is valid in the common strip of analyticity $-a < \text{Re}(\zeta) < 0$, where

$$R(\zeta) = 4\zeta^2 \alpha(\zeta)\beta(\zeta) + (b^2 - 2\zeta^2)^2. \qquad (2.5.12)$$

The function R is analytic everywhere in the complex plane except at the branch points $\zeta = \pm a$ and $\zeta = \pm b$. For the branches of α and β that were selected above, R is single valued in the ζ-plane cut along $a \le |\text{Re}(\zeta)| \le b$, $\text{Im}(\zeta) = 0$. The function R is usually called the Rayleigh wave function because the two real roots of $R = 0$, say $\zeta = \pm c$, are the inverse wave speeds of free surface Rayleigh waves traveling in opposite directions, each with absolute speed $c_R = 1/c$. In terms of Poisson's ratio ν, the value of c is given approximately by

$$\frac{c}{b} = \frac{1+\nu}{0.862 + 1.14\nu}. \qquad (2.5.13)$$

There may be other roots of $R = 0$ when other branches of the multiple valued function R are considered (that is, roots may exist on other sheets of the Riemann surface of the multiple valued function), but this possibility is of no consequence here.

The principal features of the complex ζ-plane pertinent to the solution of (2.5.11) are identified in Figure 2.5. The branch points at $\zeta = \pm a, \pm b$ and the corresponding branch cuts are shown, the zeros of the Rayleigh function at $\zeta = \pm c$ are shown, and the pole at $\zeta = 0$ associated with the applied loading in the problem is indicated. The common strip of analyticity is between the dashed line and the imaginary axis, and the overlapping half planes in which the functions labeled with subscripts $(+)$ or $(-)$ are analytic are also indicated with arrows.

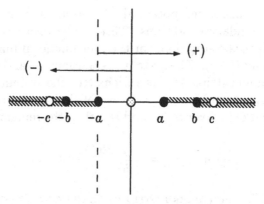

Figure 2.5. The complex ζ-plane showing the location of the singularities of the analytic functions involved in the Wiener-Hopf equation (2.5.11), which applies in the strip $-a < \text{Re}(\zeta) < 0$.

2.5.2 The Wiener-Hopf factorization

The equation (2.5.11) is typical of problems analyzed by means of the Wiener-Hopf method. It is a single functional equation that holds in a strip of the complex plane and that involves two unknown functions, each analytic in a half plane with the strip being the region of overlap of the two half planes. The essence of the Wiener-Hopf method is the process of determining *both* of the sectionally analytic functions from this single equation by applying certain theorems from the theory of analytic functions. So as not to obscure the elegance and simplicity of the steps involved with the details of the problem at hand, the procedure is first outlined in general form.

Equation (2.5.11) has the form

$$A(\zeta) + \Sigma_+(\zeta) = B(\zeta)U_-(\zeta) \qquad (2.5.14)$$

for all ζ in a strip of the complex plane parallel to the imaginary axis, say $\xi_+ < \text{Re}(\zeta) < \xi_-$. The specified functions $A(\zeta)$ and $B(\zeta)$ are analytic in the strip and are defined in the entire ζ-plane, with unique branches identified for multiple valued functions. The unknown functions $\Sigma_+(\zeta)$ and $U_-(\zeta)$ are analytic in the overlapping half planes $\text{Re}(\zeta) > \xi_+$ and $\text{Re}(\zeta) < \xi_-$, respectively. The Wiener-Hopf procedure requires that $B(\zeta)$ must be factored into the product of sectionally analytic functions

$$B(\zeta) = B_+(\zeta)B_-(\zeta), \qquad (2.5.15)$$

where $B_+(\zeta)$ is analytic and nonzero in $\mathrm{Re}(\zeta) > \xi_+$, and $B_-(\zeta)$ is analytic in $\mathrm{Re}(\zeta) < \xi_-$. Thus,

$$\frac{A(\zeta)}{B_+(\zeta)} + \frac{\Sigma_+(\zeta)}{B_+(\zeta)} = B_-(\zeta)U_-(\zeta). \qquad (2.5.16)$$

The only "mixed" function is $A(\zeta)/B_+(\zeta)$, and this must be decomposed into a sum of sectionally analytic functions

$$\frac{A(\zeta)}{B_+(\zeta)} = L_+(\zeta) + L_-(\zeta), \qquad (2.5.17)$$

where L_+ and L_- are analytic in $\mathrm{Re}(\zeta) > \xi_+$ and $\mathrm{Re}(\zeta) < \xi_-$, respectively. These factors are unique up to an additive entire function, that is, any entire function may be added to one factor and subtracted from the other factor. With this additive decomposition done, (2.5.16) takes the form

$$L_+(\zeta) + \Sigma_+(\zeta)/B_+(\zeta) = B_-(\zeta)U_-(\zeta) - L_-(\zeta) \qquad (2.5.18)$$

which is valid in the strip $\xi_+ < \mathrm{Re}(\zeta) < \xi_-$. Each side of (2.5.18) is analytic in one of the overlapping half planes, and the sides obviously coincide in the strip of overlap. Consequently, according to the identity theorem for analytic functions introduced in Section 1.3.1, each side of (2.5.18) is the analytic continuation of the other into its complementary half plane, so that the two sides together represent one and the same entire function, say $E(\zeta)$. Normally, the entire function is determined by its behavior as $|\zeta| \to \infty$, which is related to constraints on the variables representing physical quantities near $x = 0$. The most common situation is that the sectionally analytic functions have algebraic behavior at infinity, so that

$$\left| L_+(\zeta) + \Sigma_+(\zeta)/B_+(\zeta) \right| = O\left(|\zeta|^{q_+}\right), \quad \mathrm{Re}(\zeta) > \xi_+,$$
$$\left| B_-(\zeta)U_-(\zeta) - L_-(\zeta) \right| = O\left(|\zeta|^{q_-}\right), \quad \mathrm{Re}(\zeta) < \xi_-, \qquad (2.5.19)$$

as $|\zeta| \to \infty$. In order to deal with such circumstances, Liouville's theorem on bounded entire functions has been generalized to the case of entire functions with no essential singularities, even at infinity.

The result, known commonly as the generalized Liouville's theorem, requires that the entire function $E(\zeta)$ is a polynomial of degree less than or equal to the largest integer that is less than or equal to the smaller of q_+ and q_- (Whittaker and Watson 1927). Consequently, Σ_+ and U_- are determined from (2.5.18) up to a finite number of constants, the polynomial coefficients.

In some cases, the additive factorization into sectionally analytic functions indicated in (2.5.17) can be accomplished by observation. In other cases, the factorization must be done formally. The essential idea of the formal procedure follows from the statement of Cauchy's integral formula,

$$L(\zeta) = L_+(\zeta) + L_-(\zeta) = \frac{1}{2\pi i} \int_\Gamma \frac{L(\gamma)}{\gamma - \zeta}\, d\gamma, \qquad (2.5.20)$$

where ζ is a point in the strip and Γ is a small simple closed contour around ζ in the strip, executed in the counterclockwise sense. The contour Γ is then "expanded to infinity" by means of Cauchy's theorem (1.3.4). So as to continue to satisfy the conditions of Cauchy's theorem, the contour leaves in its wake clockwise contours embracing all the poles and branch cuts of $L(\zeta)$ in the whole plane as the radius of the contour is increased to indefinitely large values. If the collection of such contours in $\mathrm{Re}(\zeta) > \xi_-$ is called Γ_- and the collection in $\mathrm{Re}(\zeta) < \xi_+$ is called Γ_+, then

$$L_\pm(\zeta) = \frac{1}{2\pi i} \int_{\Gamma_\pm} \frac{L(\gamma)}{\gamma - \zeta}\, d\gamma. \qquad (2.5.21)$$

The Cauchy integral defining $L_\pm$ is analytic everywhere except for ζ approaching certain points on $\Gamma_\pm$. In view of the fact that these paths are restricted to the appropriate half planes, the indicated regions of analyticity are obvious, and the factorization is complete. It has been tacitly assumed that $|L(\zeta)| \to 0$ as $|\zeta| \to \infty$ so that the contribution to the integral from the contour at infinity vanishes, but this does not pose a serious restriction on the application of the idea.

The factorization into a product of sectionally analytic functions indicated in (2.5.15) can be accomplished in the same way by observing that

$$\log B(\zeta) = \log B_+(\zeta) + \log B_-(\zeta) = \frac{1}{2\pi i} \int_{\Gamma_+ + \Gamma_-} \frac{\log B(\gamma)}{\gamma - \zeta}\, d\gamma \qquad (2.5.22)$$

so that

$$B_\pm(\zeta) = \exp\left\{\frac{1}{2\pi i}\int_{\Gamma_\pm}\frac{\log B(\gamma)}{\gamma - \zeta}\,d\gamma\right\}. \qquad (2.5.23)$$

In writing (2.5.22), it is tacitly assumed that $B(\zeta) \to 1$ as $|\zeta| \to \infty$, and that $B_\pm(\zeta)$ are not only analytic but also nonzero in the respective half planes. The latter condition ensures that $B_\pm(\zeta)$ and $\log B_\pm(\zeta)$ have a common domain of analyticity. Again, as will be seen, these conditions do not pose serious restrictions on the application of the idea.

The functional equation (2.5.11) is now considered as a particular case of (2.5.14). The first step in determining the unknown functions Σ_+ and U_- is the factorization of $R(\zeta)$ and $\alpha(\zeta)$ into products of sectionally analytic functions. The requisite factorization of α is simple, and it is given by

$$\alpha_\pm(\zeta) = (a \pm \zeta)^{1/2}, \qquad \alpha = \alpha_+\alpha_-. \qquad (2.5.24)$$

These factors satisfy all requirements.

The Rayleigh wave function $R(\zeta)$ is more complicated than α, and the formal procedure must be followed for it. To this end, it is essential to know the behavior of $R(\zeta)$ for large $|\zeta|$. Through application of the binomial expansion theorem to the square root terms in (2.5.12), it can be shown that

$$R(\zeta) = -2(b^2 - a^2)\zeta^2 + O(1) \qquad \text{as} \qquad |\zeta| \to \infty. \qquad (2.5.25)$$

Furthermore, it is essential to know the locations of *all* zeros of the relevant branch of $R(\zeta)$, not just the locations of the real zeros corresponding to free surface Rayleigh waves. This information may be extracted by application of the zero counting formula of analytic function theory.

Consider a simple closed curve Z in the ζ-plane such that $R(\zeta)$ is analytic everywhere inside and on Z. Then, according to the zero counting formula, the number N of roots of $R(\zeta) = 0$ inside Z is

$$N = \frac{1}{2\pi i}\oint_Z \frac{R'(\zeta)}{R(\zeta)}\,d\zeta, \qquad (2.5.26)$$

where the prime denotes a derivative. The formula (2.5.26) takes proper account of the multiplicity of roots. Furthermore, Z may be a set of closed curves that together bound a multiply connected region in which $R(\zeta)$ is analytic. In either case, the sense of traversal of Z is such that the region is on the left. An operational procedure for calculating N is evident if the mapping from the complex ζ-plane to the complex ω-plane defined by $\omega = R(\zeta)$ is considered. If the integral in (2.5.26) is considered in the ω-plane, then

$$N = \frac{1}{2\pi i} \oint_\Omega \frac{d\omega}{\omega}, \qquad (2.5.27)$$

where Ω is the image of Z according to the mapping. Thus, N is simply the *net* number of times that the path Ω encircles the origin in the ω-plane in a counterclockwise sense as ζ moves in a counterclockwise sense once around Z. A convenient choice of Z, consisting of the three closed paths ABC, DEF, and GHI in the ζ-plane, and its image in the ω-plane, are shown in Figure 2.6. Evidently, the path Ω encircles the origin in the ω-plane two times in a counterclockwise sense for the choice of Z in the ζ-plane. Consequently, the two real zeros of $R(\zeta)$ in the cut ζ-plane are the only zeros of the function in the entire ζ-plane.

Recall that the factorization (2.5.15) is accomplished most directly for functions that approach unity as $|\zeta| \to \infty$ and that have neither zeros nor poles in the finite plane. A function having these properties is

$$S(\zeta) = \frac{R(\zeta)}{\kappa(c^2 - \zeta^2)}, \qquad \kappa = 2(b^2 - a^2), \qquad (2.5.28)$$

so the general result (2.5.15) is interpreted for this function. The only singularities of $S(\zeta)$ are the branch points at $\zeta = \pm a$, $\pm b$ which are shared with $R(\zeta)$, and it is single valued in the ζ-plane cut along $a \leq |\text{Re}(\zeta)| \leq b$, $\text{Im}(\zeta) = 0$. The clockwise contours Γ_+ and Γ_- are closed paths embracing the branch cuts in the left and right half planes, respectively. The property that $S(\bar\zeta) = \overline{S(\zeta)}$ can be exploited to show that

$$\log S_\pm(\zeta) = -\frac{1}{\pi} \int_a^b \tan^{-1}\left[\frac{4\eta^2 \sqrt{(\eta^2 - a^2)(b^2 - \eta^2)}}{(b^2 - 2\eta^2)^2} \right] \frac{d\eta}{\eta \pm \zeta},$$

$$(2.5.29)$$

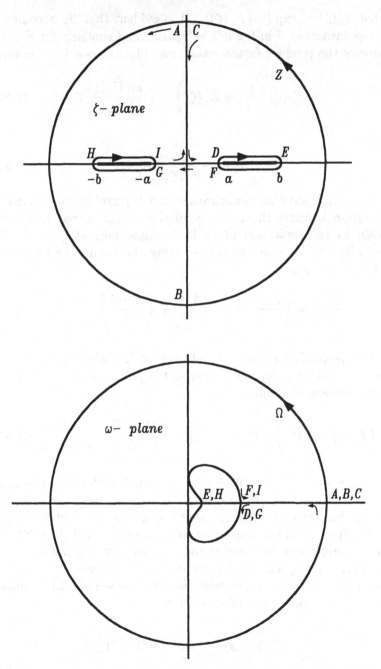

Figure 2.6. Mapping of the contour Z in the ζ-plane into the contour Ω in the ω-plane according to $\omega = R(\zeta)$.

so that $S_{\pm}(\zeta) = \exp\{\log S_{\pm}(\zeta)\}$. It is evident that S_+ accounts for the singularities of S in the left half plane, and similarly for S_-. This completes the product factorization, and (2.5.11) now has the form

$$F_+(\zeta)\left(\frac{\sigma^*}{\zeta} + \Sigma_+(\zeta)\right) = -\frac{\mu\kappa}{b^2}\frac{U_-(\zeta)}{F_-(\zeta)}, \qquad (2.5.30)$$

where

$$F_{\pm}(\zeta) = \frac{\alpha_{\pm}(\zeta)}{(c \pm \zeta)S_{\pm}(\zeta)}. \qquad (2.5.31)$$

The final additive factorization (2.5.17) can be carried out by observation because the only singularity of the mixed function in (2.5.30) in the right half plane is a simple pole at $\zeta = 0$. This singularity can be removed by requiring the residue to be zero, so the factorization is

$$\frac{F_+(\zeta)}{\zeta} = \left\{\frac{F_+(\zeta) - F_+(0)}{\zeta}\right\}_+ + \left\{\frac{F_+(0)}{\zeta}\right\}_-. \qquad (2.5.32)$$

The first (second) term on the right side of (2.5.32) is analytic in the appropriate right (left) half plane, so the particular form of (2.5.18) for the problem at hand is

$$\frac{\sigma^*}{\zeta}\left(F_+(\zeta) - F_+(0)\right) + F_+(\zeta)\Sigma_+(\zeta) = -\frac{\sigma^* F_+(0)}{\zeta} - \frac{\mu\kappa}{b^2}\frac{U_-(\zeta)}{F_-(\zeta)} \quad (2.5.33)$$

in the strip $-a < \operatorname{Re}(\zeta) < 0$. As already noted, each side is the unique analytic continuation of the other into a complementary half plane, and together the two sides represent a single entire function $E(\zeta)$.

As $|\zeta| \to \infty$ in the respective half planes, $|F_{\pm}(\zeta)| = O\left(|\zeta|^{-1/2}\right)$. Furthermore, $\hat{\sigma}(x, s)$ is expected to be square root singular as $x \to 0^+$, and $\hat{u}_-(x, s)$ is expected to vanish as $x \to 0^-$ to ensure continuity of displacement. Consequently, from the Abel theorem (1.3.18) concerning asymptotic properties of transforms,

$$\lim_{x \to 0^+} x^{1/2}\hat{\sigma}(x, s) \sim \lim_{\zeta \to +\infty} \zeta^{1/2}\Sigma_+(\zeta),$$

$$\lim_{x \to 0^-} |x|^{-q}\hat{u}_-(x, s) \sim \lim_{\zeta \to -\infty} |\zeta|^{1+q}U_-(\zeta)$$

$$(2.5.34)$$

for some $q > 0$. Therefore, each side of (2.5.33) vanishes as $|\zeta| \to \infty$ in the corresponding half plane. According to Liouville's theorem, a bounded entire function is a constant. In this case, $E(\zeta)$ is bounded in the finite plane and $E(\zeta) \to 0$ as $|\zeta| \to \infty$ so that the constant must have the value zero; thus, $E(\zeta) = 0$.

This completes the solution of the Wiener-Hopf equation (2.5.11). The two unknown functions in that equation have been determined to be

$$\Sigma_+(\zeta) = \frac{\sigma^*}{\zeta}\left(\frac{F_+(0)}{F_+(\zeta)} - 1\right),$$

$$U_-(\zeta) = -\frac{b^2}{\zeta\mu\kappa}\sigma^* F_+(0) F_-(\zeta).$$

(2.5.35)

2.5.3 Inversion of the transforms

At this point in the analysis, it would be possible to solve for $P(\zeta)$ and $Q(\zeta)$, in addition to the functions in (2.5.35), and to invert the double transforms for $\phi(x, y, t)$ and $\psi(x, y, t)$ by means of the Cagniard-de Hoop method. Attention is focused on the time history of stress ahead of the crack tip in this chapter, however, so that only the inversion of the double transform of $\sigma_+(x, t)$ will be outlined. Inversion of the two-sided Laplace transform is considered first, and the formal inverse is given by the integral

$$\hat{\sigma}_+(x, s) = \frac{1}{2\pi i}\int_{\xi_0 - i\infty}^{\xi_0 + i\infty} \frac{1}{s}\Sigma_+(\zeta) e^{s\zeta x} d\zeta,$$

(2.5.36)

where ξ_0 is a real number in the interval $-a < \xi_0 < 0$. The central idea of the Cagniard-de Hoop scheme and, likewise, the central idea here is to convert the integral in (2.5.36) to a form which will allow inversion of the one-sided Laplace transform *by observation*. To this end, the path of integration is augmented by two circular arcs of indefinitely large radius in the left half plane plus a path running from the remote point on the negative real axis inward along the upper side of the branch cut to $\zeta = -a$, around the branch point, and then outward along the lower side of the cut to the remote point; see Figure 2.7. These paths, together with the inversion path itself, form a closed contour, and the integrand of (2.5.36) is analytic inside and on this contour.

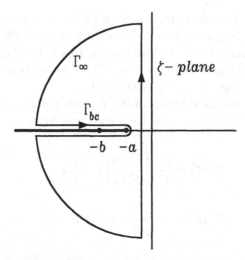

Figure 2.7. Integration path for evaluation of the integral in (2.5.37), consisting of Γ_∞, two remote quarter-circular arcs, and Γ_{bc}, a path embracing the branch cut along the negative real axis.

According to Cauchy's theorem, the integral of the function along the closed contour is zero. An immediate inference is that the integral in (2.5.36) is equal to the integral of the same function taken along the alternate path consisting of the two arcs of large radius Γ_∞ and the branch cut integral Γ_{bc},

$$\hat{\sigma}_+(x,s) = \frac{1}{2\pi i} \int_{\Gamma_\infty} \frac{1}{s} \Sigma_+(\zeta)\, e^{s\zeta x} d\zeta + \frac{1}{2\pi i} \int_{\Gamma_{bc}} \frac{1}{s} \Sigma_+(\zeta)\, e^{s\zeta x} d\zeta .$$

$$(2.5.37)$$

The function $\Sigma_+(\zeta) \to 0$ uniformly as $|\zeta| \to \infty$ in the left half plane. Consequently, according to Jordan's lemma introduced in Section 1.3.2, the value of the first integral in (2.5.37) is zero. Furthermore, $\Sigma_+(\bar{\zeta}) = \overline{\Sigma_+(\zeta)}$ so that the second integral in (2.5.37) can be written as the real integral

$$\hat{\sigma}_+(x,s) = \frac{1}{\pi s} \int_{-a}^{-\infty} \mathrm{Im}\Big\{ \Sigma_+(\zeta) \Big\}\, e^{s\zeta x} d\zeta , \qquad (2.5.38)$$

where the limiting value of the integrand as $\mathrm{Im}(\zeta) \to 0^+$ is understood. If the change of variable $\eta = -x\zeta$, $\eta \geq 0$, is introduced, the integral takes the form

$$\hat{\sigma}_+(x,s) = -\frac{1}{\pi s x}\int_{ax}^{\infty} \text{Im}\left\{\Sigma_+\left(-\frac{\eta}{x}\right)\right\} e^{-s\eta}d\eta \qquad (2.5.39)$$

and, indeed, the inversion of the one-sided transform becomes obvious. That is, (2.5.39) is a product of two transforms, so that $\sigma_+(x,t)$ is a convolution of the inverses of the two transforms,

$$\sigma_+(x,t) = -\frac{1}{\pi x}\int_{ax}^{t} \text{Im}\left\{\Sigma_+\left(-\frac{\eta}{x}\right)\right\} d\eta\, H(t-ax). \qquad (2.5.40)$$

This completes the process of inversion of the double transform for the stress distribution on the crack plane ahead of the tip. The result is in the form of a real integral, and some of its features are now examined.

The integrand of (2.5.40) has the property that

$$\Sigma_+(\zeta) = -\frac{i\sigma^*F_+(0)}{(-\zeta)^{1/2}} + O\left(|\zeta|^{-3/2}\right), \qquad |\zeta|\to -\infty \qquad (2.5.41)$$

along the upper side of the cut in the left half plane. Thus, the integral is divergent as $(x/c_d t)\to 0^+$, reflecting the anticipated singularity in stress at the crack tip. The strength of this singularity may be determined in a number of ways. For example, application of the Abel theorem (1.3.18) on asymptotic properties of transforms in the present situation implies

$$\lim_{x\to 0^+}(\pi x)^{1/2}\hat{\sigma}_+(x,s) = \lim_{\zeta\to +\infty}(s\zeta)^{1/2}\frac{1}{s^2}\Sigma_+(\zeta). \qquad (2.5.42)$$

If the indicated limits are taken for the functions at hand, then

$$\hat{K}_I(s) = \sqrt{2}\,\sigma^* F_+(0)/s^{3/2}. \qquad (2.5.43)$$

Thus, by inverting the Laplace transform, the stress intensity factor history defined in (2.4.2) is found to be

$$K_I(t) = 2\sigma^*\frac{\sqrt{c_d t(1-2\nu)/\pi}}{(1-\nu)}, \qquad (2.5.44)$$

where the coefficient in (2.5.43) has been evaluated in terms of material parameters. This evaluation is carried out by noting that $S_+(0) = S_-(0) = \sqrt{S(0)} = b^2/c\sqrt{\kappa}$ and $\alpha_+(0) = \sqrt{a}$. If the ratio of wave speeds b/a is then written in terms of Poisson's ratio ν as in (1.2.58), it is found that $F_+(0) = (1 - \nu)^{-1}\sqrt{c_d(1 - 2\nu)/2}$. The dimensionless constant C_I that was introduced in (2.4.2) has the value $C_I = \sqrt{2(1 - 2\nu)}/\pi(1 - \nu)$. Equation (2.5.44) is the main result of this section. Note that the stress intensity factor is zero for elastic incompressibility, or $\nu \to 1/2$. In this case, the stress state is apparently a uniform hydrostatic compression for $t > 0$.

The stress intensity factor (2.5.44) has been obtained for the case of crack face loading. Just as in the case of antiplane shear loading, however, it may be given another interpretation. Consider once again the same configuration, but the crack faces are now free of traction. Suppose that a plane tensile pulse propagates toward the crack plane at speed c_d. The front of the pulse is parallel to the crack plane, and it carries a jump in stress σ_{yy} from its initial value of zero to σ^*. The pulse arrives at the crack plane at time $t = 0$, and it is partially reflected and partially scattered, or diffracted, upon reaching the crack.

The complete deformation field of this stress pulse diffraction problem differs from the field due to crack face loading only by the pulse propagating in an *uncracked* solid. Because the governing equations are linear, and because the plane pulse, by itself, results in no stress singularity at the crack tip, it follows that (2.5.44) is also the stress intensity factor history for the case of stress pulse diffraction. Furthermore, the complete stress distribution on the crack plane ahead of the tip is given by (2.5.40) plus the constant σ^*, that is,

$$\sigma_{yy}(x, 0, t) = \sigma_+(x, t) + \sigma^*. \qquad (2.5.45)$$

Unlike the case of antiplane shear in Section 2.3, the expression for the complete stress distribution on $y = 0$, $0 < x < c_d t$ cannot be written in terms of elementary functions. The definite integral (2.5.40) can be evaluated numerically in a straightforward manner, however, to obtain a graphical representation of $\sigma_+(x, t)$. At first sight, it may appear that the Cauchy integral in the definition of $S_+(\zeta)$ must be evaluated for points on its path of integration. This can be done by invoking the Plemelj formula for the limiting value of a Cauchy integral as the field point approaches the path of integration

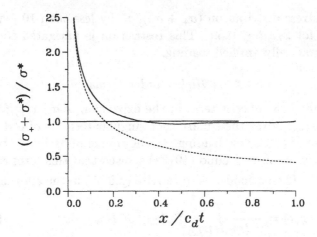

Figure 2.8. Normal stress on the plane $y = 0$ ahead of the crack tip versus normalized distance $x/c_d t$ for $\nu = 0.3$. The solid curve is the complete stress distribution (2.5.45) and the dashed curve is the singular field (2.4.1) only.

from either side (see Carrier et al. 1966, for example). If $S_+(\zeta)$ is replaced by $S(\zeta)/S_-(\zeta)$, however, then this special consideration is unnecessary because the Cauchy integral in the definition of $S_-(\zeta)$ is analytic for all points on the path of integration in (2.5.40). The result of numerical evaluation of the ratio $(\sigma_+ + \sigma^*)/\sigma^*$ versus $x/c_d t$ is shown in Figure 2.8 for $\nu = 0.3$. Also shown in the same figure as a dashed line is a graph of the singular term (2.4.1).

The size of the region, as a fraction of $x/c_d t$ or $x/c_s t$, over which the stress intensity factor field approximates the complete stress distribution is slightly smaller than in the case of antiplane shear. The reason is connected with the complexity of wave phenomena in plane strain as compared to antiplane shear deformation. In the case of plane strain, the initial wave motion radiated out from the crack faces as the pressure is applied is a negative dilation or compression. The initial tendency of material on the crack plane ahead of the crack tip is to compress slightly, as shown in Figure 2.8. The tensile stress on this plane becomes slightly *negative* with the first wave arrival, as reported in the experimental study of this configuration by Kim (1985b). It is only after the shear wavefront passes a material point on this plane that the tensile stress tends to become positive. It does not become sufficiently positive for the singular term to dominate that distribution until about $x/c_d t \leq 0.09$, at least for the case of suddenly applied loads. That is, the stress intensity factor field differs from the

complete stress distribution $(\sigma_+ + \sigma^*)/\sigma^*$ by less than 10 percent of
the latter for $x/c_d t \leq 0.09$. This restriction is mitigated somewhat
for more gradually applied loading.

2.5.4 Higher order terms

The next most significant term in the expansion of $\sigma_+(x,t)$ for small
values of $x/c_d t$, after the square root singular term, is of order unity
for the case of crack face loading. In the course of deriving the result
by an alternate route, Freund (1973a) showed that this term is simply
$\sigma_{yy}^{(1)} = -\sigma^*$. The basic idea is to rewrite (2.5.40) in the alternate form

$$\sigma_+(x,t) = -\frac{1}{2\pi i} \int_{\Gamma(t/x)} \Sigma_+(-\zeta)\, d\zeta\, H(t - ax), \qquad (2.5.46)$$

where the path $\Gamma(t/x)$ embraces the cut in the complex plane running
from $\zeta = t/x - i0$, and around the branch point at $\zeta = a$, to $\zeta =
t/x + i0$. By means of Cauchy's theorem, the integral in (2.5.46) can
be shown to be equal to an integral of the same function around a circle
of radius t/x plus the residue of the pole at $\zeta = 0$. It is the residue
which leads to the term of order unity in the near tip expansion.
Thus, the graph of the first *two terms* of the near tip expansion may
be obtained from the dashed curve in Figure 2.8 by a unit reduction
in the ordinate of each point. Likewise, the graph of $\sigma_+(x,t)/\sigma^*$ may
be obtained by a unit reduction in the ordinate of each point of the
solid curve in Figure 2.8. An immediate inference is that the stress
intensity factor field, by itself, differs from the complete distribution
σ_+/σ^* by less than 10 percent of the latter only if $x/c_d t \leq 0.0025$.
This observations shows the significance of the terms of order unity
in the near tip expansion in some cases.

The stress component σ_{xx} is also square root singular at the
crack tip. The main contribution to this stress component for points
near the tip is given by the universal spatial dependence (2.1.1) and
the stress intensity factor (2.5.44). Like the stress component σ_{yy},
the component σ_{xx} also has a term of order unity if the complete
distribution is expanded in powers of $r/c_d t$, where r is the radial
distance from the crack tip. By following the method leading to
(2.5.46), Ma and Burgers (1987) showed that the term of order unity
is

$$\sigma_{xx}^{(1)} = -\frac{\sigma^* \nu}{1 - \nu} - \frac{\sigma^* (2 - b^2/c^2)}{S_+(0) S_-(c) \sqrt{1 + c/a}}. \qquad (2.5.47)$$

The magnitude of $\sigma_{xx}^{(1)}$ depends on σ^*, of course, and on Poisson's ratio. The algebraic sign is negative for all admissible values of ν and $\sigma^* > 0$. For Poisson's ratio of $\nu = 0.3$, for example, $\sigma_{xx}^{(1)}$ has the value $-0.85\,\sigma^*$, with the contributions from the two terms in (2.5.47) being about equal. The term of order unity in the expansion of the in-plane shear stress $\sigma_{xy}^{(1)}$ is zero for this problem.

Consider once again the plane tensile stress pulse propagating in the y direction through the solid. If the pulse carries a jump in the stress component σ_{yy} from zero to σ^*, then the stress component σ_{xx} jumps from zero to $\sigma^*\nu/(1-\nu)$. But this is precisely the magnitude of the first term in (2.5.47). Consequently, for the case of diffraction of the stress pulse by a traction-free crack in the plane $y = 0$, the term of order unity in the expansion of σ_{xx} would be just the second term in (2.5.47).

2.6 Suddenly applied in-plane shear traction

Consider once again the body of elastic material that contains a half plane crack but that is otherwise unbounded; see Figure 2.1. In the present instance, the crack faces are subjected to suddenly applied, spatially uniform shear traction of magnitude τ^*. For the boundary traction, $T_y(x,z,t) = T_z(x,z,t) = 0$ and $T_x(x,z,t) = \pm\tau^*H(t)$ on $y = \pm 0$, where $\tau^* > 0$ and $H(t)$ is the unit step function. Thus, the crack faces are traction-free for time $t < 0$, and they are subjected to uniform shear traction for $t \geq 0$. The material is stress free and at rest everywhere for $t < 0$.

The components of displacement in the x, y, z-coordinate directions are u_x, u_y, u_z, respectively. Arguments similar to those used in Section 2.2 can be applied here to conclude that $u_z = 0$ for all time if it is zero initially, and that u_x and u_y are independent of the z-coordinate, hence the state of deformation is plane strain. Furthermore, the nonzero components of displacement satisfy the symmetry conditions $u_x(x, -y, t) = -u_x(x, y, t)$ and $u_y(x, -y, t) = u_y(x, y, t)$. In particular, these conditions imply that both $u_x(x, 0, t) = 0$ and $\sigma_{yy}(x, 0, t) = 0$ for $x > 0$ due to symmetry.

Before application of the loads, the material is stress free and at rest everywhere. The traction is suddenly applied to the crack faces at time $t = 0$. The anticipated wavefronts are shown in a plane $z = $ constant in Figure 2.9. For points near to the crack face compared to the distance to the crack edge, the transient field consists only of

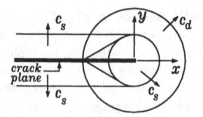

Figure 2.9. Wavefronts generated by sudden application of a spatially uniform in-plane shear traction on the faces of a crack. The state of deformation is plane strain.

a plane wave parallel to the crack face and traveling away from it at speed c_s. The shear loading does not generate a plane dilatational wave. As the plane shear wavefront passes a material point in its path, the component of stress σ_{xy} (or σ_{12}) changes discontinuously from zero to $-\tau^*$ and the particle velocity $\partial u_x/\partial t$ (or $\partial u_1/\partial t$) changes discontinuously from zero to $\pm\tau^*/\rho c_s$ for $\pm y > 0$. Near the crack edge, a nonuniform scattered field radiates out from the crack edge behind cylindrical wavefronts of radius $c_d t$ and $c_s t$ that are centered on the crack edge. The wave pattern includes the associated plane fronted headwaves traveling at speed c_s.

The arguments presented in connection with (2.2.1) can be applied here as well to conclude that the components of stress and particle velocity are homogeneous functions of x, y, t of degree zero. An immediate consequence is that, for points very close to the crack edge,

$$\sigma_{xy} \sim C_{II}\tau^*\sqrt{\frac{c_d t}{x}}, \tag{2.6.1}$$

where C_{II} is an undetermined dimensionless constant. In view of (2.6.1), the shear stress intensity factor is also known up to the constant C_{II}, that is,

$$K_{II}(t) = \lim_{x \to 0^+} \sqrt{2\pi x}\, \sigma_{xy}(x,0,t) = C_{II}\tau^*\sqrt{2\pi c_d t}. \tag{2.6.2}$$

The mathematical problem that leads to a solution with the properties outlined is now stated. The problem differs only in detail from the formulation for pressure loading in Section 2.4. The Helmholtz representation of the displacement vector in terms of the scalar dilatational potential ϕ and the single nonzero component ψ of the vector shear potential is introduced. Thus, two functions

$\phi(x, y, t)$ and $\psi(x, y, t)$ are sought that satisfy the wave equations in two space dimensions and time (2.4.3) in the half plane $-\infty < x < \infty$, $0 < y < \infty$ for time in the range $0 < t < \infty$. These wave functions are subject to the boundary conditions that

$$\sigma_{yy}(x, 0, t) = 0, \qquad -\infty < x < \infty,$$

$$\sigma_{xy}(x, 0, t) = -\tau^* H(t), \qquad -\infty < x < 0, \qquad (2.6.3)$$

$$u_x(x, 0, t) = 0, \qquad 0 < x < \infty,$$

for all time, where stress and displacement components are understood to be their representations in terms of displacement potentials ϕ and ψ. The initial conditions are given by (2.4.5), as before. These conditions ensure that the body is stress free and at rest until the crack face traction is applied. Finally, only solutions that result in stress components that vary inversely with the square root of distance from the crack tip are admitted.

The solution of this problem can be obtained by means of the Wiener-Hopf method, following the steps outlined in Section 2.5. First, the statement of each of the boundary conditions is extended to apply for the full range of x by introducing two unknown functions. The function $u_-(x, t)$ is defined to be the displacement of the crack face $y = 0^+$ in the x-direction for $-\infty < x < 0$, $0 < t < \infty$, and to be identically zero for $0 < x < \infty$. Likewise, the function $\tau_+(x, t)$ is defined to be the component of shear stress σ_{xy} along the plane $y = 0$ for $0 < x < \infty$, $0 < t < \infty$, and to be identically zero for $-\infty < x < 0$. With the introduction of these two functions

$$\sigma_{yy}(x, 0^+, t) = 0,$$

$$\sigma_{xy}(x, 0^+, t) = \tau_+(x, t) - \tau^* H(t) H(-x), \qquad (2.6.4)$$

$$u_x(x, 0^+, t) = u_-(x, t)$$

for the full range $-\infty < x < \infty$, $0 < t < \infty$.

Application of the one-sided Laplace transform on time and then the two-sided Laplace transform leads to a standard Wiener-Hopf equation, similar to (2.5.11). The steps are the same as those followed in Section 2.5, and only the final equation is included here,

$$\frac{\tau^*}{\zeta} + T_+(\zeta) = -\frac{\mu}{b^2} \frac{R(\zeta)}{\beta(\zeta)} U_-(\zeta). \qquad (2.6.5)$$

The relationship of T_+ and U_- to τ_+ and u_- is exactly the same as the relationship of Σ_+ and U_- to σ_+ and u_- established in Section 2.5. The solution of (2.6.5) for the doubly transformed shear stress ahead of the crack tip is

$$T_+(\zeta) = \frac{\tau^*}{\zeta} \left(\frac{G_+(0)}{G_+(\zeta)} - 1 \right), \qquad (2.6.6)$$

where

$$G_\pm(\zeta) = \frac{\beta_\pm(\zeta)}{(c \pm \zeta)S_\pm(\zeta)}, \qquad \beta_\pm(\zeta) = (b \pm \zeta)^{1/2}. \qquad (2.6.7)$$

The shear stress on the crack line ahead of the crack tip is given by the real integral

$$\tau_+(x,t) = -\frac{1}{\pi x} \int_{ax}^{t} \mathrm{Im}\left\{ T_+\left(-\frac{\eta}{x}\right) \right\}\, d\eta\, H(t - ax). \qquad (2.6.8)$$

It follows from this integral that the stress intensity factor history for the case of suddenly applied shear traction on the crack faces is

$$K_{II}(t) = 2\tau^* \sqrt{\frac{2c_s t}{\pi(1 - \nu)}}. \qquad (2.6.9)$$

Furthermore, the dimensionless constant that appears in (2.6.1) has the value $C_{II} = 2\pi^{-1}\sqrt{2c_s/(1 - \nu)c_d}$. The integral (2.6.8) giving the complete stress distribution on the crack line may be evaluated numerically, and the result is shown in Figure 2.10 for $\nu = 0.3$. The stress intensity factor field along $y = 0$ is shown as a dashed line in the same figure. The terms of order unity in the near tip expansions of stress components in the present case are $\sigma_{xx}^{(1)} = \sigma_{yy}^{(1)} = 0$, $\sigma_{xy}^{(1)} = -\tau^*$.

2.7 Loading with arbitrary time dependence

Transient crack tip fields were derived in the foregoing sections under the assumption that the crack face loading was suddenly applied, and thereafter held at a constant magnitude. This is reflected in the step

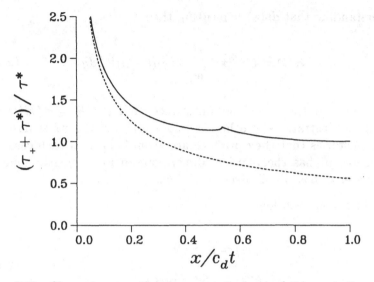

Figure 2.10. Shear stress on the plane $y = 0$ ahead of the crack tip versus normalized distance $x/c_d t$ for $\nu = 0.3$. The solid curve represents the complete stress distribution and the dashed curve shows only the singular term (2.6.9).

function time dependence of the crack face tractions in the boundary conditions (2.2.5), (2.4.4), and (2.6.4). Once the time dependence of the stress intensity factor is established for suddenly applied loading, the corresponding result for any time history of crack face traction, say $p(t)$ for $t > 0$, is obtainable by superposition. The boundary condition (2.2.5)$_1$, (2.4.4)$_1$, or (2.6.4)$_2$ is replaced by

$$\sigma_{\beta y}(x, 0^+, t) = -p(t), \qquad -\infty < x < 0, \qquad (2.7.1)$$

where β is z, y, or x, respectively. The stress intensity factor resulting from the boundary loading (2.7.1) may be considered as follows. If step loading is applied to the crack faces at time $t = \eta$ instead of time $t = 0$, and if the infinitesimal magnitude of the step depends on η according to $dp(\eta)$ instead of being τ^* or σ^*, then the stress intensity factor is

$$K(t) = dp(\eta) \, C \sqrt{2\pi c(t - \eta)}, \qquad t > \eta, \qquad (2.7.2)$$

where the values of the dimensionless parameter C and the wave speed c appropriate for the particular system being considered are presumed. If (2.7.2) is then integrated with respect to η in the range $0 < \eta < t$, understanding that $dp(\eta) = p'(\eta)\, d\eta$, then

$$K(t) = C\sqrt{2\pi c} \int_{0^-}^{t} p'(\eta)(t - \eta)^{1/2}\, d\eta. \qquad (2.7.3)$$

The result for the step function is recovered if $p(t) = \tau^* H(t)$, for example. Illustrations of other functions $p(t)$ and the stress intensity factor histories that they produce are given below. In each case, the parameter p^* has the physical dimensions of force/length^2 and the parameter t^* has the dimension of time.

i. Suddenly applied load:

$$p(t) = p^* H(t),$$

$$K(t) = Cp^* \sqrt{2\pi ct}\, H(t). \qquad (2.7.4)$$

ii. Load increasing linearly in time:

$$p(t) = p^* \left(\frac{t}{t^*}\right) H(t),$$

$$K(t) = \frac{2}{3} Cp^* \frac{t\sqrt{2\pi ct}}{t^*} H(t). \qquad (2.7.5)$$

iii. Impulsively applied load:

$$p(t) = p^* t^* \delta(t),$$

$$K(t) = \frac{1}{2} Cp^* \sqrt{2\pi ct^*} \left(\frac{t^*}{t}\right)^{1/2} H(t). \qquad (2.7.6)$$

iv. Step load of finite duration t^* as a superposition of simple steps:

$$p(t) = p^* \left\{ H(t) - H(t - t^*) \right\},$$

$$K(t) = \frac{2}{3} Cp^* \sqrt{2\pi c} \left\{ t^{1/2} H(t) - (t - t^*)^{1/2} H(t - t^*) \right\}. \qquad (2.7.7)$$

v. Step load with finite rise time t^* as a superposition of linearly increasing loads:

$$p(t) = p^* \left\{ \left(\frac{t}{t^*} \right) H(t) - \left(\frac{t}{t^*} - 1 \right) H(t - t^*) \right\},$$

$$K(t) = \frac{2}{3} C p^* \frac{\sqrt{2\pi c}}{t^*} \left\{ t^{3/2} H(t) - (t - t^*)^{3/2} H(t - t^*) \right\}. \tag{2.7.8}$$

vi. Steady state response to load varying sinusoidally in time with period t^*, after transients have decayed $(t \gg t^*)$:

$$p(t) = p^* \sin \left(\frac{2\pi t}{t^*} \right) H(t),$$

$$K(t) \approx \frac{1}{2} C p^* \sqrt{\pi c t^*} \sin \left(\frac{2\pi t}{t^*} - \frac{\pi}{4} \right). \tag{2.7.9}$$

The complete solutions of the problems described above can be obtained directly for arbitrary time dependence of the applied loading by the methods introduced in Sections 2.3 and 2.5. In the case of Green's method, the term τ^* in the integrand of the right side of the integral equation (2.3.11) would be replaced by $p(t)$. Otherwise, the solution procedure is unchanged. In the case of the Wiener-Hopf method, the factor σ^*/s in (2.5.11) would be replaced by $\hat{p}(s)$. Likewise, the presence of this factor does not alter the basic solution procedure.

3

FURTHER RESULTS FOR A
STATIONARY CRACK

3.1 Introduction

In the preceding chapter, the transient stress intensity factor history resulting from the application of a spatially uniform crack face traction was examined. The particular situations analyzed represent the simplest cases of transient loading of a stationary crack for each mode of crack opening. There are many similar situations that can be analyzed by the methods outlined in Chapter 2. For example, consider the plane strain situation of a plane tensile stress pulse propagating through the material toward the edge of the crack, which lies in the half plane $-\infty < x < 0$, $y = 0$. Suppose that the x and y components of the unit vector normal to the wavefront are $-\cos\theta$ and $-\sin\theta$, respectively, and that the pulse front reaches the crack edge at time $t = 0$. Suppose further that the incident pulse carries a jump in the normal stress component from the initial value of zero to σ_{inc}. If the solid is uncracked, then the incident pulse will induce a tensile traction of magnitude $\sigma^* = \sigma_{inc}\{1 - (1 - 2\nu)\cos^2\theta/(1 - \nu)\}$ and a shear traction of magnitude $\tau^* = \sigma_{inc}(1 - 2\nu)\sin\theta\cos\theta/(1 - \nu)$ on the plane $y = 0$ over the interval $-c_d t/\cos\theta < x < 0$. This particular stress wave diffraction problem was studied by de Hoop (1958) in his pioneering work on the subject.

The stress intensity factor history can then be determined by analyzing the situation in which equal but opposite tractions are applied to the faces of the crack over an expanding interval. The

104

analysis differs only in minor details from that presented in Section 2.5, and the results for the mixed mode stress intensity factor histories are

$$K_I(t) = 2\sigma_{inc}\left(\sin^2\theta + \frac{\nu}{1-\nu}\cos^2\theta\right)F_+(\cos\theta/c_d)\sqrt{\frac{2t}{\pi}}\,,$$

$$K_{II}(t) = 2\sigma_{inc}\frac{1-2\nu}{1-\nu}\sin\theta\cos\theta\, G_+(\cos\theta/c_d)\sqrt{\frac{2t}{\pi}}\,,$$

(3.1.1)

where $F_+(\cdot)$ and $G_+(\cdot)$ are given in (2.5.31) and (2.6.7), respectively. The formulas in (3.1.1) can be evaluated in a straightforward manner, but the matter is not pursued here. However, a comment on the special case $\theta = 0$ is in order. It is well known that if a uniform tension is applied in the direction of the crack line (in the x-direction in the present instance) and if no constraint is imposed on the material in the orthogonal direction (the y-direction here), then the stress intensity factor is zero under conditions of equilibrium. The same is not true in the case of dynamic stress wave loading, as is seen from (3.1.1). The inertial resistance of the material to Poisson contraction in the y-direction upon arrival of the stress pulse is sufficient to generate a tensile stress on the plane $y = 0$ ahead of the crack tip.

A similar plane strain situation involving an opposed pair of concentrated forces moving along the crack faces was considered by Ang (1960). The forces, each of magnitude p^*, begin to move from the crack tip at time $t = 0$ and at a constant speed $v < c_s$. The resulting mode I stress intensity factor, which can be obtained by differentiation with respect to time of the moving step load problem briefly discussed above, is

$$K_I(t) = p^*F_+(1/v)\sqrt{\frac{2}{\pi vt}}\,.$$

(3.1.2)

It was noted in the course of analysis that the problems posed in Chapter 2 have neither a characteristic length nor a characteristic time. The problems mentioned in the foregoing paragraphs are also of this type. If the crack face traction distribution is spatially nonuniform, then a characteristic length is introduced. For example, the characteristic length may be the distance of a concentrated load or a step in traction from the crack tip, or it may be a reference traction level divided by a spatial gradient of traction. The introduction of a characteristic length necessitates a reconsideration of the approach, as well as the results, for stress intensity factor histories. Two cases of plane strain deformation involving a characteristic length are studied in some detail in this chapter. One concerns the sudden loading of the faces of a crack by a pair of concentrated forces at some fixed distance from the crack tip, and the other concerns the sudden loading of a crack of finite length.

There are also several three-dimensional situations of dynamic loading of a stationary crack in an elastic solid that can be analyzed by means of extensions of methods already introduced. Two representative cases are introduced in this chapter. Finally, some implications of transient stress intensity factor results for fracture initiation from an existing crack in a body of material subjected to dynamic loading are summarized.

3.2 Nonuniform crack face traction

For the case of antiplane shear loading, Green's method of solution as introduced in Section 2.3 goes through without modification for nonuniform crack face traction, although some of the steps are more complicated. For example, consider the boundary value problem posed in Section 2.2, except that the crack faces are loaded by an opposed pair of out-of-plane point loads of magnitude q^* at a distance l from the crack tip. More precisely, q^* is the force per unit length in the z-direction of the opposed line loads acting on the crack faces. The statement of the mathematical problem is then identical to that in Section 2.2, except that the boundary condition $(2.2.5)_1$ is replaced by

$$\tau_y(x, 0^+, t) = -q^*\delta(x + l)H(t), \qquad -\infty < x < 0, \qquad (3.2.1)$$

where $l > 0$ and $\delta(\cdot)$ is the Dirac delta function, representing a distribution with respect to x in this case. The argument of the delta function has the physical dimension of length, so $\delta(x + l)$ has the dimension of length^{-1}. Thus, the multiplier q^* does indeed have dimensions of force/length.

The analysis proceeds as in Section 2.3, except that the boundary condition (3.2.1), rather than $(2.2.5)_1$, is substituted into (2.3.6). The solution of the resulting Abel integral equation leads to the transient stress intensity factor

$$K_{III}(t) = q^* \sqrt{\frac{2}{\pi l}} \, H(c_s t - l) \qquad (3.2.2)$$

for the case of concentrated antiplane shear crack face loading. The steps followed in obtaining (3.2.2) are not included here. The interpretation of (3.2.2) is straightforward, but the result is atypical for a two-dimensional wave field. The stress intensity factor is zero until the wavefront generated at the instant of load application at time $t = 0$ reaches the crack tip; this time is $t = l/c_s$. The stress intensity factor then takes on a value consistent with the equilibrium stress field arising from the applied concentrated forces, and *this value is maintained for all later times.* There is no gradual transition to the long-time value $K_{III}(\infty)$ which would be more typical for a two-dimensional wave field. It will now be demonstrated that the solution of the equivalent plane strain crack problem possesses this same remarkable feature.

The means of establishing the properties for the equivalent plane strain situation are not so transparent. In particular, a straightforward application of the Wiener-Hopf method, as outlined in Section 2.5, is not successful. Indeed, the exact solution to this problem was originally obtained by an indirect approach based on the superposition of moving dislocations (Freund 1974b). The transient stress intensity factor history solution has also been obtained by the weight function method which is introduced in Section 5.7. The original approach is the more intuitive, and the general viewpoint represented by it has been found useful in attacking a number of other problems in dynamic fracture mechanics. Consequently, this approach is outlined here. The mathematical problem is first stated, and the general approach is motivated by means of certain observations on the physical process. Next, a fundamental moving dislocation problem, which forms a key element in the approach, is solved exactly by the Wiener-Hopf technique. Finally, the exact stress intensity factor history is extracted by means of a superposition argument.

3.2.1 Suddenly applied concentrated loads

Once again, consider the elastic solid depicted in Figure 2.1 which is cut by a half plane crack but which is otherwise unbounded. In the

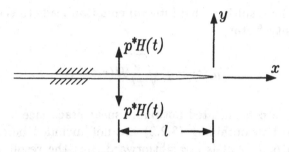

Figure 3.1. Sudden application of concentrated loads on the faces of a crack under plane strain conditions. The distance l from the load points to the crack tip is constant.

present instance, each crack face is subjected to a suddenly applied line load of intensity p^* per unit length in the z-direction. The lines of loading are parallel to the crack edge, and at a distance l from the edge. The loads on the two faces are opposed, tending to open the crack. For the boundary tractions, $T_x(x, z, t) = T_z(x, z, t) = 0$ and $T_y(x, z, t) = \pm p^* \delta(x + l) H(t)$, where $p^* > 0$, $H(\cdot)$ is the unit step function, and $\delta(\cdot)$ is the Dirac singular distribution function. The material is stress free and at rest everywhere for time $t < 0$, and the loading is applied at time $t = 0$. The two-dimensional configuration is depicted in Figure 3.1. The resulting wave field produces a state of plane strain deformation at each point.

The statement of the mathematical problem to be solved is similar to that introduced in Section 2.4. In view of the symmetries of the fields, attention can again be limited to the region $y \geq 0$. A solution of the Navier equation (1.2.51) is sought for $-\infty < x < \infty$, $0 < y < \infty$, $0 < t < \infty$, subject to the conditions that the material is stress free and at rest at time $t = 0$, and subject to the boundary conditions

$$\sigma_{yy}(x, 0, t) = -p^* \delta(x + l) H(t), \quad -\infty < x < 0,$$

$$\sigma_{xy}(x, 0, t) = 0, \quad -\infty < x < \infty, \tag{3.2.3}$$

$$u_y(x, 0, t) = 0, \quad 0 < x < \infty,$$

for all time. The form of the boundary conditions (3.2.3) is similar to (2.4.4), and this observation suggests the Wiener-Hopf method as a means of solving the problem. If integral transforms are applied in

the manner outlined in Section 2.5, however, the steps to be followed
to find a solution are not evident. A perusal of the physical process
being simulated suggests an alternate approach.

If the boundary conditions $(3.2.3)_1$ and $(3.2.3)_2$ are imposed over
the *entire* boundary $y = 0^+$, $-\infty < x < \infty$, then the problem reduces
to the classical problem treated by Lamb (1904). The displacement in
the y-direction of the surface $y = 0^+$ for $x + l > 0$ in Lamb's problem
is

$$u_y^L = -\frac{p^* b^2}{\pi \mu} \int_a^{t/(x+l)} \mathrm{Im}\left\{\frac{\alpha(\zeta)}{R(\zeta)}\right\} d\zeta \; H\big(t - a(x+l)\big), \qquad (3.2.4)$$

where $\alpha(\cdot)$ is defined in (2.5.10), the Rayleigh wave function $R(\cdot)$ is
defined in (2.5.12), and the other notation is introduced in Section
2.5. The limiting value of the integrand as $\mathrm{Im}(\zeta) \to 0^-$ is understood
in (3.2.4). The integrand has a pole singularity at $\zeta = c = 1/c_R$ along
the integration path, so the integral is interpreted in the sense of the
Cauchy principal value. In the derivation of (3.2.4), the integration
path includes a small indentation around the pole at $\zeta = c$ on both
sides of the branch cut of $\alpha(\zeta)$. Each indentation gives rise to a
contribution to the contour integral equal in magnitude to one-half
the residue of the integrand, but the contributions cancel each other
because the integrand is equal in magnitude but opposite in sign on
opposite sides of the branch cut.

In view of the boundary condition $(3.2.3)_3$, the problem at hand
is equivalent to Lamb's problem with a concentrated normal load at
$x = -l$, but with the normal displacement u_y^L of the surface $y = 0^+$
negated for $x > 0$. The necessary cancellation of normal displacement
can be accomplished by means of the fundamental moving dislocation
solution obtained in Section 3.2.2. Note that u_y^L is a homogeneous
function of degree zero of t and $x + l$. This implies that any given
displacement level radiates out along the x-axis at a constant speed.
In particular, the displacement level $u_y^L(x,t)$ propagates in the x-
direction at the speed $(x + l)/t$ for $t > 0$. This feature is exploited in
constructing the requisite moving dislocation solution. The complete
solution satisfying the boundary conditions (3.2.3) and the appropriate symmetry conditions is then obtained by superposition.

Before constructing a solution, some of the features of the transient stress intensity factor history due to the concentrated loads can
be anticipated on the basis of an examination of u_y^L in (3.2.4). For the

surface point at $x = 0$, the displacement u_y^L is zero until time $t = al$ when the cylindrical dilatational wavefront generated at the load point first arrives. Consequently, the transient stress intensity factor will be identically zero for $0 < t < al$. For $a < \zeta < c$, the integrand of (3.2.4) is positive. Therefore, the displacement u_y^L is negative for time in the range $a < t/l < c$, which means that the surface bulges *outward*. In terms of the crack problem under study, the crack tends to close or the stress intensity factor tends to become negative. This would require the unrealistic interpenetration of the crack faces, and this point is reconsidered once quantitative results are in hand. For $c < \zeta < \infty$, the integrand of (3.2.4) is negative. Thus, for time $t > cl$, the crack faces tend to move away from each other according to (3.2.4) and a positive stress intensity factor is anticipated. Finally, it is expected that the transient stress intensity factor will approach the equivalent long-time stress intensity factor, that is, $K_I(t) \rightarrow p^* \sqrt{2/\pi l}$ as $c_d t/l \rightarrow \infty$ (Tada, Paris, and Irwin 1985). These qualitative features are indeed exhibited by the solution to be derived next.

3.2.2 Fundamental solution for a moving dislocation

Consider plane strain deformation of the unbounded elastic solid containing a half plane crack that was introduced in Section 3.2.1. In the present instance, the faces of the crack are free of traction. The material is initially stress free and at rest. At time $t = 0$, an elastic dislocation with Burgers vector of magnitude 2Δ in the y-direction begins to move at constant speed v in the positive x-direction from the crack tip. In the terminology of elastic dislocation theory, this is an edge dislocation climbing in the x-direction. The speed is assumed to be in the range $0 \leq v \leq c_d$ for present purposes, although the solution can be obtained by the same procedure if the speed exceeds the dilatational wave speed of the material. As was the case in the original problem, the situation is symmetric with respect to the plane $y = 0$. Thus, a solution is sought for $-\infty < x < \infty$, $0 < y < \infty$, $0 < t < \infty$ subject to initial conditions appropriate for the material being stress free and at rest. The boundary conditions are

$$\sigma_{yy}(x, 0, t) = \sigma_+(x, t),$$

$$\sigma_{xy}(x, 0, t) = 0, \qquad (3.2.5)$$

$$u_y(x, 0, t) = u_-(x, t) + \Delta H(vt - x)$$

for all x and $t > 0$, where $\Delta > 0$ and $H(\cdot)$ is the unit step function. The interpretation of the functions σ_+ and u_- is precisely the same as in (2.5.1), that is, σ_+ is the unknown normal stress distribution on the plane $y = 0$ for $x > 0$ and $\sigma_+ = 0$ for $x < 0$. Likewise, u_- is the unknown normal displacement distribution on $y = 0$ for $x < 0$ and $u_- = 0$ for $x > 0$.

The particular moving dislocation problem posed can be solved by means of the Wiener-Hopf technique, following the steps outlined in Section 2.5. The details are not included here; only the main results are given, using the notation introduced in Section 2.5 unless a specific statement to the contrary is made. The Wiener-Hopf equation in the present case, corresponding to (2.5.11), is

$$-\frac{b^2 \alpha(\zeta)}{\mu R(\zeta)} \Sigma_+(\zeta) = U_-(\zeta) + \frac{\Delta}{\zeta + d}, \qquad (3.2.6)$$

where $d = 1/v$ is in the range $a \leq d < \infty$. This relationship between sectionally analytic functions is valid in the strip $-a < \text{Re}(\zeta) < a$ of the complex ζ-plane. Application of the factorization procedures outlined in Section 2.5 leads to the solution for the double transform of σ_+ in the form

$$\Sigma_+(\zeta) = -\frac{\mu \kappa \Delta}{b^2} \frac{1}{(\zeta + d) F_+(\zeta) F_+(d)}, \qquad (3.2.7)$$

where $F_+(\zeta)$ is defined in (2.5.31). The transform inversion procedure outlined in Section 2.5.3 leads to the expression

$$\sigma_+(x,t) = -\frac{1}{\pi x} \text{Im} \left\{ \Sigma_+ \left(-t/x + i0 \right) \right\} H(t - ax). \qquad (3.2.8)$$

As could have been anticipated from the formulation, the displacement components in this problem are homogeneous functions of x, y, t of degree zero, and the stress components are homogeneous of degree -1. The stress intensity factor history due to a moving dislocation of unit magnitude $\Delta = 1$, say $k_I(t, v)$, can be deduced from the asymptotic properties of $\Sigma_+(\zeta)$ with the result

$$k_I(t;v) = -\sqrt{\frac{2}{\pi}} \frac{\mu \kappa}{b^2 F_+(d) t^{1/2}}, \qquad t > 0. \qquad (3.2.9)$$

The stress intensity factor is identically zero for $t \leq 0$. In anticipation of the superposition scheme to be introduced in the next subsection, the dependence of the stress intensity factor on dislocation speed v has been indicated explicitly and k_I has been normalized to be the stress intensity factor for $\Delta = 1$. The fact that k_I is negative is consistent with the viewpoint of the dislocation as an extra layer of material of thickness equal to two length units inserted in the interval $0 \leq x \leq vt$ along $y = 0$.

3.2.3 The stress intensity factor history

Once again, attention is directed to the problem involving the sudden application of opposed point loads on the crack faces that was introduced in Section 3.2.1. Upon application of the loads at $x = -l$ at time $t = 0$, a transient field radiates outward from the load point, as described by Lamb (1904). When the leading dilatational wavefront reaches the crack tip at $x = 0$ and time $t = l/c_d$, the constraints on the solid surface change and the complete field can no longer be described by Lamb's solution. The viewpoint adopted here is that Lamb's solution persists for $t > l/c_d$, but that the normal displacement predicted by Lamb's solution for $x > 0$ is exactly canceled by emitting dislocations from the crack tip along $x > 0$, $y = 0$ in just the appropriate sequence.

It is observed that u_y^L is a homogeneous function of degree zero of t and $x' = x + l$, that is, it depends on these two independent variables only through the ratio t/x'. As noted above, this implies that any given displacement level radiates out along the x-axis at a constant speed. In particular, the displacement level $u_y^L(x'/t)$ propagates in the x-direction at the speed x'/t for $t > 0$. The speed is in the range between zero and the dilatational wave speed c_d. Consider a particular value of speed v in this range. The time at which the corresponding displacement level $u_y^L(v)$ arrives at $x = 0$ is $\tau(v) = l/v$ in this case. As observed above, the displacement level $u_y^L(c_d)$ arrives at time $\tau = l/c_d$. The speed of the displacement level arriving at $x = 0$ at any instant of time t is l/t. Based on these observations, a superposition integral that has a value exactly equal but opposite to the normal displacement (3.2.4) of Lamb's problem can be constructed.

Let $\Delta g(x, y, t; v)$ represent any scalar field of the fundamental dislocation solution in Section 3.2.2, such as a stress component or a displacement component, where the linear dependence on Δ is made explicit. The fundamental solution is based on the condition that an elastic dislocation with Burgers vector in the y-direction of magnitude 2Δ begins to extend from the crack tip at time $t = 0$ at speed v. If the dislocation begins moving at time $t = \tau$, instead of at $t = 0$, then the solution is $\Delta g(x, y, t - \tau; v)$. Furthermore, suppose that the strength of the dislocation is $2\,du_y^L$, instead of 2Δ. The solution of this modified problem is $g(x, y, t - \tau; v)\,du_y^L$. Finally, suppose that u_y^L and τ are prescribed functions of v. Then, the result can be summed over a range of values of v, and the appropriate continuous range for the problem at hand has already been identified in the foregoing discussion. If G and g^L denote the corresponding fields for the crack face loading problem and Lamb's problem, respectively, then

$$ G(x, y, t) = g^L(x, y, t) - \int_{c_d}^{l/t} g\big(x, y, t - \tau(v); v\big) \frac{du_y^L(v)}{dv}\, dv\,. \quad (3.2.10) $$

As a check, note that if the displacement component u_y is considered for $x > 0$, then $g(x, 0, t; v) = H(vt - x)$ and $g^L = u_y^L$, and the obvious result that $u_y(x, 0, t) = 0$ is obtained. Thus, the boundary condition $(3.2.3)_3$ is satisfied by solutions constructed according to the procedure represented by (3.2.10). With this essential fact established, attention is focused on the stress intensity history for the case of concentrated forces acting on the crack faces.

The procedure embodied in (3.2.10) is now applied to determine the stress intensity factor history according to the boundary conditions (3.2.3). The normal stress $\sigma_{yy}^L(x, 0, t)$ is not singular at $x = 0$, so that the first term on the right side of (3.2.10) need not be considered. Thus,

$$ K_I(t) = - \int_{c_d}^{l/t} k_I\big(t - \tau(v); v\big) \frac{du_y^L(v)}{dv}\, dv\,, \quad (3.2.11) $$

where the stress intensity factor history for the fundamental dislocation solution $k_I(t; v)$ is given in (3.2.9) and the surface displacement from Lamb's problem u_y^L is given in (3.2.4). If a change of variable of integration from v to η, defined by $\eta = 1/v$, is made, the integral defining the stress intensity factor history (3.2.11) takes the form

$$ K_I(t) = -\frac{p^*}{\pi} \sqrt{\frac{2}{\pi l}} \int_a^{\eta^*} \mathrm{Im} \left\{ \frac{\alpha_-(\eta)}{(c - \eta) S_-(\eta)(\eta^* - \eta)^{1/2}} \right\} d\eta \quad (3.2.12) $$

for $t \geq l/c_d$, where $\eta^* = t/l$. The limiting value of the integrand as $\text{Im}(\eta) \to 0^-$ is understood in (3.2.12). For $t < l/c_d$, no waves generated by the applied loads have arrived at the crack tip, hence the stress intensity factor is identically zero. The integrand of (3.2.12) has a pole singularity at $\eta = c$, so the integral must be interpreted in the Cauchy principal value sense for $t > cl$.

It is helpful to view the integral in (3.2.12) as a line integral in the complex η-plane. The singularities of the integrand are a simple pole at $\eta = c$ and branch points at $\eta = a, b, \eta^*$. For $a < \eta^* < b$, the integrand is analytic in the entire η-plane cut along $a < \text{Re}(\eta) < b$, $\text{Im}(\eta) = 0$, except for the pole at $\eta = c$. On the other hand, for $\eta^* > b$ the integrand is analytic in the entire η-plane cut along $a < \text{Re}(\eta) < \eta^*$, $\text{Im}(\eta) = 0$, except for the pole. In the latter case, the path of integration can be closed around the upper side of the branch cut, assuming a suitable branch of $(\eta^* - \eta)^{1/2}$, and the integral (3.2.12) can be evaluated by observation. Thus, for $\eta^* > b$, (3.2.12) takes the form

$$K_I(t) = -\frac{p^*}{2\pi i}\sqrt{\frac{2}{\pi l}} \oint_{\Gamma(\eta^*)} \frac{\alpha_-(\eta)}{(c - \eta)S_-(\eta)(\eta^* - \eta)^{1/2}}\, d\eta\,, \qquad (3.2.13)$$

where $\Gamma(\eta^*)$ denotes a closed counterclockwise contour embracing the branch cut of the integrand. Cauchy's integral formula can now be applied to show that the value of the integral in (3.2.12) is equal to the value of the integral taken along a closed circular path of indefinitely large radius in the η-plane plus, if $b < \eta^* < c$, the negative of the residue of the integrand at the pole $\eta = c$, or

$$\frac{K_I(t)}{p^*}\sqrt{\frac{\pi l}{2}} = 1 - \begin{cases} \dfrac{(c - a)^{1/2}}{S_-(c)(c - \eta^*)^{1/2}} & \text{if } b \leq \eta^* < c, \\ 0 & \text{if } c \leq \eta^* < \infty. \end{cases} \qquad (3.2.14)$$

The first term on the right side of (3.2.14) is the contribution from the closed contour of large radius and the second term is the residue contribution.

For the time range for which $a < \eta^* < b$, the direct evaluation procedure cannot be applied to this integral. Consequently, the integral (3.2.12) has been evaluated numerically in this range, and the composite result is shown in Figure 3.2 for Poisson ratio $\nu = 0.3$ in

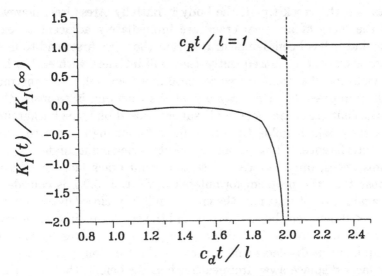

Figure 3.2. Normalized stress intensity factor (3.2.12) versus normalized time $c_d t/l$. The stress intensity factor assumes the large-time equilibrium value $K_I(\infty)$ $= p^* \sqrt{2/\pi l}$ after the Rayleigh wave generated at the load point reaches the crack tip.

the form of a graph of the stress intensity factor history normalized by the long time limit $K_I(\infty) = p^* \sqrt{2/\pi l}$ versus normalized time $c_d t/l$. The history shows the qualitative features anticipated in Section 3.1.1. Upon arrival of the dilatational wavefront generated by the point load, the stress intensity factor takes on a small negative value, reflecting the tendency for the crack faces to move toward each other for this sort of loading. The small negative values persist until the shear wavefront arrives at time $t = l/c_s$. Thereafter, the stress intensity factor decreases rapidly to a negative square root singularity at time $t = l/c_R$, which is the instant of arrival of the Rayleigh wave traveling along the crack faces from the load points to the crack tip. For time $t > l/c_R$, the remarkable result is found that the transient stress intensity factor takes on the constant value $p^* \sqrt{2/\pi l}$, which is the *equilibrium stress intensity factor* for the specified applied loading. This last observation, which is distinctly uncharacteristic of a two-dimensional wave field, is perhaps the most significant result from the practical point of view.

The analysis of this section has been carried out under the assumption that the crack faces remain traction-free between the load points and the crack tip. If the body is initially stress free, however, then the faces of an ideal crack are immediately adjacent to each other, and it has been demonstrated that the faces first tend to move toward each other. Consequently, they will interfere with each other, thus violating the traction-free condition, or they will interpenetrate, which is impossible. One way out of the dilemma is to insist that the transient crack face loading is superimposed on some background stress state which holds the crack faces far enough apart to avoid their interference. This is an acceptable viewpoint under certain circumstances, but more can be stated about crack face interaction. Suppose that the problem formulated in Section 3.2.1 is considered once again, except that now the crack is initially *closed* in the interval $-l < x < 0$ between the load points and the crack tip. In this interval, either the normal traction on the crack faces is compressive and the y-component of displacement is zero, or the traction is zero and the crack faces displace away from each other. As before, the crack tip is at $x = 0$, the opposed forces of magnitude p^* are applied at $x = -l$, and the crack faces are traction-free for $x < -l$. The exact solution of this modified problem is presented in Section 7.3, but the relevant results can be stated here. Upon application of the loads at time $t = 0$, the crack faces initially press against each other. However, the crack faces are separated behind a point traveling at the Rayleigh wave speed of the material c_R along the crack faces from the load point to the crack tip. The stress intensity factor remains at its initial value of zero until this point arrives at the crack tip at time $t = l/c_R$, whereupon it takes on its appropriate equilibrium value $K_I(\infty)$. Thus, the main result that the stress intensity factor takes on its equilibrium value a short time after the load is applied is independent of whether the crack is originally open or closed. The intermediate situation of interference over a part of the interval $-l < x < 0$ is unresolved. From the practical point of view, a crack artificially cut into a material will normally be slightly open even in the absence of applied loading, whereas a crack formed by natural growth in a material without extensive plastic deformation will be closed, or very nearly so.

Finally, some observations on more general crack face loadings are made on the basis of the foregoing results. For example, suppose that the crack faces are loaded by a pair of normal forces applied at $x = -l$ with time dependent magnitude $p^*h(t)$, where $h(t)$ is a dimensionless differentiable function of time and $h(t) = 0$ for $t < 0$.

The stress intensity factor history for this modified problem is then

$$K_I^h(t) = \int_{0^-}^{t} K_I(t-s)\dot{h}(s)\,ds\,, \qquad (3.2.15)$$

where $K_I(t)$ is understood to be the stress intensity factor for the case of step loading of constant magnitude p^*. Likewise, if a normal traction distribution $p^* f(l)$ is suddenly applied over the interval $l_1 < l < l_2$ on each crack face, where f has the physical dimension of length^{-1}, then the stress intensity factor history of the modified problem is

$$K_I^f(t) = \int_{l_1}^{l_2} K_I(t;l)f(l)\,dl\,, \qquad (3.2.16)$$

where the dependence of $K_I(t)$ on l has been indicated explicitly. A wide range of transient two-dimensional stress intensity factor solutions may be constructed on the basis of the results derived in this section, either directly or by superposition.

3.3 Sudden loading of a crack of finite length

In the previous section, a transient crack loading configuration involving a characteristic length was considered, where the characteristic length was associated with the traction distribution on the crack faces. A situation in which the configuration of the solid body itself possesses a characteristic length is considered in this section.

Consider an elastic solid containing a planar slit crack of constant width l in one direction and of indefinite length in the perpendicular direction; the body is otherwise unbounded. A right-handed rectangular x, y, z coordinate system is introduced so that $y = 0$ is the plane of the crack, the z-axis coincides with one of the edges of the crack, and the x-axis is in the direction of prospective advance of that edge. The configuration is invariant under translation in the z-direction, and a typical section $z = $ constant is shown in Figure 3.3.

The loading condition to be discussed in some detail is the sudden application of a spatially uniform pressure of magnitude σ^* to the crack faces. For the boundary traction, $T_x(x, z, t) = T_z(x, z, t) = 0$ and $T_y(x, z, t) = \pm\sigma^* H(t)$, where $\sigma^* > 0$ and $H(t)$ is the unit step function. Thus, the crack faces are traction-free for time $t < 0$, and they are subjected to uniform normal pressure for $t \geq 0$. The material

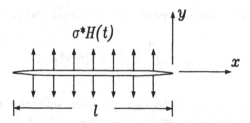

Figure 3.3. A crack of finite length l subjected to a suddenly applied crack face pressure of magnitude σ^* under plane strain conditions.

is stress free and at rest everywhere for $t < 0$. The wave field resulting from this loading produces a state of plane strain deformation at each point in the body.

The statement of the mathematical problem to be solved is similar to that introduced in Section 2.4. In view of the symmetries of the fields, attention can again be limited to the region $y \geq 0$. A solution of the Navier equation (1.2.51) is sought for $-\infty < x < \infty$, $0 < y < \infty$, $0 < t < \infty$, subject to the condition that the material is stress free and at rest at time $t = 0$, and subject to the boundary conditions

$$\sigma_{yy}(x,0,t) = -\sigma^*H(t), \quad -l < x < 0,$$

$$\sigma_{xy}(x,0,t) = 0, \quad -\infty < x < \infty, \qquad (3.3.1)$$

$$u_y(x,0,t) = 0, \quad -\infty < x < -l \quad \text{or} \quad 0 < x < \infty,$$

for all time.

The form of the boundary conditions (3.3.1) is similar to (2.4.4), and this observation suggests the Wiener-Hopf method as a means of solving the problem. If integral transforms are applied in the manner outlined in Section 2.5, a relationship among sectionally analytic functions is obtained which is somewhat more complicated in form than the standard Wiener-Hopf equations introduced in Chapter 2. In principle, the generalized Wiener-Hopf equations can be solved iteratively to obtain the complete solution at any time in the transient loading process. Essentially, each step in the iteration involves the solution of a particular semi-infinite crack problem for which the loading depends on the results of the previous iterative steps. In practice, however, only the first step in the iteration process has been carried out for the problem at hand by Thau and Lu (1971), following the work of Kostrov (1964b) and Flitman (1963), and the result is limited to the time interval $0 < t < 2l/c_d$, that is, up until a dilatational wave has traveled the length of the crack twice. An examination of the physical wave propagation process being simulated suggests an alternate approach based on the moving dislocation solution in Section 3.2.2. The alternate approach provides a solution that is valid for the same time range.

Upon sudden application of the crack face pressure, identical stress wave fields develop around each edge of the crack as though that edge was the edge of a semi-infinite half plane crack. In particular, the stress intensity factor is precisely the same as that derived for a semi-infinite crack subjected to the same pressure loading up until the time at which each crack edge becomes aware of the presence of the other edge. This information is communicated by stress waves. That time is $t = l/c_d$, the time of arrival of the cylindrical dilatational wave generated at the opposite edge at $t = 0$. Thus, from (2.5.44),

$$K_I^{(0)}(t) = 2\sigma^* \frac{\sqrt{1 - 2\nu}}{(1 - \nu)} \sqrt{\frac{c_d t}{\pi}}, \qquad 0 < t < l/c_d, \qquad (3.3.2)$$

where the superscript (0) indicates that (3.3.2) is the lowest order contribution to the total stress intensity factor history, with higher order contributions corresponding to subsequent reflections of waves between the edges of the crack.

When the dilatational wavefront generated at time $t = 0$ at the left edge of the crack $x = -l$ arrives at the right edge of the crack $x = 0$, it carries with it a displacement of the surface in the y-direction which violates the boundary condition $(3.3.1)_3$. The viewpoint adopted here is that this surface displacement persists for $x > 0$ and $t > l/c_d$. To satisfy the boundary condition $(3.3.1)_3$, however, an appropriate sequence of dislocations is emitted from the crack edge at $x = 0$ in the x-direction to negate the surface displacement.

It turns out that the surface displacement to be negated is already in hand. It is precisely the inverse of the double transform $s^{-3}U_-(\zeta)$, where $U_-(\zeta)$ is given explicitly in $(2.5.35)_2$, with the coordinate x replaced by $x' = -(x + l)$. In this case, the inversion steps outlined in Section 2.5.3 yield

$$u_-(x', t) = \frac{\sigma^* b^2 |x'| F_+(0)}{\mu\pi} \int_a^{t/|x'|} \text{Im}\{F_-(\zeta)\} \left[\frac{t}{|x'|} - \zeta\right] d\zeta, \quad (3.3.3)$$

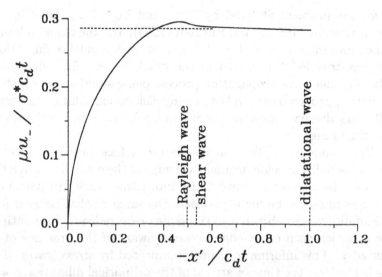

Figure 3.4. Normal displacement of the crack face (3.3.3) versus normalized distance $-x/c_d t$. The arrival time of the individual wavefronts at the observation distance is indicated on the abscissa.

where $F_\pm(\cdot)$ is given in (2.5.31) and the other notation is established in Section 2.5. The limiting value of the integrand as $\mathrm{Im}(\zeta) \to 0^+$ is understood in (3.3.3).

    The integral in (3.3.3) has been evaluated numerically for a Poisson ratio of $\nu = 0.3$ and the result is shown in Figure 3.4. For points along the crack face beyond the cylindrical dilatational wave generated at the crack edge, that is, for $|x'| > c_d t$, the surface is moving under the action of only the applied pressure. The speed of the surface particles is $\sigma^* c_s^2/\mu c_d$, and the displacement is $\sigma^* c_s^2 t/\mu c_d$. Thus, $u_- \mu/\sigma^* c_d t = (c_s/c_d)^2 = 0.286$ for $\nu = 0.3$. With the arrival of the scattered dilatational wave at $t = |x'|/c_d$, the crack faces begin to move apart at an even greater rate than the uniform motion under pressure loading. This seems to be a by-product of the large hydrostatic tension developed ahead of the crack tip. Retardation of the motion of the surface begins with the arrival of the Rayleigh wave at an observation point. When the wave motion represented in Figure 3.4 reaches the opposite crack edge, it will first incrementally *increase* the stress intensity factor there until the arrival of the Rayleigh wave. Thereafter, it will tend to *decrease* the stress intensity factor.

    The displacement u_- in (3.3.3) is a homogeneous function of x'

and t of degree one. However, the dislocation superposition scheme is most easily applied for homogeneous functions of degree zero. Consequently, the particle velocity distribution $\dot{u}_- = \partial u_-/\partial t$, which is homogeneous of degree zero, is canceled. The cancellation is achieved by superposition over a certain fundamental moving dislocation solution. In the present instance, the formulation of the fundamental moving dislocation problem is the same as that in Section 3.2.2, except that the boundary condition $(3.2.5)_3$ is replaced by

$$u_y(x,0,t) = u_-(x,t) + \dot{\Delta}\,(t - x/v)H(vt - x) \qquad (3.3.4)$$

for all x and $t > 0$. The constant $2\dot{\Delta}$ is the magnitude of a particle velocity discontinuity, or a *particle velocity dislocation*, growing from the crack tip in the x-direction at speed v. This fundamental solution was introduced by Freund (1976a) in the analysis of a particular crack arrest problem.

Analysis of the moving velocity dislocation problem could proceed by means of the Wiener-Hopf technique as in Section 3.2.2. There is no need to repeat the analysis, however. The inhomogeneous term in the present case, which is the second term on the right side of (3.3.4), is obtained by integrating over time from $t = 0$ of the inhomogeneous term in the corresponding boundary condition $(3.2.3)_3$ in Section 3.2.2. Because both problems are linear, it follows that the complete solution in the present case is the time integral of the solution obtained in Section 3.2.2. In particular, the specific stress intensity factor here for a dislocation of unit magnitude $\dot{\Delta} = 1$ is

$$k_I(t;v) = -2\sqrt{\frac{2t}{\pi}}\left[\frac{\mu\kappa}{b^2 F_+(d)}\right], \qquad t > 0, \qquad (3.3.5)$$

analogous to (3.2.9). Again, the stress intensity factor is identically zero for $t \le 0$.

From this point onward, the superposition procedure follows that developed in Section 3.2.3, except that the superposition is over a particle velocity distribution rather that over a displacement distribution. Thus, the stress intensity factor history over the time interval of interest is given by

$$K_I^{(1)}(t) = -\int_{c_d}^{l/t} k_I\left(t - \tau(v);v\right)\frac{d\dot{u}_-(v)}{dv}\,dv, \qquad (3.3.6)$$

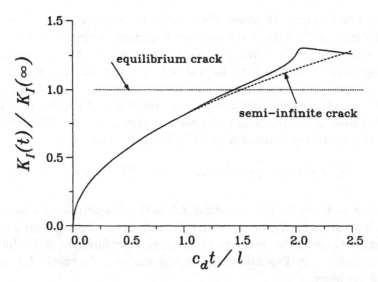

Figure 3.5. Transient stress intensity factor (3.3.7) versus normalized time $c_d t/l$. The dashed curve is the corresponding result for a semi-infinite crack. The long-time limit is $K_I(\infty) = \sigma^* \sqrt{\pi l/2}$, which is the result for the corresponding equilibrium situation.

where $k_I(t; v)$ is given in (3.3.5), $\tau(v) = l/v$, and $\dot{u}_-$ is the time derivative of (3.3.3). If the change of variable $\eta = 1/v$ is incorporated, then (3.3.6) becomes

$$K_I^{(1)}(t) = -\frac{2\sigma^* F_+(0)}{\pi} \sqrt{\frac{2l}{\pi}} \int_a^{\eta^*} \text{Im}\left\{ \frac{(\eta^* - \eta)^{1/2} F_-(\eta)}{\eta F_+(\eta)} \right\} d\eta, \quad (3.3.7)$$

where $\eta^* = t/l$, which is valid for $l/c_d < t < 2l/c_d$. The limiting value of the integrand as $\text{Im}(\eta) \to 0^+$ is understood in (3.3.7). The integrand has a pole singularity at $\eta = c$, so the integrand must be interpreted in the Cauchy principal value sense for $\eta^* > c$.

The integral (3.3.7) has been evaluated numerically for a Poisson ratio of $\nu = 0.3$, and the net stress intensity factor is shown in Figure 3.5. The transient stress intensity factor has been normalized with respect to $K_I(\infty) = \sigma^* \sqrt{\pi l/2}$, which is the equilibrium stress intensity factor for the specified loading. The variation of $K_I^{(0)}(t)$ is shown as a dashed curve, and the variation of $K_I^{(0)}(t) + K_I^{(1)}(t)$ is shown as a solid curve. The stress intensity factor history $K_I^{(0)}(t)$ prevails for time in the range $0 < t < l/c_d$, that is, before any interaction between the edges of the crack occurs and, consequently, the dashed and solid curves are identical in this time range. Upon arrival of the dilatational wavefront from the opposite edge of the crack, $K_I^{(1)}(t)$ becomes nonzero and the net stress intensity factor history rises *above* the value for a semi-infinite crack. This outcome was anticipated in the examination of Figure 3.4, which showed that the effect of the dilatational wave is to cause the faces to move away from each other at an even greater rate than is produced by the pressure loading on the free surface. This trend continues until the Rayleigh wave arrives at time $t = l/c_R$, whereupon the trend is reversed and the net stress intensity factor reaches a peak shortly thereafter. For a Poisson ratio of $\nu = 0.3$, the stress intensity factor peak is $1.30 \, K_I(\infty)$, or 30 percent above the long-time limit. This is an illustration of the phenomenon of *dynamic overshoot*. Unfortunately, the stress intensity factor solution is exact only up to time $t = 2l/c_d$. That is the time at which the effect of the scattering, at the opposite edge, of the leading wavefront generated at one edge of the crack returns to influence fields at its edge of origin. It is evident from Figure 3.5 that the scattered dilatational wave has little effect on the local stress intensity factor, however, and it is reasonable to expect that the present analysis provides the correct maximum stress intensity factor for the process. Numerical solutions due to Chen and Wilkins (1976) and Chen and Sih (1977) suggest that, after achieving the first peak shown in Figure 3.5, the net stress intensity factor history experiences oscillations of diminishing amplitude about the eventual limit of $K_I(\infty)$.

3.4 Three-dimensional scattering of a pulse by a crack

Up to this point in the study of the transient stress intensity factors that arise from the scattering of a plane stress pulse by a crack, it has been assumed that the edge of the half plane crack is *parallel* to the plane wavefront of the incident pulse. In this section, mathematical procedures are generalized to permit consideration of the scattering of a stress pulse which is *obliquely* incident on a crack edge in an elastic solid. The general approach to analysis of three-dimensional problems of this type was introduced by Freund (1971) in a study of the oblique reflection of Rayleigh surface waves from the edge of a half plane crack. The scattering of plane harmonic waves from the edge

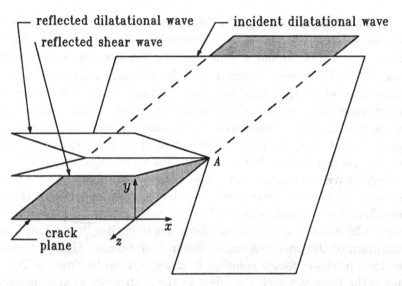

Figure 3.6. Schematic representation of a plane dilatational wave obliquely incident on a half plane crack with traction-free faces. The reflected plane dilatational wave and the reflected plane shear wave are also shown, but the scattered waves are not included in the diagram. Point A is the point of intersection of the crack edge with the incident plane wave.

of a half plane crack was also considered by Achenbach and Gautesen (1977).

Consider once again the elastic body depicted in Figure 2.1 that is cut by a half plane crack but that is otherwise unbounded. In the present instance, the body is stress free and at rest except for the influence of the remotely generated stress pulse incident on the crack. The pulse is scattered and, at each point along the crack edge, transient stress intensity factors are generated. The purpose here is to determine exact stress intensity factor histories for a particular problem of this type.

A right-handed rectangular coordinate system x, y, z is introduced in the body, oriented so that the z-axis coincides with the crack edge, and the half plane crack occupies $y = 0$, $x < 0$. Suppose that the incident stress pulse carries a jump in tensile stress of magnitude σ_{inc} in the direction of the normal to the pulse front. Consequently, the pulse is dilatational and it travels with the characteristic speed c_d. In addition, suppose that the unit vector normal to the pulse front in the direction of propagation has components $(0, -\sin\theta, -\cos\theta)$. Thus, the x-axis is parallel to the pulse front, and θ is the acute

angle between the negative z-axis and the normal to the front; see
Figure 3.6. Finally, suppose that the pulse front reaches the origin
$(0,0,0)$ at time $t = 0$. Under these conditions, the pulse front is in the
plane $c_d t + y \sin \theta + z \cos \theta = 0$. The fact that the subsequent analysis
is limited to the case when the pulse front is parallel to the x-axis
does not reflect a limitation of the analytical procedure. The main
consequence of the limitation is that the discussion of the analytical
approach is not obscured by the algebraic complexity of the case of
general oblique incidence.

The diffraction process, which is assumed to have become steady
in the z-direction, leads to a fairly complicated pattern of wavefronts.
For $x < 0$, the incident dilatational pulse is geometrically reflected
from the traction-free crack surface as a combination of a plane dilatational pulse and a plane shear pulse, both of which are shown
in Figure 3.6. For $x > 0$, on the other hand, the incident pulse
propagates past the plane $y = 0$ without modification. The movement
of the intersection point A of the incident pulse front with the crack
edge is determined as the motion of the intersection point of the
three planes $c_d t + y \sin \theta + z \cos \theta = 0$, $x = 0$, and $y = 0$, that is,
$c_d t + z \cos \theta = 0$. Thus, this point moves with the apparent speed
$v = c_d / \cos \theta$ along the crack edge and its instantaneous coordinates
are $(0, 0, -c_d t / \cos \theta)$. A scattered field trailing this point (not shown
in Figure 3.6) is established around the crack edge. The wavefronts
in the scattered field include a conical dilatational wavefront and a
conical shear wavefront, both with apex at $(0, 0, -c_d t / \cos \theta)$. Also
included is a pair of plane head wavefronts. These head wavefronts
intersect the dilatational conical wavefronts on the surfaces $y = \pm 0$,
and they extend into the material from each surface, terminating on
their lines of tangency with the conical shear wavefronts.

The same sort of superposition arguments discussed in Sections
2.5 and 3.1 can be applied here to convert the problem from one
involving a wave incident on a traction-free crack to one involving
crack face tractions but no other loading. In the present case, if the
crack were absent, then the incident wave would induce a normal
component of stress $\sigma^* = \sigma_{inc}\{\sin^2 \theta + \nu \cos^2 \theta / (1 - \nu)\}$ and a z-
component of shear stress $\tau^* = \sigma_{inc}(1 - 2\nu) \sin \theta \cos \theta / (1 - \nu)$ over
the part of the crack plane $y = 0$ for which $c_d t + z \cos \theta > 0$. The
stress intensity factor histories can then be determined by analyzing
the situation in which equal but opposite tractions are applied to the
faces of the crack. The application of the normal stress produces a

transient mode I stress intensity factor for each z, and the shear stress produces a mode III stress intensity factor for each z. The two modes are analyzed in the same way, and the steps leading to the mode I stress intensity factor history are outlined here. The corresponding result for the mode III stress intensity factor history is stated without derivation.

If the same normal pressure distribution is applied to both faces of the crack, then the wave fields are expected to have reflective symmetry with respect to the plane $y = 0$, and attention can be limited to the region $y \geq 0$. The boundary conditions imposed on the half space $y > 0$ for $x > 0$ are conditions that result from the symmetry of the fields under reflection in the plane $y = 0$, namely, $\sigma_{xy}(x, 0, z, t) = \sigma_{yz}(x, 0, z, t) = 0$ and $u_y(x, 0, z, t) = 0$.

The Helmholtz representation of the displacement vector in terms of the scalar dilatational potential $\phi(x, y, z, t)$ and the vector shear potential $\psi(x, y, z, t)$ is introduced as in (1.2.60). In general, ϕ and each rectangular component of ψ must satisfy a scalar wave equation in three space dimensions and time. In addition, it is required that the vector potential ψ must be divergence free. In any particular plane $z = $ constant, the material is stress free and at rest until time $t = -z/v$. That is, because the applied loads are moving supersonically on the crack faces in the negative z-direction, no wave motion reaches points in the plane $z = $ constant before the moving load step reaches that plane.

It is now observed that the applied loading (which is the only inhomogeneous term in the mathematical formulation) depends on z and t only through the combination $\tau = t + z/v$. If the scattered fields have become steady, then ϕ and ψ depend on z and t only through τ, as well. Consequently, $\phi(x, y, z, t) = \phi(x, y, \tau)$ and $\psi(x, y, z, t) = \psi(x, y, \tau)$. No confusion should result from the fact that obviously different functions have been identified by the same symbol. For steady fields, the wave equations governing the displacement potentials, or their components, take the form

$$\frac{\partial^2 \phi}{\partial x^2} + \frac{\partial^2 \phi}{\partial y^2} - \left(\frac{1}{c_d^2} - \frac{1}{v^2}\right)\frac{\partial^2 \phi}{\partial \tau^2} = 0,$$

$$\frac{\partial^2 \psi}{\partial x^2} + \frac{\partial^2 \psi}{\partial y^2} - \left(\frac{1}{c_s^2} - \frac{1}{v^2}\right)\frac{\partial^2 \psi}{\partial \tau^2} = 0$$

(3.4.1)

in the half plane $y > 0$. The condition that ψ must be divergence free becomes

$$\frac{\partial \psi_x}{\partial x} + \frac{\partial \psi_y}{\partial y} + \frac{1}{v}\frac{\partial \psi_z}{\partial \tau} = 0 \qquad (3.4.2)$$

in $y > 0$, where ψ_x, ψ_y, ψ_z are the rectangular components of ψ. The conditions that the material is stress free and at rest in any plane $z = $ constant until the moving load step penetrates that plane take the form

$$\phi(x,y,0) = \frac{\partial \phi}{\partial \tau}(x,y,0) = 0\,, \quad \psi(x,y,0) = \frac{\partial \psi}{\partial \tau}(x,y,0) = 0 \quad (3.4.3)$$

in $y > 0$.

The boundary value problem represented by the differential equations (3.4.1) and (3.4.2), the initial conditions (3.4.3), and the boundary conditions is similar to the plane strain problem formulated in Section 2.4. There are four unknown dependent variables here, as opposed to two in the case of plane strain, and the timelike independent variable τ has replaced time t. Furthermore, the inverse characteristic wave speeds c_d^{-2} and c_s^{-2} in Section 2.4 are replaced here by $(c_d^{-2}-v^{-2})$ and $(c_s^{-2}-v^{-2})$, respectively. There are no essential differences in the two formulations, however, and the approach developed in Section 2.5 based on the Wiener-Hopf method can be applied here without modification.

The first and last boundary conditions equivalent to $(2.5.1)_1$ and $(2.5.1)_3$ are rewritten in a form that is valid over the full range of x,

$$\sigma_{yy}(x,0,\tau) = \sigma_+(x,\tau) - \sigma^* H(\tau) H(-x)\,,$$
$$u_y(x,0,\tau) = u_-(x,\tau) \qquad (3.4.4)$$

for $-\infty < x < \infty$ and $0 < \tau$. As before, σ_+ is the unknown normal traction distribution on $x > 0$ and $\sigma_+ = 0$ for $x < 0$. Likewise, u_- is the unknown displacement in the y-direction on $x < 0$ and $u_- = 0$ for $x > 0$. The one-sided Laplace transform with transform parameter s is first applied to suppress dependence on τ. Next, the two-sided Laplace transform with parameter $s\zeta$ is applied over x. The bounded solutions of the resulting ordinary differential equations in y for the doubly transformed potentials are

$$\Phi(\zeta,y,s) = \frac{1}{s^4}P(\zeta)e^{-s\alpha y}, \qquad \Psi(\zeta,y,s) = \frac{1}{s^4}\mathbf{Q}(\zeta)e^{-s\beta y}, \quad (3.4.5)$$

where α and β are defined by

$$\alpha(\zeta, d) = (a^2 - d^2 - \zeta^2)^{1/2}, \qquad \beta(\zeta, d) = (b^2 - d^2 - \zeta^2)^{1/2}, \quad (3.4.6)$$

and $a = 1/c_d$, $b = 1/c_s$, and $d = 1/v$. Note that $d < a$ for any angle of incidence $\theta > 0$. Branch cuts are introduced and branches are selected so that $\mathrm{Re}(\alpha) \geq 0$ and $\mathrm{Re}(\beta) \geq 0$ in the entire ζ-plane. The condition (3.4.2) that the vector potential must be divergence free takes the form

$$\zeta Q_x - \beta Q_y + d Q_z = 0, \qquad (3.4.7)$$

where the functions Q_x, Q_y, Q_z are the rectangular components of the vector function $\mathbf{Q}$. At this point, the functions P and $\mathbf{Q}$ are unknown and, as in Section 2.5, the coefficients of these functions have been selected in (3.4.5) so that the functions themselves will not depend on s.

If the Laplace transformed boundary conditions (3.4.4) are imposed on (3.4.5), the result is a system of five linear equations for the six unknown functions P, Q_x, Q_y, Q_z, U_-, and Σ_+, where U_- and Σ_+ have precisely the same interpretation here as in (2.5.8). If P, Q_x, Q_y, and Q_z are eliminated, the remaining equation is a Wiener-Hopf equation of the standard type,

$$-\frac{\mu}{b^2} \frac{R(\zeta, d)}{\alpha(\zeta, d)} U_-(\zeta) = \frac{\sigma^*}{\zeta} + \Sigma_+(\zeta), \qquad (3.4.8)$$

where

$$R(\zeta, d) = 4(\zeta^2 + d^2)\alpha(\zeta, d)\beta(\zeta, d) + (b^2 - 2d^2 - 2\zeta^2)^2. \quad (3.4.9)$$

The function defined in (3.4.9) is recognized as a modified form of the Rayleigh wave function. In particular, it possesses only two zeros in the complex ζ-plane at $\zeta = \pm\sqrt{c^2 - d^2}$, where c is defined by $R(\pm c, 0) = 0$. Note that (3.4.9) is identical to (2.5.12) when $d \to 0$ (or $v \to \infty$). The strip of analyticity of (3.4.8) is $-\sqrt{a^2 - d^2} < \mathrm{Re}(\zeta) < 0$.

The remaining steps in solving the Wiener-Hopf equation and extracting the transient stress intensity factor history are identical to those followed in Section 2.5. Therefore, these steps need not be repeated here, and the stress intensity factor analogous to (2.5.44) is stated without further calculation. In terms of the apparent speed v of the step load moving along the crack faces,

$$K_I(t) = 4\sigma^* \frac{(1 - c_d^2/v^2)^{1/4}}{\sqrt{\pi(1 - 2\nu)c_d^4 R(0, 1/v)}} \sqrt{c_d(t + z/v)} \qquad (3.4.10)$$

for any fixed value of z, where $\sigma^* = \sigma_{inc}\{\sin^2\theta + \nu\cos^2\theta/(1 - \nu)\}$ and $v = c_d/\cos\theta$. It can be verified that when the stress pulse is normally incident on the crack plane, or $\theta = \pi/2$, the expression in (3.4.10) is exactly the same as that in (2.5.44). The situation of grazing incidence, when the apparent speed of the incident wave along the crack face approaches the dilatational wave speed $v = c_d$ or when $\theta = 0$, is a degenerate case. For this situation, the strip of analyticity of the Wiener-Hopf equation vanishes and the solution procedure breaks down. That this case could lead to some difficulty could have been anticipated from the theory of geometrical reflection of plane waves from a traction-free surface. For the case of grazing incidence of a dilatational pulse, the reflected dilatational pulse cancels the incident pulse, and the reflected shear pulse has zero amplitude. This dilemma arises from the combination of assumptions that the reflected field has become steady and that the excitation moves exactly at the dilatational wave speed. In fact, if the load step on the traction-free surface moves at the dilatational wave speed c_d, it is impossible for dilatational radiation from the applied load to advance into the material to establish a steady reflected plane dilatational wave. The resolution of this dilemma must be sought in the study of the phenomenon of grazing incidence as a transient process (Wright 1968; Freund and Phillips 1968).

The mode III stress intensity factor history resulting from the negation of the shear stress of magnitude τ^* induced on the crack plane by the incident wave is briefly considered. In this case, the fields are antisymmetric under reflection in the plane $y = 0$, and again attention can be limited to the region $y \geq 0$. The boundary conditions imposed on the half space are

$$\sigma_{yz}(x, 0, z, t) = -\tau^* H(t + z/v), \qquad -\infty < x < 0,$$

$$u_z(x, 0, z, t) = 0, \qquad 0 < x < \infty, \qquad (3.4.11)$$

for $-\infty < z < \infty$ and $0 < t$, where $v = c_d/\cos\theta$. The condition on $x > 0$ results from antisymmetry of the fields under reflection in the plane $y = 0$.

Although the steps in the derivation of the transient stress intensity factor history are not included here, the Wiener-Hopf equation itself is presented to illustrate an interesting feature that is not observed in two-dimensional problems of the same general type. With the functions U_- and T_+ interpreted as in Section 2.6, the Wiener-Hopf equation is

$$-\frac{\mu\beta(\zeta,d)R(\zeta,d)}{W(\zeta,d)}U_-(\zeta) = \frac{\tau^*}{\zeta} + T_+(\zeta), \qquad (3.4.12)$$

where β is defined in (3.4.6), R is defined in (3.4.9), and

$$W(\zeta,d) = 4\zeta^2\alpha\beta + 2d^2(b^2 - d^2 - \zeta^2)$$
$$+ (b^2 - 2d^2 - 2\zeta^2)(b^2 - d^2 - 2\zeta^2). \qquad (3.4.13)$$

The function W has the interesting properties that $W(\zeta,d)/R(\zeta,d) \to 1$ as $|\zeta| \to \infty$ for any admissible value of d, $W(\zeta,0)/R(\zeta,0) = 1$ for any value of ζ, and $W(\zeta,d) = 0$ has two and only two roots in the complex ζ-plane, with both roots real. The interpretation of these properties in terms of the wave propagation process follows. For values of $d = 1/v$ greater than zero, the applied shear loading gives rise to a surface wave traveling obliquely away from the crack edge on each crack face. In this case, the zeros of R provide the poles of the sectionally analytic functions corresponding to surface waves. In the limit as $v \to \infty$ or $d \to 0$, the fields reduce to the simpler antiplane shear fields considered in Sections 2.2 and 2.3. In this case, no surface waves are present. The role of the function W is to provide an automatic switch for eliminating the Rayleigh wave function from the formulation when $d \to 0$.

If the Wiener-Hopf equation (3.4.12) is solved and the stress intensity factor is extracted following the procedures developed in Section 2.5, the result is

$$K_{III}(t) = \tau^* \frac{2\sqrt{2}\,(1 - c_s^2/v^2)^{1/4}}{\sqrt{\pi c_s^3 R(0, 1/v)}}\sqrt{t + z/v} \qquad (3.4.14)$$

for any fixed value of z, where

$$v = \frac{c_d}{\cos\theta} \quad \text{and} \quad \tau^* = \sigma_{inc}\left(\frac{1-2\nu}{1-\nu}\right)\sin\theta\cos\theta. \qquad (3.4.15)$$

It can be verified that the expression in (3.4.14) reduces to the corresponding two-dimensional mode III result in (2.3.18) in the limit as $v \to \infty$ for a fixed value of τ^*.

3.5 Three-dimensional stress intensity factors

The transient mechanical fields described in Section 3.4 are three-dimensional fields in the sense that the components of stress and deformation are functions of all three spatial coordinates at any instant. Because of the assumption that the fields have become steady with respect to an observation frame moving in the negative z-direction at speed v, however, the mathematical formulation of the stress analysis problem could be cast in a form indistinguishable from the formulation of an equivalent two-dimensional problem. Three-dimensional phenomena have received relatively little attention in dynamic fracture mechanics, although certain results obtained in the study of self-similar expansion of flat circular or elliptic cracks will be noted in Section 6.3. In this section, a procedure is described for determining stress intensity factor histories for a class of three-dimensional elastodynamic crack problems. The geometrical configuration studied is again a half plane crack in an otherwise unbounded solid, with the crack faces subjected to tractions that result in variation of the transient stress intensity factors along the crack edge.

Consider again the elastic body containing a half plane crack depicted in Figure 2.1. The body is initially stress free and at rest. At a certain instant, tractions begin to act on the crack faces, resulting in a three-dimensional stress wave field in the material. The purpose here is to determine exact expressions for stress intensity factors as functions of time and position along the crack edge. To illustrate the approach, the same transient normal pressure distribution is applied to each crack face, and the tangential or shear traction components are zero. For this type of loading, the mode of deformation is mode I for each point along the crack edge. Next, a mathematical formulation of a problem corresponding to the foregoing qualitative description is given.

A right-handed rectangular coordinate system is introduced in the body, oriented so that the z-axis coincides with the crack edge, and the half plane crack occupies $y = 0$ for $x < 0$. The normal traction on the crack faces begins to act at time $t = 0$. For the time being, consider some general traction variation on $y = 0^{\pm}$, say $T_y(x, z, t) = \mp\sigma_-(x, z, t)$, where $\sigma_-(x, z, t)$ is a given function of position on the faces and of time and $\sigma_- > 0$ corresponds to tensile traction. It is assumed that there exists a distance, say z_0, such that σ_- is zero for $|z| > z_0$. The subscript $(-)$ carries the same interpretation as it did in the introduction of the Wiener-Hopf method in Section 2.5. The other components of traction are zero, that is, $T_x(x, z, t) = T_z(x, z, t) = 0$.

In view of the symmetry of the configuration and of the applied traction, the wave fields are expected to have reflective symmetry with respect to the plane $y = 0$, and attention can be limited to the region $y \geq 0$. The displacement fields satisfy $u_x(x, -y, z, t) = u_x(x, y, z, t)$, $u_y(x, -y, z, t) = -u_y(x, y, z, t)$, and $u_z(x, -y, z, t) = u_z(x, y, z, t)$. These, in turn, imply the conditions $\sigma_{xy}(x, 0, z, t) = \sigma_{yz}(x, 0, z, t) = 0$ and $u_y(x, 0, z, t) = 0$ for $x > 0$ and for all z, t. Thus, the complete set of boundary conditions to be satisfied by the stress wave fields is

$$\sigma_{yy}(x, 0, z, t) = \sigma_-(x, z, t) + \sigma_+(x, z, t),$$

$$\sigma_{xy}(x, 0, z, t) = 0,$$

$$\sigma_{yz}(x, 0, z, t) = 0, \tag{3.5.1}$$

$$u_y(x, 0, z, t) = u_-(x, z, t)$$

for $-\infty < x, z < \infty$ and $t \geq 0$. The definition of the function σ_-, which describes the imposed traction, is extended so that $\sigma_- = 0$ for all $x > 0$. As in the study of two-dimensional problems, σ_+ is the unknown normal component of stress σ_{yy} on $x > 0$ and $\sigma_+ = 0$ for $x < 0$. Likewise, u_- is the unknown y-component of displacement of material particles on the crack faces for $x < 0$ and $u_- = 0$ for all $x > 0$.

The Helmholtz representation of the displacement vector $\mathbf{u} = \nabla\phi + \nabla \times \boldsymbol{\psi}$ is adopted, as introduced in Section 1.2.3, where ∇ is the three-dimensional gradient operator, ϕ is the scalar dilatational wave potential, and $\boldsymbol{\psi}$ is the vector shear wave potential. The wave potential functions are governed by the partial differential equations

$$c_d^2 \nabla^2 \phi - \phi_{,tt} = 0, \qquad c_s^2 \nabla^2 \boldsymbol{\psi} - \boldsymbol{\psi}_{,tt} = 0, \qquad \nabla \cdot \boldsymbol{\psi} = 0 \tag{3.5.2}$$

in $y > 0$. The conditions that the material is stress free and at rest everywhere for $t \leq 0$ are expressed in terms of the potential functions by

$$\phi = \frac{\partial \phi}{\partial t} = 0, \quad \psi = \frac{\partial \psi}{\partial t} = 0 \qquad (3.5.3)$$

at $t = 0$ for all points in $y > 0$. Finally, it is noted that the boundary conditions (3.5.1) are interpreted as though the stress and displacement components were replaced by their expressions in terms of the potentials ϕ and ψ.

The mathematical boundary value problem described by the differential equations (3.5.2), the boundary conditions (3.5.1), and the initial conditions (3.5.3) is linear, and integral transforms, along with certain powerful theorems from the theory of analytic functions of a complex variable, are again used to extract stress intensity factor histories. In most respects, the procedure differs very little from that introduced in Section 2.5 for two-dimensional analysis. Certain interpretations of the transformed fields at key points in the analysis, however, make it possible to extract exact results for this class of three-dimensional problems. The first step is to apply the one-sided Laplace transform on time (1.3.7) to the differential equations (3.5.2) and the boundary conditions (3.5.1), taking into account the initial conditions (3.5.3). As before, the transform parameter is s and the transform of any function, say $\phi(x, y, z, t)$, is denoted by a superposed hat $\widehat{\phi}(x, y, z, s)$. The parameter s is considered to be a positive real parameter for the time being. Next, a two sided Laplace transform is applied to suppress the dependence on z. The complex transform parameter is $s\xi$, and the transform of any function, say $\widehat{\phi}(x, y, z, s)$, is denoted by the corresponding upper case symbol with a superposed hat

$$\widehat{\Phi}(x, y, \xi, s) = \int_{-\infty}^{\infty} \widehat{\phi}(x, y, z, s) \, e^{-s\xi z} \, dz . \qquad (3.5.4)$$

From the condition that σ_- vanishes for $|z| > z_0$, it can be expected that

$$\widehat{\phi}(x, y, z, s) = o\left(e^{-s(\pm z - z_0 - \epsilon)/c_d}\right) \qquad (3.5.5)$$

as $z \to \pm\infty$ for any $\epsilon > 0$. This condition, in turn, implies that the transform integral in (3.5.4) converges for $-a < \text{Re}(\xi) < a$, where $a = 1/c_d$. Consequently, the integral defines an analytic function in

this strip of the complex ξ-plane. From the property of uniqueness
of an analytic function in its domain of analyticity noted in Section
1.3.1, the analytic function in the strip is completely specified by
the function defined only on the portion of the real axis in the strip
$-a < \text{Re}(\xi) < a$. Furthermore, it turns out to be of great advantage
to restrict the range of ξ to this real interval in proceeding with the
analysis. At an appropriate later stage, the definition of the functions
of ξ can be extended to the entire complex ξ-plane by invoking the
relevant theorems concerning uniqueness of analytic functions and
analytic continuation from Section 1.3.

Finally, the two-sided Laplace transform that suppresses dependence on x is applied. The complex transform parameter is $s\zeta$, and
the transform of any function, say $\widehat{\Phi}(x,y,\xi,s)$, is denoted by the same
upper case symbol but without the superposed hat, $\Phi(\zeta,y,\xi,s)$. To be
able to continue with the solution procedure, some statement on the
domain of convergence of the transform integrals over x is required at
this point. If the applied loading is nonzero for indefinitely large distances from the crack edge along the crack faces, then the arguments
made in Section 2.5 can be invoked here, as well, to conclude that
the integral over negative values of x in the definition of $\Phi(\zeta,y,\xi,s)$
will converge for $\text{Re}(\zeta) < 0$. For $x > 0$, the wave fields are zero
beyond a cylindrical wavefront expanding from the y-axis at speed c_d
with instantaneous radius $z_0+\sqrt{x^2+z^2}$. To be more specific, consider
the elementary wave field $\phi(x,y,z,t) = H(c_d t - z_0 - \sqrt{x^2+z^2})$, where
$H(\cdot)$ denotes the unit step function. If the Laplace transform integrals
over t, then z, and then x are formed, it is easily demonstrated that
the triple integral converges if $\{\text{Re}(\zeta)\}^2 + \xi^2 < a^2$. This condition,
when coupled with the convergence condition $\text{Re}(\zeta) < 0$, provides a
basis for expecting that all transforms will be analytic in the strip
$-\sqrt{a^2-\xi^2} < \text{Re}(\zeta) < 0$ in the complex ζ-plane, with ξ real and in
the range $-a < \xi < a$.

It is not possible to consider completely arbitrary distributions
of traction $\sigma_-(x,z,t)$ applied to the crack faces. From this point
onward, attention is limited to those distributions for which the triple
transform has the form

$$\int_{-\infty}^{\infty}\int_{-\infty}^{\infty} \widehat{\sigma}_-(x,z,s)\,e^{-s\xi z-s\zeta x}dz\,dx = \frac{1}{s^m}\Sigma_-(\zeta,\xi), \qquad (3.5.6)$$

where m is a real number and $\Sigma_-(\zeta,\xi)$ does not depend on s. The condition (3.5.6) essentially restricts $\sigma_-(x, z, t)$ to be within a class of certain separable or homogeneous functions of x, z, t.

The steps in the solution procedure now follow those outlined in Section 2.5, although the algebraic details are somewhat more complicated here. The transforms are applied to the partial differential equations (3.5.2) to reduce them to ordinary differential equations in y. The transformed boundary conditions (3.5.1) are imposed on the solutions of the ordinary differential equations to determine the unknown parameters of integration. As usual in a problem of this general type, there are more unknown functions than there are algebraic equations to determine them. However, certain of the unknown functions are sectionally analytic functions (of ζ in this case), and the Wiener-Hopf factorization makes it possible to complete the solution by determining two unknown functions from a single equation. Only a few of the intermediate steps are included here.

The solutions of the ordinary differential equations obtained by application of transforms to the partial differential equations that are bounded as $y \to \infty$ are

$$\Phi = \frac{1}{s^{m+2}} P(\zeta, \xi) \, e^{-s\alpha y}, \qquad \Psi = \frac{1}{s^{m+2}} \mathbf{Q}(\zeta, \xi) \, e^{-s\beta y}, \qquad (3.5.7)$$

where the parameters of integration P and $\mathbf{Q}$ are unknown functions of their arguments, the number m follows from (3.5.6), and

$$\alpha(\zeta, \xi) = (a^2 - \xi^2 - \zeta^2)^{1/2}, \qquad \beta(\zeta, \xi) = (b^2 - \xi^2 - \zeta^2)^{1/2}. \qquad (3.5.8)$$

The complex ζ-plane is cut along $\sqrt{a^2 - \xi^2} < |\mathrm{Re}(\zeta)| < \infty$, $\mathrm{Im}(\zeta) = 0$, so that $\mathrm{Re}(\alpha) \geq 0$ in the entire cut ζ-plane for each value of ξ, and likewise for β. The transformation of the condition $\nabla \cdot \psi = 0$ yields

$$\zeta Q_x - \beta Q_y + \xi Q_z = 0, \qquad (3.5.9)$$

where Q_x, Q_y, Q_z are the rectangular components of $\mathbf{Q}$. The relation (3.5.9) is a linear algebraic equation for the unknown functions. The Laplace transformation of the boundary conditions provides four additional linear equations for the six unknown functions P, Q_x, Q_y,

Q_z, U_-, and Σ_+, where

$$U_-(\zeta,\xi) = s^{m+1} \int_{-\infty}^{\infty} \int_{-\infty}^{\infty} \widehat{u}_-(x,z,s)\, e^{-s(\xi z + \zeta x)} dz\, dx,$$

$$\Sigma_+(\zeta,\xi) = s^m \int_{-\infty}^{\infty} \int_{-\infty}^{\infty} \widehat{\sigma}_+(x,z,s)\, e^{-s(\xi z + \zeta x)} dz\, dx. \tag{3.5.10}$$

These four equations are

$$(b^2 - 2\zeta^2 - 2\xi^2)P - 2\xi\beta Q_x + 2\zeta\beta Q_z = \frac{\Sigma_- + \Sigma_+}{\mu},$$

$$-2\zeta\alpha P + \xi\zeta Q_x + \xi\beta Q_y + (b^2 - \xi^2 - 2\zeta^2)Q_z = 0,$$

$$2\xi\alpha P + (b^2 - 2\xi^2 - \zeta^2)Q_x + \zeta\beta Q_y + \xi\zeta Q_z = 0, \tag{3.5.11}$$

$$-\alpha P + \xi Q_x - \zeta Q_z = U_-.$$

If P, Q_x, Q_y, and Q_z are eliminated from the five equations (3.5.9) and (3.5.11), the result is a single equation involving the two remaining unknown functions U_- and Σ_+, namely,

$$-\frac{\mu}{b^2} \frac{R(\zeta,\xi)}{\alpha(\zeta,\xi)} U_-(\zeta,\xi) = \Sigma_-(\zeta,\xi) + \Sigma_+(\zeta,\xi), \tag{3.5.12}$$

where $\Sigma_-(\zeta,\xi)$ is given, α is defined in (3.5.8), and R is defined in (3.4.9). Furthermore, Σ_+ is analytic in $\mathrm{Re}(\zeta) > -\sqrt{a^2 - \xi^2}$ and U_- is analytic in $\mathrm{Re}(\zeta) < 0$. The equation (3.5.12) holds in the strip $-\sqrt{a^2 - \xi^2} < \mathrm{Re}(\zeta) < 0$. Thus, for any fixed value of ξ, (3.5.12) can be solved by factorization in much the same way that an equation of the standard Wiener-Hopf type is solved.

To make further progress, it is necessary to choose a particular applied traction distribution. A convenient case for purposes of demonstrating the general approach is that of a uniform line load which suddenly begins to act along a line *perpendicular* to the crack edge. Thus, it is assumed that

$$\sigma_-(x,z,t) = -p^* \delta(z) H(-x) H(t), \tag{3.5.13}$$

where $H(\cdot)$ is the unit step function and $\delta(\cdot)$ is the Dirac delta function. The amplitude p^* has physical dimensions of force/length, and $p^* > 0$ corresponds to traction that tends to separate the crack faces. The ensemble of wavefronts that results from application of the traction specified by (3.5.13) includes a variety of space surfaces. Among the wavefronts are cylindrical dilatational and shear wavefronts centered on the load lines in $x < 0$, plus planar head wavefronts that intersect the dilatational wavefronts on the surfaces $y = \pm 0$ and terminate at the line of tangency with the cylindrical shear wavefronts. There are also spherical dilatational and shear wavefronts centered at the origin of the coordinates, plus conical headwaves, each with its apex at a point where the spherical dilatational wave meets the z-axis and its terminus along a circle of tangency with the spherical shear wavefront. There are also conical headwaves that intersect the spherical wavefront on the surfaces $y = \pm 0$ for $x < 0$ and extend to circles of tangency with the spherical shear wave.

If the relevant integral transforms indicated in (3.5.10) are applied to (3.5.13), it is found that $m = 2$ and that

$$\Sigma_-(\zeta, \xi) = \frac{p^*}{\zeta}. \qquad (3.5.14)$$

Consequently, the relationship between the sectionally analytic functions stated in (3.5.12) is virtually identical to the relationship given in (2.5.11), except for the parametric dependence on ξ in the present case. Though the dependence on ξ must be taken into account in the solution procedure, it requires no special consideration, and the solution for $\Sigma_+(\zeta, \xi)$ can be written immediately as

$$\Sigma_+(\zeta, \xi) = \frac{p^*}{\zeta} \left(\frac{F_+(0, \xi)}{F_+(\zeta, \xi)} - 1 \right), \qquad (3.5.15)$$

where

$$\log S_\pm(\zeta, \xi) = -\frac{1}{\pi} \int_a^b \tan^{-1} \left[\frac{4\eta^2 \sqrt{(b^2 - \eta^2)(\eta^2 - a^2)}}{(b^2 - 2\eta^2)^2} \right]$$

$$\times \frac{\eta \, d\eta}{\sqrt{\eta^2 - \xi^2} \left(\sqrt{\eta^2 - \xi^2} \pm \zeta \right)} \qquad (3.5.16)$$

and

$$F_{\pm}(\zeta,\xi) = \frac{\left(\sqrt{a^2 - \xi^2} \pm \zeta\right)^{1/2}}{\left(\sqrt{c^2 - \xi^2} \pm \zeta\right) S_{\pm}(\zeta,\xi)}. \tag{3.5.17}$$

If the development in Section 2.5 is followed one step further, it is concluded that the double transform of $K_I(t, z)$, the variation of stress intensity factor history along the edge of the crack, is

$$\widehat{K}_I(s,\xi) = \sqrt{2\kappa}\,\frac{p^*}{s^{3/2}}\sqrt{\frac{(a^2 - \xi^2)^{1/2}}{R(0,\xi)}}, \tag{3.5.18}$$

where $\kappa = 2(b^2 - a^2)$, as before. Note that the notational convention on naming Laplace transforms is not followed for the stress intensity factor itself. In any case, the procedure for inversion of a double transform like that in (3.5.18) has been described in detail in Section 2.5.3. The result of applying this procedure to (3.5.18) is (Freund 1987a)

$$K_I(t, z) = \frac{p^*}{\pi}\sqrt{\frac{2\kappa}{\pi}} \int_a^{t/z} \mathrm{Im}\left\{\sqrt{\frac{\alpha(0,\xi)}{R(0,\xi)}}\right\} \frac{d\xi}{\sqrt{t - \xi z}} H(t - az) \tag{3.5.19}$$

for $z > 0$, and $K_I(t, -z) = K_I(t, z)$. The limiting value of the integrand as $\mathrm{Im}(\xi) \to 0^+$ is understood in (3.5.19).

The integral in (3.5.19) cannot be evaluated in terms of elementary functions. It can be evaluated numerically, however, and the result of numerical evaluation for a Poisson ratio of $\nu = 0.3$ is shown in Figure 3.7. Knowing the asymptotic behavior of the generalized Rayleigh wave function, namely, that $R(0,\xi) \to -\kappa\xi^2$ as $\xi \to +\infty$, the integral (3.5.19) implies that

$$\lim_{t \to \infty} K_I(t, z) = K_I(\infty, z) = \frac{p^*}{\sqrt{\pi z}} \tag{3.5.20}$$

for any fixed $z > 0$. The result of the limiting process is consistent with the independently calculated equilibrium stress intensity factor distribution along the crack edge for this configuration (Tada et al.

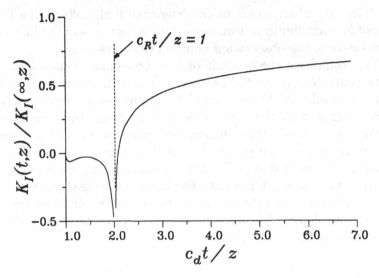

Figure 3.7. The stress intensity factor (3.5.19) versus the dimensionless time $c_d t/z$. The singularity corresponds to the arrival of the Rayleigh wave. The long-time limiting value in this case is $K_I(\infty, z) = p^*/\sqrt{\pi z}$.

1985). The stress intensity factor history in Figure 3.7 for any fixed value of z is normalized by its long-time limiting value $K_I(\infty, z)$.

The general features of $K_I(t, z)$ derive from the nature of the wave fields. For any value of $z > 0$, the stress intensity factor is zero up until the arrival of the first dilatational wave at time $t = z/c_d$. Upon sudden application of the compressive line loads on the faces of the crack, the initial response is dilatational and the surfaces at points adjacent to the line loads tend to bulge outward. The first wave arriving at the observation point along the crack edge carries this outward bulge. The crack faces tend to move toward each other, and this feature is reflected in the tendency for the stress intensity factor to become *negative* following arrival of the leading dilatational wave. This effect persists until the arrival of the Rayleigh wave at time $t = z/c_R$ ($c_d/c_R = 2.02$ for $\nu = 0.3$). The stress intensity factor history is logarithmically singular at this instant, and it tends to increase thereafter. These features are qualitatively similar to those observed for the two-dimensional configuration analyzed in Section 3.2. The transient stress intensity factor history $K_I(t, z)$ approaches its long-time limit $K_I(\infty, z)$ quite slowly, with the ratio of the former to the latter being about 0.81 at $c_d t/z = 20$ and about 0.92 at $c_d t/z =$

100. The rate of approach to the long-time limit reflects the time required for contributions from more distant segments of the line load to propagate to the observation point on the crack edge.

This completes the analysis of the three-dimensional stress intensity factor history for the particular applied traction distribution (3.5.13). Results could be derived for a number of other traction distributions in a similar way. This analysis opens the way for the examination of other three-dimensional phenomena, but few such problems have been considered. A three-dimensional configuration of particular interest is that of a pair of opposed collinear concentrated loads acting on the crack faces at a fixed distance from the crack edge. This problem involves a characteristic length, and a solution is not yet available. An application of the method to a growing crack problem is discussed by Champion (1988).

3.6 Fracture initiation due to dynamic loading

In the foregoing sections, elastodynamic analysis of cracked solids is described which leads to the dependence of the time history of the crack tip stress intensity factor on the applied loading and crack length. The analysis was motivated in Section 2.1 by the observation that the stress intensity factor provides a one-parameter representation of the mechanical load level on material near the edge of the crack for each mode of crack opening, provided that certain length requirements are met. Thus, the stress intensity factor concept provides a basis for quantifying the resistance of materials to the onset of growth of a preexisting crack, as well as for predicting the initiation of fracture in a cracked, nominally elastic structure.

3.6.1 The Irwin criterion

The engineering science of linear elastic fracture mechanics (LEFM), which has evolved from the notion of the stress intensity factor as a field characterizing parameter, has been profitably extended to situations in which material inertia plays a significant role. Given the significance of the stress intensity factor as a single characterizing parameter for each mode of crack opening, a simple criterion for the onset of crack growth is the following: A crack will begin to extend when the stress intensity factor has been increased to a material specific value, usually called the *fracture toughness* of the material and commonly denoted by K_{Ic} for mode I plane strain deformation. For values of the stress intensity factor smaller than the critical value there is no growth, and values larger than the critical value are inaccessible. This is the Irwin criterion of LEFM in its simplest form. It should be noted that such a criterion is a physical postulate on material response, on the same level as the stress–strain relation or other physical postulates on which the mathematical formulation is based. In the statement of this criterion, it should be emphasized that K_{Ic} is a material parameter and that $K_I(t)$ is a feature of the stress field. The American Society for Testing and Materials has adopted a standard procedure for establishing the fracture toughness of a metal (Annual Book of Standards, Part 10, Section E399); according to the specification in this standard, the notation K_{Ic} is reserved for an inferred measure of fracture resistance only when certain testing conditions are met.

The foregoing statement of the Irwin criterion may be applied without modification to the study of fracture initiation in nominally elastic bodies subjected to stress wave loading. That is, if a stress intensity factor field exists under rapidly applied loading then it can be postulated that crack growth will begin if the value of the stress intensity factor is increased to a certain value. This does not imply that the value of dynamic fracture toughness will be independent of the rate of loading or that dynamic effects do not influence the inferred value of fracture resistance in other ways. Several cases of dynamic loading are considered in this section. The question of whether or not the postulated criterion is valid for a range of materials and/or dynamic testing conditions can only be established through experiment.

3.6.2 Qualitative observations

Attention is limited to mode I deformation for the time being. The size of the crack is viewed as the physical feature of the configuration of prime importance, and this feature is characterized by the length l. The transient applied loading is characterized by some applied traction or incident stress wave magnitude σ^* and some temporal duration t^* of the applied loading. The bulk response of the isotropic elastic material is characterized by the Poisson ratio ν and by a wave speed, say the Rayleigh wave speed c_R. The fracture resistance is specified by the fracture toughness K_{Ic}. A complete system is shown schematically in Figure 3.8. The way in which the system responds

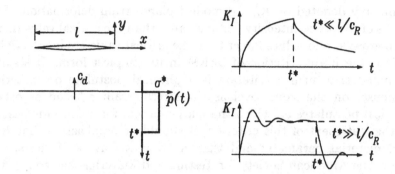

Figure 3.8. Schematic diagram of a crack of finite length loaded by means of a rectangular stress pulse under plane strain conditions. On the right, the stress intensity factor response is indicated qualitatively for the case when the pulse duration is very long or very short compared to a wave transit time along the crack length.

depends on the relative magnitudes of the system parameters, and this point is pursued on a qualitative basis in an attempt to achieve some degree of unification in this subsection.

For definiteness, consider a rectangular loading pulse of arbitrary but fixed duration t^*. Suppose that, for a given crack length l, the pulse magnitude $\sigma^* = \sigma_c^*$ is the *smallest* magnitude that will result in the Irwin criterion $K_I(t) = K_{Ic}$ being satisfied at some instant. Attention is focused on the relationship between σ_c^* and l in the dimensionless form of $\sigma_c^* \sqrt{l}/K_{Ic}$ versus $t^* c_R/l$. First, consider the case $t^* c_R/l \gg 1$, that is, the pulse is long in duration compared to the transit time of Rayleigh waves over the distance l. The *maximum* stress intensity factor achieved during the dynamic loading process is about $1.3 \sigma^* \sqrt{\pi l/2}$ according to the analysis in Section 3.3. Consequently, if the pulse amplitude σ^* is such that $\sigma^* \sqrt{l}/K_{Ic} < 0.6$, the initiation criterion cannot be satisfied and fracture will not be initiated. Thus, for $t^* c_R/l \gg 1$, it appears that $\sigma_c^* \sqrt{l}/K_{Ic} = 0.6$ is the critical value of the stress pulse amplitude. The critical stress level for initiation varies as $1/\sqrt{l}$, and the dimensionless amplitude is independent of $t^* c_R/l$. The criterion yields the same result, except for the numerical factor, as if it were applied for an equilibrium situation. The result is shown as the horizontal line in Figure 3.9.

Consider next the other extreme situation when the duration of the loading stress pulse is short compared to the transit time for Rayleigh waves over the distance l, that is, $t^* c_R/l \ll 1$. The *maximum*

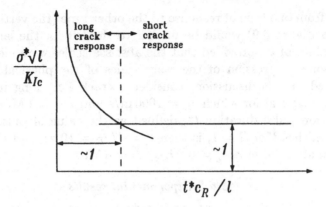

Figure 3.9. Schematic diagram of the normalized stress pulse amplitude σ^* required to satisfy a stress intensity factor fracture criterion versus stress pulse duration t^*.

stress intensity factor which can be achieved under these conditions is approximately $1.6\,\sigma^*\sqrt{t^*c_R}$ according to the analysis in Sections 2.5 and 2.7. Consequently, if the pulse amplitude σ^* is such that $\sigma^*\sqrt{l}/K_{Ic} < 0.6\,\sqrt{l/t^*c_R}$, the initiation criterion cannot be satisfied and fracture will not initiate. (The common factor $\sqrt{l}$ is not canceled in this expression so that key dimensionless parameter groups can be retained for establishing important qualitative results.) Thus, for $t^*c_R/l \ll 1$, it appears that $\sigma_c^*\sqrt{l}/K_{Ic} = 0.6\,\sqrt{l/t^*c_R}$ is the critical value of the stress pulse amplitude for a given value of t^*c_R/l. This result is also shown in Figure 3.9 as the curved line. A solid curve could easily be added to the sketch, providing a smooth transition between the two extreme cases considered. No particular significance should be attached to the precise values of numerical coefficients in the expressions for the horizontal and vertical lines in Figure 3.9, although it is noteworthy that they are both of order unity. If the extension of a three-dimensional crack due to stress wave loading were to be studied in the same way, it is expected that the results would be qualitatively similar. The coefficients would surely have different values, but it is expected that they would also be of order unity.

Recall that pulse duration t^* is fixed at some arbitrary value in the diagram in Figure 3.8. Suppose that the curve separating the growth conditions from the no-growth conditions in the plane of $\sigma_c^*\sqrt{l}/K_{Ic}$ versus t^*c_R/l is considered for two distinct values of t^*, with the parameters l and K_{Ic} held fixed. Then the transition point from one type of response to the other type (the vertical dashed line in Figure 3.9) would be further to the left for the larger of the two values of t^*, provided that the abscissa scales were identical. To give some impression of the magnitudes of the physical quantities involved in this discussion, consider a crack 1 cm long in a brittle plastic material for which $c_R = 1000$ m/s and $K_{Ic} = 1$ MPa$\sqrt{\text{m}}$. The transition pulse duration t_t^*, defined as the value of pulse duration that satisfies $t_t^* c_R/l = 1$, is then $t_t^* = l/c_R = 10\,\mu$s, and the plateau critical stress level is $\sigma_c^* = 0.6\, K_{Ic}/\sqrt{l} = 6$ MPa.

3.6.3 Experimental results

An experimental study of fracture initiation under stress wave loading conditions was reported by Kalthoff and Shockey (1977) and Shockey et al. (1983). The material selected was a brittle epoxy with a fracture toughness under slow loading conditions of $K_{Ic} = 1.1$ MPa$\sqrt{\text{m}}$. The dilatational wave speed for the material was $c_d = 2.6 \times 10^3$ m/s. Shallow cylindrical disk specimens, 50 mm in diameter and 9 mm thick, were cast with a distribution of small circular sheets of mylar embedded within each specimen. Five sizes of the mylar sheets with diameters l ranging from 0.4 mm to 12.7 mm were used. The tensile strength of the epoxy-mylar interface was measured to be less than 5 percent of the tensile strength of the epoxy. The specimens were loaded in a plate impact gas gun apparatus. The loading produced a compressive wave which propagated as a uniform plane pulse through the specimen, and then reflected from the traction-free back face as a tensile pulse of duration 2.04 μs. The amplitude of the tensile pulse was varied by varying the impact loading velocity, resulting in the range 14.9 MPa $\leq \sigma^* \leq$ 33.7 MPa. The mylar disks were viewed as penny-shaped cracks, and the experimental situation is as suggested in the two-dimensional schematic diagram in Figure 3.8. The spacing of the disks was such that there was no interaction between them during application of the loading pulse, and each acted like a penny-shaped crack in an unbounded solid during the time interval of interest.

The epoxy was transparent, allowing visual examination of the specimens after loading to determine whether or not each embedded disk/crack experienced growth. The data based on such examination are summarized in Figure 3.10. These data are taken directly from the work by Shockey et al. (1983) and are replotted in terms of the dimensionless variables introduced in Figure 3.9. The dashed curve is an estimate of the locus of points for which $\sigma^* = \sigma_c^*$, based

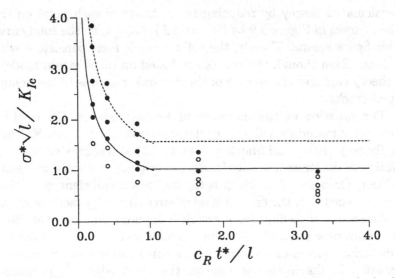

Figure 3.10. Data on fracture initiation in epoxy due to short stress pulse loading reported by Shockey et al. (1983). The filled symbols indicate that crack growth was observed, whereas the open symbols represent cases of no growth. The curves represent possible interpretations of the result of Figure 3.9 for this experiment.

on the slow loading fracture toughness value of $1.1\,\mathrm{MPa}\sqrt{\mathrm{m}}$. The horizontal portion is determined from Sneddon's (1946) result that $\sigma^*\sqrt{l}/K_{Ic} = \sqrt{\pi/2}$ for an equilibrium field, amplified by a dynamic overshoot of 25 percent. The remaining portion is estimated from the numerical calculation described by Chen and Sih (1977) for a penny-shaped crack loaded axisymmetrically by a tensile pulse. The stress overshoot for the three-dimensional penny-shaped crack is about the same as for the plane strain crack. However, the numerical results indicate that the stress maximum occurs sooner after loading for the three-dimensional case than for the two-dimensional case. The dashed curves in Figure 3.10 capture the trend of the data, but they overestimate the values of σ_c^*. It was noted by Shockey et al. (1983) that the slow loading fracture toughness $K_{Ic} = 1.1\,\mathrm{MPa}\sqrt{\mathrm{m}}$ appeared to substantially over-estimate the stress intensity factor level at which fracture was initiated in this material under rapid loading. They estimated that the fracture toughness under the dynamic loading conditions imposed was about $K_{Id} = 0.72\,\mathrm{MPa}\sqrt{\mathrm{m}}$, where the subscript d is used to indicate the special interpretation of this value. If the theoretical estimate of the locus of points for which $\sigma^* = \sigma_c^*$

is recalculated simply by reducing the ordinate of each point on the dashed curves in Figure 3.9 by the ratio K_{Id}/K_{Ic}, then the solid curve in this figure results. Clearly, the solid curve is more consistent with the data. Even though the curves are based on plane strain models, the theory captures the essence of the dynamic response of the penny-shaped cracks.

The question of the influence of loading rate on the fracture toughness of materials will be considered later in this section. Kalthoff and Shockey (1977) and Shockey et al. (1983) also suggested that the elevation of the stress intensity factor to a critical level was a necessary condition for onset of crack growth, but not a sufficient condition. They proposed that the critical level of stress intensity factor must be maintained (or exceeded) for a certain minimum time interval. Such a minimum time has no fundamental significance, and it must depend on material properties and on the operative mechanism of material separation in the nonlinear zone at the crack edge. Experiments similar to those reported by Shockey et al. (1983) were carried out on specimens of an aluminum alloy, a high strength steel, and a structural steel by Homma, Shockey and Murayama (1983). In these experiments, edge cracked strip specimens were used, and the general conclusions were similar to those reported in the earlier studies.

Ravi-Chandar and Knauss (1984a, 1984b) carried out an extensive experimental study of fracture initiation under dynamic loading conditions in thin sheets of Homalite-100, a brittle plastic material. Their experimental setup is closely modeled as a semi-infinite crack in an otherwise unbounded solid, with the crack faces subjected to uniform normal pressure. This was accomplished in a large rectangular sheet of lateral dimensions 500 mm by 300 mm and thickness 4.86 mm by cutting a crack through the sheet along the longer symmetry axis to roughly the center of the sheet. The crack tip was artificially sharpened. A copper ribbon was then inserted in the crack, running from the edge of the sheet to the crack edge along one face. At the crack edge, the ribbon was folded back onto itself, and it extended back to the edge of the specimen along the other crack face. The space within the fold of the ribbon, between the crack faces, was filled with a dielectric. A charged capacitor bank was then discharged through the ribbon, producing a large current. The pressure on the crack faces was generated by the interaction of the electric current in each leg of the ribbon with the magnetic field induced by the current in the other leg. The crack face pressure increased within about 25 μs to a plateau

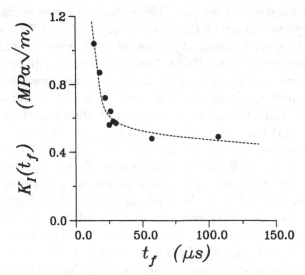

Figure 3.11. Data on the value of the stress intensity factor at initiation of crack growth versus time to fracture for a loading pulse of fixed amplitude (Ravi-Chandar and Knauss 1984a).

value, which was maintained at least until fracture initiation. The time history of the crack tip stress intensity factor was monitored by means of the optical shadow spot method. Among the data presented by Ravi-Chandar and Knauss (1984a) is the variation of the value of the stress intensity factor at the instant of fracture versus the time elapsed between the application of the crack face loading and the onset of crack growth. This elapsed time varies inversely with the magnitude of the loading pressure. These data are reproduced in Figure 3.11, along with data on the slow loading fracture toughness of Homalite-100. The data indicate that if the time to fracture is greater than about 50 μs, then the critical stress intensity factor is independent of the time to fracture, and, therefore, independent of the magnitude of the applied pressure. Furthermore, this plateau level of stress intensity factor is approximately equal to K_{Ic}. On the other hand, for high amplitude loading pulses which produce fracture in times less than 50 μs, the critical level of stress intensity factor depends strongly on the time to fracture. Ravi-Chandar and Knauss (1984a) discuss this point qualitatively on the basis of a mechanistic model involving viscous growth and coalescence of microcracks by means of a viscous mechanism.

Ravi-Chandar and Knauss (1984a) also reported measurements
of the complete history of the stress intensity factor for application
of loading to (and beyond) the onset of crack growth. This matter
was also pursued by Kim (1985a, 1985b) who used a different optical
method known as the stress intensity factor tracer method. He also
used large sheet specimens of Homalite-100, as well as the same electromagnetic loading apparatus that was introduced by Ravi-Chandar
and Knauss (1984a) to apply pressure over part of the crack faces.
With reference to the analysis in Section 3.2, the crack occupied the
half line $y = 0, x < 0$ and pressure was uniformly applied over the
crack faces for $x < -l$. By means of the superposition integrals
(3.2.15) and (3.2.16), and the fundamental result in Figure 3.2, he
determined the expected stress intensity factor history for this applied
loading. The predicted history showed nearly all of the features shown
in Figure 3.2, including the fall-off in stress intensity factor between
arrival of the shear wavefront and arrival of the Rayleigh surface wave
at the crack tip. The measured $K_I(t)$ followed the prediction very
closely. There was even some evidence of crack face interference in
the early stages of loading, as discussed in Section 3.2.

An experimental approach to the study of dynamic fracture initiation in metals was developed by Costin, Duffy, and Freund (1977).
In the experiment, a tensile wave was generated in a precracked round
bar by means of an explosion behind a loading head bolted to one end
of the bar. The wave produced a fracture at the cracked section within
about 25 μs. The load on the ligament at the cracked section and the
crack opening displacement were measured during the course of the
experiment. Because the ligament dimension was small compared to
the length of the incident loading pulse, the mechanical fields at the
cracked section were deduced from the measured quantities on the
basis of the assumption that the distribution of field quantities was
the same as if the loading had been statically applied. The assumption
has been verified by means of a detailed calculation of the transient
fields by Nakamura, Shih, and Freund (1986).

It was noted by Shockey et al. (1983) that the fracture toughness under dynamic loading conditions (termed K_{Id} in the foregoing
discussion) was substantially lower than K_{Ic} for the epoxy used in
their experiments. Indeed, it is frequently found that the rate of
loading on a fracture specimen has an influence on the level of stress
intensity factor at which fracture initiates. This was indeed the case
for structural steels in the work reported by Costin et al. (1977).

In some cases, the toughness appears to increase with the rate of loading whereas in other cases the opposite dependence is found. The explanation for a shift in toughness, either upward or downward, must be sought in the mechanisms of inelastic deformation and material separation in the highly stressed region at the edge of the crack in the loaded body.

To illustrate the fundamental difference in fracture response of materials which gives rise to the two general types of behavior, consider two ideal fracture mechanisms. Suppose that materials separate within the crack tip zone when either a critical stress is reached at a material element near the crack edge or when a critical strain is reached there. The critical stress criterion could apply for the case of cleavage fracture initiation in a polycrystalline metal with intrinsically cleavable grains, such as polycrystalline iron with brittle grain boundary carbides at or below room temperature. Due to the combined effects of plastic flow which is inhomogeneous on a grain-to-grain scale and the high degree of stress triaxiality at points ahead of the crack tip in mode I, fracture can initiate in such a material when the local stress becomes large enough to drive a crack from the brittle phase into a favorably oriented adjacent grain *as a cleavage crack*. The cleavage microcrack is presumed to precipitate macroscopic fracture in the ideal case. Thus, only a critical level of stress in the crack tip region is required for any rate of loading. The critical strain criterion, on the other hand, could apply for the case of locally ductile fracture initiation in a two-phase material consisting of hard brittle particles embedded in a ductile matrix, for example, carbon steel at temperature above room temperature. Under load, the large strains near the crack edge cause separations at the particle-matrix interface or cracking of individual particles, and cause subsequent ductile growth of these cavities to coalescence. Given a distribution of particles to serve as cavity nucleation sites, a critical level of crack tip strain is required for fracture initiation (that is, ductile cavity coalescence) for any rate of loading. Inelastic deformation over some small region at the crack edge is essential in either case.

An effect of increased rate of deformation on the stress–strain response of the material is to *increase* the stress necessary for continued plastic flow at a *given* level of strain. The magnitude of this effect can have enormous variation, depending on the material and its microstructural condition. Consider now two loading situations, one "rapid" and the other "slow." A critical level of stress at a point

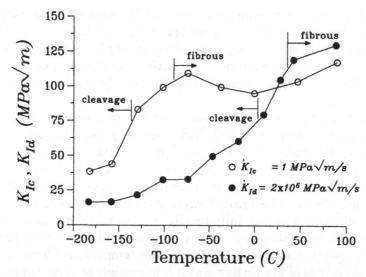

Figure 3.12. Data due to Wilson et al. (1980) showing the influence of loading rate on the fracture toughness of a 1018 cold-rolled structural steel.

in the crack tip region will be reached at a lower level of strain for rapid loading than for slow loading. Consequently, the stress work density (1.2.38) will be less in the case of rapid loading. If less work has to be done on the material element ahead of the crack tip to bring it to the point of incipient fracture, then the material will appear to be more brittle. Therefore, for materials in which fracture initiates by a stress activated mechanism and the flow stress increases with increasing deformation rate, the effect of increase of loading rate is to reduce the apparent fracture toughness. If the same argument is applied in the case of the critical strain criterion, it is clear that the stress work density at a point in the crack tip region will be greater for rapid loading than for slow loading when the critical level of strain is reached. The material appears to be tougher in the former case than in the latter. Therefore, for materials in which fracture initiates by a critical deformation mechanism, the effect of increase of loading rate is to increase the apparent fracture toughness, if it is changed at all.

This brief discussion of ideal fracture initiation mechanisms is intended only to suggest the wide range of responses which are possible in real materials. The use of energy concepts is somewhat premature here; this point will be pursued at greater depth in later discussion. Physical mechanisms of separation are commonly far more complex than the ideal mechanisms indicated in this discussion. Furthermore, it is possible that the main influence of high loading rate is not to change the toughness level, but instead to completely change the mechanism of fracture. This is evident in the data reported by Wilson et al. (1980) on fracture initiation in structural steels, which was obtained by means of the method introduced by Costin et al. (1977); see Figure 3.12. Their data showed that, at low testing temperature where the fracture initiation mechanism is essentially stress induced cleavage, the influence of increased loading rate is to reduce the toughness level by a small amount. Likewise, at high testing temperature where the fracture mechanism is essentially deformation-controlled ductile cavity growth, the influence of increased loading rate is to increase by a small amount the toughness level. The main effect of increased rate of loading, however, was to shift the temperature of transition from cleavage to ductile behavior to a position about 100 C higher on the temperature scale. This shift reflects a change in fracture mechanism induced by an increase in rate of loading.

4

ASYMPTOTIC FIELDS
NEAR A MOVING CRACK TIP

4.1 Introduction

In Section 2.1, it was argued that the fields at an interior point on the edge of a stationary crack in an elastic solid are asymptotically two-dimensional. The convention of resolving the local deformation field into the in-plane opening mode (mode I), the in-plane shearing mode (mode II), and the antiplane shearing mode (mode III) was adopted. It was pointed out that the components of stress have an inverse square root dependence on normal distance from the crack edge and a characteristic variation with angular position around the edge. This variation is specified through the functions $\Sigma_{ij}(\cdot)$ for each mode in Section 2.1. These general features are common to all configurations and all loading conditions. The influence of configuration and loading are included in the asymptotic description of stress only through scalar multipliers, one for each mode, which are the elastic stress intensity factors. The role of the stress intensity factor as a crack tip field characterizing parameter has been discussed in Sections 2.1 and 3.6.

The existence of similar universal fields for growing cracks is considered next. The same convention for categorizing local deformation modes will be followed. Except where noted explicitly to the contrary, the asymptotic crack tip fields will be determined for *variable* crack tip speed. Mode III will be considered first, because it is the simplest case to analyze, and the in-plane modes will be analyzed subsequently.

A rough estimate of the upper limit of the range of crack speeds for which inertial effects can be *ignored* in the description of an asymptotic crack tip field for a given material can be obtained on the basis of an equilibrium field for quasi-static crack growth. To illustrate this idea, two estimates are included here. Consider first the case of mode I crack growth in an elastic material at speed v for which the quasi-static crack tip field is given in Section 2.1. As the crack tip approaches a material particle near its path, the magnitude of the particle velocity increases sharply and the kinetic energy density increases accordingly. The magnitude of the stress and strain fields also increase sharply as the crack tip approaches, resulting in a rapid increase in strain energy density. These two energy changes take place during comparable intervals of time, so a relative comparison of the two energy density values when the material particle is very close to the crack tip can shed some light on the degree of influence of material inertia on the local fields.

At a radial distance r from the crack tip, the particle velocity is proportional to $vK_I/E\sqrt{r}$, so the kinetic energy density is

$$T \sim \frac{\rho v^2}{2} \frac{K_I^2}{rE^2} \tag{4.1.1}$$

to within a multiplier of order unity, where ρ is the material mass density and E is Young's modulus. Likewise, the stress components are proportional to $K_I/\sqrt{r}$, so the stress work density (strain energy density in this case) at the same point is

$$U \sim \frac{1}{2}\frac{K_I^2}{rE} \tag{4.1.2}$$

to within a multiplier of order unity. The ratio T/U is then

$$\frac{T}{U} \sim \frac{v^2}{E/\rho}. \tag{4.1.3}$$

The ratio is independent of r, and the result (4.1.3) implies that inertial effects may not be important in determining the structure of the crack tip field as long as the crack speed v is less than about one-third of the elastic wave speed $c_o = \sqrt{E/\rho}$.

The asymptotic structure of the equilibrium fields for quasi-static mode I crack advance in an elastic-ideally plastic material under plane strain conditions is also known (Rice et al. 1980), so a similar estimate can be obtained for this case. Within the angular sectors around the crack tip in which the plastic strain rate is most singular, the particle velocity and plastic shear strain in crack tip polar coordinates vary with radial distance as

$$\dot{u}_r, \dot{u}_\theta \sim v\epsilon_o \ln\left(\frac{\bar{r}_p}{r}\right) , \quad \epsilon^p_{r\theta} \sim \epsilon_o \ln\left(\frac{\bar{r}_p}{r}\right) \qquad (4.1.4)$$

where ϵ_o is the tensile yield strain and $\bar{r}_p$ is a constant with dimension of length that is usually identified with the size of the crack tip plastic zone. If expressions for the kinetic energy density T and the stress work density U are derived from the fields given by Rice et al. (1980) for $\theta = \pi/2$, say, then the ratio of the two energy measures is

$$\frac{T}{U} \sim 10\frac{v^2}{c_o^2}\ln\left(\frac{\bar{r}_p}{r}\right) \qquad (4.1.5)$$

to within a multiplier of order unity. The ratio in this case depends on r, and it can be made indefinitely large simply by taking r to be small enough. This observation may have important mathematical consequences in asymptotic analysis. However, it is inconsequential from a practical point of view unless the value of the ratio T/U becomes significant compared to unity for values of $r/\bar{r}_p$ that represent distances of relevance on the scale of continuum fracture analysis. For example, if $v/c_o = 0.1$ then the ratio T/U is greater than one-tenth if $r/\bar{r}_p \leq 0.3$. While the estimate is based on a number of approximations, it raises the possibility that the influence of inertial effects on the local fields may become significant at lower crack speeds in elastic-plastic materials than in elastic materials.

There are several reasons for the study of the asymptotic crack tip field for dynamic growth of a crack in a material. The influence of material inertia on the distribution of stress and deformation near the crack edge is of interest in assessing mechanisms of crack advance because these fields represent the environment in which the mechanisms are operative. Furthermore, numerical methods are often the only means for obtaining full field solutions within this problem class. For points very close to the crack edge where gradients are most severe the accuracy of numerical solutions is difficult to assess. The ability to match computed fields to asymptotic fields valid in this region establishes confidence in the numerical results.

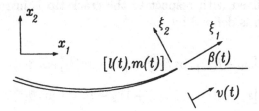

Figure 4.1. Crack growing along a smooth curved path under two-dimensional conditions. The instantaneous crack tip position is $x_1 = l(t)$, $x_2 = m(t)$, and the instantaneous crack speed is $v(t)$ in the local ξ_1-direction.

4.2 Elastic material; antiplane shear

Suppose that the elastic plane in Figure 4.1 is subjected to loading that results in antiplane shear deformation. A rectangular coordinate system labeled x_1, x_2, x_3 is introduced so that the only nonzero component of displacement $u_3(x_1, x_2, t)$ is in the x_3-direction. A crack grows along a curved path $x_1 = l(t)$, $x_2 = m(t)$ where l and m are continuous functions of time. The instantaneous speed of the crack tip in the plane is $v(t) = \sqrt{\dot{l}^2 + \dot{m}^2}$, where the superposed dot is used to denote differentiation with respect to time. The crack tip speed is restricted to the range $0 < v < c_s$ and it is assumed to be continuous for the time being. A local rectangular coordinate system ξ_1, ξ_2 is introduced in such a way that the crack tip velocity is always in the ξ_1-direction. A local polar coordinate system r, θ is also defined as shown in Figure 4.1, with $r = \sqrt{\xi_1^2 + \xi_2^2}$ and $\tan \theta = \xi_2/\xi_1$. Suppose that the crack faces are subjected to equal but opposite shear traction and that the finite limiting value of the component of the applied stress σ_{23} (in the local coordinates) at the crack tip is $\sigma_{23}^{(1)}$.

The near tip stress distribution is derived as an interior asymptotic expansion for the boundary value problem by means of standard asymptotic methods (Freund and Clifton 1974). From Section 1.2.3, out-of-plane displacement $u_3(x_1, x_2, t)$ is governed by the partial differential equation

$$\frac{\partial^2 u_3}{\partial x_1^2} + \frac{\partial^2 u_3}{\partial x_2^2} - \frac{1}{c_s^2}\frac{\partial^2 u_3}{\partial t^2} = 0 \qquad (4.2.1)$$

in regions where it is differentiable. Next, a transformation of coordinates from the fixed rectangular system x_1, x_2 to the rectangular system ξ_1, ξ_2 fixed with respect to the crack tip is introduced. This transformation is defined by

$$\xi_1 = \left[x_1 - l(t)\right] \cos \beta(t) + \left[x_2 - m(t)\right] \sin \beta(t),$$
$$\xi_2 = -\left[x_1 - l(t)\right] \sin \beta(t) + \left[x_2 - m(t)\right] \cos \beta(t), \tag{4.2.2}$$

where $\beta(t)$ is the angle from the x_1-axis to the ξ_1-axis, measured positively in the counterclockwise sense. Thus, $\tan \beta = \dot{m}/\dot{l}$. The displacement u_3 is then viewed as a function of position ξ_1, ξ_2 in the moving coordinate system and of time, that is, $u_3(x_1, x_2, t) = w(\xi_1, \xi_2, t)$. Under the transformation of coordinates (4.2.2), the partial differential equation (4.2.1) becomes

$$\frac{\partial^2 w}{\partial \xi_1^2} + \frac{\partial^2 w}{\partial \xi_2^2} - \frac{1}{c_s^2} \left\{ \frac{\partial^2 w}{\partial \xi_i \partial \xi_j} \dot{\xi_i} \dot{\xi_j} + \frac{\partial w}{\partial \xi_i} \ddot{\xi_i} + 2 \frac{\partial^2 w}{\partial \xi_i \partial t} \dot{\xi_i} + \frac{\partial^2 w}{\partial t^2} \right\} = 0, \tag{4.2.3}$$

where $\dot{\xi_1} = \dot{\beta} \xi_2 - v$ and $\dot{\xi_2} = -\dot{\beta} \xi_1$. Summation over the range 1,2 is implied for repeated indices. Obviously, $\partial u_3 / \partial t$ is distinct from $\partial w / \partial t$ because x_1, x_2 is held fixed in the former operation whereas ξ_1, ξ_2 is held fixed in the latter.

To derive the first term in the asymptotic expansion of stress components as $r \to 0$, a standard device is employed whereby the region around the crack tip is expanded so that it fills the entire field of observation. To this end, rescaled coordinates $\eta_i = \xi_i / \epsilon$ are introduced, where ϵ is a small parameter. If ϵ is taken to be indefinitely small then all points in the ξ_1, ξ_2-plane except those near the crack tip are pushed out of the field of observation in the η_1, η_2-plane. Furthermore, as ϵ becomes indefinitely small, the crack line occupies the negative η_1-axis in the field of observation.

It is assumed that w has an expansion in powers of ϵ of the form

$$w(\xi_1, \xi_2, t) = w(\epsilon \eta_1, \epsilon \eta_2, t)$$
$$= \epsilon^{p_0} W_0(\eta_1, \eta_2, t) + \epsilon^{p_1} W_1(\eta_1, \eta_2, t) + \dots, \tag{4.2.4}$$

where W_0 represents the dominant asymptotic solution, W_1 represents the first order correction to the asymptotic solution, and so on. This implies that the exponents are ordered so that $p_0 < p_1 < p_2 < \cdots$. The exponent p_0 is expected to be in the range $0 < p_0 < 1$ if the displacement is to be bounded at the crack tip but the stress is to be singular. Each term in the expansion (4.2.4) is expected to be bounded as $\epsilon \to 0^+$ for any fixed point $\xi_i \xi_i \neq 0$. The expansion (4.2.4) is essentially an assumption that the near tip displacement field can be represented as a series of homogeneous functions of increasing degree, a feature shared with the more familiar approach based on separation of variables in local polar coordinates. The assumed form of the solution is substituted into the transformed wave equation (4.2.3) and the coefficient of each power of ϵ is set equal to zero. The coefficient of the lowest power of ϵ vanishes if W_0 satisfies

$$\left(1 - \frac{v^2}{c_s^2}\right) \frac{\partial^2 W_0}{\partial \eta_1^2} + \frac{\partial^2 W_0}{\partial \eta_2^2} = 0, \qquad (4.2.5)$$

where $v = v(t)$ is the instantaneous crack tip speed, as before. This equation is recognized as Laplace's equation with the coordinate η_2 rescaled by the factor $\alpha_s = \sqrt{1 - v^2/c_s^2}$. Thus, a general solution is

$$W_0 = \mathrm{Im}\{F(\zeta)\}, \qquad (4.2.6)$$

where $\zeta = \eta_1 + i\alpha_s \eta_2$ and $F(\zeta)$ is an analytic function in the complex ζ-plane. The function can depend parametrically on t. A displacement field that is symmetric with respect to the crack line $\xi_2 = 0$ results in zero stress on the crack line, so there is no loss in generality by assuming that F is further restricted by the symmetry condition $F(\bar{\zeta}) = \overline{F}(\zeta)$. Note that W_0 could equally well be expressed as the real part of an analytic function. In that case, the analytic function would be subject to a different symmetry condition and the final result for W_0 would be the same as that obtained below.

In light of the expansion (4.2.4), if the limiting value of the crack face traction at the crack tip, say $\sigma_{23}^{(1)}$, is bounded and if $0 < p_0 < 1$ then

$$\frac{\partial W_0(\eta_1, \pm 0, t)}{\partial \eta_2} = \mathrm{Re}\{\alpha_s F'_{\pm}(\eta_1)\} = 0 \qquad (4.2.7)$$

for $\eta_1 < 0$, where the prime denotes differentiation of F with respect to its argument. In view of the assumed symmetry condition on $F(\cdot)$,

$$F'_+(\eta_1) + \overline{F}'_+(\eta_1) = F'_+(\eta_1) + F'_-(\eta_1) = 0 \qquad (4.2.8)$$

on $\eta_1 < 0$, where the subscript plus or minus indicates a limiting value of the function as the negative real axis is approached through positive or negative values of η_2, respectively. The equation (4.2.8) represents a standard Hilbert problem in analytic function theory (Muskhelishvili 1953).

The most general analytic function that satisfies (4.2.8) and that leads to a deformation field with integrable mechanical energy density is

$$F'(\zeta) = \frac{A(\zeta)}{\zeta^{1/2}}, \qquad (4.2.9)$$

where $A(\zeta)$ is an entire function and a branch cut is provided along the negative real axis to render $F'(\zeta)$ single valued in the cut ζ-plane. The only feature of the function $A(\zeta)$ that influences the asymptotic field as $\zeta \to 0$ is $A(0)$, hereafter denoted simply by A (which can still depend on time t). The condition $F(\overline{\zeta}) = \overline{F(\zeta)}$ further requires that A must be a real parameter. Note that if $F(\overline{\zeta}) = -\overline{F(\zeta)}$, then the condition equivalent to (4.2.8) leads to the conclusion that either $F'(\zeta)$ is entire or $F'(\zeta)$ has a pole at the origin. In the former case, the boundedness condition leads only to the trivial solution, whereas the latter case is ruled out because the singularity is too strong to be admissible. Neither possibility is of interest in this development. Consequently, the dominant contribution to the near tip field has the symmetry $F(\overline{\zeta}) = \overline{F(\zeta)}$.

In view of (4.2.4) and (4.2.6), the result (4.2.9) implies that the dominant contribution to the near tip displacement field is

$$w \sim 2Ar_s^{1/2}\sin\tfrac{1}{2}\theta_s, \qquad (4.2.10)$$

where $r_s = \sqrt{\xi_1^2 + \alpha_s^2\xi_2^2}$, $\tan\theta_s = \alpha_s\xi_2/\xi_1$, and A is an undetermined parameter that is independent of ξ_1, ξ_2. The parameter p_0 in (4.2.4) evidently has the value one-half. From the stress–strain relation for the material it follows that

$$\sigma_{23} = \mu\frac{\partial w}{\partial \xi_2} \sim \frac{A\mu\alpha_s}{r_s^{1/2}}\cos\tfrac{1}{2}\theta_s. \qquad (4.2.11)$$

If the dynamic stress intensity factor for mode III is defined by

$$K_{III}(t) = \lim_{\xi_1\to 0}\sqrt{2\pi\xi_1}\,\sigma_{23}(\xi_1,0,t) \qquad (4.2.12)$$

then $A = K_{III}(t)/\mu\alpha_s\sqrt{2\pi}$. Thus, the feature that the near crack tip field has universal spatial dependence carries over to the case of nonsteady mode III crack growth as well.

The next term in the expansion (4.2.4) will satisfy the boundary conditions if $p_1 = 1$. Then, equating coefficients of like powers of ϵ, the boundary condition equivalent to (4.2.7) is

$$\frac{\partial W_1}{\partial \eta_2} = \frac{\sigma_{23}^{(1)}}{\mu} \qquad (4.2.13)$$

and the differential equation governing the function W_1 is

$$\left(1 - \frac{v^2}{c_s^2}\right)\frac{\partial^2 W_1}{\partial \eta_1^2} + \frac{\partial^2 W_1}{\partial \eta_2^2} = 0. \qquad (4.2.14)$$

Like W_0, the function W_1 can be expressed in terms of an analytic function. The relationship between limiting values of this analytic function corresponding to (4.2.8) has the nonzero right side $2\sigma_{23}^{(1)}/\mu\alpha_s$. It follows immediately that $\mu W_1 = \eta_1\sigma_{13}^{(1)} + \eta_2\sigma_{23}^{(1)}$, where $\sigma_{13}^{(1)}$ is the value of the stress component σ_{13} at the crack tip. Consequently, the asymptotic stress distribution can be summarized as

$$\sigma_{i3} = \frac{K_{III}(t)}{\sqrt{2\pi r}}\Sigma_{i3}(\theta, v) + \sigma_{i3}^{(1)} + o(1) \quad \text{as} \quad r \to 0 \qquad (4.2.15)$$

in the presence of crack face tractions. The physical coordinates r, θ are related to r_s, θ_s by

$$r_s = r\sqrt{1 - (v\sin\theta/c_s)^2} = r\gamma_s, \qquad \tan\theta_s = \alpha_s\tan\theta. \qquad (4.2.16)$$

The functions Σ_{ij} in (4.2.15) are

$$\Sigma_{13} = -\frac{\sin\frac{1}{2}\theta_s}{\alpha_s\sqrt{\gamma_s}}, \qquad \Sigma_{23} = \frac{\cos\frac{1}{2}\theta_s}{\sqrt{\gamma_s}}. \qquad (4.2.17)$$

Upon comparison of (4.2.17) with the corresponding results for a stationary crack tip, it is evident that the former reduce to the latter as $v \to 0$. The distribution of particle velocity in the crack tip region is also recorded here for future reference. The asymptotic form of particle velocity as $r \to 0$ is

$$\dot{u}_3 \sim \frac{v K_{III}(t)}{\sqrt{2\pi r}} \frac{\sin \frac{1}{2}\theta_s}{\alpha_s \sqrt{\gamma_s}}. \qquad (4.2.18)$$

The asymptotic forms (4.2.15) and (4.2.18) represent the main results of this section. Several observations can be made concerning these results. For example, the dominant singular term in the expansion of stress or particle velocity does not depend on the fact that the crack path is (or can be) curved. Certainly, the influence of crack path curvature will enter through the higher order terms in the expansion. Indeed, examination of the partial differential equation (4.2.3) makes it clear that the terms of order $r^{1/2}$ in expansions of stress and particle velocity components will include a contribution with its coefficient proportional to the angular velocity of the crack propagation direction $\dot{\beta}$. The instantaneous radius of curvature of the crack path is $v/\dot{\beta}$. The asymptotic expansion can be continued to higher order terms by equating the coefficient of the next highest power of ϵ in the differential equation obtained by substituting (4.2.4) into (4.2.3). This step is pursued for a straight tensile crack (that is, for $\beta = 0$) in the next subsection. For a crack following a curved path, the construction of the expansions is conceptually straightforward but the details are not yet available.

4.3 Elastic material; in-plane modes of deformation

Suppose that a planar crack is extending in an elastic body in such a way that the state of deformation is two-dimensional plane strain. The case of generalized plane stress is identical except for the interpretation of the elastic constants. A rectangular x_1, x_2, x_3 coordinate system is introduced so that the crack edge is parallel to the x_3-axis and the displacement vector of each particle is parallel to the x_1, x_2-plane. For definiteness, suppose the crack grows in the x_1-direction so that the coordinates of the crack tip in the x_1, x_2-plane are $x_1 = l(t)$, $x_2 = 0$, where l is a continuous function of time. The instantaneous speed of the crack tip is $v(t) = \dot{l}(t)$. The crack tip speed is restricted to the range $0 < v < c_R$ and to be continuous for the time being. A local rectangular coordinate system ξ_1, ξ_2 with its origin at the crack tip is introduced, where $\xi_1 = x_1 - l(t)$, $\xi_2 = x_2$. A local polar coordinate system r, θ is also introduced, where $r = \sqrt{\xi_1^2 + \xi_2^2}$, $\tan \theta = \xi_2/\xi_1$. The near tip stress distribution is derived for modes I and II, following the procedure outlined for mode III in the preceding section.

4.3.1 Singular field for mode I

Suppose that the crack faces are subjected to tractions consistent with the mode I symmetry conditions, and that the finite limiting value of the component of applied crack face stress σ_{22} as $\xi_1 \to 0^-$ at the crack tip is $\sigma_{22}^{(1)}$. The representation of crack tip fields in terms of the displacement potentials ϕ and $\psi = \psi_3$ is adopted as in Section 1.2.3. If ϕ is viewed as a function of position (ξ_1, ξ_2) in the local coordinate system and of time t, then it is governed by the differential equation

$$\left(1 - \frac{v^2}{c_d^2}\right) \frac{\partial^2 \phi}{\partial \xi_1^2} + \frac{\partial^2 \phi}{\partial \xi_2^2} + \frac{\dot{v}}{c_d^2} \frac{\partial \phi}{\partial \xi_1} + 2\frac{v}{c_d^2} \frac{\partial^2 \phi}{\partial \xi_1 \partial t} - \frac{1}{c_d^2} \frac{\partial^2 \phi}{\partial t^2} = 0. \quad (4.3.1)$$

In light of the derivation of (4.2.3) in the previous section, the interpretation of the terms in (4.3.1) is obvious. The shear wave potential $\psi(\xi_1, \xi_2, t)$ must satisfy the same equation with c_d replaced by c_s.

The development follows that presented in Section 4.2 for mode III. First, the rescaled coordinates $\eta_i = \xi_i/\epsilon$ are introduced, where ϵ is an arbitrary small parameter. It is then assumed that ϕ has an expansion in powers of ϵ of the form

$$\phi(\xi_1, \xi_2, t) = \phi(\epsilon\eta_1, \epsilon\eta_2, t)$$
$$= \epsilon^{p_0} \Phi_0(\eta_1, \eta_2, t) + \epsilon^{p_1} \Phi_1(\eta_1, \eta_2, t) + \cdots, \quad (4.3.2)$$

where Φ_0 represents the main contribution to the asymptotic solution, Φ_1 represents the first order correction, and so on. The powers of ϵ are ordered so that $p_0 < p_1 < p_2 < \cdots$, and p_0 is expected to be in the range $1 < p_0 < 2$ if displacement is to be bounded but stress is to be singular at the crack tip. The assumed expansion (4.3.2) is substituted into the governing equation (4.3.1) and the coefficient of each power of ϵ is set equal to zero. The coefficient of the lowest power of ϵ vanishes if Φ_0 satisfies

$$\alpha_d^2 \frac{\partial^2 \Phi_0}{\partial \eta_1^2} + \frac{\partial^2 \Phi_0}{\partial \eta_2^2} = 0, \quad (4.3.3)$$

where $\alpha_d = \sqrt{1 - v^2/c_d^2}$. Thus, Φ_0 is viewed as a function of η_1, η_2 with the possibility that it also depends on time t as a parameter. A general solution of this equation is

$$\Phi_0 = \text{Re}\{F(\zeta_d)\}, \qquad (4.3.4)$$

where $\zeta_d = \eta_1 + i\alpha_d\eta_2$ and $F(\zeta)$ is an analytic function everywhere in the complex ζ-plane except along the nonpositive real axis. For mode I deformation, F must satisfy $F(\bar{\zeta}) = \overline{F(\zeta)}$. Similarly, the leading term in the expansion for ψ has the form

$$\Psi_0 = \text{Im}\{G(\zeta_s)\}, \qquad (4.3.5)$$

where $\zeta_s = \eta_1 + i\alpha_s\eta_2$ and $G(\zeta)$ has the same domain of analyticity and symmetry as does $F(\zeta)$. The powers of ϵ in the expansion for ψ must be the same as for ϕ because neither type of wave function alone can satisfy the boundary conditions on the crack faces.

If the expansions of ϕ and ψ are substituted into the stress boundary conditions on the crack faces, and the coefficients of independent powers of ϵ are set equal to zero, then

$$(1 + \alpha_s^2)\left[F_+''(\eta_1) + F_-''(\eta_1)\right] + 2\alpha_s\left[G_+''(\eta_1) + G_-''(\eta_1)\right] = 0\,,$$
$$2\alpha_d\left[F_+''(\eta_1) - F_-''(\eta_1)\right] + (1 + \alpha_s^2)\left[G_+''(\eta_1) - G_-''(\eta_1)\right] = 0\,. \qquad (4.3.6)$$

The solution of this pair of equations that leads to singular stress but integrable mechanical energy density is

$$F''(\zeta) = -\frac{1 + \alpha_s^2}{D}\frac{A}{\zeta^{1/2}}\,, \qquad G''(\zeta) = \frac{2\alpha_d}{D}\frac{A}{\zeta^{1/2}}\,, \qquad (4.3.7)$$

where A is independent of ζ (although it can depend on time) and

$$D = 4\alpha_d\alpha_s - (1 + \alpha_s^2)^2\,. \qquad (4.3.8)$$

Note that $D \to 0$ as $v \to c_R$, where c_R is the Rayleigh wave speed from Section 2.5. In light of (4.3.7), the exponent p_0 in (4.3.2) has the value $3/2$.

If the dynamic stress intensity factor for mode I crack growth is defined by

$$K_I(t) = \lim_{\xi_1 \to 0^+} \sqrt{2\pi\xi_1}\, \sigma_{22}(\xi_1, 0, t) \qquad (4.3.9)$$

then $A = K_I(t)/\mu\sqrt{2\pi}$. The dominant contribution to the near tip field is therefore completely determined up to the scalar multiplier $K_I(t)$ for *nonuniform* crack tip motion.

The next term in the expansion (4.3.2) can be determined by following steps similar to those for mode III. The details are not included here, but it is noted that the exponent p_1 takes on the value 2 and the stress components derived are locally uniform. Consequently, the leading terms in the asymptotic expansion of the crack tip stress field can be summarized by

$$\sigma_{ij} = \frac{K_I(t)}{\sqrt{2\pi r}}\Sigma_{ij}^I(\theta, v) + \sigma_{ij}^{(1)} + o(1) \quad \text{as} \quad r \to 0. \qquad (4.3.10)$$

The functions $\Sigma_{ij}^I(\theta, v)$ that represent the angular variation of stress components for any value of instantaneous crack tip speed v are

$$\Sigma_{11}^I = \frac{1}{D}\left\{(1+\alpha_s^2)(1+2\alpha_d^2-\alpha_s^2)\frac{\cos\frac{1}{2}\theta_d}{\sqrt{\gamma_d}} - 4\alpha_s\alpha_d\frac{\cos\frac{1}{2}\theta_s}{\sqrt{\gamma_s}}\right\},$$

$$\Sigma_{12}^I = \frac{2\alpha_d(1+\alpha_s^2)}{D}\left\{\frac{\sin\frac{1}{2}\theta_d}{\sqrt{\gamma_d}} - \frac{\sin\frac{1}{2}\theta_s}{\sqrt{\gamma_s}}\right\},$$

$$\Sigma_{22}^I = -\frac{1}{D}\left\{(1+\alpha_s^2)^2\frac{\cos\frac{1}{2}\theta_d}{\sqrt{\gamma_d}} - 4\alpha_d\alpha_s\frac{\cos\frac{1}{2}\theta_s}{\sqrt{\gamma_s}}\right\},$$

$$(4.3.11)$$

where

$$\gamma_d = \sqrt{1 - (v\sin\theta/c_d)^2}\,, \quad \tan\theta_d = \alpha_d\tan\theta\,,$$

$$\gamma_s = \sqrt{1 - (v\sin\theta/c_s)^2}\,, \quad \tan\theta_s = \alpha_s\tan\theta\,. \qquad (4.3.12)$$

The terms of order unity denoted by $\sigma_{ij}^{(1)}$ in (4.3.10) have relatively simple interpretations. The term $\sigma_{12}^{(1)} = 0$ to conform with the symmetry of a mode I deformation field. The term $\sigma_{22}^{(1)}$ is determined from the crack face loading conditions as the limiting value of normal crack face traction at the crack tip. If the crack faces are traction-free during crack growth, then this term has value zero. Finally, the term $\sigma_{11}^{(1)}$ cannot be determined from the asymptotic solution. It must be determined as a feature of the complete stress distribution in any particular stress analysis problem. An elastic field consisting *only* of a uniform tensile stress in the direction of the crack line (the x_1-direction, in this case) can always be added to any other stress field without violating the boundary conditions on the crack faces, provided that the deformation in the transverse direction is completely unconstrained.

It can be verified by direct calculation that the terms in the expansion (4.3.10) do indeed reduce to the equivalent results for equilibrium crack tip fields given in Section 2.1 as the crack speed becomes vanishingly small. Some care is required in the calculation because both the term D in the denominator of each expression and each of the quantities enclosed in brackets in (4.3.11) vanish as $v \to 0$. However, in each case both numerator and denominator vanish as v^2 so the ratio has an unambiguous limit as $v \to 0$ and the Irwin-Williams asymptotic field (2.1.3) is recovered.

It is noteworthy that the mode I crack tip field for nonuniform motion of the crack tip involves the crack tip motion only through the instantaneous crack tip speed $v(t)$. An immediate consequence is that the near tip field for nonuniform motion is identical to that for steady state crack growth in the same material, at least for the singular and locally uniform contributions investigated up to this point. This was demonstrated by Freund and Clifton (1974), Nilsson (1974b), and Achenbach and Bazant (1975), each of whom compared asymptotic solutions for nonuniform crack growth with earlier results based on the assumption of steady growth obtained by Cotterell (1964), Rice (1968), and Sih (1970). It is shown in the next subsection that the expansions do indeed differ if terms beyond those included in (4.3.10) are examined. Such higher order terms are used to interpret experimental data obtained by optical sensing methods applied over some region around the tip of a rapidly growing crack, so the interpretation of such higher order terms is of practical significance.

The variation of stress components near the crack tip can be represented in any number of ways. Here, the circumferential tensile stress (the "hoop" stress or the tensile stress in the θ-direction), the

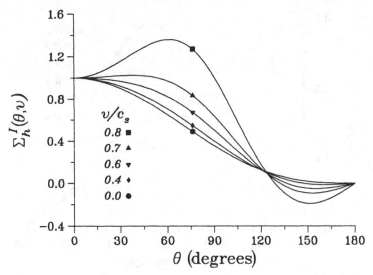

Figure 4.2. Variation of the circumferential tensile stress $(4.3.13)_1$ with angle θ around the crack edge for several values of normalized crack speed. Crack growth is in mode I.

largest principal stress, and the maximum shear stress are plotted against the angular position θ around the crack tip for a fixed value of $K_I(t)/\sqrt{2\pi r}$. These three stress measures are denoted by

$$\sigma_h = \frac{K_I(t)}{\sqrt{2\pi r}}\Sigma_h^I(\theta, v) \quad \text{(circumferential tensile stress)},$$

$$\sigma_1 = \frac{K_I(t)}{\sqrt{2\pi r}}\Sigma_1^I(\theta, v) \quad \text{(largest principal stress)}, \qquad (4.3.13)$$

$$\tau_{max} = \frac{K_I(t)}{\sqrt{2\pi r}}\Sigma_{max}^I(\theta, v) \quad \text{(maximum shear stress)},$$

respectively. The angular variations of the circumferential tensile stress, the largest principal stress, and the maximum shear stress are shown in Figures 4.2 through 4.4. Note that the definition of stress intensity factor (4.3.9) implies that $\Sigma_{22}^I = 1$.

For crack speeds less than about 60 percent of the shear wave speed of the material, the hoop stress or transverse tensile stress shown in Figure 4.2 has a maximum value along $\theta = 0$, that is, in the direction of crack growth. For crack speeds greater than about

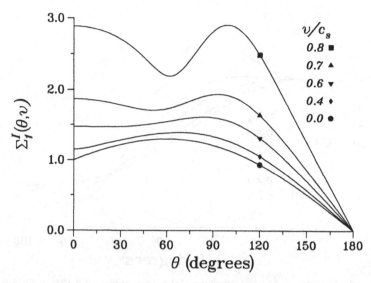

Figure 4.3. Variation of the maximum principal stress $(4.3.13)_2$ with angle θ around the crack edge for several values of normalized crack speed. Crack growth is in mode I.

$0.6\, c_s$, however, this stress measure develops a maximum in a direction inclined at about 60° to the direction of crack growth. The feature was first observed by Yoffe (1951) who suggested that this inertia induced modification of the local singular stress field could account for the tendency of rapidly growing cracks in very brittle materials to develop roughened fracture surfaces and to bifurcate into several branched cracks. Though this feature does provide a mechanism by which a growing crack can sample propagation directions inclined to the straight ahead direction, it is not able to fully explain crack branching and bifurcation behavior in brittle materials. In general, bifurcation half angles (one-half the total angle between the two branches of a bifurcated crack) tend to be substantially smaller than 60° and the attainment of a critical speed appears to be neither a necessary nor a sufficient condition for a fast running crack to bifurcate. The bifurcation behavior of fast running fractures appears to depend on properties of the stress field not included in the singular term in (4.3.10) alone (Cotterell 1964; Ramulu and Kobayashi 1983). This point will be discussed further in connection with crack propagation behaviors.

The same crack tip singular field that was found by Yoffe (1951) in a study of a steady crack growth process was also found in a study of a transient crack growth problem by Baker (1962). It was only later in the development of the subject that this field was established as a *universal* crack tip field for any instantaneous stress intensity factor and instantaneous crack tip speed. Baker suggested that it was not the hoop stress or transverse tensile stress that was of fundamental significance in analyzing crack advance in brittle materials. Instead, he suggested that if the local condition for fracture is the attainment of a critical tensile stress on some plane, then the maximum principal stress is the stress measure of significance. The angular variation of this stress measure is shown in Figure 4.3. For low crack speeds, this angular variation has a shallow maximum between about 60° and 90°. For very high speeds, on the other hand, the variation shows local maxima at both $\theta = 0$ and for some value of θ beyond 90°.

The variation of maximum shear stress for the singular field in (4.3.10) with angle θ around the crack edge is shown in Figure 4.4. It is evident from this figure that there is a monotonic increase in maximum shear stress with increasing crack speed for a fixed level of stress intensity factor. Thus, the driving force for inelastic modes of deformation that are *shear* driven, such as dislocation glide, twinning, or viscous relaxation, is increased with increasing speed. Along $\theta = 0$, the in-plane principal stresses within the singular field are equal for zero crack speed and nearly equal for low crack speeds. However, the ratio σ_{22}/σ_{11} on $\theta = 0$ is a rapidly decreasing function of speed for values of v greater than about 50 percent of the shear wave speed. The change from the corresponding equilibrium result in the nature of the asymptotic field accounts for the dramatic increase in maximum shear stress along $\theta = 0$ with crack speed. Thus, at high crack speeds, the larger in-plane principal stress ahead of the crack edge acts on planes perpendicular to the direction of growth, and the largest in-plane principal stress in the regions near the crack edge acts at points off of the crack plane. These effects may underlie the observed tendency for rapidly growing cracks in brittle materials to develop rough fracture surfaces.

Finally, it is noted that the elastodynamic particle velocity field corresponding to the square root singular stress field in (4.3.10) is also square root singular at the crack edge. The dominant spatial dependence of the particle velocity components near the crack edge is

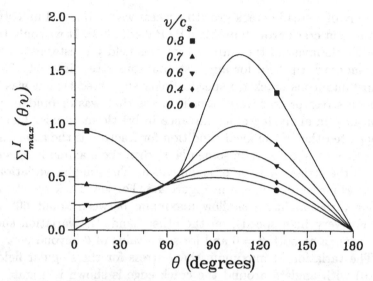

Figure 4.4. Variation of the maximum shear stress $(4.3.13)_3$ with angle θ around the crack edge for several values of normalized crack speed. Crack growth is in mode I.

given by

$$\dot{u}_1 \sim -\frac{vK_I(t)}{\mu D\sqrt{2\pi r}}\left\{(1+\alpha_s^2)\frac{\cos\frac{1}{2}\theta_d}{\sqrt{\gamma_d}} - 2\alpha_d\alpha_s\frac{\cos\frac{1}{2}\theta_s}{\sqrt{\gamma_s}}\right\},$$

$$\dot{u}_2 \sim -\frac{v\alpha_dK_I(t)}{\mu D\sqrt{2\pi r}}\left\{(1+\alpha_s^2)\frac{\sin\frac{1}{2}\theta_d}{\sqrt{\gamma_d}} - 2\frac{\sin\frac{1}{2}\theta_s}{\sqrt{\gamma_s}}\right\}.$$

(4.3.14)

The limiting velocity field as the crack tip speed becomes vanishingly small is

$$\lim_{v\to 0}\left(\frac{\dot{u}_j}{v}\right) = -\left\{\frac{\partial u_j}{\partial x_1}\right\}_{equil},$$

(4.3.15)

where the right side is intended to represent a feature of the Irwin-Williams equilibrium asymptotic field.

It is evident from (4.3.10) and (4.3.14) that for crack tip speeds in the range $0 < v < c_s$ the local stress and particle velocity vary as the inverse square root of radial distance from the crack tip. Due to the appearance of the factor D in the expressions for these fields, however, the algebraic sign of the coefficient of the square root singularity depends on whether the crack tip speed is less than or greater than the Rayleigh wave speed of the material c_R. Suppose that the crack faces separate behind the advancing crack tip. Inspection of (4.3.10) and (4.3.14) leads to the conclusion that the traction on the prospective fracture plane ahead of the tip is tensile if $0 < v < c_R$ but that it is *compressive* if $c_R < v < c_s$. This feature implies that there is a net flux of energy out of the advancing crack tip for $c_R < v < c_s$, whereas there is a net flux of energy into the crack tip for crack tip speeds less than the Rayleigh wave speed. This result is established in Section 5.3.

4.3.2 Higher order terms for mode I

The next term in the expansion (4.3.2) for ϕ which can yield a non-trivial result has exponent $p_2 = 5/2$. If the expansion is substituted into the governing equation (4.3.1) and the coefficient of each power of ϵ is set equal to zero, then

$$\alpha_d^2 \frac{\partial^2 \Phi_2}{\partial \eta_1^2} + \frac{\partial^2 \Phi_2}{\partial \eta_2^2} = -\frac{\partial}{\partial t}\left(\frac{v}{c_d^2}\frac{\partial \Phi_0}{\partial \eta_1}\right), \qquad (4.3.16)$$

where the function $\Phi_0(\eta_1, \eta_2, t)$ was determined in the previous subsection. Again, Φ_2 is viewed as a function of η_1, η_2 that possibly depends on t as a parameter. The equation is more convenient to consider in the distorted polar coordinates $\rho_d = \sqrt{\eta_1^2 + \alpha_d^2\eta_2^2}$, $\omega_d = \arctan(\alpha_d\eta_2/\eta_1)$, so that for constant crack speed

$$\frac{\partial^2 \Phi_2}{\partial \rho_d^2} + \frac{1}{\rho_d}\frac{\partial \Phi_2}{\partial \rho_d} + \frac{1}{\rho_d^2}\frac{\partial^2 \Phi_2}{\partial \omega_d^2} = A^* \rho_d^{1/2} \cos\tfrac{1}{2}\omega_d, \qquad (4.3.17)$$

where

$$A^* = \frac{\sqrt{v}}{\alpha_d^2}\frac{d}{dt}\left[\frac{4}{\sqrt{2\pi}}\frac{\sqrt{v}}{c_d^2}\frac{1+\alpha_s^2}{\mu D}K_I(t)\right]. \qquad (4.3.18)$$

Note that A^* is the rate of change of a quantity that depends on the instantaneous stress intensity factor $K_I(t)$ and on the crack tip speed v. A general solution of (4.3.17) for mode I deformation is

$$\Phi_2 = \rho_d^{5/2}\left(\tfrac{1}{6}A^* \cos\tfrac{1}{2}\omega_d + A_2 \cos\tfrac{5}{2}\omega_d\right), \qquad (4.3.19)$$

where A_2 is a constant of integration to be determined from the boundary conditions.

Likewise, the next term in the asymptotic expansion of ψ is given by

$$\Psi_2 = -\rho_s^{5/2} \left(\tfrac{1}{6} B^* \sin \tfrac{1}{2}\omega_s + B_2 \sin \tfrac{5}{2}\omega_s \right) \qquad (4.3.20)$$

for mode I deformation, where B_2 is a constant to be determined from the boundary conditions, $\rho_s = \sqrt{\eta_1^2 + \alpha_s^2 \eta_2^2}$, $\omega_s = \arctan(\alpha_s \eta_2 / \eta_1)$, and

$$B^* = \frac{\sqrt{v}}{\alpha_s^2} \frac{d}{dt} \left[\frac{8}{\sqrt{2\pi}} \frac{\sqrt{v}}{c_s^2} \frac{\alpha_d}{\mu D} K_I(t) \right] . \qquad (4.3.21)$$

The condition of traction-free crack faces will be satisfied to corresponding order if A_2 and B_2 are related by $5(30A_2 + A^*) = 8(15B_2 + B^*)$. The coefficients A_2 and B_2 are otherwise unrestricted by the near tip conditions, and a second relationship between them can only follow from remote conditions. The mean in-plane stress derived from (4.3.19) and (4.3.20) is

$$\frac{\sigma_{11}^{(2)} + \sigma_{22}^{(2)}}{\mu} = \frac{2}{1 - 2\nu} \left\{ A^* r_d^{1/2} \left[\left(1 - \frac{3}{8} \frac{v^2}{c_d^2} \right) \cos \tfrac{1}{2}\theta_d \right. \right.$$

$$\left. \left. - \frac{1}{4} \frac{v^2}{c_d^2} \sin \tfrac{1}{2}\theta_d \sin \theta_d \right] + A_2 \frac{15}{4} \frac{v^2}{c_d^2} r_d^{1/2} \cos \tfrac{1}{2}\theta_d \right\} . \qquad (4.3.22)$$

Only the second term in each of (4.3.19) and (4.3.20) is commonly viewed as the next most significant term, after the term corresponding to locally uniform stress, in the expansion of ϕ or ψ. While this is indeed the case for steady state crack growth, it is not correct when either the crack speed or the stress intensity factor is changing with time. From their definitions in (4.3.18) and (4.3.21), it is evident that A^* and B^* are nonzero, in general, when either $\dot{v} \neq 0$ or $\dot{K}_I(t) \neq 0$. When this is the case, the first term in each of (4.3.19) and (4.3.20) must be included in the expansion. The existence of these contributions associated with nonsteady situations has not been generally recognized in dynamic fracture studies, and their importance relative to the other terms of the same order in the expansion of crack tip fields has not been studied systematically.

4.3.3 Singular field for mode II

The asymptotic field for mode II deformation can be analyzed by following the steps outlined for the case of mode I deformation. The results for the leading term in the crack tip field are given here; details of the analysis are omitted. For mode II deformation, the stress distribution near the crack edge has the form

$$\sigma_{ij} = \frac{K_{II}(t)}{\sqrt{2\pi r}} \Sigma_{ij}^{II}(\theta, v) + o(1) \qquad \text{as} \qquad r \to 0^+ , \qquad (4.3.23)$$

where $K_{II}(t)$ is the instantaneous mode II stress intensity factor. In this case, the functions Σ_{ij}^{II} representing angular variation of stress components for any value of instantaneous crack tip speed v are

$$\Sigma_{11}^{II} = -\frac{2\alpha_s}{D} \left\{ (1 + 2\alpha_d^2 - \alpha_s^2) \frac{\sin \frac{1}{2}\theta_d}{\sqrt{\gamma_d}} - (1 + \alpha_s^2) \frac{\sin \frac{1}{2}\theta_s}{\sqrt{\gamma_s}} \right\} ,$$

$$\Sigma_{12}^{II} = \frac{1}{D} \left\{ 4\alpha_d \alpha_s \frac{\cos \frac{1}{2}\theta_d}{\sqrt{\gamma_d}} - (1 + \alpha_s^2)^2 \frac{\cos \frac{1}{2}\theta_s}{\sqrt{\gamma_s}} \right\} , \qquad (4.3.24)$$

$$\Sigma_{22}^{II} = \frac{2\alpha_s(1 + \alpha_s^2)}{D} \left\{ \frac{\sin \frac{1}{2}\theta_d}{\sqrt{\gamma_d}} - \frac{\sin \frac{1}{2}\theta_s}{\sqrt{\gamma_s}} \right\} .$$

The angular variations of the circumferential tensile stress, the largest principal stress, and the maximum shear stress as determined from (4.3.24) are shown in Figures 4.5 through 4.7. The definition of stress intensity factor adopted in (4.3.23) implies that $\Sigma_{12}^{II}(0, v) = 1$.

The elastodynamic particle velocity field corresponding to the singular stress field (4.3.24) is given by

$$\dot{u}_1 \sim \frac{v\alpha_s K_{II}(t)}{\mu D \sqrt{2\pi r}} \left\{ 2\frac{\sin \frac{1}{2}\theta_d}{\sqrt{\gamma_d}} - (1 + \alpha_s^2) \frac{\sin \frac{1}{2}\theta_s}{\sqrt{\gamma_s}} \right\} ,$$

$$\qquad (4.3.25)$$

$$\dot{u}_2 \sim -\frac{v K_{II}(t)}{\mu D \sqrt{2\pi r}} \left\{ 2\alpha_d \alpha_s \frac{\cos \frac{1}{2}\theta_d}{\sqrt{\gamma_d}} - (1 + \alpha_s^2) \frac{\cos \frac{1}{2}\theta_s}{\sqrt{\gamma_s}} \right\} .$$

4.3.4 Supersonic crack tip speed

In the foregoing analyses of crack tip singular fields, it has been assumed that the crack speed is subsonic, that is, $v < c_s$ in all cases. The observation of crack speeds greater than c_s or even greater than

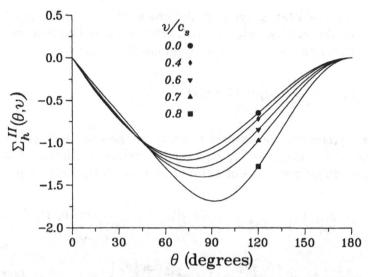

Figure 4.5. Variation of the circumferential tensile stress with angle θ around the crack edge for five values of normalized crack speed. Crack growth is in mode II.

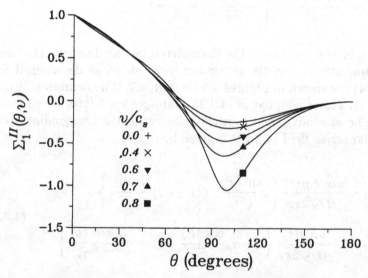

Figure 4.6. Variation of the maximum principal stress with angle θ around the crack edge for five values of normalized crack speed. Crack growth is in mode II.

c_d in structural materials has been limited to cases where the loading is applied directly at the crack tip by means of a fluid under high

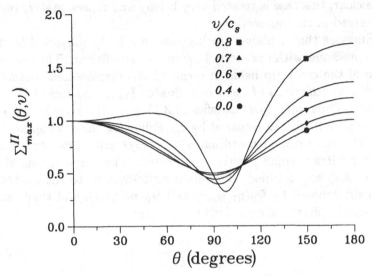

Figure 4.7. Variation of the maximum shear stress with angle θ around the crack edge for several values of normalized crack speed. Crack growth is in mode II.

pressure flowing into the opening crack or by some other extreme condition. Winkler, Curran, and Shockey (1970a, 1970b) focused an intense laser beam at a point in the interior of a sodium chloride crystal. The vaporized material expanding in the resulting cavity induced supersonic crack growth. Analysis of seismic data taken during crustal earthquakes has led to the conclusion that a shear fracture on a preexisting fault can propagate at a speed greater than the shear wave speed of the crustal material (Archuleta 1982). Few analytical results are available for crack growth at a speed greater than c_s or c_d. In the truly supersonic case where $v > c_d$ the crack is growing into material that has no advance warning that the crack is approaching. Thus, it seems that such a process in a brittle-elastic material must be driven by crack face loading that acts up to the crack tip. The crack tip field in this case consists of plane wavefronts trailing the crack edge, with the deformation distribution determined by the distribution of traction on the crack faces. Indeed, the solution is exactly that for a supersonically moving pressure step load on the surface of an elastic half space (Cole and Huth 1958). The case of a supersonic step pressure load on the surface of an elastic-plastic half space was analyzed by Bleich and Matthews (1967). The situation for the intermediate case with $c_s < v < c_d$ (sometimes called transonic) is less clear; this case is treated very briefly and representative results are derived in this subsection.

Suppose that a planar crack grows in a body subjected to tractions consistent with the mode I symmetry conditions. The representation of the crack tip fields in terms of the displacement potentials ϕ and ψ as functions of local coordinates ξ_1, ξ_2 and time t is again adopted. Thus, $\phi(\xi_1, \xi_2, t)$ satisfies (4.3.1) and $\psi(\xi_1, \xi_2, t)$ satisfies the same equation with c_d replaced by c_s. Following the development for mode III, the rescaled coordinates $\eta_i = \xi_i/\epsilon$ are introduced, where ϵ is an arbitrary small positive parameter. Then, expansions of the form (4.3.2) are assumed. The main contributions to the potentials are again denoted by $\Phi_0(\eta_1, \eta_2, t)$ and $\Psi_0(\eta_1, \eta_2, t)$, and they satisfy the partial differential equations (4.3.3) and

$$\widehat{\alpha}_s^2 \frac{\partial^2 \Psi_0}{\partial \eta_1^2} - \frac{\partial^2 \Psi_0}{\partial \eta_2^2} = 0, \qquad (4.3.26)$$

respectively, where $\widehat{\alpha}_s = \sqrt{v^2/c_s^2 - 1}$. Equation (4.3.26) is the elementary wave equation in two independent variables. The general solution of (4.3.3) is given in (4.3.4) and the general solution of (4.3.26) which satisfies the principle of causality and which has the symmetries consistent with mode I deformation is

$$\Psi_0 = \begin{cases} G(\eta_1 + \widehat{\alpha}_s \eta_2) & \text{if } \eta_2 \geq 0, \\ -G(\eta_1 - \widehat{\alpha}_s \eta_2) & \text{if } \eta_2 \leq 0, \end{cases} \qquad (4.3.27)$$

where G is an arbitrary function with the property that $G(s) = 0$ if $s > 0$. If the expansions of ϕ and ψ are substituted into the stress boundary conditions on the crack faces, and the coefficients of independent powers of ϵ are set equal to zero, then

$$\left(2 - v^2/c_s^2\right) \left[F_+''(\eta_1) + F_-''(\eta_1)\right] - 4\widehat{\alpha}_s G''(\eta_1) = 0,$$

$$\alpha_d \left[F_+''(\eta_1) - F_-''(\eta_1)\right] + \left(2 - v^2/c_s^2\right) G''(\eta_1) = 0 \qquad (4.3.28)$$

for $-\infty < \eta_1 < 0$. The solution of this pair of equations that corresponds to singular stress but integrable mechanical energy density is

$$F''(\zeta) = \frac{A}{\zeta^q}, \qquad G''(\eta) = \frac{A\left(v^2/c_s^2 - 2\right)\cos(q\pi)}{2\widehat{\alpha}_s |\eta|^q}, \qquad (4.3.29)$$

where A is independent of position in the plane and

$$q = \frac{1}{\pi} \tan^{-1} \left[\frac{4\widehat{\alpha}_s \alpha_d}{(2 - v^2/c_s^2)^2} \right].$$ (4.3.30)

The exponent q depends on crack speed v in the following simple way. It has the value $q = 0$ when $v = c_s$, it increases monotonically to $q = 1/2$ when $v = c_s\sqrt{2}$, and it decreases monotonically to $q = 0$ when $v = c_d$.

In terms of crack tip coordinates ξ_1, ξ_2, a representative pair of stress component and particle velocity component is

$$\sigma_{22} \sim \left(v^2/c_s^2 - 2\right) A \left\{ \frac{\cos(q\theta_d)}{r_d^q} - \frac{H(-\xi_1 - \widehat{\alpha}_s \xi_2)\cos(q\pi)}{|\xi_1 + \widehat{\alpha}_s \xi_2|^q} \right\},$$

$$\dot{u}_2 \sim vA \left\{ \alpha_d \frac{\sin(q\theta_d)}{r_d^q} - \frac{H(-\xi_1 - \widehat{\alpha}_s \xi_2)\left(1 - \widehat{\alpha}_s^2\right)\cos(q\pi)}{2\widehat{\alpha}_s |\xi_1 + \widehat{\alpha}_s \xi_2|^q} \right\},$$

(4.3.31)

where $H(\cdot)$ is the unit step function.

Thus, A is also a stress intensity factor but the singularity in (4.3.31) is weaker, in general, than the inverse square root singularity in the case of subsonic crack growth. The shear wave contribution to the above expressions is fundamentally different from that in the case of $v < c_s$. Because the crack speed is greater than the shear wave speed, no shear wave radiated from the crack edge can propagate ahead of the running crack tip. Instead, shear wave motion can exist only behind plane wavefronts trailing from the advancing crack edge. In the local coordinate system, these plane wavefronts coincide with the lines $\xi_1 + \widehat{\alpha}_s |\xi_2| = 0$ and the step function in (4.3.31) ensures that the shear wave fields are confined to the material points behind the wavefronts. Some particular aspects of transonic crack growth in elastic materials were considered by Burridge (1973), Burridge, Conn, and Freund (1979), Freund (1979a), and Simonov (1983).

4.4 Elastic-ideally plastic material; antiplane shear

Suppose that the plane in Figure 4.8 is subjected to loading that results in antiplane shear deformation. A rectangular coordinate system

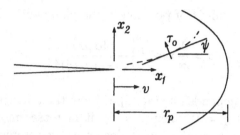

Figure 4.8. Schematic diagram of the near tip region for steady crack growth through an elastic-plastic material at speed v in the x_1-direction. The length r_p is the extent of the active plastic zone in the direction of crack growth. The dashed line is a shear line with local slope ψ.

labeled x_1, x_2, x_3 is introduced so that the only nonzero component of displacement $u_3(x_1, x_2, t)$ is in the x_3-direction. A crack in the plane $x_2 = 0$ grows in the x_1-direction with steady crack tip speed v in the range $0 < v < c_s$. The crack faces are free of traction, and the driving force is assumed to be applied remotely.

The equation ensuring momentum balance is

$$\frac{\partial \sigma_{13}}{\partial x_1} + \frac{\partial \sigma_{23}}{\partial x_2} = \rho \frac{\partial^2 u_3}{\partial t^2}, \tag{4.4.1}$$

where ρ is the material mass density. The nonzero strain components are ϵ_{13} and ϵ_{23}, and these strain components are subject to the compatibility condition

$$2\frac{\partial \epsilon_{13}}{\partial x_2} = \frac{\partial^2 u_3}{\partial x_1 \partial x_2} = 2\frac{\partial \epsilon_{23}}{\partial x_1}. \tag{4.4.2}$$

The total strain is assumed to be the sum of elastic strain and plastic strain,

$$\epsilon_{ij} = \epsilon_{ij}^e + \epsilon_{ij}^p. \tag{4.4.3}$$

The material time rate of change of elastic strain is assumed to be proportional to the material stress rate

$$\dot{\epsilon}_{ij}^e = \frac{\dot{\sigma}_{ij}}{2\mu}, \tag{4.4.4}$$

where μ is the elastic shear modulus. In plastically deforming regions of the plane, the stress state at each point is assumed to satisfy the Mises yield criterion

$$\sigma_{13}^2 + \sigma_{23}^2 = \tau_o^2, \tag{4.4.5}$$

where τ_o is the flow stress in homogeneous plastic shearing. For the case of antiplane shear, the Tresca yield criterion takes on the same simple form. The crack tip field for the case of a yield condition represented by a diamond-shaped yield locus in stress space has been considered by Nikolic and Rice (1988).

The plastic strain rate is assumed to depend on stress at each point through the associated flow rule

$$\dot{\epsilon}^p_{ij} = \lambda s_{ij} , \qquad (4.4.6)$$

where s_{ij} is the deviatoric stress $s_{ij} = \sigma_{ij} - \frac{1}{3}\sigma_{kk}\delta_{ij}$ and λ is a nonnegative factor proportional to the plastic work rate. For the case of antiplane shear, the stress state is deviatoric. Furthermore, if (4.4.6) is written for the two nonzero strain rate components and the unknown parameter λ is eliminated from the resulting equations, then

$$\frac{2\mu\dot{\epsilon}_{13} - \dot{\sigma}_{13}}{\sigma_{13}} = \frac{2\mu\dot{\epsilon}_{23} - \dot{\sigma}_{23}}{\sigma_{23}} \qquad (4.4.7)$$

must hold in a plastically deforming region.

Finally, for steady growth in the x_1-direction at speed v, any field quantity, say $f(x_1, x_2, t)$, depends on x_1 and t only through the combination $x_1 - vt$. Consequently, the material time derivative of f can be replaced by the spatial gradient $-v(\partial f/\partial x_1)$. The spatial gradient form for the time rates will be assumed in the remainder of this section. Without loss of generality, it is assumed that the crack tip passes the point $x_1 = 0$ at time $t = 0$.

The yield condition (4.4.5) is satisfied identically at each point within a plastically deforming region if the nonzero deviatoric stress components are represented in terms of a single function $\psi(x_1, x_2, t)$, commonly called the local shear angle, according to

$$\sigma_{13} = -\tau_o \sin\psi , \qquad \sigma_{23} = \tau_o \cos\psi . \qquad (4.4.8)$$

If σ_{13} and σ_{23} are viewed as the components of the shear traction vector in the x_1, x_2-plane, (4.4.8) implies that this vector has magnitude τ_o and direction at a counterclockwise angle ψ from the x_2-axis. The shear angle field is assumed to have the symmetry $\psi(x_1, -x_2, t) = -\psi(x_1, x_2, t)$. Thus, if ψ is continuous across the crack line $x_2 = 0$, then $\psi = 0$ on the crack line. If (4.4.8) is introduced into both the momentum equation (4.4.1) and the incremental constitutive equation (4.4.7), and ϵ_{23} is eliminated by means of the compatibility equation (4.4.2), then the first order system of partial differential equations governing the remaining dependent variables ψ and $\gamma = 2\mu\epsilon_{13}/\tau_o$ is

$$\frac{\partial \psi}{\partial x_1} + \cos \psi \frac{\partial \gamma}{\partial x_1} + \sin \psi \frac{\partial \gamma}{\partial x_2} = 0 \,,$$

$$\cos \psi \frac{\partial \psi}{\partial x_1} + \sin \psi \frac{\partial \psi}{\partial x_2} + m^2 \frac{\partial \gamma}{\partial x_1} = 0 \,,$$

(4.4.9)

where $m = v/c_s$ is the ratio of the crack tip speed to the elastic shear wave speed. The system of partial differential equations (4.4.9) is quasi-linear and hyperbolic with two families of real characteristic curves in a plastically deforming region. Because the equations are quasi-linear (the nonlinear equations are linear in the derivatives), the characteristic network cannot be established without knowing the solution beforehand. Characteristic curves will be discussed in Section 8.3; they are not required to extract asymptotic fields in this section.

4.4.1 Asymptotic fields for steady dynamic growth

To aid in the construction of an asymptotic solution, a polar coordinate system r, θ centered at the crack tip in the plane is introduced, with $r = \sqrt{(x_1 - vt)^2 + x_2^2}$ and $\theta = \tan^{-1}[x_2/(x_1 - vt)]$. In the moving crack tip coordinate system, fields depend only on the relative position r, θ. Because the stress components σ_{13} and σ_{23} must be bounded and single valued at the crack tip, the shear angle $\psi(r, \theta)$ must have the same properties. Thus, there exists a function of θ, denoted by $p(\theta)$, such that $\psi(r, \theta) \to p(\theta)$ as $r \to 0^+$. The derivatives of ψ in (4.4.9)$_2$ are no more singular in r than r^{-1} as $r \to 0$, so that the derivatives of γ must have the same property. It follows that

$$\gamma(r, \theta) \sim A \ln r + g(\theta)$$

(4.4.10)

for small r, where A is a constant and $g(\theta) = -g(-\theta)$ is a function to be determined. For antiplane deformation with reflective symmetry with respect to the crack plane, $\gamma(r, 0) = 0$ so that $A = 0$.

The pair of partial differential equations (4.4.9) is now reduced to a pair of ordinary differential equations for $p(\theta)$ and $g(\theta)$. This is accomplished by writing the partial derivatives as their polar coordinate equivalents, multiplying the terms in each equation by r, and taking the limit as $r \to 0$. The result is

$$p'(\theta)\sin\theta - g'(\theta)\sin(p - \theta) = 0,$$

$$p'(\theta)\sin(p - \theta) - m^2 g'(\theta)\sin\theta = 0,$$

$$(4.4.11)$$

where the prime denotes differentiation with respect to θ. There are two solutions of the pair of equations. Either

$$p'(\theta) = 0, \qquad g'(\theta) = 0 \tag{4.4.12}$$

separately, or the determinant of coefficients of p' and g' vanishes so that

$$p(\theta) = \theta \pm \sin^{-1}(m\sin\theta). \tag{4.4.13}$$

The former solution leads to an asymptotically uniform field; the latter solution represents a nonuniform field similar to a centered fan. The choice of sign in the latter case is dictated by the requirement of positive plastic dissipation.

An asymptotic expansion for λ in a region where (4.4.13) holds can be obtained as follows. Form the product of s_{jk} with each term in (4.4.6) and contract over the indices i and k, with the result that

$$\lambda \tau_o^2 = -v\frac{\partial \epsilon_{13}}{\partial x_1}\sigma_{13} - v\frac{\partial \epsilon_{23}}{\partial x_1}\sigma_{23}, \tag{4.4.14}$$

where (4.4.3) has been used to express the difference between total strain and plastic strain. Next, the partial derivatives in (4.4.14) are transformed to their polar coordinate equivalents, both sides of (4.4.14) are multiplied by r, and then the equation is examined as r becomes indefinitely small. The result is

$$\lambda r \sim -\frac{v}{2\mu}g'(\theta)\cos(p - \theta). \tag{4.4.15}$$

The factor $g'(\theta)$ can be obtained from (4.4.11) in light of (4.4.13), and $\cos(p-\theta) = \mp\sqrt{1 - m^2\sin^2\theta}$, where the upper (lower) algebraic sign corresponds to the upper (lower) sign in (4.4.13). Thus,

$$\lambda r\frac{2\mu}{v} \sim \mp\frac{1}{m}\sqrt{1 - m^2\sin^2\theta} - \cos\theta \tag{4.4.16}$$

as $r \to 0$. The magnitude of the first term is greater than or equal to the magnitude of the second term for any $m > 0$ and for any value of θ in the range $0 < \theta < \pi$, so that the condition of nonnegative plastic dissipation is ensured if the lower (positive) sign is selected in (4.4.16) and, consequently, the lower (negative) sign must be selected in the definition of $p(\theta)$ in (4.4.13).

The asymptotic stress distribution in a plastically deforming region corresponding to the asymptotic form of the stress function ψ in (4.4.13) is

$$\sigma_{13} \sim -\tau_0 \left(\sqrt{1 - m^2 \sin^2 \theta} - m \cos \theta \right) \sin \theta ,$$

$$\sigma_{23} \sim \tau_0 \left(\cos \theta \sqrt{1 - m^2 \sin^2 \theta} + m \sin^2 \theta \right) . \tag{4.4.17}$$

This stress distribution satisfies the symmetry condition with respect to the plane $\theta = 0$, but it does not satisfy the boundary condition of zero traction on the crack faces at $\theta = \pi$. The stress distribution is uniform for the other type of region (4.4.12). The particular choice

$$\sigma_{13} \sim -\tau_0 , \qquad \sigma_{23} \sim 0 \tag{4.4.18}$$

does satisfy the boundary condition of vanishing traction on the crack face. Furthermore, if (4.4.17) holds for $0 < \theta < \theta^*$ and (4.4.18) holds for $\theta^* < \theta < \pi$, then the two stress distributions satisfy continuity of traction across the radial line $\theta = \theta^*$ provided that $p(\theta^*) = \pi/2$ or

$$\theta^* = \tan^{-1} (-1/m) . \tag{4.4.19}$$

(The question of continuity of traction in the asymptotic field will be examined in some detail in Section 4.5.) Thus, the stress distribution described satisfies all conditions imposed, and it therefore represents a possible asymptotic stress distribution. This is the distribution determined originally by Slepyan (1976).

The corresponding asymptotic deformation field can be established in a similar manner. From (4.4.12) and (4.4.13), the distribution of the strain component ϵ_{13} is given by

$$\epsilon_{13} \sim -\frac{\tau_0}{2\mu m} \begin{cases} \theta - \sin^{-1}(m \sin \theta) & \text{if } 0 < \theta < \theta^* , \\ \pi/2 & \text{if } \theta^* < \theta < \pi . \end{cases} \tag{4.4.20}$$

This distribution satisfies the symmetry condition on $\theta = 0$, and it implies continuity of particle velocity across $\theta = \theta^*$. Furthermore, because $2\epsilon_{13} = \partial u_3 / \partial x_1$, it implies that the crack tip opening displacement is zero, that is,

$$\lim_{x_1 \to vt^-} u_3(x_1, 0, t) = 0 \qquad (4.4.21)$$

and the crack tip opening angle is

$$\lim_{x_1 \to vt^-} \frac{\partial u_3}{\partial x_1}(x_1, 0, t) = -\frac{\pi \tau_0}{2m\mu}. \qquad (4.4.22)$$

A particularly significant feature of the deformation field is the distribution of shear strain ϵ_{23} ahead of the advancing crack tip. For present purposes, it suffices to extract the variation of ϵ_{23} with $r = x_1 - vt$ on $\theta = 0$. From (4.4.20) and the compatibility equation (4.4.2), it is evident that

$$\frac{\partial \epsilon_{13}}{\partial x_2} = \frac{\partial \epsilon_{23}}{\partial x_1} \sim \frac{\tau_0}{2\mu} \frac{\cos \theta}{r} \left[-\frac{1}{m} + \frac{\cos \theta}{\sqrt{1 - m^2 \sin^2 \theta}} \right] \qquad (4.4.23)$$

for $0 < \theta < \theta^*$. Therefore, on $\theta = 0$,

$$\frac{\partial \epsilon_{23}}{\partial x_1} \sim -\frac{\tau_0(1 - m)}{2\mu m (x_1 - vt)}, \qquad x_1 > vt, \qquad (4.4.24)$$

so that

$$\epsilon_{23} \sim \frac{\tau_0}{2\mu} \frac{1 - m}{m} \ln \frac{1}{(x_1 - vt)} = \frac{\tau_0}{2\mu} \frac{1 - m}{m} \ln \frac{1}{r} \qquad (4.4.25)$$

as $r \to 0$ on $\theta = 0$ for any speed $m > 0$. Thus, the shear strain on the prospective fracture plane is logarithmically singular. The equation (4.4.23) can be integrated to determine the asymptotic form of ϵ_{23} as a function of θ; the result is given by Slepyan (1976) who showed that the strain is logarithmically singular throughout the sector of active plastic flow. This point will be dealt with in greater detail when more complete solutions of elastic-plastic crack growth problems are considered.

4.4.2 Comparison with equilibrium results

For purposes of comparison, some features of the crack tip solutions for a stationary crack or a steadily growing crack in the same elastic-plastic material but under equilibrium conditions are included here. The governing equations once again are (4.4.1) through (4.4.6). The main point for comparison is the strength of the singularity in plastic strain ϵ_{23}^p ahead of the crack tip because of its significance for crack advance studies. Other features of the fields are also noted.

Consider first a stationary crack in a body subjected to monotonically increasing loading resulting in antiplane shear deformation with the symmetry $u_3(r, -\theta) = -u_3(r, \theta)$ in terms of crack tip polar coordinates. This situation was analyzed by Hult and McClintock (1956) and Rice (1968). In the absence of inertial effects, the equilibrium equation and the yield condition lead to the result that the shear lines are straight. Furthermore, in anticipation of a strain singularity ahead of the crack tip within the plastic zone, the shear lines are taken to be radial. Actually, choices are severely restricted by the conditions that no two shear lines may intersect (some stress component would exceed τ_0 in magnitude at the intersection point) and shear lines may intersect the traction-free crack faces only at right angles. If the shear lines are radial, then the stress distribution in the active plastic zone is $\sigma_{13} = -\tau_0 \sin\theta$ and $\sigma_{23} = \tau_0 \cos\theta$. From (4.4.8) it is evident that

$$\psi = \theta, \qquad 0 < \theta < \pi/2, \qquad (4.4.26)$$

over the plastic sector ahead of the tip, and the material remains elastic for $\pi/2 < \theta < \pi$. Along any radial shear line, $\epsilon_{r3} = 0$ because the stress vector in the x_1, x_2-plane is always transverse to the radial line. Thus, particle displacement is uniform, that is,

$$u_3 = u_3(\theta), \qquad 0 < \theta < \pi/2. \qquad (4.4.27)$$

The shear strain within the plastic zone is

$$\epsilon_{13} = -\frac{\tau_0}{2\mu} \frac{R(\theta)\sin\theta}{r}, \qquad \epsilon_{23} = \frac{\tau_0}{2\mu} \frac{R(\theta)\cos\theta}{r}, \qquad (4.4.28)$$

where $R(\theta)$ is the distance from the crack tip to the elastic-plastic boundary in the direction from the tip defined by the angle θ. Thus, the singularity in plastic strain ϵ_{23}^p ahead of the crack tip is proportional to r^{-1}, which is much stronger than that for the steadily growing dynamic crack in the same material.

To complete the summary of results for the stationary crack problem, it is noted that the crack tip opening displacement is *not* zero, as it is in the case of a growing crack. Instead, it has the value

$$\lim_{x_1 \to 0^-} u_3(x_1, 0^+) = \frac{\tau_0}{\mu} \int_0^{\pi/2} R(\theta)\, d\theta. \qquad (4.4.29)$$

The crack tip opening angle or slope in this case has the value

$$\lim_{x_1 \to 0^-} \frac{\partial u_3}{\partial x_1}(x_1, 0^+) = -\frac{\tau_0}{2\mu}. \qquad (4.4.30)$$

For a crack growing steadily under equilibrium conditions in the same material, (4.4.9) can be applied with $m = 0$. Then $(4.4.9)_2$ implies that the gradient of ψ in the direction defined by the angle ψ is zero, or that the shear lines are straight. For the reasons outlined in the foregoing paragraphs, therefore, the shear lines are taken to be radial lines focused at the crack tip in the region ahead of the advancing crack, so that $\psi = \theta$. Then $(4.4.9)_1$ implies that the gradient of γ in the radial direction is equal to $-\partial\psi/\partial x_1$ or

$$\frac{\partial \gamma}{\partial r} = \frac{\sin \theta}{r}. \qquad (4.4.31)$$

The result of integrating with respect to r and imposing the condition that $\epsilon_{13} = -(\tau_0/2\mu)\sin\theta$ on the elastic-plastic boundary at $r = R(\theta)$ is

$$\gamma(r,\theta) = -\sin\theta \left[1 + \ln \frac{R(\theta)}{r} \right]. \qquad (4.4.32)$$

Then, upon differentiation with respect to x_2 and incorporation of the compatibility equation (4.4.2),

$$\frac{\partial \epsilon_{23}}{\partial x_1} = \frac{\tau_0}{2\mu r} \left\{ 1 - \cos^2\theta \left[2 + \ln \frac{R(\theta)}{r} \right] - \sin\theta\cos\theta \frac{R'(\theta)}{R(\theta)} \right\}. \qquad (4.4.33)$$

Consider the strain distribution along $\theta = 0$ directly ahead of the crack tip. Along this line, (4.4.33) can be integrated subject to the condition that $\epsilon_{23} = \tau_o/2\mu$ when $r = R(0) = R_0$, with the result that

$$\epsilon_{23}(r,0) = \frac{\tau_o}{2\mu} \left\{ 1 + \ln\left(\frac{R_0}{r}\right) + \frac{1}{2}\ln^2\left(\frac{R_0}{r}\right) \right\}. \qquad (4.4.34)$$

This result, which was presented by Rice (1968), provides not only the asymptotic strain distribution, but also the *full* distribution on the crack line within the active plastic zone.

An immediate observation is that the crack tip strain singularity for the case of a steadily growing crack is much weaker than that for a stationary crack subjected to monotonic loading in the same material. In the latter case, the stress state at a material particle near the crack tip deep within the plastic zone is fixed and plastic strain accumulates without a change in the direction of plastic straining. In the former case, on the other hand, the crack is advancing into plastically deformed material. The stress state, and consequently the direction of plastic straining, rotates continuously at a material point as the crack tip moves by, and a lesser strain concentration is established.

Comparison of the crack tip strain distributions along $\theta = 0$ for steady growth under equilibrium and dynamic conditions reveals a paradox at this stage of development of elastic-plastic crack growth. For steady quasi-static growth, the strain ϵ_{23} near the crack tip can be extracted from (4.4.34), whereas for steady dynamic growth the asymptotic result

$$\epsilon_{23} \sim \frac{\tau_o}{2\mu}\frac{1-m}{m}\ln\frac{1}{r} \qquad (4.4.35)$$

was obtained in (4.4.25). It would be natural to expect that the form of the former result for small r with $m = 0$ would be the same as the form of the latter result for $m \to 0$ with small r, but this is not the case. The asymptotic solutions do not provide a basis for pursuing the matter, and the paradox can be resolved only through a more complete dynamic solution. The analysis provided in Section 8.3 reveals that the range of validity of this asymptotic solution vanishes as $m \to 0$.

4.5 Elastic-ideally plastic material; plane strain

Consider a mode I tensile crack growing steadily at speed v in an elastic-ideally plastic material under plane strain conditions. The material is assumed to be fully incompressible. Both a rectangular coordinate system x_1, x_2 and a polar coordinate system r, θ are introduced with a common origin at the tip of the crack. The coordinate systems translate with the crack tip which moves steadily in the x_1-direction at speed v. The equations of momentum balance are

$$\frac{\partial \sigma_{ij}}{\partial x_j} = \rho \frac{\partial \dot{u}_i}{\partial t}, \qquad (4.5.1)$$

where σ_{ij} are the rectangular components of the Cauchy stress tensor, $\dot{u}_i$ are the rectangular components of the particle velocity, and ρ is the mass density of the material. As in the previous section, it is convenient to introduce the dimensionless crack tip speed $m = v/c_s$, where $c_s = \sqrt{\mu/\rho}$ and μ is the elastic shear modulus of the material. Furthermore, steady state conditions imply that the material time derivative of any field quantity f can be replaced by the scaled spatial derivative $\dot{f} = -v \partial f / \partial x_1$.

In terms of the particle velocity, the rectangular components of the small strain rate tensor are

$$\dot{\epsilon}_{ij} = \frac{1}{2} \left(\frac{\partial \dot{u}_i}{\partial x_j} + \frac{\partial \dot{u}_j}{\partial x_i} \right), \qquad (4.5.2)$$

and additive decomposition of strain into elastic and plastic parts is assumed as in (4.4.3). The rectangular components of the elastic strain rate, plastic strain rate, and deviatoric stress rate tensors are $\dot{\epsilon}_{ij}^e$, $\dot{\epsilon}_{ij}^p$, and $\dot{s}_{ij}$, respectively. The elastic strain components are related to the deviatoric stress components $s_{ij} = \sigma_{ij} - \frac{1}{3}\sigma_{kk}\delta_{ij}$ through Hooke's law,

$$\epsilon_{ij}^e = s_{ij}/2\mu. \qquad (4.5.3)$$

The plastic strain rate is assumed to be proportional to the deviatoric stress,

$$\dot{\epsilon}_{ij}^p = \lambda s_{ij}, \qquad (4.5.4)$$

where the proportionately factor λ is a nonnegative function of position to be determined as part of the solution. Combining (4.5.2) through (4.5.4) yields

$$\frac{1}{2} \left(\frac{\partial \dot{u}_i}{\partial x_j} + \frac{\partial \dot{u}_j}{\partial x_i} \right) = \lambda s_{ij} + \frac{\dot{s}_{ij}}{2\mu}, \qquad (4.5.5)$$

which describes the material response in plastically deforming regions. The yield condition in plane strain is assumed to be

$$\tfrac{1}{4}(\sigma_{11} - \sigma_{22})^2 + \sigma_{12}^2 = \tau_{\mathrm{o}}^2 , \tag{4.5.6}$$

where τ_{o} is the constant flow stress in shear. This condition states that in plastically deforming regions the maximum shear stress in the plane of deformation has the value τ_{o}. Such a yield condition results for an isotropic material exhibiting plastic incompressibility when the elastic deformation is also incompressible. The Tresca and Mises yield conditions are particular cases under these assumptions.

The deformation field possesses the symmetry

$$\dot{u}_1(r,\theta) = \dot{u}_1(r,-\theta) \quad , \qquad \dot{u}_2(r,\theta) = -\dot{u}_2(r,-\theta) \tag{4.5.7}$$

corresponding to mode I crack opening. The crack faces are traction-free, which is ensured if

$$\sigma_{22}(r,\pm\pi) = 0 , \qquad \sigma_{12}(r,\pm\pi) = 0 \tag{4.5.8}$$

for all $r > 0$.

It is assumed that that the out-of-plane stress σ_{33} is the intermediate principal stress, an assumption that can be verified subsequently from the solution. Then there is no plastic strain in the out-of-plane direction, that is, $\dot{\epsilon}_{33}^p = 0$. (It is unlikely that this feature can be carried over to the case of elastic compressibility.) In light of (4.5.4), this implies that

$$s_{11} = -s_{22} , \tag{4.5.9}$$

which allows the yield condition (4.5.6) to be expressed as

$$s_{22}^2 + s_{12}^2 = \tau_{\mathrm{o}}^2 . \tag{4.5.10}$$

This condition, with (4.5.9), implies that the deviatoric stress components are bounded for all values of $r \geq 0$. If, in addition, it is assumed that the hydrostatic stress $\sigma = \tfrac{1}{3}\sigma_{kk}$ is bounded, then the momentum equations can be used to discern the form of $\dot{u}_i$. A discussion of the conditions under which σ can be shown to be bounded has been presented by Leighton, Champion and Freund (1987).

4.5.1 Asymptotic field in plastically deforming regions

A momentum equation for steady crack growth is

$$\frac{\partial \sigma_{11}}{\partial x_1} + \frac{\partial \sigma_{12}}{\partial x_2} = -v\rho \frac{\partial \dot{u}_1}{\partial x_1}. \tag{4.5.11}$$

If the dependent variables are treated as functions of r and θ, derivatives with respect to the rectangular coordinates can be replaced by derivatives with respect to the polar coordinates by means of a simple change of variables, for example,

$$\frac{\partial f}{\partial x_1} = \cos\theta \frac{\partial f}{\partial r} - \frac{\sin\theta}{r} \frac{\partial f}{\partial \theta}. \tag{4.5.12}$$

If the stress components are bounded as $r \to 0$, then in view of (4.5.12), the left side of (4.5.11) is $O(1/r)$. The right side of (4.5.11) must therefore be $O(1/r)$. In this way, (4.5.11) can be used to determine the form of $\dot{u}_i$ that gives rise to the strongest allowable singularity. If the particle velocity is logarithmically singular as $r \to 0$, then the first term in (4.5.12) is $O(r^{-1})$. In order to avoid a stronger singularity arising from the second term in (4.5.12), the coefficient of the $\ln r$ term must be independent of θ. Alternately, if the particle velocity is a function of θ as $r \to 0$, then the second term of (4.5.11) is $O(r^{-1})$ whereas the first term is $o(r^{-1})$. In summary, it appears that the two possibilities leading to the strongest allowable singularity in particle acceleration are

$$\dot{u}_i \sim vA_i \ln\left(\frac{R}{r}\right), \quad \frac{dA_i}{d\theta} = 0 \quad \text{or} \quad \dot{u}_i \sim vB_i(\theta), \tag{4.5.13}$$

where R is a parameter with the dimensions of length which is assumed to be the same order of magnitude as the maximum extent of the active plastic zone. The plane strain condition implies that $B_3(\theta) = 0$ and $A_3 = 0$.

It is assumed that the behavior of $\dot{u}_i$ can vary from one angular sector to another around the crack tip. Symmetry requires that $A_2 = 0$ in the upstreammost sector, and therefore the condition that the velocity components must be continuous (to be obtained subsequently) implies that $\dot{u}_2$ cannot be logarithmically singular in any angular sector about the crack tip. Thus $(4.5.13)_2$ is the asymptotic representation of $\dot{u}_2$ for $0 \leq \theta \leq \pi$. From here on, A_1 is replaced by A.

The incompressibility condition implies that the particle velocity is divergence free, which in plane strain is expressed by

$$\frac{\partial \dot{u}_1}{\partial x_1} + \frac{\partial \dot{u}_2}{\partial x_2} = 0. \tag{4.5.14}$$

This implies that the velocity components are derivable from a stream function $\phi(r, \theta)$ according to

$$\dot{u}_1 = v\frac{\partial \phi}{\partial x_2}, \quad \dot{u}_2 = -v\frac{\partial \phi}{\partial x_1}. \tag{4.5.15}$$

If the velocity components are taken to be a combination of the forms indicated in (4.5.13),

$$\dot{u}_1 = vA\ln\left(\frac{R}{r}\right) + vB_1(\theta), \quad \dot{u}_2 = vB_2(\theta), \tag{4.5.16}$$

then the divergence-free condition (4.5.14) requires that A, $B_1(\theta)$, and $B_2(\theta)$ satisfy the relation

$$B_1'(\theta)\tan\theta = B_2'(\theta) - A, \tag{4.5.17}$$

where the prime denotes differentiation with respect to θ. This relation allows integration of (4.5.15) to obtain

$$\phi(r, \theta) = Ax_2\left\{\ln\left(\frac{R}{r}\right) + 1\right\} + x_2 B_1(\theta) - x_1 B_2(\theta) \tag{4.5.18}$$

to within an arbitrary constant.

It should be pointed out that the asymptotic form of $\dot{u}_1$ is actually

$$\dot{u}_1 \sim vA\ln\left(\frac{R}{r}\right) + F(r) + vB_1(\theta) + o(1) \quad \text{as } r \to 0, \tag{4.5.19}$$

where $F(r) = o(\ln r)$ does not contribute to the leading order terms in any of the governing equations, and therefore cannot be determined without a higher order analysis. The function $F(r)$ has been neglected here. If such a function were part of the complete asymptotic form of $\dot{u}_1$ it would not affect the results presented here for stress, strain rate, or $\dot{u}_2$. The results for the strain and $\dot{u}_1$ would require important modification, however. That there is no similar function in the asymptotic form of $\dot{u}_2$ follows from the foregoing argument that led to the exclusion of a logarithmic term from $\dot{u}_2$.

The material strain rates are obtained from (4.5.2) and (4.5.16) as

$$\dot{\epsilon}_{11} = -\dot{\epsilon}_{22} = -\frac{v}{2x_2} B_2'(\theta) \sin 2\theta \,,$$

$$\dot{\epsilon}_{12} = \frac{v}{2x_2} [B_2'(\theta) \cos 2\theta - A] \,.$$

$$(4.5.20)$$

The strain rate components are therefore $O(r^{-1})$ as $r \to 0$.

Throughout any region of the body in which the material is at a state of yielding, the yield condition (4.5.10) can be satisfied identically if the two deviatoric stress components appearing in (4.5.10) are derived from a stress function $\psi(\theta)$ according to

$$s_{22} = \tau_0 \cos [2\theta - \psi(\theta)] \,, \qquad s_{12} = -\tau_0 \sin [2\theta - \psi(\theta)] \,. \qquad (4.5.21)$$

Because the hydrostatic stress σ is taken to be bounded, it is expressed as a function of θ where

$$\sigma(\theta) = \tfrac{1}{2}(\sigma_{11} + \sigma_{22}) \,. \qquad (4.5.22)$$

In terms of the functions $B_2(\theta)$, $\psi(\theta)$, and $\sigma(\theta)$, the symmetry conditions (4.5.7) and the boundary conditions (4.5.8) are

$$\psi(0) = 0 \,, \qquad B_2(0) = 0 \,,$$

$$\psi(\pi) = \pi \,, \qquad \sigma(\pi) = \tau_0 \,,$$

$$(4.5.23)$$

respectively, for the range $0 \le \theta \le \pi$. The system of ordinary differential equations that these functions must satisfy is obtained next.

In addition to the incompressibility condition (4.5.14), the governing equations that remain to be satisfied are the two nontrivial momentum balance equations (4.5.1) and the two rate equations

(4.5.5) describing the material model. If the representations (4.5.16), (4.5.21), and (4.5.22) are substituted into these four equations then a straightforward manipulation of the results leads to

$$[\psi'(\theta) - 2]\left[\cos^2\psi(\theta) - m^2\sin^2\theta\right] = \frac{m^2}{\tau_0}A\cos(\psi(\theta) - 2\theta),$$

$$\frac{2\tau_0 r}{v}\lambda(\theta) = \tau_0\left[\psi'(\theta) - 2\right]\sin\theta\tan\psi(\theta) + 2A\cos\theta\sec\psi(\theta),$$

$$\qquad\qquad\qquad\qquad\qquad\qquad\qquad\qquad (4.5.24)$$

$$B_2'(\theta) = \frac{\tau_0}{m^2}\left[\psi'(\theta) - 2\right]\cos\psi(\theta),$$

$$B_1'(\theta)\tan\theta = B_2'(\theta) - A, \quad \sigma'(\theta) = \tau_0\left[\psi'(\theta) - 2\right]\sin\psi(\theta).$$

These differential equations now are considered for the case (4.5.13)$_2$, that is, for $A = 0$. The case of $A \neq 0$, which was considered by Gao and Nemat-Nasser (1983a), will be examined subsequently.

4.5.2 A complete solution

A solution to the system of equations (4.5.24) is sought for the case when $A = 0$. This condition is equivalent to the assumption that the particle velocity components have the asymptotic form (4.5.13)$_2$. Attention is first focused on the function $\psi(\theta)$ which, according to (4.5.24), must satisfy

$$[\psi'(\theta) - 2]\left[\cos^2\psi(\theta) - m^2\sin^2\theta\right] = 0 \qquad (4.5.25)$$

in the interval $0 \leq \theta \leq \pi$, varying from $\psi(0) = 0$ to $\psi(\pi) = \pi$. Clearly (4.5.25) is satisfied if either $\psi'(\theta) = 2$ or $\cos\psi(\theta) = \pm m\sin\theta$. The curves defined by $\cos\psi(\theta) = \pm m\sin\theta$ are shown in Figure 4.9. Any straight line in this plane with a slope of two is an integral curve of $\psi'(\theta) = 2$.

A complete solution is constructed in the following way. The curve $\psi(\theta) = 2\theta$ satisfies the boundary condition at $\theta = 0$, and the curve $\psi(\theta) = 2\theta - \pi$ satisfies the boundary condition at $\theta = \pi$. The two curves can be joined by $\cos\psi(\theta) = \pm m\sin\theta$ with either sign, and the choice is governed by the requirement that the plastic flow factor $\lambda(\theta)$ given by (4.5.24)$_2$ must be nonnegative. The factor $\psi'(\theta) - 2$ is negative along both possible joining curves. Furthermore,

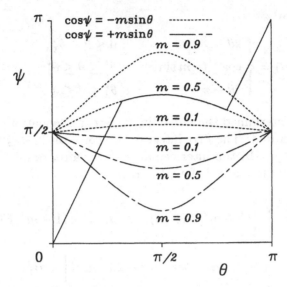

Figure 4.9. Solution curves $\psi(\theta)$ that satisfy the condition (4.5.25), including curves $\cos\psi(\theta) = \pm m\sin\theta$ and lines with slope of two.

$\tan\psi(\theta)\sin\theta = \pm m^{-1}\sin\psi(\theta)$, where $m^{-1}\sin\psi$ is nonnegative along either curve. Therefore, the curve $\cos\psi(\theta) = -m\sin\theta$ must be chosen to render λ nonnegative. For the case $m = 0.5$, the complete solution for $\psi(\theta)$ is shown as the solid line in Figure 4.9, and it is the only continuous solution satisfying the boundary conditions. The interpretation of this solution is straightforward. The stress and deformation fields are spatially uniform within the angular sector $0 \leq \theta \leq \theta_1^*$, the fields are nonuniform within the sector $\theta_1^* \leq \theta \leq \theta_2^*$, and again the fields are uniform within the sector $\theta_2^* \leq \theta \leq \pi$. That the upstream and downstream regions are uniform follows directly from $\psi'(\theta)-2 = 0$ and (4.5.21). The transition angles bounding the region of nonuniform deformation are determined to be

$$\theta_1^* = \sin^{-1}\left[\frac{m + \sqrt{8 + m^2}}{4}\right],$$

$$\theta_2^* = \pi - \sin^{-1}\left[\frac{-m + \sqrt{8 + m^2}}{4}\right]$$

(4.5.26)

and the deviatoric stress components are determined by means of

(4.5.21) from

$$\psi(\theta) = \begin{cases} 2\theta & \text{if } 0 \le \theta \le \theta_1^*, \\ \cos^{-1}(-m\sin\theta) & \text{if } \theta_1^* \le \theta \le \theta_2^*, \\ 2\theta - \pi & \text{if } \theta_2^* \le \theta \le \pi. \end{cases} \qquad (4.5.27)$$

With $\psi(\theta)$ determined the remaining differential equations in (4.5.24) can be integrated. The first integrals are constant in $0 \le \theta \le \theta_1^*$ and in $\theta_2^* \le \theta \le \pi$. In the intermediate region, after application of the boundary conditions, it is found that

$$B_1(\theta) = \frac{k}{m^2} \left[(1 - m^2)F(\theta; m) - E(\theta; m) - (1 - m^2)F(\theta_1^*; m) \right.$$

$$\left. + E(\theta_1^*; m) + 2m\sin\theta - 2m\sin\theta_1^* \right] + B_1^o,$$

$$B_2(\theta) = \frac{k}{m^2} \left[\chi(\theta) - \chi(\theta_1^*) - 2m\cos\theta + 2m\cos\theta_1^* \right],$$

$$\sigma(\theta) = \tau_o \left[1 + m\sin\theta - 2E(\theta; m) - m\sin\theta_2^* + 2E(\theta_2^*; m) \right],$$

$$(4.5.28)$$

where $k = \tau_o/\mu$, $\chi(\theta) = \sqrt{1 - m^2 \sin\theta}$, and $F(\theta; m)$ and $E(\theta; m)$ are elliptic integrals of the first and second kind, respectively, both with parameter m. B_1^o is a constant whose value cannot be determined from the asymptotic analysis.

With the functions $B_i(\theta)$ determined by (4.5.28), the expressions for strain rate components in terms of $B_i(\theta)$ in (4.5.20) can be integrated along a line $x_2 = $ constant from $x_1 = x_2 \cot\theta_1^*$ through decreasing values of x_1 to determine the strain distribution in the nonuniform region $\theta_1^* \le \theta \le \theta_2^*$. The results of this integration, after application of the boundary conditions, are

$$\epsilon_{11}(\theta) = -\epsilon_{22}(\theta) = -B_1(\theta),$$

$$\epsilon_{12}(\theta) = \frac{k}{m^2} \left[\frac{m^2}{2} \ln \left(\frac{[1 - \chi(\theta)]\sin\theta_1^*}{[1 - \chi(\theta_1^*)]\sin\theta} \right) + \chi(\theta) \right.$$

$$\left. - 2m\cos\theta - \chi(\theta_1^*) + 2m\cos\theta_1^* - m\ln\left(\frac{\tan\frac{1}{2}\theta}{\tan\frac{1}{2}\theta_1^*} \right) \right]. \qquad (4.5.29)$$

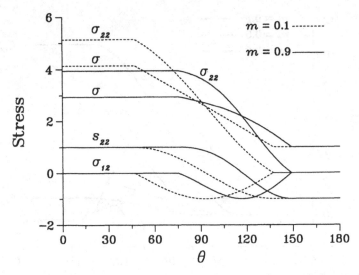

Figure 4.10. Variation of stress components with angle θ around the crack edge for $m = 0.1$ and $m = 0.9$.

The strain components are uniform in the angular sectors $0 \leq \theta \leq \theta_1^*$ and $\theta_2^* \leq \theta \leq \pi$.

The angular variations of the stress components for dimensionless crack speeds $m = 0.1$ and 0.9 are shown in Figure 4.10. The influence of inertia, aside from increasing the transition angles θ_1^* and θ_2^* somewhat, is merely to cause a moderate reduction of the hydrostatic stress σ in the upstream region. Of course, this implies that both σ_{11} and σ_{22} are similarly reduced. Figure 4.11 shows the variation of the strain components for $m = 0.1$, 0.5, and 0.9. Inertial effects are more significant here than for stress; the downstream value of ϵ_{12} is reduced almost to the upstream level, and the variations of both ϵ_{12} and ϵ_{22} are reduced in the transition region. This results in very little strain variation for large m. Note that the unknown constant B_1^o has been chosen to be unity for displaying ϵ_{22}. B_1^o can only be determined by matching the asymptotic solution to the outer solution. However, it is possible that $B_1^o \to 0$ as $m \to 1$.

The solution presented here includes no elastic unloading in any angular sector about the crack tip, and therefore it cannot reduce to the corresponding quasi-static solution of Rice, Drugan, and Sham (1980). A similar situation arises in mode III, where the dynamic steady state solution fails to reduce to the corresponding quasi-static

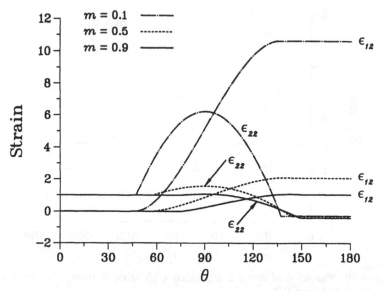

Figure 4.11. Variation of strain components with angle θ around the crack edge for $m = 0.1$, 0.5, and 0.9.

solution. However, this issue can be resolved only by a complete solution of the problem. For the case of antiplane shear crack growth, a complete crack line solution is discussed in Chapter 8. In this case, the domain of validity of the dynamic asymptotic solution vanishes with vanishing crack tip speed. Carrying this observation over to the mode I problem, it appears that here, too, the domain of validity of the dynamic asymptotic field probably vanishes with vanishing crack tip speed. The stress components reduce to those observed in the Prandtl punch field as $m \to 0$, as noted by Achenbach and Dunayevsky (1981). The plastic strain components are nonsingular as $r \to 0$ for a fixed, positive value of m, whereas the corresponding plastic strain components in the quasi-static analysis by Rice et al. (1980) are logarithmically singular at the crack tip. The components of plastic strain are unbounded as $m \to 0$, although ϵ_{12} remains zero in the upstream sector. The region where ϵ_{22} is singular as $m \to 0$ depends on the behavior of B_1^o as $m \to 0$.

4.5.3 Other possible solutions

Next, the possibility of constructing an asymptotic solution with logarithmic dependence of $\dot{u}_1$ on radial coordinate r is considered, that is, the constant A in (4.5.16) is taken to be positive. With reference to (4.5.24), the stress function $\psi(\theta)$ must satisfy the nonlinear ordinary differential equation

$$[\psi'(\theta) - 2] \left[\cos^2 \psi(\theta) - m^2 \sin^2 \theta\right] = \frac{m^2}{k} A \cos(\psi(\theta) - 2\theta) , \quad (4.5.30)$$

with boundary conditions

$$\psi(0) = 0 \quad , \quad \psi(\pi) = \pi . \qquad (4.5.31)$$

In this case, the plastic flow factor is

$$\frac{2\tau_o r}{v} \lambda(\theta) = A \left[\frac{2 \cos\theta \cos\psi(\theta) + m^2 \sin\theta \sin(\psi(\theta) - 2\theta)}{\cos^2 \psi(\theta) - m^2 \sin^2 \theta} \right] . \qquad (4.5.32)$$

Recall that the conditions (4.5.30)–(4.5.32) correspond to a stress state satisfying the yield condition over the full angular range around the crack tip. A continuous solution for $\psi(\theta)$ is inadmissible on the grounds that it would necessarily pass through a range of θ for which the plastic work rate is negative. This fact is demonstrated next.

From (4.5.32) it is noted that the sign of the flow parameter $\lambda(\theta)$ depends on the relative signs of the functions

$$N(\theta) = 2 \cos\theta \cos\psi + m^2 \sin\theta \sin(\psi - 2\theta) ,$$
$$D(\theta) = \cos^2 \psi - m^2 \sin^2 \theta . \qquad (4.5.33)$$

The curves $N(\theta) = 0$, denoted by (I,I'), and the curves $D(\theta) = 0$, denoted by (II,II'), are shown in Figure 4.12; the algebraic sign of the plastic flow parameter is indicated in each region.

Consider (4.5.30) and the first boundary condition (4.5.31). The solution curve $\psi(\theta)$ is well defined up to its point of intersection c with the curve $\cos\psi = m\sin\theta$ (curve II in Figure 4.13), at which point $\psi' \to \infty$. Equation (4.5.30) can be used to show that

$$\psi'(\theta) > 2 , \qquad \psi''(\theta) > 0 \qquad (4.5.34)$$

for $0 \le \theta < \theta_c$, where θ_c is the value of θ at the point c. This indicates that the solution of (4.5.30) up to the point c is of the form

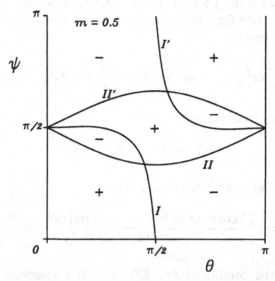

Figure 4.12. Diagram of the regions in which the plastic flow factor λ is positive (+) or negative (−) for $m = 0.5$ and $A \neq 0$.

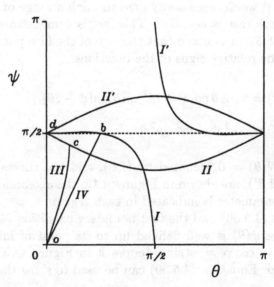

Figure 4.13. Curves that separate regions where the plastic flow factor λ is either positive or negative (see Figure 4.12), and a possible solution curve III. The curve III was obtained by numerical integration of the governing equation for $\psi(\theta)$ for $m = 0.5$ and $A = 1$. Curve IV is the line $\psi = 2\theta$.

shown by curve III in Figure 4.13, where the line $\psi = 2\theta$ (curve IV in Figure 4.13) is shown for comparison. In particular, (4.5.34) shows that the solution curve must lie above the line $\psi = 2\theta$, at least until point c, where it encounters the curve $\cos \psi = m \sin \theta$ (curve II' in Figure 4.13) with $\psi'(\theta) \to \infty$. The point c will always have the property that $\psi(\theta_c) < \pi/2$ and, therefore, the segment of the solution curve III must always lie inside the triangular region Odb defined by the lines $\theta = 0$, $\psi = \pi/2$, and $\psi = 2\theta$. Point b in Figure 4.13 is the intersection of line IV with $\psi = \pi/2$. The only way for the solution to be extended continuously as a plastic state beyond the point c is through a region of negative plastic work rate, which is inadmissible. A possible remedy for this difficulty exists in the form of a discontinuity in $\psi(\theta)$ initiating at the point c. However, this approach is shown later in this section to be inconsistent with the principle of maximum plastic work, so that discontinuities in field variables are inadmissible within the context of the present mechanical problem.

Seemingly, the only way to construct a *continuous* asymptotic solution for which the velocity components depend logarithmically on r and which satisfies the maximum plastic work principle would involve the insertion of an elastically deforming sector starting at some point on the curve III. This possibility is also considered below, where it is shown that an elastic sector cannot initiate at any point inside the region Odb, and therefore on curve III. The consequence is that a solution with logarithmic behavior in the velocity components is not admissible, and the velocity components must therefore be bounded.

4.5.4 Discontinuities

Possible discontinuities in field variables across radial lines from the crack tip are considered next. For quasi-statically moving surfaces of strong discontinuity in elastic-plastic solids, Drugan and Rice (1984) showed that all stress components are continuous, although certain velocity components in the plane of the discontinuity may suffer jumps. The following analysis leads to the conclusion that for the dynamic problem considered in this section all field quantities must be continuous.

Consider a hypothetical line discontinuity Σ propagating with the crack and inclined at an angle θ_Σ to the crack line, as shown in Figure 4.14. Define local rectangular coordinates x, y such that the y-direction is along the line of discontinuity. The jump in a quantity

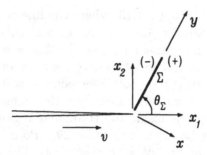

Figure 4.14. A radial line of discontinuity Σ with local rectangular coordinates x, y. The upstream and downstream regions are identified by $(+)$ and $(-)$, respectively.

f is denoted by $[\![f]\!] = f^+ - f^-$, where the superscripts plus and minus denote limiting values of f on the leading and trailing sides of the discontinuity, respectively. Application of momentum balance across a moving surface of discontinuity gives rise to the following jump conditions (Truesdell and Toupin 1960):

$$[\![\sigma_{ij}]\!]n_j + \rho c_\Sigma [\![\dot{u}_i]\!] = 0 \ , \qquad (4.5.35)$$

where n_j is the normal to the surface of discontinuity, and c_Σ is the speed of the surface of discontinuity in the direction of the normal. From Figure 4.14, $c_\Sigma = v \sin \theta_\Sigma$ and (4.5.35) can be rewritten as

$$[\![\sigma_{xx}]\!] + \rho v \sin \theta_\Sigma [\![\dot{u}_x]\!] = 0 \ , \qquad [\![\sigma_{xy}]\!] + \rho v \sin \theta_\Sigma [\![\dot{u}_y]\!] = 0 \ . \quad (4.5.36)$$

In terms of the particle displacement components u_x and u_y, the velocity components are

$$\dot{u}_x = v \left(\sin \theta_\Sigma \frac{\partial u_y}{\partial y} - \cos \theta_\Sigma \frac{\partial u_x}{\partial y} \right) ,$$

$$\dot{u}_y = - v \left(\sin \theta_\Sigma \frac{\partial u_y}{\partial x} + \cos \theta_\Sigma \frac{\partial u_y}{\partial y} \right) , \qquad (4.5.37)$$

where the incompressibility condition $\partial u_x / \partial x + \partial u_y / \partial y = 0$ has been incorporated.

Displacement is continuous across Σ so the gradient of displacement along Σ is also continuous, or

$$\left[\!\!\left[\frac{\partial u_x}{\partial y} \right]\!\!\right] = \left[\!\!\left[\frac{\partial u_y}{\partial y} \right]\!\!\right] = 0 . \qquad (4.5.38)$$

Substitution of (4.5.38) into (4.5.37) reveals that

$$[\dot{u}_x] = 0 , \qquad [\dot{u}_y] = -v \sin\theta_\Sigma \left[\!\left[\frac{\partial u_y}{\partial x}\right]\!\right] = -2v \sin\theta_\Sigma [\epsilon_{xy}] . \quad (4.5.39)$$

In view of (4.5.39), the jump relations (4.5.36) reduce to

$$[\sigma_{xx}] = 0 , \qquad [\sigma_{xy}] = 2\rho v^2 \sin^2\theta_\Sigma [\epsilon_{xy}] . \qquad (4.5.40)$$

The constitutive equations combined with (4.5.40) imply that

$$[\sigma_{xy}] = 2q\mu[\epsilon^p_{xy}] , \qquad [\sigma_{yy}] = -4\mu[\epsilon^p_{yy}] , \qquad (4.5.41)$$

where $q = (m^2 \sin^2\theta_\Sigma)/(1 - m^2 \sin^2\theta_\Sigma)$.

Suppose that material particles follow a continuous path in strain space through the discontinuity. Furthermore, suppose that if the jump in any quantity is zero, then that quantity remains constant along the strain path through the discontinuity. In the following discussion the phrase "inside the jump" is understood to mean "along the strain path from strain state ϵ^+_{ij} to ϵ^-_{ij} within the discontinuity." Then, (4.5.40) and (4.5.41) can be written in the differential form

$$d\sigma_{xx} = 0 , \qquad d\sigma_{xy} = 2q\mu d\epsilon^p_{xy} , \qquad d\sigma_{yy} = -4\mu d\epsilon^p_{yy} \qquad (4.5.42)$$

for states inside the jump.

The maximum plastic work inequality is

$$(\sigma_{ij} - \sigma^*_{ij})d\epsilon^p_{ij} \geq 0 , \qquad (4.5.43)$$

where σ^*_{ij} is any stress state on or inside the yield surface in stress space. Following the analysis of Drugan and Rice (1984) for the quasi-static case, the maximum plastic work inequality is integrated across the discontinuity to obtain

$$W = \int_{\epsilon^{p+}_{ij}}^{\epsilon^{p-}_{ij}} (\sigma_{ij} - \sigma^*_{ij})d\epsilon^p_{ij} \geq 0 . \qquad (4.5.44)$$

W is now written in the form $W = W^p - W^*$, where

$$W^p = \int_{\epsilon_{ij}^{p+}}^{\epsilon_{ij}^{p-}} \sigma_{ij} d\epsilon_{ij}^p \,, \qquad W^* = \int_{\epsilon_{ij}^{p+}}^{\epsilon_{ij}^{p-}} \sigma_{ij}^* d\epsilon_{ij}^p \,. \qquad (4.5.45)$$

The incremental relations (4.5.42) imply that

$$W^p = \sigma_{xx}^\Sigma \int_{\epsilon_{xx}^{p+}}^{\epsilon_{xx}^{p-}} d\epsilon_{xx}^p + \frac{1}{q} \int_{\sigma_{xy}^+}^{\sigma_{xy}^-} \sigma_{xy} d\sigma_{xy} - \frac{1}{4} \int_{\sigma_{yy}^+}^{\sigma_{yy}^-} \sigma_{yy} d\sigma_{yy}$$

$$= -\sigma_{xx}^\Sigma [\![\epsilon_{xx}^p]\!] + (\sigma_{xy}^+ + \sigma_{xy}^-) \left[\frac{1}{8\mu} [\![\sigma_{yy}]\!] - \frac{1}{2q\mu} [\![\sigma_{xy}]\!] \right] , \qquad (4.5.46)$$

where the superscript Σ denotes the value of a continuous field quantity at the discontinuity. Use of (4.5.41) in (4.5.46) results in

$$W^p = -\tfrac{1}{2}(\sigma_{ij}^+ + \sigma_{ij}^-) [\![\epsilon_{ij}^p]\!] \,. \qquad (4.5.47)$$

The integral W^* can also be evaluated with the result that

$$W^* = -\sigma_{ij}^* [\![\epsilon_{ij}^p]\!] \,. \qquad (4.5.48)$$

Finally, combining (4.5.47) and (4.5.48) yields

$$W = \tfrac{1}{2}(2\sigma_{ij}^* - \sigma_{ij}^+ - \sigma_{ij}^-) [\![\epsilon_{ij}^p]\!] \geq 0 \,. \qquad (4.5.49)$$

Because the material is nonhardening, any stress state in the body can be chosen for σ_{ij}^*. Choosing $\sigma_{ij}^* = \sigma_{ij}^+$ and $\sigma_{ij}^* = \sigma_{ij}^-$ in (4.5.49) gives $[\![\sigma_{ij}]\!][\![\epsilon_{ij}^p]\!] \geq 0$ and $[\![\sigma_{ij}]\!][\![\epsilon_{ij}^p]\!] \leq 0$, respectively. These relations are satisfied for a plastically incompressible material if and only if

$$[\![s_{ij}]\!][\![\epsilon_{ij}^p]\!] = 0 \,. \qquad (4.5.50)$$

Using (4.5.40) and (4.5.41) with (4.5.50) implies that

$$[\![s_{xy}]\!] = \pm\sqrt{q} \, [\![s_{yy}]\!] \,. \qquad (4.5.51)$$

As noted above, any state inside the jump is a possible candidate for σ_{ij}^* in equation (4.5.43) and the limits of integration in (4.5.49) may be any pair of states between ϵ_{ij}^+ and ϵ_{ij}^- inclusive. Thus, (4.5.51) must hold for any pair of states inside the jump. This implies that

$$s_{xy} = \pm\sqrt{q}\, s_{yy} + C \qquad (4.5.52)$$

inside the jump, where C is an unknown constant. In addition, (4.5.40) and (4.5.41) show that no part of the strain path through the discontinuity may be purely elastic, and hence the yield condition must be satisfied at every state inside the jump.

Use of (4.5.52) in the yield condition (4.5.6) shows that

$$[\![\sigma_{yy}]\!] = [\![\sigma_{xy}]\!] = 0 \,. \qquad (4.5.53)$$

These results, together with (4.5.39) through (4.5.41), show that all field quantities must be continuous across the discontinuity, that is,

$$[\![\sigma_{ij}]\!] = 0 \,, \qquad [\![\dot{u}_i]\!] = 0 \,, \qquad [\![\epsilon_{ij}^p]\!] = 0 \,. \qquad (4.5.54)$$

In particular, (4.5.54) requires that $[\![s_{ij}]\!] = 0$, and therefore that $\psi(\theta)$ is continuous. In a related study of the motion of a smooth punch along the surface of a rigid-perfectly plastic material, Spencer (1960) concluded that the velocity field in the immediate vicinity of the punch edge must be continuous. His dynamic solution was obtained in the form of a perturbation from the equilibrium slip line field.

To summarize, if it is required that the strain histories must abide by the principle of maximum plastic work through the jump, then discontinuities are ruled out. Consequently, only asymptotic solutions that exhibit continuous angular variations in stress and particle velocity are in full compliance with the maximum plastic work principle. A general result on the existence of discontinuities of field quantities in the material model considered here has been presented by Drugan and Shen (1987).

The question of whether or not the strain paths through the jumps *should* be required to satisfy the principle of maximum plastic work could be raised, of course. Within the context of gas dynamics, Courant and Friedrichs (1948) show that, for a mechanical analysis of the propagation of discontinuities, the sequence of mechanical states experienced by a material particle as a discontinuity propagates across it must be the same as though the transition from the initial state to the final state has occurred in a smooth simple wave. The reason for this outcome, in a somewhat simplistic form, is that in a mechanical description of the phenomenon (as opposed to a more complete thermodynamic description) the behavior of the material does not have the flexibility afforded by thermodynamic properties to depart from the mechanical constitutive law imposed. These observations can be carried over to the present case of elastic-plastic crack growth. The model is strictly mechanical so it may be expected that the behavior of the material in a jump discontinuity must also be describable as the limit of a smooth wave as the slope of the smooth wave becomes indefinitely steep. If this is so, then it follows immediately that the sequence of states in the jump must be consistent with the principle of maximum plastic work.

It should be pointed out that Courant and Friedrichs (1948) only state that for a mechanical description of the transition through a discontinuity, the initial and final states are the same as if the transition had occurred in a simple wave. However, their proof of this statement demonstrates that the sequence of states within the discontinuity must also be the same as if the transition had occurred in a simple wave. In order to describe the state of a material particle passing through the discontinuity, they used essentially the assumption that was used by Drugan and Rice (1984) and that was used in this section: If the jump in some quantity is zero, then that quantity remains constant along the path through the discontinuity.

For the case of antiplane shear crack growth in a material with a *diamond-shaped* yield locus, Nikolic and Rice (1988) showed that the near tip solution consists of sectors which carry constant stress at yield levels corresponding to adjacent vertices on the diamond-shaped yield locus, and which are joined along an elastic-plastic discontinuity. All plastic flow occurs within the discontinuity. Plastic strains and particle velocity are found to be finite at the crack tip. The special features of this asymptotic solution arise from the existence of vertices on the yield locus in stress space. The corresponding results for dynamic plane strain crack growth are not yet available.

4.5.5 Elastic sectors

The possibility of including an elastic sector in the asymptotic field, initiating at some angle θ^* on curve III in Figure 4.13, is considered in this section. In an elastic sector, $\dot{\epsilon}_{ij}^p = 0$ and, consequently, the stress rate and strain rate are related by

$$\dot{\epsilon}_{ij} = \frac{1}{2\mu}\dot{s}_{ij}. \tag{4.5.55}$$

Substitution of the incompressibility condition and (4.5.55) into the momentum equations reveals that the stream function $\phi(r,\theta)$ must satisfy

$$\nabla^2 \left[\beta^2 \frac{\partial^2 \phi}{\partial x_1^2} + \frac{\partial^2 \phi}{\partial x_2^2} \right] = 0 \tag{4.5.56}$$

where $\beta^2 = 1 - m^2$. As can be seen from (4.5.18), the stream function has the general asymptotic form

$$\phi(r,\theta) = rg(\theta) - Ar\sin\theta\ln r. \tag{4.5.57}$$

The representation (4.5.57) implies that

$$\beta^2 \frac{\partial^2 \phi}{\partial x_1^2} + \frac{\partial^2 \phi}{\partial x_2^2} = \frac{G(\theta)}{r}, \tag{4.5.58}$$

where

$$G(\theta) = (1 - m^2\sin^2\theta)\left[g(\theta) + g''(\theta) - 2A\cos(\theta)\right]. \tag{4.5.59}$$

Using (4.5.58) in (4.5.56) shows that $G(\theta)$ must satisfy the ordinary differential equation $G''(\theta) + G(\theta) = 0$ which, with (4.5.59), shows that $g(\theta)$ satisfies the inhomogeneous equation

$$g''(\theta) + g(\theta) = \frac{a_0\sin\theta + b_0\cos\theta}{1 - m^2\sin^2\theta} + 2A\cos\theta, \tag{4.5.60}$$

where a_0 and b_0 are constants to be determined. The general solution of this equation has the form

$$g(\theta) = \sum_{i=1}^{5} a_i g_i(\theta), \tag{4.5.61}$$

where the a_i are constants, and $a_3 = -A$. The functions $g_i(\theta)$ are given by

$$g_1(\theta) = \cos\theta, \quad g_2(\theta) = \sin\theta, \quad g_3(\theta) = \theta\cos\theta,$$

$$g_4(\theta) = \chi(\theta)\cos\theta + \sin\theta\left[\theta - \beta\tan^{-1}(\beta\tan\theta)\right], \qquad (4.5.62)$$

$$g_5(\theta) = -\chi(\theta)\sin\theta + \cos\theta\left[\theta - \beta^{-1}\tan^{-1}(\beta\tan\theta)\right],$$

where $\chi(\theta) = \ln\sqrt{1 - m^2\sin^2\theta}$. The strains can be obtained from the stream function (4.5.57) by integration along $x_2 = $ constant. However, ϵ_{12} can be determined only to within an arbitrary function of x_2. This unknown function is determined by the boundary conditions at θ^*. Expressions for the deviatoric stress components are obtained from the constitutive equations

$$s_{ij}/\mu = 2\left[\epsilon_{ij} - \epsilon_{ij}^p(x_2)\right], \qquad (4.5.63)$$

and the hydrostatic stress σ is obtained from the momentum equations (4.5.1). The resulting expressions are

$$s_{11}/\mu = -2a_2 + 2a_4\left[\beta\tan^{-1}(\beta\tan\theta) - \theta\right] + 2a_5\chi(\theta)$$
$$+ 2A(1 + \ln r) - 2\epsilon_{11}^p(x_2),$$

$$s_{12}/\mu = s_0/\mu + a_4\left[m^2\ln r + (2 - m^2)\chi(\theta)\right] \qquad (4.5.64)$$
$$+ a_5\left[2\theta - (\beta + \beta^{-1})\tan^{-1}(\beta\tan\theta)\right] - 2A\theta,$$

$$\sigma/\mu = \sigma_0/\mu - m^2a_4\theta + m^2(A - a_5)\ln r - 2\epsilon_{11}^p(x_2),$$

where s_0 and σ_0 are constants.

By means of (4.5.59), boundedness of s_{11} in the elastic sector requires that

$$\epsilon_{11}^p(x_2) \sim -A\ln(r\sin\theta) + \epsilon_0, \qquad (4.5.65)$$

where ϵ_0 is a constant. Also, because s_{12} must be bounded, $(4.5.64)_2$ shows that $a_4 = 0$. Continuity of the hydrostatic stress at an elastic-plastic boundary implies that σ is bounded in elastic regions, and, therefore, $a_5 = -A(2 - m^2)/m^2$. The resulting expressions for the stress components are

$$s_{11}/\mu = c_1 - A\left[2\ln(\sin\theta) + \frac{2 - m^2}{m^2}\ln(1 - m^2\sin^2\theta)\right] ,$$

$$s_{12}/\mu = c_2 - \frac{A}{m^2}\left[4\theta - \frac{2 - m^2}{\beta}\tan^{-1}(\beta\tan\theta)\right] , \qquad (4.5.66)$$

$$\sigma/\mu = c_3 - 2A\ln(\sin\theta) ,$$

where c_1, c_2, and c_3 are constants.

With reference to Figure 4.13, the possibility of an elastic sector initiating on curve III at some angle $\theta = \theta^*$, with $0 < \theta^* < \theta_c$, is considered. The point is to determine if it is possible to avoid the region of negative plastic work at $\theta = \theta_c$ through introduction of an elastic sector. In any such elastic sector the unloading condition

$$\frac{d}{d\theta}\left(s_{22}^2 + s_{12}^2\right) \leq 0 \qquad (4.5.67)$$

must hold in some neighborhood of $\theta = \theta^*$. Hence, the inequality

$$s_{22}^e(\theta^*)s_{22}^{e\prime}(\theta^*) + s_{12}^e(\theta^*)s_{12}^{e\prime}(\theta^*) \leq 0 \qquad (4.5.68)$$

must hold, where the superscript e indicates that quantities are evaluated on the elastic side of the elastic-plastic boundary.

From continuity of stress at an elastic-plastic boundary, (4.5.21) implies

$$s_{22}^e(\theta^*) = \tau_0\cos\left(2\theta^* - \psi^*\right) ,$$
$$s_{12}^e(\theta^*) = -\tau_0\sin\left(2\theta^* - \psi^*\right) , \qquad (4.5.69)$$

where ψ^* denotes $\psi(\theta^*)$. The use of (4.5.66) and (4.5.69) in the inequality (4.5.68) shows that ψ^* must satisfy

$$\tan(\psi^* - 2\theta^*) \leq \frac{2\cot\theta^*\cos 2\theta^*}{4\cos^2\theta^* - m^2} . \qquad (4.5.70)$$

It can be shown that this inequality is not satisfied anywhere inside the triangle *Odb* of Figure 4.13. But, as was mentioned earlier in this section, the point (ψ^*, θ^*) must lie on curve III, which in turn must lie inside triangle Obd. Hence an elastic sector cannot be inserted into the plastic field to circumvent the region of negative plastic work. This implies that there is no solution curve if the region of negative plastic work, indicated in Figure 4.13, is present. Therefore, ways to eliminate it must be sought. Apparently, this can be accomplished only by choosing $A = 0$, and therefore the velocity component $\dot{u}_1$ may not have a logarithmic singularity in r.

For $A = 0$, it can be seen from (4.5.66) that the only possible elastic sector is a region of constant stress which must satisfy the yield condition in light of full continuity at an elastic-plastic boundary. But such constant state regions are included in the solution presented above. It is concluded that the solution above is unique within the solution class considered. Finally, it is noted once again that the range of dominance of an asymptotic solution cannot be established without having a more complete solution in hand. It must be recognized that the size of the region near the crack tip in which these asymptotic results provide an accurate description of the local fields may be too small to be of practical significance.

4.6 Elastic-viscous material

In this section, some results on the nature of near tip fields for dynamic growth of a crack in an elastic-viscous material are considered. The study is based on a material model that is a special type of viscoelastic material commonly called a generalized Maxwell material. The essential properties of the model are that the total strain rate is a sum of the elastic strain rate and the viscous strain rate, the elastic strain rate depends on the stress rate through the rate form of Hooke's law, and the viscous strain rate depends on the current stress but is independent of stress history. A stress–strain relation having these properties is

$$\dot{\epsilon}_{ij} = \frac{1+\nu}{E}\dot{\sigma}_{ij} - \frac{\nu}{E}\dot{\sigma}_{kk}\delta_{ij} + \frac{g(\sigma)}{\sigma}s_{ij}, \qquad (4.6.1)$$

where E and ν are Young's modulus and Poisson's ratio of the elastically isotropic material, s_{ij} is the deviatoric stress, $\sigma = \sqrt{3s_{ij}s_{ij}/2}$ is the effective stress, and $g(\sigma)$ is the uniaxial viscous response function for isotropic, incompressible viscous behavior. This function is assumed to have the properties that

$$g'(\sigma) > 0 \qquad \text{for} \qquad \sigma > 0,$$

$$g(\sigma) \to 0 \qquad \text{as} \qquad \sigma \to 0^+, \qquad (4.6.2)$$

$$g(\sigma) = O(\sigma^n) \qquad \text{as} \qquad \sigma \to \infty,$$

where n is a nonnegative real number. Some special cases of interest include the situation with g proportional to σ, which is a linear Maxwell material, and g proportional to σ^n, which is the nonlinear power law viscous material. Condition $(4.6.2)_2$ ensures that the viscous strain rate is zero when the effective stress is zero, and $(4.6.2)_3$ provides information on the strength of the singularity in viscous strain rate as the stress becomes unbounded, such as in a crack tip singular field.

Some features of the stress and deformation distributions on the crack line ahead of the crack tip are considered in the subsections to follow. Several assumptions are made in order to obtain specific results. For one, crack tip fields are assumed to be steady as seen by an observer moving with the crack tip. Though this assumption results in simpler mathematical expressions than those for nonsteady motion, the results are asymptotically correct for nonsteady motion as well. The reason for this situation is that, in a local crack tip coordinate system, the material time derivative is a spatial gradient in local coordinates plus a time rate of change at a fixed point in local coordinates. For any singular crack tip field, the former contribution is sensitive to local gradients whereas the latter is not, so that the former dominates the latter. The assumption that fields are steady leads to the result that the latter contribution is not merely negligible but that it is zero. A second assumption is that the asymptotic stress distribution is a *separable* function in crack tip polar coordinates. Though it cannot be established that the crack tip fields must have a separable form, it is nonetheless possible to satisfy all conditions imposed on the basis of this assumption.

4.6.1 Antiplane shear crack tip field

Suppose that a two-dimensional elastic-viscous solid is subjected to loading that results in antiplane shear deformation. A rectangular x_1, x_2, x_3 coordinate system is introduced so that the only nonzero component of displacement $u_3(x_1, x_2, t)$ is in the x_3-direction. A crack grows in the plane $x_2 = 0$ in the x_1-direction with steady crack tip speed v in the range $0 < v < c_s$. The crack faces are free of traction, and the driving force is assumed to be applied remotely and to result in a field with the symmetry $u_3(x_1, -x_2, t) = -u_3(x_1, x_2, t)$.

The equation ensuring momentum balance, in a form appropriate for steady antiplane shear deformation fields, is

$$\frac{\partial \sigma_{13}}{\partial x_1} + \frac{\partial \sigma_{23}}{\partial x_2} = 2\rho v^2 \frac{\partial \epsilon_{13}}{\partial x_1}, \tag{4.6.3}$$

where ρ is the material mass density. For steady fields, the stress–strain relation (4.6.1) simplifies to

$$-v\frac{\partial \epsilon_{i3}}{\partial x_1} = -\frac{v}{2\mu}\frac{\partial \sigma_{i3}}{\partial x_1} + \frac{g(\sigma)}{\sigma}\sigma_{i3} \tag{4.6.4}$$

for $i = 1, 2$, where $\mu = E/2(1 + \nu)$ is the elastic shear modulus and $\sigma = \sqrt{3(\sigma_{13}^2 + \sigma_{23}^2)}$.

Polar coordinates r, θ centered at the crack tip and translating with it are also introduced. The crack tip is at $r = 0$ and the direction of crack growth is $\theta = 0$. For the asymptotic analysis, suppose that the stress components have the separable form

$$\sigma_{i3} \sim \frac{1}{r^p}\Sigma_i(\theta, v), \qquad i = 1, 2, \tag{4.6.5}$$

as $r \to 0$, where p is a positive real number and $\Sigma_i(\theta, v)$ represents the angular variation of the singular stress field. Order arguments are next outlined which lead to conclusions on the value of p. The reasoning follows that introduced by Hui and Riedel (1981) for crack growth under equilibrium conditions and by Lo (1983) and Brickstad (1983b) for crack growth under dynamic conditions in an elastic-viscous material. According to the assumption (4.6.5), the stress components are singular at the crack tip. Consequently, only the form of $g(\sigma)$ for large σ, say

$$g(\sigma) \sim b_\infty \sigma^n, \tag{4.6.6}$$

where b_∞ is a constant, is required.

First, suppose that the elastic strain rate dominates the viscous strain rate in the near tip region, that is, suppose that

$$\lim_{r \to 0} \sqrt{\dot{\epsilon}^v_{ij}\dot{\epsilon}^v_{ij}} \bigg/ \sqrt{\dot{\epsilon}^e_{ij}\dot{\epsilon}^e_{ij}} = 0. \tag{4.6.7}$$

The implication of (4.6.7) is that the asymptotic fields are governed by (4.6.3) and (4.6.4) with the second term on the right side of (4.6.4) being of negligible magnitude. The remaining equations are essentially the elasticity equations, which implies that

$$\sigma_{i3} \sim \frac{1}{r^{1/2}} \Sigma_i(\theta, v) \tag{4.6.8}$$

or that $p = 1/2$. But with this result in hand, consistency with the original hypothesis (4.6.7) must be checked. With $p = 1/2$, the elastic strain rate is singular as $r^{-3/2}$ whereas from (4.6.6) the viscous strain rate is singular as $r^{-n/2}$. Thus, (4.6.7) holds only if $n < 3$. Otherwise, the elastic strain rate cannot dominate the viscous strain rate.

Suppose now that the viscous strain rate is much larger than the elastic strain rate in the near tip region. Then (4.6.4) implies that

$$v \left| \frac{\partial \epsilon_{i3}}{\partial x_1} \right| \sim b\sigma^{n-1}|\sigma_{i3}| \gg \frac{v}{2\mu} \left| \frac{\partial \sigma_{i3}}{\partial x_1} \right| \tag{4.6.9}$$

for $i = 1$ or 2. But if this is true, then (4.6.3) implies that

$$\frac{\partial \epsilon_{13}}{\partial x_1} \sim 0 \tag{4.6.10}$$

which is inconsistent with (4.6.9). The hypothesis leads to a contradiction and it is therefore false. Thus, the viscous strain rate cannot dominate the elastic strain rate in the near tip region for any value of n.

Having shown that the viscous strain rate cannot dominate the elastic strain rate for any value of n, and that the elastic strain rate cannot dominate the viscous strain rate unless $n < 3$, it must be concluded that the two strain rates are singular of the same order if $n \geq 3$. In view of the fact that the elastic strain rate is singular as

$r^{-(p+1)}$ and that the viscous strain rate is singular as r^{-np}, it follows that

$$p = 1/(n-1) \qquad (4.6.11)$$

for $n \geq 3$.

With the strength of the singularity in stress determined, it remains to determine the angular variation $\Sigma_i(\theta, v)$ of the stress components. The momentum equation and the compatibility condition for antiplane shear deformation are given by (4.4.1) and (4.4.2), respectively. The particular forms of the momentum equation and the stress–strain relation for steady crack growth in the x_1-direction are given in (4.6.3) and (4.6.4), respectively. For the present calculation, it is assumed that

$$g(\sigma) = b\sigma^n \qquad (4.6.12)$$

for $n \geq 1$ and all $\sigma \geq 0$ in the general constitutive equation (4.6.4). The parameter b has the physical dimensions of length$^{2n}\times$time$^{-1}\times$ force^{-n}. The qualitative features of material response included in such a power law dependence of inelastic strain rate on stress are evident. The apparent stiffness of the material increases with the rate of deformation. The coefficient b is a measure of the viscosity of the material.

It is convenient to express the governing equations in terms of dimensionless quantities. Thus, the normalized coordinates and field quantities

$$\xi_\alpha = 2(3)^{\frac{n-1}{2}} b\mu^n x_\alpha / v, \qquad m = v/c_s,$$
$$\gamma = 2\epsilon_{13}, \qquad \tau_\alpha = \sigma_{\alpha 3}/\mu, \qquad \tau = \sqrt{\tau_\alpha \tau_\alpha} \qquad (4.6.13)$$

are introduced, where c_s is the elastic shear wave speed. Greek indices range over 1,2 throughout, and repeated indices are summed. In light of the notation conventions established, the constitutive equation takes the form

$$\frac{\partial \tau_\alpha}{\partial \xi_1} - \tau^{n-1} \tau_\alpha - \frac{\partial \gamma}{\partial \xi_\alpha} = 0, \qquad \alpha = 1,2, \qquad (4.6.14)$$

where ϵ_{32} has been eliminated by means of the compatibility equation. The balance of linear momentum takes the form

$$\frac{\partial \tau_1}{\partial \xi_1} + \frac{\partial \tau_2}{\partial \xi_2} - m^2 \frac{\partial \gamma}{\partial \xi_1} = 0. \qquad (4.6.15)$$

The auxiliary conditions in this problem include boundary conditions to ensure traction-free crack faces, continuity conditions on certain fields, and remote asymptotic conditions to render the crack tip fields consistent with the outer fields. The conditions that the crack faces are traction-free and that the displacement field is antisymmetric with respect to the crack plane are

$$\tau_2(\xi_1, 0) = 0 \qquad \text{for} \qquad \xi_1 < 0,$$

$$\tau_1(\xi_1, 0) = \gamma(\xi_1, 0) = 0 \qquad \text{for} \qquad \xi_1 > 0,$$

(4.6.16)

where it has been assumed that the fields are continuous across $\xi_2 = 0$ for $\xi_1 > 0$. It can be argued that the regularity requirement of Hui and Riedel (1981) is equivalent to the second of these conditions. All fields are assumed to be continuous.

This completes a formulation of the problem of determining the crack tip asymptotic field. In general, the nature of the solution will depend on three parameters, namely, the crack tip speed v, the rate exponent n, and the remote loads. Both the rate coefficient b, which is the material property representing the ratio of inelastic strain rate to τ^n, and the shear modulus μ have already been absorbed into the dimensionless physical quantities. The remote asymptotic conditions will not be considered for the time being, but any indeterminate parameters in the description of the local crack tip fields will be interpreted as parameters whose value can be determined only through coupling with the remote mechanical fields.

Attention is now focused on the near tip fields for steady crack propagation. A local polar coordinate system r, θ is introduced, with the origin of coordinates moving with the crack tip and $r = \sqrt{\xi_1^2 + \xi_2^2}$, $\theta = \tan^{-1}(\xi_2/\xi_1)$. For $n \geq 3$, it was shown above that, for small values of r,

$$\tau_1(r, \theta) \sim r^{-p} T_1(\theta), \qquad \tau_2(r, \theta) \sim r^{-p} T_2(\theta) \qquad (4.6.17)$$

plus terms with less singular dependence on r, where $p = 1/(n - 1)$. The strength of the singularity was obtained by balancing the dominant singularities of elastic and plastic strain rate. The stress field is anticipated to be uniform with respect to r for any crack speed m or rate exponent n. In contrast, the dependence of the deformation field on r is expected to be nonuniform in n and m, but this can be demonstrated only if a full solution is known. Likewise, the strain component appearing in the reduced equations (4.6.14) and (4.6.15) is assumed to have the asymptotic form

$$\gamma(r,\theta) \sim r^{-p}G(\theta) \qquad (4.6.18)$$

plus terms less singular in r as $r \to 0$.

If the assumed asymptotic forms are substituted into the field equations and dominant terms in r are collected, then a quasi-linear system of ordinary differential equations is obtained in the matrix form

$$\begin{pmatrix} -\sin\theta & \cos\theta & m^2\sin\theta \\ -\sin\theta & 0 & \cos\theta \\ 0 & -\sin\theta & -\cos\theta \end{pmatrix} \begin{pmatrix} T_1' \\ T_2' \\ G' \end{pmatrix}$$

$$= \begin{pmatrix} p\cos\theta & p\sin\theta & pm^2\cos\theta \\ p\cos\theta + T^{n-1} & 0 & -p\cos\theta \\ 0 & p\cos\theta + T^{n-1} & -p\sin\theta \end{pmatrix} \begin{pmatrix} T_1 \\ T_2 \\ G \end{pmatrix}, \qquad (4.6.19)$$

where $T(\theta) = \sqrt{T_1(\theta)^2 + T_2(\theta)^2}$ and a prime denotes differentiation with respect to θ. This system is subject to the boundary conditions

$$T_1(0) = 0, \quad G(0) = 0, \quad T_2(\pi) = 0. \qquad (4.6.20)$$

The first two conditions in (4.6.19) result from symmetry of the fields plus continuity across $\theta = 0$, and the third equation ensures a traction-free crack face at $\theta = \pi$.

Note that the three differential equations (4.6.19) are subject to three boundary conditions (4.6.20). Consequently, it appears that a solution of the equations will be completely determined without free parameters for any specific values of m and n. In other words, the somewhat surprising result is obtained (for $n \geq 3$) that the asymptotic field is *completely* determined by the imposed local boundary conditions for a given crack speed v. The asymptotic field involves no free parameter to be determined from remote conditions and to represent the influence of the applied loads. In this sense, the asymptotic solution is *independent* of the applied loads. Suppose that some crack growth criterion is applied to this asymptotic solution. In view of the character of the asymptotic field, it appears that a crack in an elastic-viscous material will grow at just the right speed v so that a crack

growth criterion is satisfied within the near tip region, no matter what
the applied loading might be. However, it must be kept in mind that
the asymptotic field has limited implication for crack growth unless
the asymptotic solution is an accurate representation of the actual
field over a region of size greater than some characteristic dimension
of the physical separation process. This matter cannot be addressed
on the basis of an asymptotic analysis alone.

At first sight, the boundary value problem represented by (4.6.19)
and its boundary conditions seems to be of a standard type for which
a solution can be obtained by numerical methods. Upon closer examination, however, it is seen that the coefficient matrix on the left
side of (4.6.19) is singular, or its determinant vanishes, at $\theta = 0$.
Consequently, evaluation of the differential equation at $\theta = 0$ does
not provide values of $T_1'(0)$ and $G'(0)$ which are essential in starting
a numerical procedure, say a finite difference method. Thus, the
features of the solution near $\theta = 0$ are examined in order to find
a way around this difficulty.

Consider the local fields in the asymptotic power series form

$$\tau_2 \sim \frac{T_2(0)}{\xi_1^p}\left(1 + \kappa\frac{\xi_2^2}{\xi_1^2}\right),$$

$$\tau_1 \sim \frac{T_1'(0)}{\xi_1^p}\left(\frac{\xi_2}{\xi_1}\right), \quad \gamma \sim \frac{G'(0)}{\xi_1^p}\left(\frac{\xi_2}{\xi_1}\right) \tag{4.6.21}$$

for $\xi_1 > 0$ and $\xi_2 \ll \xi_1$, where κ is a constant. These functions have
the symmetry and continuity properties anticipated in any solution
and they satisfy the boundary conditions $(4.6.20)_1$ and $(4.6.20)_2$ on
$\theta = 0$. Note that $T_2'(0)$ is expected to be zero from symmetry of the
fields. If these functions are substituted into the differential equation
(4.6.19), and if the lowest order terms in powers of ξ_2 are required to
balance, then it is found that

$$G'(0) = -\left[pT_2(0) + T_2(0)^n\right],$$

$$T_1'(0) = -\frac{(p+1)\left[p + T_2(0)^{n-1}\right]T_2(0)}{\left[1 + p + T_2(0)^{n-1}\right]}. \tag{4.6.22}$$

An expression for κ in terms of $T_2(0)$ can also be found, but it is of
no consequence.

With the above values of $T_1'(0)$ and $G'(0)$ in terms of the initial
value of $T_2(0)$ in hand, an approximate solution can be obtained by
application of a standard numerical integration algorithm. Initial
values are assigned as $(0, T_2(0), 0)$ and the solution can then proceed
by means of (4.6.19) from $\theta = 0$ up to $\theta = \pi$. For an arbitrary
choice of $T_2(0)$, it is found that any procedure leads to $T_2(\pi) \neq 0$, in
general. Thus, to satisfy the boundary condition at $\theta = \pi$, a shooting
method for solving a two-point boundary value problem is adopted
to systematically adjust the value of $T_2(0)$ for the next numerical
integration from $\theta = 0$ to $\theta = \pi$ on the basis of the value of $T_2(\pi)$
obtained from the previous integration, until a value is found for which
$T_2(\pi) = 0$. An asymptotic solution has been found when this is the
case. This numerical analysis has been carried out by Lo (1983).

The distribution of viscous strain on the crack line $\theta = 0$ ahead
of the crack tip was found by Lo (1983) to have the form

$$2\epsilon_{23}^v(r, 0) \sim \frac{n-1}{\mu} \left(\frac{v}{b\mu x_1} \right)^{1/(n-1)} T_2(0)^n, \qquad (4.6.23)$$

where $T_2(0)$ is the dimensionless constant introduced above. By numerical calculation, he found that the amplitude factor $T_2(0)$ decreased from 0.662 for $m = 0$ to 0.470 for $m = 0.7$ if $n = 4$, and
that it decreased from 1.08 to 0.890 over the same speed range for
$n = 10$. The result (4.6.23) illustrates features of the asymptotic field
that have already been discussed.

4.6.2 Plane strain crack tip field

The arguments presented in Section 4.6.1 for antiplane shear deformation apply equally well in considering the case of crack growth in an
elastic-viscous material under plane strain opening mode conditions.
For the plane strain form of the stress–strain relation (4.6.1), the near
tip stress distribution in crack tip polar coordinates is found to be

$$\sigma_{ij} \sim \left(\frac{v}{b\mu r} \right)^{1/(n-1)} \Sigma_{ij}(\theta, v), \qquad (4.6.24)$$

where i or $j = 1, 2$. Numerical results showing the dependence of the
angular variation of stress components with angle θ are presented by
Lo (1983).

4.7 Elastic-viscoplastic material; antiplane shear

The general discussion on crack growth in rate-dependent materials begun in the previous section is continued here by considering the situation of steady crack growth in an elastic-viscoplastic material. In particular, an elastic-viscoplastic material is considered that shows the same power law dependence of inelastic strain rate on stress for large stress magnitudes as suggested in (4.6.2), but that retains the properties of existence of an elastic region in stress space and a yield condition on stress state for moderate stress levels. With reference to the stress–strain relation (4.6.1), the distinction is made more precise by noting that the dependence of effective viscous strain rate $\dot{\epsilon}^v = \sqrt{2\dot{\epsilon}_{ij}^v \dot{\epsilon}_{ij}^v / 3}$ on effective stress σ indicated in Section 4.6 has the form

$$\dot{\epsilon}^v = \tfrac{2}{3}g(\sigma) = \tfrac{2}{3}b\sigma^n \qquad (4.7.1)$$

for an elastic-viscous material, where the exponent n is a material parameter indicating stress sensitivity of viscous strain rate. On the other hand, an elastic-viscoplastic material model having essentially the same dependence of inelastic strain rate on stress for large stress magnitude, but retaining the characteristics of plastic yield at moderate stress levels, is $\dot{\epsilon}^v = \tfrac{2}{3}g(\sigma)$, where

$$g(\sigma) = \begin{cases} 0 & \text{for } \sigma \le \sigma_0, \\ b(\sigma - \sigma_0)^n & \text{for } \sigma > \sigma_0. \end{cases} \qquad (4.7.2)$$

The material parameter n has the same interpretation as in (4.7.1) but the reference stress σ_0 can now be interpreted as a flow stress for slow deformations. This constitutive assumption was introduced by Malvern (1951) in a study of one-dimensional stress waves in an elastic-viscoplastic material, and it was later generalized to three-dimensional stress states by Perzyna (1963). The general structure of constitutive laws of this form was considered by Rice (1970).

    The notation established in Section 4.6 is followed here unless noted otherwise. The momentum equation and the compatibility condition for antiplane shear are given by (4.4.1) and (4.4.2), respectively. For steady crack growth, the governing equations are most conveniently expressed in terms of the normalized convected coordinates and functions of position as

$$\xi_\alpha = 2(3)^{\frac{n-1}{2}}\mu\tau_0^{n-1}x_\alpha/v\,, \qquad m = v/c_s\,,$$

$$\gamma = 2\mu\epsilon_{13}/\tau_0\,, \qquad \tau_\alpha = \sigma_{\alpha3}/\tau_0\,, \tag{4.7.3}$$

where c_s is the elastic wave speed and $\tau_0 = \sigma_0/\sqrt{3}$ is the flow stress in shear for very slow deformation. In addition, the normalized effective stress $\tau = \sqrt{\tau_\alpha\tau_\alpha}$ is introduced. The constitutive equation takes the form

$$\frac{\partial\tau_\alpha}{\partial\xi_\alpha} - \frac{(\tau-1)^n}{\tau}\tau_\alpha - \frac{\partial\gamma}{\partial\xi_\alpha} = 0\,, \qquad \alpha = 1, 2\,, \tag{4.7.4}$$

where ϵ_{32} has been eliminated by means of the compatibility relation. The momentum balance is identical to (4.6.15) but it must be remembered that the coordinates and fields have been normalized in a slightly different way here. Finally, the conditions that the crack faces are traction-free and that the displacement field is antisymmetric with respect to the crack plane are

$$\tau_2(\xi_1, 0) = 0 \qquad \text{for} \qquad \xi_1 < 0\,,$$

$$\tau_1(\xi_1, 0) = \gamma(\xi_1, 0) = 0 \qquad \text{for} \qquad \xi_1 > 0\,. \tag{4.7.5}$$

It is instructive to consider the elastic-viscoplastic constitutive relationship in a very simple uniaxial form so that the qualitative features of the response become clearer. To this end, let $\tau_2 = 0$ in (4.7.4), so that $|\tau_1| = \tau$. Furthermore, $-\xi_1$ is identified as a timelike parameter and γ is the magnitude of the normalized x_1-component of strain. Thus, with τ and γ as the measures of stress and strain, respectively, the relationship is

$$\dot\gamma = \dot\tau + \langle\tau - 1\rangle^n\,, \tag{4.7.6}$$

where the superimposed dot denotes differentiation with respect to the timelike parameter and the angle brackets $\langle\cdot\rangle$ are used to indicate that the value of the term is zero if the enclosed quantity is nonpositive.

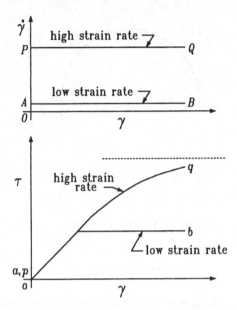

Figure 4.15. Schematic diagram of material response according to (4.7.6) under conditions of imposed uniform strain rate from zero initial strain.

Consider material response according to (4.7.6) under conditions of imposed strain rate from zero initial strain. If $\dot{\gamma}$ has a very small constant value greater than zero, then $\dot{\tau}$ is also very small. The stress rate $\dot{\tau}$ is proportional to $\dot{\gamma}$ until $\tau = 1$, and τ is close to one thereafter. The response is essentially an elastic-perfectly plastic response, as indicated in Figure 4.15. On the other hand, if $\dot{\gamma}$ has a very large constant value, then $\dot{\tau}$ is equal to $\dot{\gamma}$ up to $\tau = 1$, and thereafter τ asymptotically approaches the stress level $1 + \dot{\gamma}^{1/n}$ as γ becomes very large. This behavior is also indicated in Figure 4.15.

Consider now the response to a nonuniform strain rate having the general continuous variation with strain shown in Figure 4.16. The dependence has a maximum at some strain level, and it has a zero at some higher strain level. The strain necessarily begins to decrease once the strain rate becomes negative. The response to such an imposed strain rate according to (4.7.6) is shown in the lower part of Figure 4.16 where the algebraic signs of the elastic rate $\dot{\gamma}^e = \dot{\tau}$ and the viscoplastic strain rate $\dot{\gamma}^p = \langle \tau - 1 \rangle^n$ vary along the stress strain path. For example, $\dot{\gamma}^e = \dot{\gamma} > 0$ and $\dot{\gamma}^p = 0$ in oa; $\dot{\gamma} > 0$, $\dot{\gamma}^e > 0$, and $\dot{\gamma}^p > 0$ in ab; $\dot{\gamma} > 0$, $\dot{\gamma}^e < 0$, and $\dot{\gamma}^p > 0$ in bc; $\dot{\gamma} < 0$, $\dot{\gamma}^e < 0$,

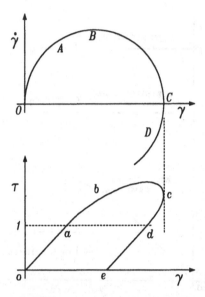

Figure 4.16. Schematic diagram of material response according to (4.7.6) under conditions of imposed nonuniform strain rate from zero initial strain.

and $\dot{\gamma}^p > 0$ in cd; and $\dot{\gamma} = \dot{\gamma}^e < 0$ and $\dot{\gamma}^p = 0$ in de. A material particle close to the crack path in the crack growth problem under study is expected to experience a stress–strain history having these general features, but under multiaxial conditions. The stress may be singular in a region of elastic-viscoplastic response, but the material can unload elastically only from a stress level of $\tau = 1$.

The understanding of the asymptotic crack tip field for the situation under discussion is far from complete, so this section is concluded with some general observations in light of the results described previously for the case of elastic-viscous response. Suppose that the term $(\tau - 1)^n$ in the constitutive equation is expanded by means of the binomial theorem, so that

$$(\tau - 1)^n = \tau^n - n\tau^{n-1} + \frac{n(n-1)}{2}\tau^{n-2} - \ldots . \qquad (4.7.7)$$

In any region near the crack tip in which $\tau \gg 1$ the first term in this expansion is much larger than any of the succeeding terms. Indeed, if $(\tau - 1)^n$ is approximated by τ^n in such a region, then the constitutive equation (4.7.4) is identical to the equation (4.6.14) for the elastic-viscous case. The elastic-viscous material admits an asymptotic solution that is singular as r^{-p}, where $p = 1/(n-1)$, for $n \geq 3$, so the elastic-viscoplastic material admits a singular solution of exactly the same strength over a sector of some angular extent around the crack tip. For the elastic-viscous material, the asymptotic solution was found to be singular over the entire range of θ in the interval $0 \leq \theta < \pi$. Is the same true for the elastic-viscoplastic case?

For the case of elastic-viscoplastic response, suppose that

$$\tau(r, \theta) \sim r^{-p} T(\theta) \qquad (4.7.8)$$

near $\theta = 0$, where the notation of the previous section has been adopted, and suppose that $T(0) > 0$. This seems to be reasonable because plastic strain is concentrated ahead of the tip for mode III deformation. As θ increases from zero, it appears that (4.7.8) provides the correct asymptotic form for the solution as long as $T(\theta) > 0$. The form (4.7.8) could give way to an unloading sector or some other type of region, say at $\theta = \theta^*$, only if $T(\theta^*) = 0$. This matter has not been thoroughly studied, and a complete description of the asymptotic field which accounts for all possibilities is not yet available.

It was observed by Hui and Riedel (1981) for the case of crack growth through an elastic-viscous material under quasi-static conditions and by Lo (1983) for dynamic crack growth through the same material that the computed asymptotic fields do not seem to converge to the commonly accepted asymptotic solution for steady crack growth through a rate independent elastic-plastic material reported by Chitaley and McClintock (1971) as $n \to \infty$. In a sense, this is not too surprising because the limiting response of the elastic-viscous material model as n becomes indefinitely large is not the same as the rate independent elastic-perfectly plastic response. For example, it does not exhibit an elastic region in stress space. This elastic region plays a significant role in the rate independent case. It is also interesting to consider this general question from the point of view of asymptotic analysis. For either the elastic-viscous or elastic-viscoplastic models considered here, the dominant term in the crack tip asymptotic solution has the radial dependence r^{-p}. The next term in the expansion in either case appears to be of order unity as $r \to 0$. However, the observation that the equations determining this contribution are slightly different for the two cases, due to the second term in the expansion (4.7.7), is potentially significant. In either case, the term that is $O(r^{-p})$ as $r \to 0$ will dominate the second term of order unity only over a region with radial coordinates r for which $r^{-p} \gg 1$. As $n \to \infty$ or as $p \to 0$, the size of this region of dominance vanishes. In other words, as n becomes large, the second term in the asymptotic expansion appears to be as significant as the first term. Perhaps a study of the higher order terms in the crack tip asymptotic solution will help to resolve the apparent discrepancy among limiting cases as $n \to \infty$. Results on this question are not yet available.

5

ENERGY CONCEPTS
IN DYNAMIC FRACTURE

5.1 Introduction

Analytical methods based on the work done by applied loads and the changes in the energy of a system that accompany a real or virtual crack advance have been of central importance in the development of fracture mechanics. These methods have provided a degree of unification of seemingly diverse ideas in fracture mechanics, and they have led to procedures of enormous practical significance for the characterization of the fracture behavior of materials. In addition, some of the most elegant theoretical analyses in the field have been those associated with energy methods. In this chapter, energy concepts that are particularly relevant to the study of dynamic fracture processes are considered.

The importance of the variation of energy measures during crack growth was recognized by Griffith (1920) in his pioneering discussion of brittle fracture, as outlined in Section 1.1.2. The extension of a crack requires the formation of new surface, he reasoned, with its associated surface energy. Consequently, a crack in a brittle solid should advance when the reduction of the total potential energy of the body during a small amount of crack advance equals the surface energy of the new surface thereby created. For an elastic body containing a crack, the negative of the rate of change of total potential energy with respect to crack dimension is called the energy release rate. This quantity, which is usually denoted by the symbol G, is a function of crack size, in general. From its definition, G is the amount of energy, per unit length along the crack edge, that is supplied by the elastic energy in the body and by the loading system in creating the new fracture surface. In this sense, G is a generalized force that is work-conjugate to the amount of crack advance. Consequently, G has also become known as the crack driving force. These ideas have been exploited and extended in a host of ways in fracture mechanics, including dynamic fracture mechanics.

Irwin (1957) recognized the role of the elastic stress intensity factor as a characterizing parameter for the elastic crack tip field and, on this basis, he proposed the critical stress intensity factor fracture criterion as discussed in Section 3.6. The energy criterion and the stress intensity factor criterion were then shown to be equivalent for an elastic-brittle material under equilibrium conditions by means of an energy analysis presented by Irwin (1957, 1960). The essential features of the analysis are summarized here. Consider a planar crack in a homogeneous and isotropic elastic solid under plane strain mode I conditions. If x is a spatial coordinate in the direction of prospective crack advance with its origin at the crack tip position, then the tensile stress distribution on the prospective fracture plane for points very close to the crack tip compared to any physical dimension of the configuration is

$$\sigma(x) = \frac{K_I}{\sqrt{2\pi x}}, \qquad (5.1.1)$$

where K_I is the mode I stress intensity factor; see (2.1.1). Imagine that the crack then advances a small distance Δx under equilibrium conditions. After advance, the outward normal displacement of either crack face near the crack tip is

$$u(x) = \frac{2(1 - \nu^2)(K_I + \Delta K_I)}{E} \sqrt{\frac{\Delta x - x}{2\pi}} \qquad (5.1.2)$$

for $x < \Delta x$, where E and ν are the elastic constants and ΔK_I is the change in stress intensity factor during the crack advance. The energy that was released from the body and loading system in going from the configuration before crack advance to the configuration after advance is the work that must be supplied to return the latter configuration to the former. In view of the linearity of the system, that work is

$$\Delta W = \int_0^{\Delta x} \sigma(x) u(x)\, dx = \frac{(1 - \nu^2)}{E} K_I^2 \left(1 + \frac{\Delta K_I}{K_I}\right) \Delta x. \quad (5.1.3)$$

The energy release rate is then obtained in terms of the stress intensity factor by means of its definition above as

$$G = \lim_{\Delta x \to 0} \frac{\Delta W}{\Delta x} = \frac{(1 - \nu^2)}{E} K_I^2, \qquad (5.1.4)$$

where it has been assumed that $\Delta K_I / K_I \to 0$ as $\Delta x \to 0$. The relationship (5.1.4), which is commonly known as the Irwin relationship, is of fundamental importance in fracture mechanics. It provides an unambiguous link between stress intensity factor and energy release rate for elastic crack problems. Furthermore, if crack growth criteria are postulated in terms of critical values of these parameters, then (5.1.4) implies that values of these material specific parameters cannot be independent. The form of (5.1.4) with $(1 - \nu^2)$ replaced by unity, which is appropriate for two-dimensional plane stress, was given as (1.1.8).

The relationship (5.1.4) is also at the root of the so-called compliance methods of fracture testing introduced by Irwin and Kies (1954). As noted above, the energy release rate is the change in potential energy of a system with respect to crack length. This change can be expressed in terms of the overall compliance of the body and the applied loading. Likewise, the energy release rate is related to the crack tip stress intensity factor through (5.1.4). Thus, if the compliance versus crack length can be estimated or measured separately, then the energy approach provides a direct relationship among load level, crack length, and stress intensity factor. For example, for the split beam configuration used to motivate the energy approach in Section 1.1.2, the relation $G = -\partial\Omega/\partial l$ implies that

$$K_I = \sqrt{\frac{3h}{2}} \frac{Ehq^*}{l^2} \qquad (5.1.5)$$

in view of the Irwin relationship.

The path-independent J-integral introduced by Eshelby (1956) and Rice (1968) provides a more general means of relating crack tip fields to remote loading conditions in cracked solids. The path integral method was generalized to the case of nonlinear elastic material response for which Hutchinson (1968) and Rice and Rosengren (1968) showed that the value of J provides a crack tip field characterizing parameter, similar to the stress intensity factor in linear fracture mechanics. These key ideas involving energy concepts in equilibrium fracture mechanics provide the basis for the development of similar ideas in dynamic fracture, as described in the sections to follow.

5.2 The crack tip energy flux integral

A crack tip contour integral expression for elastodynamic energy release rate was proposed by Atkinson and Eshelby (1968) who argued that the form for dynamic crack growth should be the same as for quasi-static growth with the elastic energy density replaced by the total mechanical energy density, that is, the elastic energy plus the kinetic energy. The integral expression for dynamic energy release rate in terms of crack tip stress and deformation fields was subsequently derived directly from the field equations for general material behavior by Kostrov and Nikitin (1970) and for elastodynamics by Freund (1972c). They enforced an instantaneous energy rate balance for the time-dependent volume of material bounded by the outer boundary of the solid, the crack faces, and small loops surrounding each moving crack tip and translat ng with it. By application of the generalized transport theorem (1.2.41) and the divergence theorem, an expression for crack tip energy flux in the form of a path integral of field quantities along the crack tip loops was obtained. A result that is applicable to a broad range of material response and that contains the earlier elastodynamic result as a special case is derived here. Although a variety of seemingly different path-independent energy integrals have been proposed, each can be extracted from this general result by invoking the appropriate restrictions on material response and crack tip motion.

5.2.1 The energy flux integral for plane deformation

In the absence of body forces, the equation of motion in terms of rectangular components of stress σ_{ij} and displacement u_i is

$$\frac{\partial \sigma_{ji}}{\partial x_j} - \rho \frac{\partial^2 u_i}{\partial t^2} = 0, \tag{5.2.1}$$

where ρ is mass density. If the inner product of (5.2.1) with the particle velocity $\partial u_i / \partial t$ is formed, then the resulting equation can be

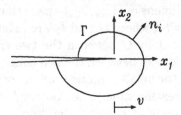

Figure 5.1. A crack tip moving with the instantaneous speed v. The contour Γ translates with the crack tip.

written in the form

$$\frac{\partial}{\partial x_j}\left(\sigma_{ji}\frac{\partial u_i}{\partial t}\right) - \frac{\partial}{\partial t}(U+T) = 0, \qquad (5.2.2)$$

where U is the stress work density and T is the kinetic energy density, that is,

$$U = \int_{-\infty}^{t} \sigma_{ji}\frac{\partial^2 u_i}{\partial t'\partial x_j}\,dt', \qquad T = \tfrac{1}{2}\rho\frac{\partial u_i}{\partial t}\frac{\partial u_i}{\partial t} \qquad (5.2.3)$$

at any material point, as defined in Section 1.2.2. There are no constitutive assumptions implicit in the definition of stress work density.

If the time coordinate is now treated in the same way as a spatial coordinate, then (5.2.2) is the requirement that a certain vector expression is divergence-free in a particular "volume" of space-time, that is, it represents a mechanical conservation law. Thus, if the expression is integrated over this volume and the divergence theorem is applied, the integral of the inner product of the divergence-free vector with the surface unit normal vector, taken over the bounding surface of this volume, vanishes. This idea is developed below.

Consider a two-dimensional body that contains an extending crack. Suppose that the outer boundary of the body is the plane curve C. Rectangular coordinates are introduced so that the plane of the body is the x_1, x_2-plane, the crack plane is $x_2 = 0$, and crack growth is along the x_1-axis. The instantaneous crack tip speed is v. A small contour Γ begins on one traction-free face of the crack, surrounds the tip, and ends on the opposite traction-free face. This crack tip contour is fixed in both size and orientation in the crack tip reference frame as the crack grows; see Figure 5.1. Consider the volume in the three-dimensional x_1, x_2, t-space that is bounded by the planes $t = t_1$ and $t = t_2$, where t_1 and t_2 are arbitrary times, by the right cylinder swept out by C between the two times, by the planes swept out by the crack faces between the two times, and by the tubular surface swept out by the small contour Γ as the crack advances in the x_1-direction with increasing time. The "bottom" of the volume is the configuration of the body at time t_1, the "top" of the volume is the configuration of the body at time t_2, and the lateral surface is the time history of C. The outward unit normal vector to the bounding surface of the volume of space-time is denoted by $\boldsymbol{\nu}$ which has components ν_1, ν_2, ν_t in the coordinate directions. A second unit normal vector is defined for the plane body at any fixed time. This vector is denoted by **n** and it is directed out of the material for points on C or on the crack faces, and it is directed away from the crack tip for points on Γ. Note that $n_1 \neq \nu_1$ and $n_2 \neq \nu_2$, in general, but these components are equal at points where $\nu_t = 0$. The general approach is similar to that followed in deriving energy integrals in the theory of hyperbolic partial differential equations (Courant and Hilbert 1962).

If the quantity (5.2.2) is integrated over the volume in space-time and the divergence theorem is applied to the result, then

$$\int_{t_1}^{t_2} \int_C \sigma_{ji} n_j \frac{\partial u_i}{\partial t} \, dC \, dt - \int_{A(t_2)} (U + T) \, dA + \int_{A(t_1)} (U + T) \, dA$$

$$= -\int_{\text{Tube}} \left[\sigma_{ji} \nu_j \frac{\partial u_i}{\partial t} - (U + T) \nu_t \right] dS, \qquad (5.2.4)$$

where $A(t)$ is the cross-sectional area of the volume on any plane $t = $ constant and the integral on the right-hand side is the surface integral over the portion of the tube between times t_1 and t_2. Because the tube is formed by translating the fixed contour Γ in the x_1-direction at instantaneous speed v, the components of $\boldsymbol{\nu}$ are related by $\nu_t = -v\nu_1$. Thus, the right side of (5.2.4) takes the form

$$-\int_{\text{Tube}} \left[\sigma_{ji} \frac{\partial u_i}{\partial t} + (U + T) v \delta_{1j} \right] \nu_j \, dS, \qquad (5.2.5)$$

where δ_{ij} is the Kronecker delta operator (1.2.6). Suppose that the surface of the tube is parameterized in terms of surface coordinates t and s, where the arclength s is measured along Γ at constant t

from some arbitrary point. Then, from the theory of surfaces, the area element $\nu_j\, dS$ is the element $-n_j\, ds\, dt$ in terms of the surface coordinates. The relationship (5.2.4) becomes

$$\int_{t_1}^{t_2} \int_C \sigma_{ji} n_j \frac{\partial u_i}{\partial t}\, dC\, dt = \int_{A(t_2)} (U+T)\, dA - \int_{A(t_1)} (U+T)\, dA$$

$$+ \int_{t_1}^{t_2} \int_{\Gamma} \left[\sigma_{ji} n_j \frac{\partial u_i}{\partial t} + (U+T)vn_1 \right] ds\, dt. \qquad (5.2.6)$$

The term on the left side of the equal sign in (5.2.6) is the total work done on the body by tractions applied to either C or the crack faces in the time interval of interest, and the first two terms on the right side represent the net stress work done plus kinetic energy added during the same time. Consequently, the last term in (5.2.6) is the total energy lost from the body due to flux through Γ during the interval $t_1 \le t \le t_2$. Because this time interval is completely arbitrary, it follows immediately that the crack tip line integral

$$F(\Gamma) = \int_{\Gamma} \left[\sigma_{ji} n_j \frac{\partial u_i}{\partial t} + (U+T)vn_1 \right] ds \qquad (5.2.7)$$

is the *instantaneous* rate of energy flow through Γ toward the crack tip. From its definition, F is the flux of total energy. It is the flux of kinetic energy plus free energy only under the special conditions of adiabatic and isentropic deformation. The energy flux has the physical dimensions of energy per unit time per unit width, or force/time. This result has also been obtained by Willis (1975).

5.2.2 Some properties of $F(\Gamma)$

The interpretation of the terms in (5.2.7) is straightforward. The first term is the rate of work done on the material inside of Γ by the tractions acting on Γ. If the curve Γ were a material line, then this would be the only contribution to the energy flux. The curve is moving through the material, however, and the second term in (5.2.7) represents the contribution to the energy flux due to transport of material through Γ, with its associated energy density $U + T$.

The energy flux integral expression (5.2.7) was derived within the framework of infinitesimal deformations, and no distinction was made

between the deformed and undeformed configurations of the body.
In fact, the integral is valid for finite deformation if the terms are
properly interpreted. Suppose that a deformable body is undergoing
a deformation of arbitrary magnitude with respect to some reference
configuration. A common rectangular coordinate system is assumed,
and components of vectors and tensors are defined in reference to the
coordinate directions. Suppose that u_i denotes the time-dependent
displacement of the particle at location x_i in the reference configura-
tion, so that $\partial u_i/\partial t$ is the velocity of the particle at x_i in the reference
configuration. Suppose that Γ is a surface with unit normal vector n_i
in the reference configuration that moves with respect to the material
particles instantaneously on it with normal speed v. If ρ is the mass
density of the material in the reference configuration and σ_{ij} is the
transpose of the nominal stress (or the first Piola-Kirchhoff stress)
then (5.2.7) represents the instantaneous energy flux through Γ. The
work rate balance for a continuum undergoing finite deformation is
discussed in a form useful for considering the energy flux by Ogden
(1984), as outlined in Section 1.2.2. The energy flux integral for the
particular case of finite deformation of a hyperelastic material was
discussed by Gurtin and Yatomi (1980).

The integral defining F itself is not path-independent, in general,
as can be seen by means of a direct check. Consider the closed path
formed by two crack tip contours Γ_1 and Γ_2, and by the segments of
the crack faces that connect the ends of these contours. Application
of the divergence theorem to the energy flux integral for this entire
closed path leads to the result that

$$F(\Gamma_2) - F(\Gamma_1) = \int_{A_{12}} \left[\left(\frac{\partial U}{\partial t} + v\frac{\partial U}{\partial x_1} \right) \right.$$
$$\left. + \rho\frac{\partial u_i}{\partial t}\left(\frac{\partial^2 u_i}{\partial t^2} + v\frac{\partial^2 u_i}{\partial t\partial x_1} \right) \right] dA, \quad (5.2.8)$$

where A_{12} is the area within the closed path. Without loss of generality, the algebraic signs in (5.2.8) are chosen to be consistent with
the normal vector to Γ_1 directed into A_{12} and the normal vector to
Γ_2 directed away from A_{12}. Except in special circumstances, the
integrand of (5.2.8) is not zero, so that $F(\Gamma_1) \neq F(\Gamma_2)$ in general.
The path dependence can be understood quite readily in terms of the
wave character of the fields represented by the integrand. For any

choice of Γ_1 and Γ_2, transient stress fields can be introduced which influence the energy flux through one contour but not through the other.

One special case for which the integral is indeed path-independent is that of steady crack growth. If the mechanical fields are invariant in a reference frame traveling with the crack tip in the x_1-direction at a uniform speed v, then any field quantity depends on x_1 and t only through the combination $\xi = x_1 - vt$. In this case the area integral in (5.2.8) vanishes at each point in the plane, implying that the energy flux integral (5.2.7) is independent of path Γ for steady crack growth situations. The energy flux integral takes the special form

$$F(\Gamma) = v \int_\Gamma \left[(U + T)n_1 - \sigma_{ji} n_j \frac{\partial u_i}{\partial x_1} \right] d\Gamma, \qquad (5.2.9)$$

where, in this case,

$$U = \int_{x_1 - vt}^{\infty} \sigma_{ji} \frac{\partial^2 u_i}{\partial x_j \partial \xi} \, d\xi, \qquad T = \tfrac{1}{2}\rho v^2 \frac{\partial u_i}{\partial x_1} \frac{\partial u_i}{\partial x_1}. \qquad (5.2.10)$$

This property of path-independence is exploited in analyzing crack growth problems in Sections 5.3, 5.4, and 8.3.

The energy flux integral can be extended to the case of three-dimensional deformation fields. If Γ is reinterpreted as a surface in space with local unit normal vector with components n_i and if the surface is translating in the direction defined by the unit vector with components m_i with instantaneous speed v, then the energy flux through this surface is

$$F(\Gamma) = \int_\Gamma \left[\sigma_{ji} \frac{\partial u_i}{\partial t} + (U + T)v m_j \right] n_j \, d\Gamma, \qquad (5.2.11)$$

where the integration is taken over the surface Γ. In this case, F has the physical dimensions of force×length/time. The result (5.2.7) or (5.2.11) is a purely mechanical result that does not depend on the existence of a thermodynamic equation of state. All applications to follow are based on this mechanical result.

An expression similar to (5.2.11), including internal energy and heat flux quantities, was presented by Kostrov and Nikitin (1970) on the basis of the first law of thermodynamics, and the result was further discussed by Nilsson and Ståhle (1988). The existence of an internal energy presumes an underlying thermodynamic equation of state (Malvern 1969). If an equation of state exists as a constitutive assumption, then an equivalence between (5.2.11) and the expression proposed by Kostrov and Nikitin (1970) can be established. In the absence of heat conduction or any other nonmechanical energy transfer modes, the two forms are the same if the quantity U in (5.2.11) is identified as the internal energy of the material per unit volume. In the presence of heat conduction (but no other nonmechanical energy transfer modes), the results are equivalent if U is related to the internal energy per unit mass e by

$$\frac{\partial U}{\partial t} = \rho \frac{\partial e}{\partial t} + \frac{\partial q_j}{\partial x_j}, \qquad (5.2.12)$$

where ρ is the material mass density and q_j are the components of the heat flux vector. This expression states that the internal energy density and the stress work density at a point differ by an amount equal to the net heat flow away from that point. When (5.2.12) applies, terms in the energy-momentum equation (5.2.2) can be regrouped as

$$\frac{\partial}{\partial x_j}\left(\sigma_{ji}\frac{\partial u_i}{\partial t} - q_j\right) - \frac{\partial}{\partial t}(\rho e + T) = 0. \qquad (5.2.13)$$

Application of the divergence theorem then yields the energy flux expression given by Kostrov and Nikitin (1970).

Finally, it is noted that the energy flux integral does not depend on the fact that the energy flux is associated with crack propagation. Rather, the expression (5.2.7) is a general expression for energy flux through a surface translating through a deforming solid. If the total deformation rate, that is, the symmetric part of $\partial^2 u_i / \partial x_j \partial t$, is viewed as the sum of an elastic deformation rate and an inelastic deformation rate, then the stress work density U can also be viewed as the sum of an elastic (recoverable) part and an inelastic (dissipated) part. Then the flux of recoverable energy through the contour Γ is given by (5.2.7) with U replaced by the elastic part. An example of this separation of U into elastic and inelastic parts appears in Section 8.3.

5.3 Elastodynamic crack growth

Applications of the general energy flux integral $F(\Gamma)$ for the case of elastodynamic crack growth are considered next. For linear elastic material response, the stress work density is the strain energy density $U = \frac{1}{2}\sigma_{ij}\epsilon_{ij} = \frac{1}{2}C_{ijkl}\epsilon_{ij}\epsilon_{kl} = \frac{1}{2}M_{ijkl}\sigma_{ij}\sigma_{kl}$, where M_{ijkl} is the array of elastic compliance parameters for the material. For isotropic response

$$M_{ijkl} = -\frac{\nu}{2\mu(1+\nu)}\delta_{ij}\delta_{kl} + \frac{1}{4\mu}(\delta_{ik}\delta_{jl} + \delta_{il}\delta_{jk}) \qquad (5.3.1)$$

in terms of the shear modulus μ and Poisson's ratio ν. The elastic wave speeds are defined in terms of elastic moduli and the mass density in Section 1.2.

5.3.1 Dynamic energy release rate

The *dynamic energy release rate* is defined as the rate of mechanical energy flow out of the body and into the crack tip per unit crack advance. The energy flux, or energy flow per unit time, must be divided by v, the crack movement per unit time, in order to obtain the energy released from the body per unit crack advance. The dynamic energy release rate is denoted by G and it is defined by

$$
\begin{aligned}
G &= \lim_{\Gamma \to 0}\left\{\frac{F(\Gamma)}{v}\right\}\\[2mm]
&= \lim_{\Gamma \to 0}\left\{\frac{1}{v}\int_{\Gamma}\left[\sigma_{ij}n_j\frac{\partial u_i}{\partial t} + (U+T)vn_1\right]d\Gamma\right\},
\end{aligned}
\qquad (5.3.2)
$$

where n_i is the unit normal vector to Γ directed away from the crack tip and the limit implies that the contour Γ is shrunk onto the crack tip. For two-dimensional fields, the quantity G is the mechanical energy released per unit crack advance per unit thickness of the body in the direction perpendicular to the plane of deformation, so that the physical dimensions of G are force/length.

For the energy release rate concept to have a fundamental significance, the value calculated according to (5.3.2) must be independent of the shape of Γ. For linear elastic material response, the integral

(5.2.8) expressing the difference in values of the energy flux for two arbitrary crack tip paths Γ_1 and Γ_2 becomes

$$F(\Gamma_1) - F(\Gamma_2) = \int_{A_{12}} \left[\sigma_{ij} \frac{\partial}{\partial x_j} \left(\frac{\partial u_i}{\partial t} + v \frac{\partial u_i}{\partial x_1} \right) \right.$$

$$\left. + \rho \frac{\partial u_i}{\partial t} \frac{\partial}{\partial t} \left(\frac{\partial u_i}{\partial t} + v \frac{\partial u_i}{\partial x_1} \right) \right] dA, \qquad (5.3.3)$$

where the notation is as established in connection with (5.2.8). It is clear from the asymptotic analysis in Section 4.3 that, in the near tip region, u_i depends on x_1 and t only through $x_1 - vt$. If both contours Γ_1 and Γ_2 are within the near tip region, then the terms in parentheses in (5.3.3) vanish and the value of $F(\Gamma)$ is indeed path-independent for all paths within the near tip region. Consequently, the value of G will not depend on the choice of path Γ used to compute it.

In view of its definition (5.3.2), G is determined by the near crack tip mechanical field. In addition, the near tip spatial distribution of mechanical fields was established in Chapter 4 for each mode of deformation. Only the scalar multiplier, the elastic stress intensity factor, was undetermined. Consequently, the terms in (5.3.2) can be expressed in terms of their near tip universal spatial descriptions and the integral can be evaluated to obtain a relationship between the energy release rate and the stress intensity factor. For mode I deformation, the near tip stress and particle velocity components have the form

$$\sigma_{ij} \sim \frac{K_I}{\sqrt{2\pi r}} \Sigma_{ij}(\theta, v), \qquad \frac{\partial u_i}{\partial t} \sim \frac{v K_I}{\mu \sqrt{2\pi r}} V_i(\theta, v) \qquad (5.3.4)$$

in terms of crack tip polar coordinates r, θ, where v is the instantaneous crack tip speed and the functions Σ_{ij} and V_i are given explicitly in (4.3.11) and (4.3.14), respectively. Thus, the integrand of (5.3.2) is $v K_I^2 B(\theta)/2\pi r \mu$, where

$$B(\theta) = \Sigma_{ij} n_j V_i + \tfrac{1}{2} n_1 \mu M_{ijkl} \Sigma_{ij} \Sigma_{kl} + \tfrac{1}{2} \frac{v^2}{c_s^2} n_1 V_i V_i. \qquad (5.3.5)$$

The integrand has several notable features. For instance, the crack speed v appears as a common factor, canceling the $1/v$ multiplier in

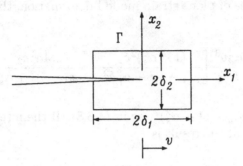

Figure 5.2. A particular choice of the contour Γ in (5.3.2) that is useful in establishing the relationship between energy release rate and stress intensity factor in (5.3.9).

the definition of G. Furthermore, the integrand depends on r only through the common factor $1/r$. If Γ is chosen to be a circular path of radius r centered at the crack tip then

$$G = \frac{K_I^2}{2\pi\mu} \int_{-\pi}^{+\pi} B(\theta)\, d\theta \,. \qquad (5.3.6)$$

Thus, G is proportional to K_I^2/μ and the dimensionless proportionality factor depends on the instantaneous crack speed v and on Poisson's ratio for the material. The integral (5.3.6) was evaluated for a particular case by Atkinson and Eshelby (1968).

   The value of the proportionality factor is obtained here by choosing the contour Γ to be the rectangle shown in Figure 5.2. This is a convenient choice because $n_1 = 0$ along the segments parallel to the x_1-axis. Furthermore, if the rectangle Γ is shrunk onto the crack tip by first letting $\delta_2 \to 0$ and then $\delta_1 \to 0$, there is no contribution to G from the segments parallel to the x_2-axis. Consequently, G can be computed by evaluating only the first term of (5.3.2) along the segments parallel to the x_1-axis, that is,

$$G = 2 \lim_{\delta_1 \to 0} \left\{ \lim_{\delta_2 \to 0} \int_{l-\delta_1}^{l+\delta_1} \sigma_{i2}(x_1, \delta_2, t) \frac{\partial u_i(x_1, \delta_2, t)}{\partial t} \, dx_1 \right\} , \qquad (5.3.7)$$

where l is the instantaneous x_1-coordinate of the crack tip and the factor 2 is introduced to account for the side of the rectangle at $x_2 = -\delta_2$ by symmetry.

For the case of plane strain mode I deformation, the integrand of (5.3.7) is

$$\sigma_{i2}\dot{u}_i = \frac{K_I^2}{4\pi\mu} \frac{\alpha_d^2 v^3}{Dc_s^2} \left[\frac{(1+\alpha_s^2)^2 \delta_2}{(x_1-l)^2 + \alpha_d^2 \delta_2^2} - \frac{4\alpha_s^2 \delta_2}{(x_1-l)^2 + \alpha_s^2 \delta_2^2} \right], \quad (5.3.8)$$

where $D = 4\alpha_d\alpha_s - (1+\alpha_s^2)^2$, as in (4.3.8). If the integral in (5.3.7) is then evaluated, the result is

$$G = \frac{v^2 \alpha_d}{2c_s^2 \mu D} K_I^2 . \quad (5.3.9)$$

If the computation is repeated for mode II (plane strain) and mode III deformations, then it is found that

$$G = \frac{1-\nu^2}{E} \left[A_I(v)K_I^2 + A_{II}(v)K_{II}^2 \right] + \frac{1}{2\mu} A_{III}(v)K_{III}^2 , \quad (5.3.10)$$

where

$$A_I = \frac{v^2 \alpha_d}{(1-\nu)c_s^2 D} , \quad A_{II} = \frac{v^2 \alpha_s}{(1-\nu)c_s^2 D} , \quad A_{III} = \frac{1}{\alpha_s} . \quad (5.3.11)$$

The functions A_I, A_{II}, and A_{III} are also universal functions, in the sense that they do not depend on the details of the applied loading or on the configuration of the body being analyzed. They do depend on the crack speed and on the properties of the material. Each function in (5.3.11) has the properties that $A \to 1$ as $v \to 0^+$, $dA/dv \to 0$ as $v \to 0^+$, and $A = O[(c_R - v)^{-1}]$ as $v \to c_R$ for plane strain, or $A = O[(c_s - v)^{-1}]$ as $v \to c_s$ for antiplane strain. The variation of each function $A(v)$ with v is monotonic. For plane stress conditions, the factor $1 - \nu^2$ in (5.3.10) is absent and the wave speeds must be adjusted to account for the lesser in-plane stiffness of the material.

As noted in Section 5.1, the relationship between stress intensity factor and energy release rate under equilibrium conditions ($v \to 0$) was first obtained by Irwin (1957, 1960). The result (5.3.10) thus generalizes the Irwin relationship to the case of elastodynamics. The quantity G introduced by Irwin, as well as the corresponding elastodynamic quantity introduced here, refers to the total mechanical energy change that is not taken into account in a continuum description of the crack advance process. Although G is sometimes called the elastic energy release rate, this terminology is strictly correct only if the crack advances under equilibrium conditions and under the fixed grip boundary condition. The fixed grip condition precludes the exchange of mechanical energy between the elastic body and its surroundings. If the deformation during crack advance is always at equilibrium, then the only global energy measure that can change its value is the elastic energy and, consequently, the energy release rate is exactly the elastic energy change due to crack growth. For crack growth with more general boundary loading conditions and with inertial resistance taken into account, energy variations may include exchanges with the surroundings and/or changes in kinetic energy. Under these general conditions, G cannot be related to elastic energy variations alone. Instead, G is the total mechanical energy release rate. For example, for dynamic crack growth and fixed grip conditions, there can be an exchange between elastic and kinetic energy of the body through wave reflection at a remote boundary. This variation in elastic energy has no immediate influence on the energy release rate G. The significance of the energy release rate concept, and on its role in crack growth criteria, will be discussed further in Chapters 7 and 8.

5.3.2 Cohesive zone models of crack tip behavior

The study of the advance of a sharp crack edge on the basis of an elastodynamic model leads naturally to the result that the stress is singular at the crack edge. The result cannot be accepted literally, of course. Instead, the result is accepted on the basis that the size of the region over which the material behavior deviates from linear elasticity is too small to be detected on the scale of the model. The cohesive zone idea provides a simple yet useful device for examining crack tip phenomena at a level of observation for which deviations from linearity in material response are admitted.

The essential idea of the cohesive zone model for plane strain mode I crack growth under the simplest circumstances is as follows. A tensile crack advances in a body under the action of some applied loading, and the configuration is such that a stress intensity factor field of the form (4.3.10) is fully established around the crack edge. The stress intensity factor due to this field alone is called the applied stress intensity factor K_{Iappl}. To simulate material response beyond the linear elastic range, the opening of the crack behind the mathematical crack tip at $x_1 = l(t)$ is resisted by a traction distribution called the *cohesive traction* or the *cohesive stress*, say $\sigma(\xi)$ for $-\Lambda < \xi < 0$, where $\xi = x_1 - l(t)$, acting on both crack faces. The cohesive traction leads to a negative stress intensity factor K_{Icoh} at $\xi = 0$. For a given cohesive stress, the total stress intensity factor is the sum $K_{Iappl} + K_{Icoh} = K_I$. It is now observed that the total stress can be rendered nonsingular by choosing the length of the cohesive zone Λ to satisfy the condition $K_I = 0$. It is noted here that for the case of mode III, the use of this device to eliminate the crack tip singularity does *not* result in all stress components being bounded everywhere.

To pursue the cohesive zone concept somewhat further, consider the specific case of crack growth by simple cleavage of a material. That is, the advance of the crack is viewed as the breaking of atomic bonds between pairs of atoms that end up on opposite faces of the advancing crack. A unique relationship between bonding force and separation distance is assumed. This situation was studied by Barenblatt (1959a, 1959b) on the basis of a cohesive zone model for equilibrium conditions, and the same idea was subsequently applied to the situation of dynamic cleavage crack growth by Barenblatt, Salganik, and Cherepanov (1962). In this case, the cohesive stress σ depends on position ξ with respect to the mathematical crack tip only through the separation of the crack faces at that position, say $\delta(\xi)$. The *crack opening displacement* is zero at the mathematical crack tip, so that $\delta(0) = 0$. The opening increases with distance behind the mathematical tip, and it reaches some final value at the end of the cohesive zone, that is, at $\xi = -\Lambda$, where the cohesive stress is effectively reduced to zero. The end of the cohesive zone is the physical crack tip in this model, and the opening displacement there has special significance. It is denoted here by $\delta(-\Lambda) = \delta_t$.

Suppose that, on a laboratory scale, the cleavage crack appears to be a perfectly sharp crack and to be growing at speed v under the action of an applied stress intensity factor K_{Iappl} as defined in Section 4.3. On a finer scale of observation, the same crack appears to open gradually from the leading edge of the cohesive zone (the mathematical crack tip at $\xi = 0$ where the limit of linearity of material behavior is presumably reached) to its full opening δ_t at the physical crack tip at $\xi = -\Lambda$. The final opening δ_t is determined by the interatomic force law assumed for the material. There must be a connection between the applied stress intensity factor K_{Iappl} and the parameters of the cohesive zone model, and it is the purpose here

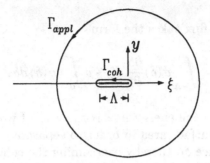

Figure 5.3. Two crack tip contours for the evaluation of the energy flux integral (5.2.7) in order to relate the crack tip parameters to the remote loading parameters.

to illustrate this connection on the basis of energy arguments. This equivalence was first examined and the conditions under which it holds were established by Willis (1967a) by means of direct stress analysis.

Consider two crack tip contours as shown in Figure 5.3. The contour Γ_{coh} embraces the cohesive zone, whereas the other contour Γ_{appl} is circular in shape, with center on the mathematical crack tip and radius large compared to Λ. Suppose that the larger contour passes through the region where the fields are accurately described by the stress intensity factor field. In view of (5.3.10), the energy flux through Γ_{appl} is

$$F(\Gamma_{appl}) = \frac{1 - \nu^2}{E} v A_I(v) K_{Ic}^2 . \qquad (5.3.12)$$

On the other hand, the fundamental energy flux integral (5.2.7) leads immediately to the result that the energy flux through the cohesive zone contour is

$$F(\Gamma_{coh}) = 2 \int_{-\Lambda}^{0} \sigma_{2j} \dot{u}_j \, d\xi , \qquad (5.3.13)$$

where the symmetry property of mode I deformation has been incorporated. But in the cohesive zone, $\delta(\xi, t) = 2u_2(\xi, 0, t)$, $\sigma_{12}(\xi, 0, t) = 0$, and $\sigma_{22}(\xi, 0, t) = \sigma(\delta)$. Therefore,

$$F(\Gamma_{coh}) = \int_{-\Lambda}^{0} \sigma(\delta) \dot{\delta}(\xi, t) \, d\xi , \qquad (5.3.14)$$

where the superposed dot denotes a material time derivative. Thus,

$$\dot{\delta}(\xi, t) = \frac{\partial \delta}{\partial t} - v \frac{\partial \delta}{\partial \xi} , \qquad (5.3.15)$$

so that the energy flux takes the form

$$F(\Gamma_{coh}) = \int_{-\Lambda}^{0} \sigma(\delta) \frac{\partial \delta}{\partial t} \, d\xi + v \int_{0}^{\delta_t} \sigma(\delta) \, d\delta . \qquad (5.3.16)$$

The second term in the energy flux is the rate of work that must be supplied per unit surface area in order to separate the crack faces to a separation distance δ_t, thereby overcoming the cohesive stress. The first term reflects the fact that the rate of work of the cohesive stress depends on internal adjustments associated with nonuniform growth within the cohesive zone.

If the crack growth process is *steady* as seen by an observer moving at uniform speed v with the crack tip, then the first term on the right side in (5.3.16) is identically zero. Furthermore, under these conditions, the energy flux integral is path-independent. Thus, in the case of steady crack growth, $F(\Gamma_{coh}) = F(\Gamma_{appl})$ or

$$\frac{1 - \nu^2}{E} A_I(v) K_{Iappl}^2 = \int_{0}^{\delta_t} \sigma(\delta) \, d\delta . \qquad (5.3.17)$$

This result establishes a relationship between the prevailing applied stress intensity factor and the cohesive property of the material, represented by the interaction law embodied in $\sigma(\delta)$. The plane stress result corresponding to (5.3.17) is obtained by a simple redefinition of the material constants.

If the notion is adopted that the cohesion arises from atomic interaction during growth of a cleavage crack, then the term on the right side of (5.3.17) is twice the surface energy of the material. This point, which was mentioned briefly in Section 1.1.2, will be pursued in Section 7.4, where crack growth criteria are considered. The matter of determining the size of the cohesive zone Λ is a stress analysis issue, and it is considered in Section 6.2. However, it is emphasized here that (5.3.17) is valid only for a cohesive zone that is small enough to be completely surrounded by a fully established stress intensity factor field characterized by K_{Iappl}. Strictly speaking, it is valid asymptotically as the length of the cohesive zone becomes vanishingly small.

A cohesive zone model of crack tip plasticity was introduced by Dugdale (1960). The physical phenomenon of interest was the highly localized plastic deformation ahead of the tip of a crack in a thin sheet of a ductile material subjected to tension in the direction perpendicular to the crack line. The mathematical model is similar to that already introduced above and, because the model circumvents the path dependent nature of plastic flow, it can be applied for growing as well as stationary cracks. The cohesive stress is taken as the tensile flow stress σ_o of an ideally plastic solid. In the case of steady dynamic crack growth under plane stress conditions the energy flow per unit crack advance into the cohesive zone is simply $\sigma_o \delta_t$. If the cohesive zone is completely surrounded by a stress intensity factor field with parameter K_{Iappl}, then

$$A_I(v)K^2_{Iappl} = E\sigma_o\delta_t, \qquad (5.3.18)$$

where the plane stress form of $A_I(v)$ is assumed.

Yet another cohesive zone model with particular relevance to shear cracks was introduced by Ida (1972, 1973) to analyze dynamic earth faulting phenomena and by Palmer and Rice (1973) to analyze stability of a slope under gravitational loading. This model also has the feature that the cohesive stress representing the interaction between the crack faces depends on the relative slip between the faces. For a plane strain mode II crack occupying the half plane $x_2 = 0$, $x_1 < vt$ advancing in the x_1-direction steadily at speed v, the relative coordinate $\xi = x_1 - vt$ is again introduced. The relative slip between the crack faces is $\delta(\xi) = u_1(\xi, 0^+) - u_1(\xi, 0^-)$ and the cohesive shear traction is denoted by $\tau(\delta)$. The essential features of the model are as follows. The shear stress σ_{12} in the unslipped portion of the plane $\xi > 0$ is assumed to increase to a characteristic value τ_{sf} as ξ decreases toward the mathematical crack tip at $\xi = 0$. The magnitude of τ_{sf} may be loosely identified with the static frictional strength of the interface along which the shear crack advances. At the mathematical crack tip the relative slip is zero, that is, $\delta(0) = 0$. As ξ decreases further, the crack faces begin to slide so that δ increases. As δ increases, the cohesive shear stress $\tau(\delta)$ decreases monotonically. When the relative slip has increased to the level δ_t, the cohesive stress has been reduced to a characteristic value τ_{df}. The latter value has been identified with the dynamic friction strength of the interface. For values of δ greater than δ_t the relative motion of the crack faces is resisted by a uniform shear traction of magnitude τ_{df}. The cohesive zone properties are depicted in Figure 5.4. This model is commonly called the *slip-weakening model* because, once the strength limit is

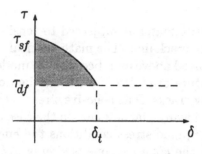

Figure 5.4. The relationship between cohesive zone traction τ and the amount of relative slip δ for the slip-weakening model. The shaded area is the energy release rate in (5.3.19).

achieved on the fracture plane, the cohesive stress magnitude falls off as slip progresses.

The excess of the shear stress over and above the residual level τ_{df} is the cohesive stress that must be overcome to advance the crack. For steady growth of a plane strain mode II shear crack under these conditions, the energy that must be supplied per unit crack advance is

$$G = \int_0^{\delta_t} [\tau(\delta) - \tau_{df}] \, d\delta \,, \qquad (5.3.19)$$

which is shown as the shaded area in Figure 5.4.

5.3.3 Special forms for numerical computation

In elastodynamic crack growth analysis based on numerical finite element or finite difference methods, it is difficult to calculate the dynamic energy release rate by evaluating the expression for G from the fundamental definition in (5.3.2). The difficulty arises from the fact that G is given in (5.3.2) in terms of fields evaluated at points arbitrarily close to the crack tip. In numerical approaches, values of fields are available only at discrete points and these values are least accurate within the high gradient region around the crack tip. Two computational algorithms that have been successfully applied for calculating G are briefly described here.

Consider first a method based on (5.2.8) which is known as the *area domain integral method* (Kishimoto, Aoki, and Sakata 1980c; Nishioka and Atluri 1983; Nakamura, Shih, and Freund 1985a). The relation (5.2.8) is valid for any contours Γ_1 and Γ_2 encircling a simply connected region free of body forces under conditions for which the various quantities can be evaluated. With a view toward computational problems, consider the particular case for which Γ_1 is shrunk onto the crack tip and $\Gamma_2 = \Gamma$ encircles the crack tip at a sufficient distance from it so that accurate numerical results can be expected for points along Γ. Then, in light of (5.3.2),

$$ G = \frac{1}{v}F(\Gamma) + \frac{1}{v}\int_{A(\Gamma)} \frac{\partial}{\partial x_j}\left[\sigma_{ij}\frac{\partial u_i}{\partial t} + v(U + T)\right]dA, \qquad (5.3.20) $$

where Γ is a contour encircling the crack tip and $A(\Gamma)$ is the area enclosed by Γ and the crack faces up to the crack tip. The integrand of the area integral appears to be singular as r^{-2} for $r \to 0$, in which case the integral would be divergent. As noted by Nakamura et al. (1985a), however, this singularity is only apparent and the integral does indeed have a finite value for boundary conditions of interest.

Many different forms of the type (5.3.20) can be obtained by applying the divergence theorem and by recognizing that $\partial f/\partial t \approx -v\partial f/\partial x_1$ for any field f very near to the crack tip, but all of these relationships have the form of a line integral along a crack tip contour Γ plus a surface integral over the area bounded by Γ and the crack faces up to the crack tip. Even though the value of G computed in this way does not depend on the particular choice of Γ, the descriptive term "path-independent integral" is not applicable due to the presence of the area term. Advantages and disadvantages of some particular forms of (5.3.20) are discussed by Nakamura et al. (1985a), who present numerical finite element applications of the area domain method of computing G. An application based on a finite difference numerical method is described by Ravichandran and Clifton (1989).

A second algorithm that has been useful in extracting values of G for elastodynamic crack growth in finite element simulations is the *node release method*. The main idea underlying this method can be seen in the expression (5.3.13) for the rate of energy flux out of the body when a crack advances by gradual reduction of the tractions restraining the crack faces as the crack opens (Keegstra 1976; Rydholm, Fredriksson, and Nilsson 1978). Consider planar crack growth in a two-dimensional body covered by a finite element mesh so that the crack path extends from node to node along straight element boundaries. A finite element node ahead of the moving crack tip is viewed as being two coincident nodes that are kinematically constrained against relative displacement until the crack tip arrives.

When the crack tip passes the material point occupied by the node pair, the nodes are allowed to separate dynamically under the action of equal but opposite nodal forces acting to resist the opening. The initial magnitude of these nodal forces is typically assumed to be the magnitude of the nodal constraint force at the instant that the crack tip passes the node pair. The kinematic constraint on the node pair is released at the instant the crack tip passes, and the nodal force magnitude is assumed to gradually diminish from its initial value to zero as the crack tip propagates to the next node pair. Several schemes for reducing the nodal force magnitude have been introduced but, typically, the magnitude is always proportional to the initial value, with the proportionality factor depending instantaneously on the fraction of the distance to the next node pair that the crack tip has traversed along an element side. For example, consider crack growth in a single mode and suppose that the nodal constraint force at a particular node pair has magnitude Q_0 when the crack tip arrives. Let H denote the distance to the next node pair along the fracture path ahead of the crack tip. If the crack tip is instantaneously at a distance h beyond the separating node pair, then the instantaneous value of the restraining force there, say Q, is assumed to have the form

$$Q = Q_0 \, q(h/H), \qquad (5.3.21)$$

where $q(\cdot)$ is a monotonically decreasing function of its argument with $q(0) = 1$ and $q(1) = 0$. Some particular choices that have been considered are $q(x) = (1 - x)^{1/2}$ by Rydholm et al. (1978), $q(x) = (1 - x)^{3/2}$ by Malluck and King (1980), and $q(x) = (1 - x)$ by Kobayashi, Emery, and Mall (1976). A comparative study was reported by Malluck and King (1980) in which some sensitivity of the results to the form of $q(\cdot)$ was found.

The two computational approaches for finding G noted above do not rely on special interpolation functions in the crack tip region for inferring field values between nodes in the finite element or finite difference discretization. There are other numerical approaches in which the knowledge of the crack tip asymptotic field for elastodynamic crack growth is exploited. Typically, the displacement interpolation functions very near to the crack tip are chosen to be precisely the displacement distributions corresponding to the asymptotic field; the common amplitude of these distributions is the stress intensity factor under single mode deformation conditions. This general approach, of which there are a number of variations, is usually called the *singular element method* in finite element analysis (Aoki et al. 1978; Kishimoto, Aoki, and Sakata 1980a; Nishioka and Atluri 1980a, 1980b). Although the idea is to compute the stress intensity factor directly, without the use of its relationship to energy release rate (5.3.10), it must be recognized that the variational work/energy principle underlying the finite element formulation must account for the flux of energy out of the body during crack growth.

5.4 Steady crack growth in a strip

Determination of the dependence of the dynamic stress intensity factor on body configuration and applied loading during rapid crack growth is a principal objective of stress analysis in dynamic fracture. Some direct methods of stress analysis for elastodynamic crack growth are introduced and illustrated in Chapter 6. One point that will become clear, if it is not already so, is that exact analytical results can be extracted for only a few situations. Furthermore, analytical results for configurations with boundaries, either constrained or free, other than the crack faces are rare. A particular situation for which exact stress intensity factor results can be obtained is that of steady crack growth along the length of a strip subjected to simple boundary loading on its edges. This configuration has been used by Nilsson (1974a) in experiments on rapid fracture in high strength steel plates. The full boundary value problem can be approached with a view toward integrating the field equations, thereby establishing the dependence of the stress intensity factor on loading, configuration, and crack speed. This approach was described by Nilsson (1972) and it is outlined in Section 6.2. Certain results can be obtained more directly by application of energy arguments, however, and the energy approach is summarized in this section.

5.4.1 Strip with uniform normal edge displacement

Consider a homogeneous and isotropic elastic strip in the region of the x_1, x_2-plane with $-\infty < x_1 < +\infty$, $-h < x_2 < +h$. Suppose the body is a thin sheet or plate, and that the elastodynamic fields are adequately described by the two-dimensional theory of generalized plane stress. The analysis for plane strain differs only in the interpretation of elastic constants. Suppose that the edges of the strip at $x_2 = \pm h$ are displaced away from each other a distance $2u_0$, that is,

$$u_2(x_1, \pm h, t) = \pm u_0. \tag{5.4.1}$$

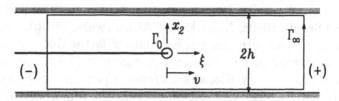

Figure 5.5. Steady crack growth at speed v in a strip of width $2h$. The edges
of the strip are given a uniform normal displacement u_0, and the relationship
between the boundary loading and the crack tip stress intensity factor is obtained
by means of the path-independent integral evaluated on the near tip and remote
contours shown.

Specification of the complementary component of traction or displacement is postponed, but the second boundary condition will be either

$$u_1(x_1, \pm h, t) = 0 \quad \text{or} \quad \sigma_{12}(x_1, \pm h, t) = 0. \qquad (5.4.2)$$

Suppose that a crack with traction-free faces grows steadily at speed
$v < c_R$ in the x_1-direction along the symmetry line $x_2 = 0$ of the
strip; see Figure 5.5. Without loss of generality, the crack is assumed
to pass $x_1 = 0$ at time $t = 0$. If the crack growth process is steady,
then the elastic fields depend on x_1 and t only through $\xi = x_1 - vt$.

Both far ahead of the crack tip and far behind the crack tip, fields
are uniform in the x_1-direction. The kinematic boundary condition
(5.4.1) requires that the extensional strain in the x_2-direction far
ahead of the tip is $\epsilon_{22} \to \epsilon_0$ as $\xi \to +\infty$, where $\epsilon_0 = u_0/h$. Far
behind the crack tip, the only stress distribution that is uniform in
the x_1-direction and that is consistent with the condition that the
crack face is traction-free has the property that

$$\sigma_{22}(\xi, x_2) \to 0, \quad \sigma_{12}(\xi, x_2) \to 0 \quad \text{as} \quad \xi \to -\infty. \qquad (5.4.3)$$

To demonstrate this for one of the stress components, consider balance
of momentum in the x_1-direction which requires that

$$\frac{\partial \sigma_{11}}{\partial x_1} + \frac{\partial \sigma_{12}}{\partial x_2} = \rho \frac{\partial^2 u_1}{\partial t^2}. \qquad (5.4.4)$$

For fields that are steady in the x_1-direction, this balance equation
reduces to $\partial \sigma_{12}/\partial x_2 = 0$, which implies that σ_{12} is a constant. The

boundary condition that $\sigma_{12} = 0$ on the crack faces then requires that the constant is zero.

Suppose that the limiting value of any field as $\xi \to \pm\infty$ is indicated by a superposed $(+)$ or $(-)$ sign. For example, $\epsilon_{22}^+(\xi, x_2) = \epsilon_0$ or $\sigma_{22}^-(\xi, x_2) = 0$. Then, the stress–strain relations imply that

$$\sigma_{22}^+ = \frac{E}{1 - \nu^2}(\epsilon_0 + \nu\epsilon_{11}^+),$$

$$\sigma_{11}^+ = \frac{E}{1 - \nu^2}(\epsilon_{11}^+ + \nu\epsilon_0), \tag{5.4.5}$$

$$\sigma_{11}^- = E\epsilon_{11}^-.$$

Consider now the particular case $(5.4.2)_1$ so that the edges of the strip are constrained against displacement in the x_1-direction. Then, $\epsilon_{11}^\pm(\xi, x_2) = 0$ and $\epsilon_{12}^+(\xi, x_2) = 0$. The value of the energy release rate for a remote contour Γ_∞ shown in Figure 5.5 is

$$G(\Gamma_\infty) = \int_{\Gamma_\infty} \left[(T + U)n_1 - \sigma_{ij}n_j \frac{\partial u_i}{\partial x_1} \right] d\Gamma = \frac{h\epsilon_0^2 E}{1 - \nu^2}. \tag{5.4.6}$$

The only contribution to the integral arises from the segment of Γ_∞ parallel to the x_2-axis at $\xi \to +\infty$. The path independence of the energy release rate integral for steady crack growth implies that

$$G(\Gamma_0) = G(\Gamma_\infty), \tag{5.4.7}$$

where Γ_0 is a small contour surrounding the crack tip within the region where the asymptotic stress intensity factor field prevails. In view of (5.3.10), the stress intensity factor is

$$K_I = \frac{u_0 E}{\sqrt{h(1 - \nu^2)A_I(v)}} \tag{5.4.8}$$

for any crack speed v. The stress intensity factor depends linearly on u_0, as it must in view of linearity of the system. The nature of the dependence on E and h could also have been anticipated on the basis of dimensional considerations.

Although the result (5.4.8) was derived through formal application of the energy release rate integral, the same result could have
been obtained even more directly by means of an intuitive argument.
The mechanical energy density of the system per unit length in the
x_1-direction far ahead of the crack tip is $\epsilon_0^2 hE/(1-\nu^2)$. Far behind the
crack tip, the energy density per unit length in the x_1-direction is zero.
During advance of the crack tip a unit distance in the x_1-direction,
a length of material with the former energy density is essentially
replaced by an equal length with zero energy density. There is no
energy exchange of the body with its surroundings and overall energy
must be balanced for a steady state situation. Consequently, the
mechanical energy lost from the body must flow out through the crack
tip. It follows immediately that

$$\frac{A_I(v)K_I^2}{E} = \frac{\epsilon_0^2 hE}{1-\nu^2}. \tag{5.4.9}$$

The argument is quite general, and it can be applied to advantage
in a variety of situations. Some equilibrium situations are considered
from this point of view by Rice (1968).

Suppose for the moment that the boundary condition $(5.4.2)_2$ applies, that is, suppose that $\sigma_{12}(\xi, \pm h) = 0$ is considered. In this case,
the body is unconstrained in the x_1-direction so that an additional
condition must be imposed to ensure overall momentum balance in
the x_1-direction. For the symmetries associated with mode I deformation and for fields uniform in the x_1-direction at remote points,
this condition takes the form

$$\int_{-h}^{+h} \left[\sigma_{11}^+ - \sigma_{11}^- - \rho v^2(\epsilon_{11}^+ - \epsilon_{11}^-)\right] dx_2 = 0. \tag{5.4.10}$$

For purposes of illustration, suppose that all limiting fields are
uniform across the strip so that the integrand of (5.4.10) vanishes
pointwise in x_2. Consider the particular case with $\sigma_{11}^+ = 0$, which
implies from (5.4.5) that $\epsilon_{11}^+ = -\nu\epsilon_0$. Thus, global momentum balance
(5.4.10) requires that $\sigma_{11}^- = \rho v^2(\epsilon_{11}^- + \nu\epsilon_0)$. In light of (5.4.5),

$$\epsilon_{11}^-(\xi, x_2) = \frac{\nu\epsilon_0}{(E/\rho v^2 - 1)}. \tag{5.4.11}$$

With this result in hand, either the energy release rate integral or the energy balance argument leads to the result that the crack tip energy release rate is

$$G = \epsilon_0^2 h E \left[\frac{1 - (1 - \nu^2) v^2 / c_0^2}{1 - v^2 / c_0^2} \right] , \qquad (5.4.12)$$

where $c_0 = \sqrt{E/\rho}$. Note that v is always less than c_R. Other particular cases, such as $\epsilon_{11}^+ = 0$, can be considered as well.

5.4.2 Shear crack with a cohesive zone in a strip

As a second example of steady dynamic crack growth in a strip, consider the propagation of a mode II shear crack in a slab of material of thickness $2h$ under plane strain conditions; see Figure 5.6. Under the action of applied displacements

$$u_1(x_1, \pm h, t) = \pm u_0 , \qquad u_2(x_1, \pm h, t) = 0 \qquad (5.4.13)$$

the mode II crack grows in the x_1-direction at speed $v < c_R$. The coordinate x_1 is eliminated in favor of ξ through $\xi = x_1 - vt$. The rectangular coordinate system ξ, x_2 has its origin at the crack tip and it translates with the mathematical tip. Although a singularity in the stress component σ_{12} at the leading edge of the crack can be considered, a linear slip-weakening cohesive zone is introduced instead. The features of the slip-weakening cohesive zone model are discussed in Section 5.3.2. The slip across the plane $x_2 = 0$ is denoted by $\delta(\xi) = u_1(\xi, 0^+) - u_1(\xi, 0^-)$. On this plane, $\sigma_{12}(0, 0) = \tau_{sf}$ with $\delta(0) = 0$, $\sigma_{12}(-\Lambda, 0) = \tau_{df}$ with $\delta(-\Lambda) = \delta_t$, and $\sigma_{12}(-\lambda, 0) = \tau_{df}$ with $\delta(-\lambda) = \delta_T > \delta_t$. According to this model, the physical process of slipping begins at $\xi = 0$, it continues throughout the interval $-\lambda < \xi < 0$, and it terminates with total relative displacement of δ_T at $\xi = -\lambda$. Relative slipping is resisted by the cohesive/frictional traction $\sigma_{12}(\xi, 0) = \tau_{sf} - (\tau_{sf} - \tau_{df})\delta(\xi)/\delta_t$ in the interval $-\Lambda < \xi < 0$, and by the uniform traction τ_{df} in the interval $-\lambda < \xi < -\Lambda$. The frictional stress might be due to Coulomb friction arising from a superimposed uniform compressive stress in the x_2-direction. If such a stress were to be introduced by specifying the alternate boundary condition $u_2(x_1, \pm h, t) = \mp v_0$, however, it would have no influence on the main results to be obtained in this subsection.

A notable feature of this particular problem is that it represents crack propagation under displacement control or very stiff loading

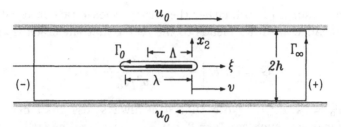

Figure 5.6. Steady growth of a shear crack at speed v in a slab of thickness $2h$. The faces of the slab are given a uniform relative tangential displacement $2u_0$. The relationship between the boundary loading and the features of a crack tip slip-weakening zone is obtained by means of the energy integral (5.2.9) evaluated on the contours shown.

conditions, in contrast to crack propagation under stress control or very compliant boundary loading as modeled by growth of a crack in an otherwise unbounded solid. Displacement controlled conditions allow for the consideration of stress relief by crack extension, that is, the shear stress on the slip plane at some point far behind the slipping region, say σ_{12}^-, will be less than the shear stress on the fault plane far ahead of the slipping region, say σ_{12}^+. In terms of the boundary displacement and the total residual offset,

$$\sigma_{12}^+ = \mu u_0/h, \qquad \sigma_{12}^- = \mu(u_0 - \delta_T/2)/h. \qquad (5.4.14)$$

The relationship among the various stress magnitudes and slip magnitudes is obtained by direct application of the path independent energy integral (5.2.9).

The two choices of contour Γ for evaluation of the energy integral are shown in Figure 5.6. The value of $F(\Gamma_\infty)$ will be considered first. For points far ahead of and far behind the crack tip the only nonzero stress component is $\sigma_{12}(\xi, x_2) = \sigma_{12}^+$ or σ_{12}^-, respectively, and the only nonzero displacement gradient is $\partial u_1(\xi, x_2)/\partial x_2 = \sigma_{12}^+/\mu$ or σ_{12}^-/μ, respectively. These components are essentially uniform over remote regions. For the portion of the remote contour along $x_2 = \pm h$, the displacement components are uniform and $n_1 = 0$ so that the integrand of (5.2.9) vanishes identically and there is no contribution to the value of F. The only contributions to F arise from the segments of Γ_∞ parallel to the x_2-axis in the remote uniform regions. Thus,

$$F(\Gamma_\infty) = v(\sigma_{12}^{+\,2} - \sigma_{12}^{-\,2})h/\mu = v(\sigma_{12}^+ + \sigma_{12}^-)\delta_T/2. \qquad (5.4.15)$$

As is indicated in Figure 5.6, the other choice of contour Γ_0 embraces the crack and $n_1 = 0$ at virtually all points of this path. Thus,

$$F(\Gamma_0) = -v \int_{-\lambda}^{-\Lambda} \tau_{df} \frac{\partial \delta}{\partial \xi} \, d\xi - v \int_{-\Lambda}^{0} \tau(\delta) \frac{\partial \delta}{\partial \xi} \, d\xi \qquad (5.4.16)$$

$$= v(\tau_{sf} - \tau_{df})\delta_t/2 + v\tau_{df}\delta_T \, .$$

In view of the property of path independence, $F(\Gamma_\infty) = F(\Gamma_0)$ or

$$\frac{\delta_T}{\delta_t} = \frac{\tau_{sf} - \tau_{df}}{\sigma_{12}^+ + \sigma_{12}^- - 2\tau_{df}} \, . \qquad (5.4.17)$$

This analysis introduces two stress differences that can be associated with *stress drops* in the seismological characterization of earthquakes. These stress differences are the dynamic stress drop $(\sigma_{12}^+ - \tau_{df})$ and the total stress drop $(\sigma_{12}^+ - \sigma_{12}^-)$. The second of these is particularly interesting because it satisfies the relationship $\sigma_{12}^+ - \sigma_{12}^- = \mu\delta_T/2h$. If h can be identified in some way with the total distance of travel of the edge of the slip region, then this expression coincides with one of the standard definitions of stress drop in seismology. It is noteworthy that σ_{12}^+ and σ_{12}^- do not represent frictional properties of the slip surface. These magnitudes are arbitrary except that $\sigma_{12}^+ + \sigma_{12}^- > 2\tau_{df}$ and $\sigma_{12}^+ > \sigma_{12}^-$ for the process to occur at all. Unfortunately, the energy integral approach provides no information on the magnitudes of the physical lengths λ and Λ in the model.

The strip problems considered in this chapter are representative of steady crack propagation problems that can be analyzed by means of energy arguments. In none of these cases have complete stress and deformation fields been determined by solution of the corresponding boundary value problems. As long as a crack edge is moving more or less steadily, a steady state analysis yields relationships among the various parameters used to characterize the process without reference to a growth criterion. For example, this approach can provide relationships among the parameters used to describe a crack growth process at different scales of observation, as noted in Section 5.3.2. The study of the way in which a crack tip or slip zone edge moves according to a particular crack propagation condition must be based on a transient analysis of the propagation process, and such analysis is considered in Chapters 7 and 8.

5.5 Elementary applications in structural mechanics

The energy flux integral (5.2.7) was derived under the assumption of a general field description of the mechanical behavior of a continuum. The concepts are more general than the derivation suggests, however. In this section, several applications of the same energy concepts are considered within the framework of elastic structural analysis. The applications illustrate the relationship between stress resultants at the level of structural analysis and crack tip parameters. Furthermore, in some particular cases, energy arguments can be applied to determine crack tip parameters in terms of applied loading for dynamic crack growth in structural elements.

Analytical models of dynamic crack growth involving a single spatial dimension have been developed in connection with dynamic fracture toughness testing by Kanninen (1974), Bilek and Burns (1974), and Freund (1977), in connection with earthquake source modeling by Knopoff, Mouton, and Burridge (1973) and Landoni and Knopoff (1981), and in connection with peeling of a bonded layer by Hellan (1978a, 1978b). These models have been remarkably successful in enhancing insight into the particular physical process of interest, and the few models for which complete mathematical solutions exist provide rare opportunities to see all aspects of a dynamic crack growth event in a common and relatively transparent framework.

5.5.1 A one-dimensional string model

The discussion here will focus on the near tip fields for a model developed to illustrate the influence of reflected waves on crack growth in a double cantilever beam fracture specimen (Freund 1977). The description can be recast in terms of the dynamics of an elastic string, an even simpler conceptual model, and it has been analyzed with numerous variations from this point of view by Burridge and Keller (1978).

Consider a stretched elastic string lying along the positive x-axis, that is, along $0 < x < \infty$. The string has mass per unit length ρ and characteristic wave speed c. The small transverse deflection is $w(x,t)$, the elastic strain is $\gamma(x,t) = \partial w/\partial x$, and the transverse particle velocity is $\eta(x,t) = \partial w/\partial t$. Initially, the string is free of transverse loading in the interval $0 < x < l_0$ and it is bonded to a rigid, flat surface for $x > l_0$. A boundary condition in the form of a condition on the deflection $w(0,t)$, the slope $\gamma(0,t)$, or a linear combination of the two is required; specific cases will be considered in a later section when the full motion of the string during a crack growth process is considered. Finally, suppose that the string is initially deflected but at rest, so that $w(x, 0) = w_0(x)$ and $\eta(x, 0) = 0$, where $w_0(x)$ is specified. At time $t = 0$, the string begins to peel away from the surface, so that at some later time $t > 0$ the free length is $0 < x < l(t)$. Thus, the field equations

$$\frac{\partial \gamma}{\partial t} = \frac{\partial \eta}{\partial x}, \qquad c^2 \frac{\partial \gamma}{\partial x} = \frac{\partial \eta}{\partial t} \qquad (5.5.1)$$

are to be satisfied in the time-dependent interval $0 < x < l(t)$ subject to the stated initial and boundary conditions. If $l(t)$ is specified, then the solution of the governing differential equation is subject to the "crack tip" condition that $w\big(l(t), t\big) = 0$ or, in rate form, that

$$\dot{l}\gamma + \eta = 0 \quad \text{at} \quad x = l(t). \qquad (5.5.2)$$

The field equations and boundary conditions can be satisfied for a host of crack motions $l(t)$. If a crack growth criterion is also specified at $x = l(t)$, then a part of the solution procedure is to find that particular crack motion $l(t)$ for which the growth criterion is satisfied pointwise in time. Some cases of crack motion will be discussed in Chapter 7 after the issues of crack tip singular field and energy release rate have been considered within the framework of the dynamic string model.

Ahead of the crack tip the strain and particle velocity are zero, whereas behind the tip they have nonzero values, in general. Thus, the crack tip carries a propagating discontinuity in strain and particle velocity. Because the equation governing motion of the string is the elementary wave equation, discontinuities propagate freely only at the characteristic wave speed of the string c. But the speed of the crack tip $\dot{l}(t)$ is not restricted to be equal to c, hence the propagating crack tip must carry a momentum source or sink. In view of the fact that the only degree of freedom is the transverse deflection, the momentum source must be a generalized transverse force that is work-conjugate to w, namely, a concentrated force acting at $x = l(t)$ and traveling with that point. This is the crack tip singular field for this simple structure. The elastic energy density and kinetic energy density for the string are

$$U(x, t) = \tfrac{1}{2}\rho c^2 \gamma^2, \qquad T(x, t) = \tfrac{1}{2}\rho \eta^2. \qquad (5.5.3)$$

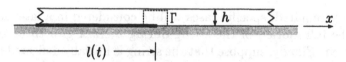

Figure 5.7. A crack tip contour for evaluation of the energy release rate in a model of crack growth based on the elastic string.

Suppose that the string is now viewed as a two-dimensional structure with the internal deformation fields constrained to be consistent with the string idealization. Thus, for example, all points on a cross section have a velocity in the direction transverse to the x-axis equal to η and no velocity in the direction of the x-axis. The component of traction on a cross section in the direction transverse to the x-axis is $\pm \rho c^2 \gamma$, where ρc^2 is the tension in the undeflected string and the algebraic sign is determined by the direction of the outward normal on the cross section in the $\pm x$-direction.

Consider now a contour Γ as shown in the sketch in Figure 5.7. The contour begins on the crack face indefinitely close to the crack tip at $x = l^-$, it traverses the cross section there, then extends along a portion of the upper traction-free surface of the material to $x = l^+$, and then traverses the cross section to the rigid surface. Ahead of the crack tip, both γ and η are zero. Thus, the portion of Γ ahead of the tip yields no contribution to the energy flux. Along the segment from $x = l^-$ to $x = l^+$ the traction is zero and the outward normal vector component n_1 is zero, so no contribution is obtained from this segment of Γ. Therefore, the only contribution to energy flux arises from the portion of Γ along the cross section at $x = l^-$. Application of the energy flux integral (5.2.7) yields

$$F = G\dot{l} = -\rho c^2 \gamma \eta - \tfrac{1}{2}\rho c^2 \gamma^2 \dot{l} - \tfrac{1}{2}\rho \eta^2 \dot{l} \qquad (5.5.4)$$

and, in light of the kinematic boundary condition (5.5.2) at the crack tip,

$$G = \tfrac{1}{2}\rho c^2 (1 - \dot{l}^2/c^2)\gamma(l^-, t)^2 . \qquad (5.5.5)$$

This expression provides the energy release rate G for an arbitrary crack growth rate $\dot{l}(t)$ in terms of the value of a crack tip parameter $\gamma(l^-, t)$. If this model is developed to simulate the double cantilever beam situation mentioned above then the factor one-half in (5.5.5) would be absent to account for energy contributions from both halves of the symmetric specimen. Note that G has the physical dimensions of energy/length for this simple one-dimensional model.

The observation made in the foregoing discussion that the traction exerted on the string by the rigid surface is a concentrated restraining force Q acting at the moving point $x = l(t)$ provides the basis for an alternate derivation of the result (5.5.5). In general, the rate of work of a concentrated force on a body is the inner product of the force with particle velocity of the point on the body at which the force instantaneously acts. (In order to account for the possibility that the force moves with respect to the body on which it acts, this definition of the rate of work of a force is slightly more elaborate than is often necessary in continuum mechanics. In the present situation, the point of application of the force moves in a direction perpendicular to the direction of the force, so a naive definition of the rate of work of a force, based on the motion of its point of application as opposed to the particle velocity at its point of application, would lead to the incorrect conclusion that the rate of work is zero.) In the present case, the velocity η is discontinuous at $x = l(t)$ so that the rate of work is $\frac{1}{2}Q\eta(l^-, t)$. The present approach requires the evaluation of the integral $\int_{-\infty}^{\infty} \delta(x)H(x)\, dx$, where δ and H are the Dirac delta function and the unit step function, respectively. The integral can be evaluated formally by noting that $\delta(x) = dH(x)/dx$. A discussion of integrals of this type is presented by Fisher (1971) within the context of the theory of distributions. Application of the principle of impulse and momentum to the portion of the string enclosed in the contour Γ of Figure 5.7 leads to the conclusion that

$$Q = -\rho c^2 \gamma(l^-, t) - \rho \dot{l} \eta(l^-, t) = -\rho c^2 (1 - \dot{l}^2/c^2)\gamma(l^-, t). \qquad (5.5.6)$$

The energy flux is, by definition, the negative rate of work of the concentrated force, and the result (5.5.4) follows immediately.

Finally, it is noted that the result can also be derived by direct application of the overall energy rate balance. That is, if P is the rate of work of the boundary load at $x = 0$ then

$$F = P - (\dot{U}_{tot} + \dot{T}_{tot}), \qquad (5.5.7)$$

where U_{tot} and T_{tot} are the integrals over the length of the string of the energy densities defined in (5.5.3). In the present notation, the work rate is $P = -\rho c^2 \eta(0,t) \gamma(0,t)$. Then

$$F = -\rho c^2 \eta(0,t) \gamma(0,t) - \int_0^l \left[\rho c^2 \gamma \frac{\partial \gamma}{\partial t} + \rho \eta \frac{\partial \eta}{\partial t} \right] dx$$

$$- \left[\tfrac{1}{2} \rho c^2 \gamma(l^-,t)^2 + \tfrac{1}{2} \rho \eta(l^-,t)^2 \right] \dot{l}. \quad (5.5.8)$$

If the terms $\partial \gamma / \partial t$ and $\partial \eta / \partial t$ in the integrand are replaced by $\partial \eta / \partial x$ and $c^2 \partial \gamma / \partial x$ from (5.5.1) then the integrand is an exact differential. The result of evaluating the integral is once again the expression (5.5.4) for energy flux.

5.5.2 Double cantilever beam configuration

A specimen configuration that has been employed for some crack propagation experiments is the split rectangular plate shown in the diagram in Figure 5.8. Because of its configuration, the specimen has been analyzed by a strength of materials approach, whereby the arms of the specimen are viewed as beams cantilevered at the crack tip end. For this reason, the configuration has become known as the double cantilever beam. The origin of the descriptive term for this specimen configuration is not clear, but Benbow and Roessler (1957), Gilman, Knudsen, and Walsh (1958), and Burns and Webb (1966) were among the first to take advantage of its simplicity. The deformation and associated stress distribution obtained by applying the usual assumptions of beam theory represent internally constrained plane strain or plane stress fields, and the general energy flux expression can be applied directly to compute the energy release rate. Suppose that the plane of deformation is the x_1, x_2-plane, that the crack is in the plane $x_2 = 0$, and that the tip is at $x_1 = l(t)$. The crack is assumed to open symmetrically in mode I. The contour Γ in the plane of deformation begins on one traction-free crack face at $x_1 = l(t) - 0$. It runs along a cross section of one beam arm to the traction-free boundary at $x_2 = h$, along this boundary until $x_1 = l(t) + 0$, and then along the cross section to $x_2 = -h$. It returns to the other crack face in such a way that Γ has reflective symmetry with respect to the crack plane.

Suppose that the energy flux integral (5.2.7) is now interpreted within the framework of elastic Bernoulli-Euler beam theory. The transverse deflection of the neutral axis of the upper beam arm at

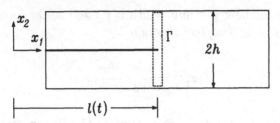

Figure 5.8. The split rectangular plate configuration, commonly known as the double cantilever beam configuration, showing a contour for evaluation of the energy release rate during crack growth.

$x_2 = h/2$ is denoted by $w(x_1, t)$. Ahead of the crack tip $x_1 > l(t)$, curvature and velocity are both zero and there is no contribution to the energy flux from this region. Along the portion of Γ with $x_2 = h$, the traction is zero and $n_1 = 0$, so there is no contribution from this portion of Γ either. Therefore, the energy flux integral is reduced to

$$F(\Gamma) = 2 \int_0^h \left[\sigma_{ij} n_j \frac{\partial u_i}{\partial t} + (U + T)\dot{l}n_1 \right]_{x_1 = l^-} dx_2 . \qquad (5.5.9)$$

For a Bernoulli-Euler beam cantilevered at $x_1 = l(t)$,

$$\frac{\partial u_1}{\partial t}(l^-, x_2, t) = - (x_2 - h/2)\frac{\partial^2 w}{\partial x_1 \partial t}(l^-, t) ,$$

$$\frac{\partial u_2}{\partial t}(l^-, x_2, t) = 0 . \qquad (5.5.10)$$

Furthermore, the slope of the neutral axis is zero there, that is, $\partial w(l^-, t)/\partial x_1 = 0$. The rate form of this condition is

$$\frac{\partial^2 w}{\partial x_1^2}\dot{l} + \frac{\partial^2 w}{\partial x_1 \partial t} = 0 \qquad (5.5.11)$$

as $x_1 \to l^-$.

The normal stress $\sigma_{11}(l^-, x_2, t)$ appearing in (5.5.9) is related to the bending moment per unit width in the beam $M(l^-, t)$ through

$$\sigma_{11}(l^-, t) = -\frac{12M(l^-, t)(x_2 - h/2)}{h^3} . \qquad (5.5.12)$$

The bending moment per unit width is related to the local curvature of the neutral axis $\partial^2 w/\partial x^2$ through

$$M(x,t) = \frac{Eh^3}{12}\frac{\partial^2 w}{\partial x^2}(x,t) \qquad (5.5.13)$$

for plane stress deformation. Each term in (5.5.9) can now be written in the form of $(x_2 - h/2)^2$ times $M(l^-,t)^2$. Because the dependence on x_2 is explicit, the integral can be evaluated with the result that the energy release rate is

$$F(\Gamma)/\dot{l} = G = \frac{12M(l^-,t)^2}{Eh^3}\left(1 - \frac{\dot{l}^2}{c_o^2}\right), \qquad (5.5.14)$$

where $M(l^-,t)$ is the bending moment per unit thickness in each beam arm at the crack tip end and $c_o^2 = E/\rho$. The second term in parentheses in (5.5.14) is the kinetic energy contribution due to the velocity distribution in (5.5.10). This "rotary inertia" contribution is not included in a derivation of the equation of motion of a Bernoulli-Euler beam, and it should be omitted if the energy release rate expression is to be interpreted strictly within the framework of Bernoulli-Euler beam theory. It is interesting to note that if the corresponding energy release rate expression is derived within the framework of Timoshenko beam theory, with both rotary inertia and shear deformation effects included, and the deformation is then constrained so that each cross section remains normal to the neutral axis as the beam deforms, the result is (5.5.14). If the result is to be applied to a state of plane strain rather than plane stress then a factor $(1 - \nu^2)$ should be inserted in the numerator of (5.5.14).

It can be argued, as in the case of compliance methods of equilibrium fracture mechanics, that the energy being supplied to the crack tip cross section is actually being absorbed at the crack tip. Then (5.3.10) and (5.5.14) yield an expression for the crack tip stress intensity factor for the specimen in terms of the internal bending moment at $x = l(t)$, namely,

$$K_I(t) = \frac{\sqrt{12}\,M(l^-,t)}{\sqrt{A_I(\dot{l})h^3(1 - \dot{l}^2/c_o^2)}}. \qquad (5.5.15)$$

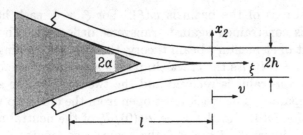

Figure 5.9. Schematic diagram of a long rectangular plate being split in half by a wedge. The process is analyzed as a steady crack growth process.

At the level of the beam approximation, the singularity at the crack tip on the fracture plane is a concentrated force. On the level of the plane elastic field, however, it is the square root singular stress distribution. A number of other connections between strength of materials models and elastic field models can be established in a similar manner.

5.5.3 Splitting of a beam with a wedge

The results of the previous subsection can be applied to study crack propagation phenomena in structures for which the beam idealization is appropriate. Consider the situation in which a long uniform elastic strip is split by driving a wedge at speed v along the symmetry line of the strip; see Figure 5.9. The stress state is assumed to be two-dimensional plane stress, the fields are steady as seen by an observer moving with the wedge or with the crack edge, and the system has the symmetries typical of mode I deformations. The total wedge angle is 2α, where $\alpha \ll 1$. The crack tip is located a distance l ahead of the point on the advancing wedge at which the arms first come into contact with the wedge. The purpose of this brief analysis is to obtain relationships among the wedge angle α, the crack length l, the speed v, and a crack tip characterizing parameter. The length l is supposed to be much greater than the beam height h so that the use of Bernoulli-Euler beam theory is suitable. Generalization of the procedure to more sophisticated beam or plate theories is possible.

The x_1-coordinate is eliminated in favor of $\xi = x_1 - vt$ and, without loss of generality, the position of the crack edge in the plane of deformation is taken to be $\xi = 0$, $x_2 = 0$. Only the half of the strip with $x_2 \geq 0$ is considered, and the influence of the other beam arm is included by symmetry. The beam arm is modeled as a beam cantilevered at the crack tip end, and the transverse deflection of the neutral axis of the beam is $w(\xi)$. For $\xi > 0$, each half of the specimen is constrained against transverse deflection. Thus, within the context of elementary beam theory, the internal bending moment vanishes for $\xi > 0$. As the crack tip passes a section of the beam, the transverse constraint is relieved and the two arms of the split strip begin to separate. The crack must open from the crack tip with both the deflection $w(0)$ and the slope $dw(0)/d\xi$ of the neutral axis being zero. In the interval $-l < \xi < 0$ the beam arms separate freely, and no transverse load acts in this interval. At $\xi = -l$, the surfaces of the beam arms come into contact with the wedge, and the arms are assumed to remain in contact with the wedge for $\xi < -l$. Thus, the beam arms have no curvature and, consequently, no internal bending moment for $\xi < -l$. However, the arms arrive at the contact point with nonzero curvature. Therefore, the contact point is a point of discontinuity in the internal bending moment per width in the beam $M(\xi)$ which, in turn, implies that the reaction force of the wedge on each beam arm is a concentrated force per unit width at $\xi = -l$. Suppose that the magnitude of the concentrated force per unit width is Q, positive in a direction tending to open the crack. The wedge is assumed to be smooth for the time being, but the effect of sliding friction is considered subsequently.

Consider the energy integral (5.2.9) along a path enclosing the deformed portion of the beam in Figure 5.9. Because the process is steady state and the path is closed, the value of the integral is zero. If the terms in the integrand are interpreted as representing two-dimensional fields constrained to be consistent with beam theory, as in the previous subsection, then the value of the integral is

$$-\tfrac{1}{2}\rho h v^2 \alpha^2 + Q\alpha - 6M(0^-)^2/Eh^3 = 0, \qquad (5.5.16)$$

where ρ is the mass density of the material, E is the elastic modulus, and the form of the energy release rate strictly applicable for a Bernoulli-Euler beam is assumed. The relationship is a statement of energy rate balance and the interpretation of the individual terms is straightforward. The first term accounts for the residual kinetic energy the material has as it slides up the wedge face beyond the contact point. The second term is the rate of work of the wedge reaction force on the material, and the last term on the left side of the equation is the energy flow into the crack tip region as given in (5.5.14).

The equation of motion for the beam governing the transverse deflection under steady state conditions is

$$w'''' + \beta^2 w'' = 0 \qquad (5.5.17)$$

subject to the boundary conditions

$$w(0) = 0, \quad w'(0) = 0, \quad w'(-l) = -\alpha, \quad w''(-l) = 0, \qquad (5.5.18)$$

as already noted, where $\beta^2 = 12\rho v^2/Eh^2$. The solution of this boundary value problem is

$$w(\xi) = \frac{\alpha[\beta\xi\cos\beta l + \sin\beta l - \sin\beta(\xi + l)]}{\beta(1 - \cos\beta l)} \qquad (5.5.19)$$

which implies that

$$Q = \frac{\alpha\rho h v^2}{(1 - \cos\beta l)}. \qquad (5.5.20)$$

The result for Q can be substituted into (5.5.16) to obtain

$$-\tfrac{1}{2}\rho h v^2 \alpha^2 + \frac{\alpha^2 \rho h v^2}{(1 - \cos\beta l)} - \tfrac{1}{2}G = 0 \qquad (5.5.21)$$

where G is the crack tip energy flux from (5.5.14).

Suppose that the crack grows in such a way that the energy release rate is a constant G_c, as in pure cleavage, for example. This point of view was adopted in a study of the cleavage strength of mica under equilibrium conditions by Obreimoff (1930), independently of the earlier work by Griffith on the connection between the energy release rate and the surface energy. With a constant energy release rate, (5.5.21) provides an expression for the length l in terms of other system parameters in the form

$$\frac{l}{h} = \frac{1}{2\sqrt{3}}\frac{c_o}{v}\cos^{-1}\left[\frac{G_c/Eh - \alpha^2(v/c_o)^2}{G_c/Eh + \alpha^2(v/c_o)^2}\right], \qquad (5.5.22)$$

where $c_o = \sqrt{E/\rho}$ is the speed of one-dimensional stress waves in an elastic bar. Some special cases can be extracted from this general result. For example, for vanishing $v/c_0 \to 0$ and any $G_c/Eh \neq 0$, (5.5.22) implies that

$$\frac{l}{h} \to \alpha \sqrt{\frac{Eh}{3G_c}} \qquad \text{as} \qquad \frac{v}{c_0} \to 0. \qquad (5.5.23)$$

If $\alpha = 10°$ and $Eh/G_c = 10^4$ then $l/h \approx 17$. On the other hand, if the fracture resistance of the material is relatively low, then it can be shown from (5.5.22) that

$$\frac{l}{h} \to \frac{\pi}{2\sqrt{3}} \frac{c_0}{v} \qquad \text{as} \qquad \frac{G_c}{Eh} \to 0. \qquad (5.5.24)$$

If $\alpha = 10°$ and $v/c_0 = 0.1$ then $l/h \approx 2.4$.

The influence of frictional interaction between the wedge and the beam arms can also be included in several ways. In the simplest approach, the transverse force Q gives rise to a force fQ in the direction of motion, where f is the coefficient of sliding friction between the wedge and beam. If buckling of the beam arms in the column mode is not considered to be important then the equations of motion can be written in the undeformed configuration for small deflections. The energy balance equation (5.5.16) is modified by including the term $-fQ$ on the left side to account for the work dissipated in frictional sliding. The solution of the relevant beam boundary value problem corresponding to (5.5.17) can also be obtained in a straightforward manner. The beam arms are assumed to be axially inextensible and the last boundary condition in (5.5.18) is replaced by

$$w''(-l) = -fQ\frac{h}{2}\frac{12}{Eh^3} \qquad (5.5.25)$$

to account for the bending moment at the contact point due to the eccentricity of the axial frictional force. The magnitude of the shear force Q corresponding to (5.5.20) is found to be

$$Q = \frac{\alpha\rho h v^2}{(1 - \cos\beta l - 0.5f\beta h \sin\beta l)}. \qquad (5.5.26)$$

The significance of the frictional effect can be seen by examining the limiting case of crack growth with constant energy release rate G_c and with vanishing speed v. Under these conditions,

$$\frac{l}{h} \rightarrow \frac{f}{2} + \sqrt{\frac{f^2}{4} + \frac{\alpha^2 Eh}{3G_c}} \qquad \text{as} \qquad \frac{v}{c_0} \rightarrow 0. \qquad (5.5.27)$$

As $f \rightarrow 0$, this result approaches the corresponding result in (5.5.23).

5.5.4 Steady crack growth in a plate under bending

Although the energy release rate idea has been applied for cases of plane strain, plane stress, or antiplane shear deformations, the concept itself can be applied equally well to analyze crack propagation in plates and shells. A few simple illustrations of the application of energy concepts are included here.

Consider a thin isotropic elastic Kirchhoff plate under bending. The undeformed mid-plane of the plate lies in the x_1, x_2-plane and it occupies the region $-\infty < x_1 < \infty$, $-b \le x_2 \le b$. Suppose that a through-crack grows steadily at speed v along the centerline of the plate in the positive x_1-direction. That is, the crack surface extends across the full thickness of the plate, and it occupies the portion of the x_1-axis with $-\infty < x_1 < vt$. The edges of the plate $x_2 = \pm b$ are constrained to remain in the plane, but the mid-plane normals on opposite edges are rotated an angle α away from each other. There is no transverse loading on the plate. The transverse deflection depends on x_1 and t only through $\xi = x_1 - vt$ and it is denoted by $w(x_1, x_2, t) = w(\xi, x_2)$. The boundary conditions imposed on the deflection are

$$w(\xi, \pm b) = 0, \qquad \frac{\partial w}{\partial x_2}(\xi, \pm b) = \mp \alpha. \qquad (5.5.28)$$

Far ahead of the crack tip the plate is at rest and in a state of pure bending across the width of the strip. The deflection there is

$$w(\xi, x_2) = \frac{b\alpha}{2}\left(1 - \frac{x_2^2}{b^2}\right). \qquad (5.5.29)$$

Far behind the crack tip, the plate is also at rest but completely stress-free. Thus, by adopting the intuitive argument on energy balance put forward in Section 5.4 the crack tip energy release rate can be determined directly. Indeed, the crack tip energy release rate is equal to the elastic energy per unit length in the x_1-direction stored in the plate far ahead of the advancing crack tip. Thus, the energy release rate is

$$G = \frac{Eh^3\alpha^2}{12(1-\nu^2)b}.$$ (5.5.30)

The stress resultants are singular at the crack tip, but the matter of crack tip stress singularities is not pursued further here. A full stress analysis of a plate problem of this type has been presented by Fossum (1978) who found the energy release rate to be identical to (5.5.30) even though he included rotary inertia effects in the model of the plate. The reason is that the energy release rate depends only on the energy stored in the plate far ahead of the crack tip, and not on wave propagation in the plate. Indeed, an analysis of this situation based on any of the standard bending theories of plates will lead to the same energy release rate. However, the crack tip stress distribution corresponding to this energy release will depend on the exact nature of the bending model.

5.5.5 Crack growth in a pressurized cylindrical shell

For propagation of a sharp crack in a cylindrical membrane shell at instantaneous crack tip speed v the rate of energy flow into the crack tip is

$$F(\Gamma) = \lim_{\Gamma \to 0} \int_\Gamma \left[\nu_\alpha N_{\alpha\beta}\dot{u}_\beta + \tfrac{1}{2}v(N_{\alpha\beta}\epsilon_{\alpha\beta} + \rho h\dot{u}_\beta\dot{u}_\beta + \rho h\dot{w}^2)\right] d\Gamma,$$

(5.5.31)

where the subscripts have the range 1,2, Γ is a contour in the middle surface passing from one crack face to the other around the crack tip, ν_β is the unit normal to Γ in the tangent plane of the middle surface, $N_{\alpha\beta}$ are the in-plane stress resultants referred to orthogonal coordinates, $\epsilon_{\alpha\beta}$ are the work-conjugate strains of the middle surface, u_β are the in-plane displacement components, w is the displacement in the direction normal to the undeformed middle surface, h is the wall thickness, and the superposed dot denotes a material time derivative. This energy flux integral can be deduced from (5.2.7) or from direct application of the overall energy rate balance to the shell configuration. For a membrane shell the crack tip stress distribution is identical to the plane stress distribution, so that (5.5.31) can be evaluated in

terms of a stress intensity factor. On the other hand, if the crack tip
region is modeled by a cohesive zone then the results of Section 5.3
can be applied.

The particular case of an axial crack propagating steadily along a
generator of the cylindrical shell under the action of internal pressure
p provides an instructive situation. Again, for steady propagation,
all field quantities including the pressure depend on the axial coordinate x and time t only through the combination $\xi = x - vt$. For
steady growth, and in the absence of internal pressure, it can be
shown that the energy integral (5.5.31) is path-independent for all
crack tip contours, analogous to the plane strain result established in
Section 5.2. In the presence of internal pressure, the integral is path
dependent, and the nature of this path dependence can be illustrated
by considering the following special case.

Far ahead of the crack tip the material is at rest and the internal
pressure is at some reference level, say p_∞. Far behind the crack
tip, the material is also at rest and the internal pressure has been
reduced to zero, perhaps due to the escape of an initially pressurized
gas through the opening crack. Then, by applying the divergence
theorem to the integral (5.5.31) for the closed path consisting of Γ
around the crack tip, the traction-free crack faces, and circumferential
circles far ahead of and far behind the crack tip, it follows that the
energy release rate is

$$G = \frac{\pi a}{E} \left(\frac{a}{h}\right)^2 p_\infty^2 - \int_A p \frac{\partial w}{\partial x} \, dA, \qquad (5.5.32)$$

where a is the shell mean radius and A is the entire middle surface
of the shell enclosed within the contour Γ. The first term in (5.5.32)
is the contribution to the energy release rate from the elastic energy
stored in the shell per unit length in the axial direction far ahead
of the advancing crack tip (free axial expansion is tacitly assumed),
whereas the second term is the contribution to the energy release rate
from the work done on the shell by the escaping gas as the crack
opens. From this perspective, the shell is treated as a closed system
and the gas as an external agent. The energy stored initially in the
pressurized gas that is carried away through the opening crack, as
well as the way in which the pressure decays behind the advancing
crack tip, must be determined by the dynamics of the gas flow and
escape along the cylinder. Some elaboration on this point is given by

Kanninen, Sampath, and Popelar (1976) and by Parks and Freund (1978). Because the gradient $\partial w/\partial x$ vanishes far ahead of the tip and the pressure p vanishes far behind the tip, the second term in (5.5.32) will exist, in general, even though the range of integration is infinite in the axial direction.

In any case, once G is computed from the pressure distribution, it can be converted to a stress intensity factor or a crack tip opening displacement. It is observed that the form of G in (5.5.32) is independent of the specific shell model assumed, but the actual value of G will depend on the details of the solution through $\partial w/\partial x$ and possibly through the influence of deformation on the pressure distribution.

5.6 A path-independent integral for transient loading

The energy flux integral and its variations which were discussed in the preceding sections of this chapter have been useful in several ways in dynamic fracture studies. The crack tip energy flux provides a useful one parameter characterization of the mechanical fields in the crack tip region in some cases. If the integral definition of such a parameter is also independent of the path used to evaluate the integral, then this property can be exploited to relate the near tip field to remote loading conditions without the need to solve the field equations for the problem.

The crack tip energy integral (5.2.7) was noted to be path dependent, in general, even for the case of linear elastodynamics. However, Nilsson (1973) introduced a path-independent integral for plane elastodynamics that can be used to advantage in certain transient crack problems. This integral is defined and applied in this section. The integral has been found useful in the study of two-dimensional plane strain, plane stress, or antiplane shear deformation problems involving the transient loading of a body containing a crack that does not extend. The integral is first introduced, and its use is then illustrated by application to a transient mode I problem that has already been analyzed in Chapter 2.

5.6.1 The path-independent integral

The path-independent integrals of equilibrium fracture mechanics, especially the J-integral due to Eshelby (1956) and Rice (1968), have played a central role in the development of fracture mechanics. The path-independent integrals can be interpreted as consequences of the conservation laws of elastostatics. These conservation laws, in turn, are consequences of the principle of stationary potential energy and the invariance of the potential energy functional over the elastic field under certain classes of transformations. These ideas are generalized and made precise by Noether's theorem on invariant variational principles, as described by Günther (1962) and Knowles and Sternberg (1972). Here, the connection between a path-independent integral and the underlying variational principle is illustrated by means of a simple approach introduced by Eshelby (1970).

Suppose that the configuration of a mechanical system is described by a vector function of position over a volume of space, say R. The rectangular components of the vector are denoted by u_i and the spatial gradient of the vector has components $\partial u_i/\partial x_j = u_{i,j}$. Suppose that the field is governed by a variational principle of the form $\delta\Phi[u_i, R] = 0$, where

$$\Phi[u_i, R] = \int_R L(u_i, u_{i,j})\, dR + \text{boundary terms} \qquad (5.6.1)$$

and R is held fixed as u_i is varied. Thus, $\Phi[u_i, R]$ is a functional of the field u_i over the region of space R, and the integrand L is assumed to be a differentiable function of its arguments. The functional Φ is stationary under variations in u_i that are consistent with all constraints on the admissible functions within the range of Φ. Under suitable assumptions, a necessary condition on u_i for Φ to be stationary is that u_i must satisfy the Euler equations

$$\frac{\partial}{\partial x_j}\left(\frac{\partial L}{\partial u_{i,j}}\right) - \frac{\partial L}{\partial u_i} = 0 \qquad (5.6.2)$$

in R. In view of its arguments, the spatial gradient of L is

$$\frac{\partial L}{\partial x_k} = \left(\frac{\partial L}{\partial u_i} - \frac{\partial}{\partial x_j}\frac{\partial L}{\partial u_{i,j}}\right)u_{i,k} + \frac{\partial}{\partial x_j}\left(\frac{\partial L}{\partial u_{i,j}}u_{i,k}\right). \qquad (5.6.3)$$

In light of (5.6.3), the terms of (5.6.2) can be rearranged to yield

$$\frac{\partial}{\partial x_j}\left(L\delta_{kj} - \frac{\partial L}{\partial u_{i,j}}u_{i,k}\right) = 0, \qquad k = 1, 2, 3. \qquad (5.6.4)$$

A particularly noteworthy feature of (5.6.4) for present purposes is that, for each value of k, the divergence of a certain vector is equal to zero. Therefore, if (5.6.4) is integrated over R or any subregion of R, an application of the divergence theorem leads immediately to the *mechanical conservation law*

$$\int_S \left[Ln_k - \frac{\partial L}{\partial u_{i,j}} u_{i,k}\, n_j \right] dS = 0, \qquad (5.6.5)$$

where S is the closed bounding surface of the region or subregion, and n_i are the components of the outward unit normal vector to S.

The result (5.6.5) applies for any three-dimensional region S in which (5.6.2) holds. The equivalent result for two-dimensional fields takes virtually the same form, where the counterpart of S is the closed contour surrounding the region of the plane in which (5.6.2) holds. To see this, introduce a rectangular coordinate system x_1, x_2, x_3 and suppose that the field u_i is independent of x_3. Take the surface in (5.6.5) to be a right cylinder with its generator in the x_3-direction. The end faces are planes of constant x_3 at an arbitrary distance apart, and the right section of the cylinder is bounded by the curve C. For $k = 1$ or 2 there is no contribution to the surface integral from the end faces, and for $k = 3$ the contributions from the two end faces cancel. Consequently, in two dimensions,

$$\int_C \left[Ln_k - \frac{\partial L}{\partial u_{i,j}} u_{i,k}\, n_j \right] dC = 0, \qquad k = 1, 2, \qquad (5.6.6)$$

where C is any closed curve in the plane bounding a region in which (5.6.2) holds.

As a simple illustration of the conservation law (5.6.6), consider the case of a two-dimensional elastodynamic field in the x_1, x_2-plane that is steady with respect to a reference frame translating in the x_1-direction at speed v. The momentum equation can be written in terms of the displacement $u_i(x_1 - vt, x_2)$ as

$$C_{ijpq} \frac{\partial^2 u_p}{\partial x_q \partial x_j} = \rho v^2 \frac{\partial^2 u_i}{\partial x_1^2}, \qquad (5.6.7)$$

where C_{ijpq} is the array of elastic constants. This is a specific example of the field equation (5.6.2). The variational statement underlying this equation is

$$\delta \left\{ \int_R \left(\tfrac{1}{2} C_{ijpq} u_{i,j}\, u_{p,q} - \tfrac{1}{2} \rho v^2 u_{i,1}\, u_{i,1} \right) dR + \ldots \right\} = 0, \qquad (5.6.8)$$

where R is now understood to be any region of the plane in which (5.6.7) holds. Consider (5.6.6) for $k = 1$; the integrand is

$$L n_1 - \frac{\partial L}{\partial u_{i,j}} u_{i,1}\, n_j = (T + U) n_1 - \sigma_{ij} u_{i,1}\, n_j \qquad (5.6.9)$$

and the conservation law

$$\int_{\mathcal{C}} \left[(T + U) n_1 - \sigma_{ij} u_{i,1}\, n_j \right] d\mathcal{C} = 0 \qquad (5.6.10)$$

follows immediately.

Consider steady growth of a crack with traction-free faces in the x_1-direction. Form a closed path consisting of two contours, say Γ_1 and Γ_2, each of which begins on one crack face, encircles the crack tip, and terminates on the other face, plus the two crack face segments necessary to complete the closed path. The first term in the integrand of (5.6.10) vanishes on the crack faces because $n_1 = 0$ there, and the second term vanishes because $\sigma_{ij} n_j = 0$ for $i = 1$ or 2 to satisfy the traction condition. It follows that

$$G(\Gamma) = \int_\Gamma \left[\tfrac{1}{2}(T + U) n_1 - \sigma_{ij} u_{i,1}\, n_j \right] d\Gamma \qquad (5.6.11)$$

is independent of path Γ for all paths beginning on one crack face, surrounding the tip, and ending on the other face. By custom, n_i is the unit normal vector to Γ directed away from the tip. This integral is a special case of (5.2.9) and both its property as an energy release rate and its property of path independence have been exploited in the preceding sections of this chapter.

To obtain the main result of this subsection as a second example of a path-independent integral, consider the case of a plane strain elastodynamic field represented by the displacement $u_i(x_1, x_2, t)$. The components of the displacement vector satisfy

$$C_{ijpq} \frac{\partial^2 u_p}{\partial x_q \partial x_j} = \rho \frac{\partial^2 u_i}{\partial t^2}, \qquad (5.6.12)$$

where C_{ijpq} is the array of elastic constants. Assume that the displacement and particle velocity components are zero initially, that is,

$$u_i(x_1, x_2, 0) = 0, \qquad \frac{\partial u_i}{\partial t}(x_1, x_2, 0) = 0. \qquad (5.6.13)$$

If the Laplace transform (1.3.7)

$$\widehat{u}_i(x_1, x_2, s) = \int_0^\infty u_i(x_1, x_2, t)e^{-st}\, dt \qquad (5.6.14)$$

is applied to (5.6.12), then

$$C_{ijpq}\frac{\partial^2 \widehat{u}_p}{\partial x_q \partial x_j} = \rho s^2 \widehat{u}_i. \qquad (5.6.15)$$

To a certain extent, the introduction of the Laplace transform can be motivated through the following argument. The hyperbolic nature of the partial differential equations governing the transient elastodynamic response precludes the existence of path-independent integrals. Furthermore, the mechanical systems for which path-independent integrals have been found are governed by elliptic partial differential equations. It is well known that the hyperbolic equations of elastodynamics can be converted to elliptic equations by application of the Laplace transform over time, and this is the conversion represented by (5.6.15). The elliptic equation (5.6.15) is the Euler equation implied by the underlying variational statement of the problem,

$$\delta\Phi\left[\widehat{u}_i, R\right] = 0, \qquad (5.6.16)$$

where

$$\Phi\left[\widehat{u}_i, R\right] = \int_R \left[\tfrac{1}{2}C_{ijpq}\widehat{u}_{p,q}\,\widehat{u}_{i,j} + \tfrac{1}{2}\rho s^2 \widehat{u}_i\widehat{u}_i\right]\, dR \qquad (5.6.17)$$

plus any boundary terms. The two-dimensional conservation law

$$\int_C \left[\tfrac{1}{2}(\widehat{\sigma}_{ij}\widehat{u}_{i,j} + \rho s^2 \widehat{u}_i\widehat{u}_i)n_1 - \widehat{\sigma}_{ij}n_j\widehat{u}_{i,1}\right]\, dC = 0 \qquad (5.6.18)$$

follows immediately, where $\hat{\sigma}_{ij} = C_{ijpq}\hat{u}_{p,q}$ are the Laplace transformed stress components. The integration path C in (5.6.18) is any closed path in the plane of deformation bounding a region in which (5.6.15) is satisfied.

Consider now a body containing a planar crack with traction-free faces that is deformed under plane strain conditions. Rectangular coordinates labeled x_1, x_2, x_3 are introduced so that the displacement is independent of x_3, and the plane of the crack is $x_2 = 0$. A crack tip contour Γ is introduced starting on one crack face, encircling the crack tip in the plane, and ending at some point on the other crack face. Then a line integral along Γ is defined as

$$J_N(\Gamma; s) = \int_\Gamma \left[\tfrac{1}{2}(\hat{\sigma}_{ij}\hat{u}_{i,j} + \rho s^2 \hat{u}_i \hat{u}_i)n_1 - \hat{\sigma}_{ij}n_j\hat{u}_{i,1} \right] d\Gamma, \qquad (5.6.19)$$

where n_i is a unit vector normal to Γ and directed away from the crack tip. Because the integrand of (5.6.19) vanishes pointwise on the crack faces, this conservation law implies that $J_N(\Gamma; s)$ is independent of the particular choice of Γ used to compute its value. Again, Γ must be a material line, and the region between Γ and the crack tip must be simply connected and free of body force. This path-independent crack tip integral was introduced by Nilsson (1973), who noted that it applies for transient viscoelastic fields as well as elastic fields. As noted by Gurtin (1976), the path-independent integral (5.6.19) has an expression in the time domain equivalent to its expression in the Laplace transformed domain.

5.6.2 Relationship to stress intensity factor

The path-independent integral $J_N(\Gamma; s)$ does not represent the Laplace transform of an energy measure or other physically significant feature of the elastic field. For the integral to be useful, it must be related to some quantity of physical significance. That quantity is the dynamic stress intensity factor, and the connection is established in this subsection. The case of mode I deformation under plane strain conditions is considered here.

Again, a rectangular coordinate system x_1, x_2, x_3 is introduced in the body. The deformation is independent of x_3, and the crack lies in the plane $x_2 = 0$. Consider a crack tip contour Γ that is arbitrarily close to the crack tip. For a mode I plane strain deformation and a stationary crack tip, the stress and displacement fields very near to the crack tip are

$$\sigma_{ij} \sim \frac{K_I(t)}{\sqrt{2\pi r}} \Sigma_{ij}(\theta),$$

$$u_i \sim \frac{K_I(t)}{2\mu} \sqrt{\frac{r}{2\pi}} U_i(\theta)$$

(5.6.20)

as $r \to 0$, in terms of polar coordinates r, θ centered at the crack tip. The direction $\theta = 0$ coincides with the direction of prospective crack advance, and μ is the elastic shear modulus. The functions $\Sigma_{ij}(\theta)$ are given in (2.1.3) and the $U_i(\theta)$ are readily determined from the limiting values of the angular variations in (4.3.14) as $v \to 0^+$. Taking the Laplace transform of (5.6.20), the near tip distributions of $\hat{\sigma}_{ij}$ and $\hat{u}_i$ are

$$\hat{\sigma}_{ij} \sim \frac{\hat{K}_I(s)}{\sqrt{2\pi r}} \Sigma_{ij}(\theta), \quad \hat{u}_i \sim \frac{\hat{K}_I(s)}{2\mu} \sqrt{\frac{r}{2\pi}} U_i(\theta). \qquad (5.6.21)$$

But these near tip fields are exactly the same as the corresponding results for equilibrium fields with $\hat{K}_I(s)$ in place of the equilibrium stress intensity factor. The term in $J_N(\Gamma; s)$ that is proportional to $\hat{u}_i \hat{u}_i$ makes no contribution to the integral along Γ as the path shrinks onto the crack tip because it is proportional to r as $r \to 0$. Each of the remaining terms is proportional to r^{-1} and, consequently, each contributes to the value of $J_N(\Gamma; s)$ as Γ is shrunk onto the crack tip. However, the remaining terms together are identical to those making up the integrand of Rice's J-integral as applied in equilibrium elastic fracture mechanics (Rice 1968). Consequently,

$$\lim_{\Gamma \to 0} J_N(\Gamma; s) = \frac{1 - \nu^2}{E} \hat{K}_I(s)^2 \qquad (5.6.22)$$

for an isotropic elastic material with Young's modulus E and Poisson's ratio ν under plane strain conditions.

   Evidently, the value of $J_N(\Gamma; s)$ for Γ very near the crack tip is related to the value of the Laplace transform of the stress intensity factor $\hat{K}_I(s)$. But $J_N(\Gamma; s)$ is path-independent. Therefore, if the value of J_N can be determined in terms of the Laplace transformed boundary loading by evaluating the integral for a contour Γ far from the crack tip, then the property of path independence can be invoked to obtain a relationship between the applied loading and the stress intensity factor. Such an application is illustrated in the next subsection.

5.6.3 An application

To see how the conservation law (5.6.19) can be applied to extract a stress intensity factor history for a transient crack problem without solving the field equations, consider the problem formulated in Section 2.4. This is the situation of a half plane crack in an otherwise unbounded elastic solid. The faces of the crack are subjected to a uniform pressure of magnitude σ^*, suddenly applied at time $t = 0$. A rectangular coordinate system x_1, x_2, x_3 is introduced with the crack occupying the half plane $x_2 = 0$, $x_1 \leq 0$. The deformation is plane strain with all field quantities being independent of the x_3-coordinate and the crack opens with mode I deformation.

As noted in Section 2.4, the transient stress distribution resulting from this loading consists of cylindrical waves diverging from the crack edge plus plane dilatational waves propagating away from the crack faces. As the plane wave passes a point with coordinate x_2, the normal stress component changes discontinuously according to

$$\sigma_{22}(x_1, x_2, t) = -\sigma^* H(c_d t - |x_2|),\qquad (5.6.23)$$

where $H(\cdot)$ is the unit step function. Likewise, the normal displacement changes according to

$$u_2(x_1, x_2, t) = \pm \frac{\sigma^*}{\rho c_d^2}(c_d t \mp x_2) H(c_d t \mp x_2)\qquad (5.6.24)$$

for $\pm x_2 \geq 0$.

The conservation law (5.6.18) is now applied to determine the stress intensity factor for this loading without solving the boundary value problem. For this purpose, the Laplace transform of the field is considered, that is, $\widehat{u}_i(x_1, x_2, s)$ for s in some small interval of the positive real axis. The path C is chosen to be the closed curve shown in Figure 5.10. The path AA' is a circular crack tip contour of indefinitely small radius. The paths AB and $A'B'$ coincide with the crack faces, and B and B' are far from the crack tip compared to

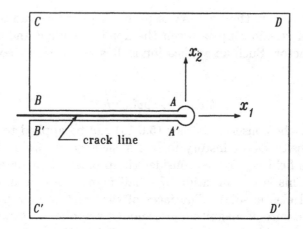

Figure 5.10. Diagram of the path C for evaluation of the integral (5.6.18) in order to relate the applied loading to the transient crack tip stress intensity factor for the case of pressure suddenly applied to the faces of the crack. A diagram of the wavefronts appears in Figure 2.4.

the distance c_d/s for any fixed s in the real interval. The remainder of C is the rectangle $BCDD'C'B'$ with all four edges far from the crack tip compared to c_d/s. The Laplace transforms of the diverging cylindrical waves decay rapidly with distance from the crack edge, as discussed in Section 2.4. In fact, the decay is rapid enough so that these transformed quantities make no contribution to the integral.

The remaining features of the wave field that could contribute are the plane waves traveling in the $\pm x_2$-directions, the crack face loading, and the crack tip field. The contributions can be worked out as follows. In light of (5.6.22), the nontrivial contributions are summarized as

$$\frac{1-\nu^2}{E}\widehat{K}_I(s)^2 + 2\int_A^B \widehat{\sigma}_{22}\widehat{u}_{2,1}\, dx_1$$

$$+ \int_B^C (\widehat{\sigma}_{22}\widehat{u}_{2,2} + \rho s^2\widehat{u}_2\widehat{u}_2)\, dx_2 = 0. \quad (5.6.25)$$

Along AB, $\widehat{\sigma}_{22}$ is independent of x_1 so the first term is proportional to the value of $\widehat{u}_2$ at point B, which is known. All points on BC are far enough from the crack edge so that the diverging cylindrical waves make no contribution, and the integral is given in terms of the plane wave traveling away from the crack faces. From (5.6.23) and (5.6.24) the relevant Laplace transforms are

$$\widehat{\sigma}_{22} = -\frac{\sigma^*}{s}e^{\mp sx_2/c_d}\,, \qquad \widehat{u}_2 = \pm\frac{\sigma^*}{\rho c_d s^2}e^{\mp sx_2/c_d} \qquad (5.6.26)$$

for $\pm x_2 > 0$. If the integrals in (5.6.25) are evaluated for the case when the x_1-coordinate of B and the x_2-coordinate of C are indefinitely large, it is found that

$$\frac{1-\nu^2}{E}\widehat{K}_I(s)^2 = \frac{\sigma^{*2}}{\rho c_d s^3}\,. \qquad (5.6.27)$$

With the identity $\rho c_d^2 = (1-\nu)E/[(1-2\nu)(1+\nu)]$ this result simplifies to

$$\widehat{K}_I(s) = \frac{\sigma^*}{s^{3/2}}\frac{\sqrt{(1-2\nu)c_d}}{1-\nu}\,, \qquad (5.6.28)$$

where only the positive root is retained on physical grounds. The inverse Laplace transform is

$$K_I(t) = 2\sigma^*\frac{\sqrt{c_d t(1-2\nu)/\pi}}{(1-\nu)}\,, \qquad (5.6.29)$$

which is identical to the result (2.5.44) obtained from a complete solution of the field equations. Another application of the path integral to an elastodynamic strip problem is given by Nilsson (1973).

Finally, it is noted that the path integral idea applies in the case of steady vibratory deformation, as well as for Laplace transformed transient fields. For example, suppose that an elastodynamic field has the form

$$u_i(x_1, x_2, t) = \widetilde{u}_i(x_1, x_2)e^{i\omega t}\,. \qquad (5.6.30)$$

Such a field might represent the steady harmonic response of a cracked elastic solid to a periodic applied traction with frequency ω. The quantity $\widetilde{u}_i$ represents the amplitude field of the resulting harmonic response. Under these conditions, the conservation law (5.6.18) applies with s replaced by $i\omega$. The stress intensity factor for a mode I crack under these conditions has the form

$$K_I(t) = \widetilde{K}_I(\omega)e^{i\omega t} \qquad (5.6.31)$$

and the value of the integral taken along a crack tip contour analogous
to (5.6.22) is

$$\frac{1-\nu^2}{E}\widetilde{K}_I(\omega)^2 \, . \tag{5.6.32}$$

5.7 The transient weight function method

Up to this point, dynamic stress intensity factor results have been
obtained on the basis of a solution of a particular boundary value
problem or the application of a path-independent integral to a particular problem. Here, the notion of an influence function for stress
intensity factor is introduced. For a body of any given shape, the
stress intensity history for a linear elastodynamic crack problem must
have the form

$$K_I(t) = \int_0^t \int_B h(x_B, t-\tau)\, p(x_B, \tau)\, dx_B\, d\tau \, , \tag{5.7.1}$$

where B is the spatial boundary of the body, including the crack faces,
p is the time-dependent loading applied to the boundary which begins
to act at time $t = 0$, and x_B symbolically represents points on the
boundary B. Thus, the stress intensity factor for *any* applied loading
p is determined by the function $h(x_B, t - \tau)$. This function is called
the *influence function*, or, alternatively, the *weight function*, for the
dynamic stress intensity factor for a particular configuration.

5.7.1 The weight function based on a particular solution

In Section 5.6 it was shown that the stationary character of the
functional $\Phi[\widehat{u}_i; R]$ given in (5.6.17), where the argument $\widehat{u}_i$ is the
Laplace transform on time of a displacement field, leads to the path-
independent crack tip integral $J_N(\Gamma; s)$ given in (5.6.19). The notation
established in that section is followed without modification. In particular, the plane of deformation is the x_1, x_2−plane and the crack opens
in the plane strain opening mode. There are two properties of this
path-independent integral that are exploited in this development. The
first is that the integral J_N has a direct relationship with the Laplace
transform of the stress intensity factor, as stated in (5.6.22), and the
second property concerns the change in Φ due to a change in crack
length. The latter property is discussed next.

Suppose that the solution to the crack problem $\hat{u}_i$ can be found for any fixed position of the crack tip, say at $x_1 = l$, $x_2 = 0$. Then $\hat{u}_i$ depends in some way on l, as well as on position in the plane, and Φ can be viewed as a function of l. Suppose that the total derivative $d\Phi/dl$ is evaluated. The region R in (5.6.17) is now considered to be the entire two-dimensional body exterior to a small but arbitrary loop around the crack tip of interest. For present purposes, it is sufficient to consider the outer boundary of R to be constrained against working displacement during the increment dl, that is, $d\hat{u}_i/dl = 0$ at all points on the outer boundary on which nonzero traction acts. (If this condition is not assumed then it is necessary to add appropriate boundary terms to the functional in (5.6.17); the final result is unaffected.) To evaluate $d\Phi/dl$, the parameter l can be viewed as a timelike parameter, and the generalized transport theorem (1.2.41) can be applied to yield

$$\frac{d\Phi}{dl} = \int_R \left[\hat{\sigma}_{ij} \frac{\partial \hat{u}_{i,j}}{\partial l} + \rho s^2 \hat{u}_i \frac{\partial \hat{u}_i}{\partial l} \right] dR - \int_\Gamma \frac{1}{2} \left[\hat{\sigma}_{ij} \hat{u}_{i,j} + \rho s^2 \hat{u}_i \hat{u}_i \right] n_1 \, d\Gamma ,$$

$$(5.7.2)$$

where n_i is the unit normal vector to Γ directed away from the crack tip. An application of the divergence theorem yields

$$\frac{d\Phi}{dl} = - \int_\Gamma \left[\frac{1}{2} \left(\hat{\sigma}_{ij} \hat{u}_{i,j} + \rho s^2 \hat{u}_i \hat{u}_i \right) n_1 + \hat{\sigma}_{ij} n_j \frac{\partial \hat{u}_i}{\partial l} \right] d\Gamma . \qquad (5.7.3)$$

The derivatives with respect to l in this expression are computed with x_i held fixed. For points very close to the crack tip,

$$\left. \frac{\partial \hat{u}_i}{\partial l} \right|_{x_1} \approx - \left. \frac{\partial \hat{u}_i}{\partial x_1} \right|_{x_1 - l} . \qquad (5.7.4)$$

Comparison of (5.6.19) with the result (5.7.3) then leads to the second main property of the path-independent integral, namely,

$$\frac{d\Phi}{dl} = - J_N . \qquad (5.7.5)$$

In a study of linear elastostatic crack tip fields, Rice (1972) employed Irwin's relation between energy release rate and the stress intensity factor to show that if the displacement field and stress intensity factor are known as functions of crack length for any one symmetrical load system acting on a linear elastic body in plane strain, then the stress intensity factor for any other symmetrical load system acting on the same body can be determined directly. As shown by Freund and Rice (1974), an analogous result can be obtained for dynamic stress fields in an elastic solid containing a crack. This approach is developed and illustrated in this section.

Consider an elastic solid containing a planar crack of fixed length under conditions of plane strain. The body is assumed to be symmetrical with respect to the plane of the crack, and only loading resulting in mode I deformation is considered. Consider any particular load system on the body and suppose that the crack length is determined by the parameter l. For time $t < 0$, the material is stress-free and at rest. At time $t = 0$, a traction distribution begins to act on the boundary B of the body. (A body force distribution can also be taken into account, but only the final result is stated with body forces included.) The boundary traction is given at any later time $t > 0$ by $\mathbf{T} = Q_1 \mathbf{T}^{(1)}(x_B, t)$ for x_B on B. As noted in Section 1.2.1, the boldface characters in this development represent vector quantities. The parameter Q_1, which appears as a scale factor, will subsequently be viewed as a generalized force. Suppose that the resulting displacement field $\mathbf{u} = Q_1 \mathbf{u}^{(1)}(x_B, t)$ and stress intensity factor $K_I = Q_1 K^{(1)}(t)$ are known as functions of l. These quantities are necessarily linear in the load scale factor Q_1. It will now be shown that if the Laplace transforms on time $\hat{\mathbf{u}}$ and $\hat{K}_I$ are known as functions of l for a range of positive real values of the Laplace transform parameter s, then the Laplace transform on time of the stress intensity factor for any other load system acting on the same body can be determined. This development hinges on the two properties of the path-independent integral J_N identified above. These properties imply that

$$\left. \frac{d\Phi}{dl} \right|_{\text{fixed displ.}} = -\frac{1-\nu^2}{E} \hat{K}_I^2 , \qquad (5.7.6)$$

where fixed displacement means that the increment in Φ due to an increment in l is evaluated with the loaded portion of the boundary held fixed in position.

Consider now a second load system characterized by the generalized force Q_2, in addition to the loading system introduced above.

The complete solution for the case of the loading system represented by the generalized force Q_1 is assumed to be known. Generalized displacements q_1 and q_2 are associated with any displacement field $\widehat{\mathbf{u}}$ by

$$q_i = \int_B \widehat{\mathbf{T}}^{(i)} \cdot \widehat{\mathbf{u}} \, dx_B . \qquad (5.7.7)$$

If both load systems act simultaneously, linear superposition yields

$$q_i = C_{ij} Q_j , \qquad C_{ij} = \int_B \widehat{\mathbf{T}}^{(i)} \cdot \widehat{\mathbf{u}}^{(j)} \, dx_B , \qquad (5.7.8)$$

where summation over the repeated index is implied and $Q_1 \widehat{\mathbf{u}}^{(1)}$ and $Q_2 \widehat{\mathbf{u}}^{(2)}$ are the separate transformed displacement fields. It is clear from (5.7.8) that the generalized displacements q_i are work-conjugate to Q_i with respect to the energylike quantity Φ.

When both loading systems are applied to the body simultaneously, then Φ can be viewed as a function of q_1, q_2, and l with

$$\frac{\partial \Phi}{\partial l} = -\frac{1-\nu^2}{E} \widehat{K}_I^2 , \qquad \frac{\partial \Phi}{\partial q_i} = Q_i , \qquad (5.7.9)$$

where $\widehat{K}_I = Q_i \widehat{K}^{(i)}$. It is then possible to write the variation of Φ due to a small variation in each of its arguments as

$$\delta \Phi = Q_1 \delta q_1 + Q_2 \delta q_2 - \left(\frac{1-\nu^2}{E}\right) \widehat{K}_I^2 \delta l . \qquad (5.7.10)$$

An interchange of the dependent and independent variables, commonly known as a Legendre transformation, yields the equivalent complete differential

$$\delta(Q_1 q_1 + Q_2 q_2 - \Phi) = q_1 \delta Q_1 + q_2 \delta Q_2 + \left(\frac{1-\nu^2}{E}\right) \widehat{K}_I^2 \delta l . \qquad (5.7.11)$$

Because (5.7.11) is an exact differential, the relationships

$$\frac{\partial q_i}{\partial l} = Q_j \frac{dC_{ij}}{dl} = \left(\frac{1-\nu^2}{E}\right) \frac{\partial \widehat{K}_I^2}{\partial Q_i} = 2 \left(\frac{1-\nu^2}{E}\right) \widehat{K}^{(i)} \widehat{K}_I \qquad (5.7.12)$$

must be valid. Furthermore, these relationships must hold for arbitrary values of Q_j so that

$$\frac{dC_{ij}}{dl} = 2 \left(\frac{1 - \nu^2}{E} \right) \widehat{K}^{(i)} \widehat{K}^{(j)} . \qquad (5.7.13)$$

But $C_{12} = C_{21}$ and $\widehat{K}^{(1)}$ are known functions of l. Thus, making use of (5.7.8), relation (5.7.13) with $i = 2$ and $j = 1$ can be solved for $\widehat{K}^{(2)}$ to yield

$$\widehat{K}^{(2)} = \frac{E}{2\widehat{K}^{(1)}(1 - \nu^2)} \int_B \widehat{\mathbf{T}}^{(2)} \cdot \frac{\partial \widehat{\mathbf{u}}^{(1)}}{\partial l} \, dx_B , \qquad (5.7.14)$$

where the derivative with respect to l is taken with x_B and s held fixed.

As in the case of equilibrium fields, the stress intensity factor for the load system represented by Q_2 should not depend in any way on the particular choice of loading system represented by Q_1. Hence, there must exist a function, say $\mathbf{h}$, which is defined in terms of any particular solution by

$$\mathbf{h} = \frac{E}{2\widehat{K}_I(1 - \nu^2)} \frac{\partial \widehat{\mathbf{u}}}{\partial l}(x_1, x_2, s) \qquad (5.7.15)$$

but which is a universal function for a cracked body of given geometrical configuration, regardless of the way in which the body is loaded. The uniqueness of $\widehat{\mathbf{h}}$ in the equilibrium case is established by Rice (1972). It is then evident that if $\widehat{\mathbf{T}}$ and $\widehat{\mathbf{F}}$ are the transforms of any particular boundary traction distribution and body force distribution, respectively, that result in a mode I deformation field then the corresponding transformed stress intensity factor is

$$\widehat{K}_I = \int_B \widehat{\mathbf{T}} \cdot \mathbf{h} \, dx_B + \int_R \widehat{\mathbf{F}} \cdot \mathbf{h} \, dx_R \qquad (5.7.16)$$

in terms of the weight function $\mathbf{h}$. In all cases, of course, the task of finding the stress intensity factor history is not complete until the Laplace transform has been inverted into the time domain. A specific case is examined next.

The weight function method as developed above is now applied to a particular plane strain case of transient loading considered in Section 3.2. A semi-infinite crack in an otherwise unbounded body is subjected to an opposed pair of concentrated loads of magnitude p^* acting at some fixed distance l from the crack tip and tending to separate the crack faces. If the crack is in the plane $x_2 = 0$ and the crack occupies the interval $-\infty < x_1 < l$, the boundary condition on normal stress on the crack faces is

$$\sigma_{22}(x_1, 0^\pm, t) = -p^*\delta(x_1)H(t), \quad -\infty < x_1 < l. \qquad (5.7.17)$$

The other boundary conditions on the region $x_2 > 0$ are given in $(3.2.3)_2$ and $(3.2.3)_3$.

The stress intensity factor for this case will be determined directly from the results for the same body subjected to a suddenly applied crack face pressure of magnitude σ^* at time $t = 0$. Except for a minor coordinate transformation, this is precisely the problem analyzed in Section 2.5. The Laplace transform of the stress intensity factor for that problem is given in (2.5.43) as

$$\widehat{K}^{(1)} = \frac{\sqrt{2}\,\sigma^* F_+(0)}{s^{3/2}} \qquad (5.7.18)$$

in terms of quantities defined in Section 2.5. The Laplace transform of the displacement of the crack face is $\widehat{u}_-(x_1, s)$ and this function is determined by inverting the transform expression in $(2.5.35)_2$. If the minor change in coordinates between the two cases is taken into account, then it is found that

$$\frac{\partial \widehat{u}_2^{(1)}}{\partial l}(x_1, 0^\pm, s) = \pm \frac{\partial \widehat{u}_-}{\partial l}(x_1, s)$$

$$= \pm \frac{\sigma^* b^2 F_+(0)}{2\pi i \mu \kappa s} \int_{\xi_0 - i\infty}^{\xi_0 + i\infty} F_-(\zeta)\, e^{s\zeta(x_1 - l)}\, d\zeta,$$

$$(5.7.19)$$

where, again, the various quantities appearing in (5.7.19) are defined in Section 2.5. The weight function is required on the boundary where traction is applied, and it is given by

$$h_2(x_1, 0^\pm, s) = \frac{E}{2\widehat{K}^{(1)}(1 - \nu^2)} \frac{\partial \widehat{u}_2^{(1)}}{\partial l}(x_1, 0^\pm, s). \qquad (5.7.20)$$

All of the entries in the general expression (5.7.14) have now been identified. If they are substituted into (5.7.14) and the resulting expression is simplified, then it is found that

$$\widehat{K}_I(s) = p^* \sqrt{\frac{2}{s}} \, \frac{1}{2\pi i} \int_{\xi_0 - i\infty}^{\xi_0 + i\infty} F_-(\zeta) \, e^{-s\zeta l} \, d\zeta \,, \qquad (5.7.21)$$

where $-a < \xi_0 < a$. The constant coefficient of this expression simplifies considerably if it is noted that $b^2/\kappa = (1 - \nu)$. Upon inversion of the Laplace transform on time, the integral expression for $K_I(t)$ is exactly the same as the expression obtained as (3.2.12). The features of the stress intensity factor history are described there, and they are interpreted in terms of the process of load transfer by means of stress waves. The purpose here has been to illustrate the application of the weight function method for a case that was already familiar.

5.7.2 A boundary value problem for the weight function

The discussion of the preceding subsection emphasized the construction of the weight function for a particular configuration from any one known solution. However, it is also possible to pose a boundary value problem for the weight function itself, as discussed within the context of elastostatic crack fields by Bueckner (1970) and Rice (1972). It is evident from (5.7.15) that the weight function vector h_i must satisfy the same field equations as does the displacement vector itself. The initial and boundary conditions on the solution of the displacement equation of motion, as well as the nature of the crack tip singularity for the weight function, are considered here. The analysis proceeds by invoking the elastodynamic reciprocal theorem (1.2.59) as it applies for the field for which the dynamic stress intensity factor is sought and the weight function field for the same geometrical configuration. The conditions under which the statement of the reciprocal theorem reduces to (5.7.16) are identified, and the weight function meeting these conditions is determined. The development presumes a mode I deformation field, but it can be carried out equally well for other modes of deformation.

The particular situation to be studied here is the transient loading of a cracked body under plane strain opening mode conditions. The plane of deformation is the x_1, x_2-plane, and the notation of the previous subsection is adopted. A crack occupies the half plane $x_2 = 0$, $-\infty < x_1 \leq 0$. The material is stress-free and at rest for time $t < 0$. At time $t = 0$ a pressure distribution $p(x_1, t)$ begins to act on the crack faces tending to open the crack, and the objective is to determine the resulting stress intensity factor history $K_I(t)$. In view of the mode I symmetry with respect to the crack plane, only the half plane $x_2 \geq 0$ is considered. The elastodynamic field within this region, represented by the displacement u_i, say, is subject to the boundary conditions

$$\sigma_{22}(x_1, 0^+, t) = -p(x_1, t), \quad -\infty < x_1 < 0,$$

$$\sigma_{12}(x_1, 0^+, t) = 0, \quad -\infty < x_1 < \infty, \qquad (5.7.22)$$

$$u_2(x_1, 0^+, t) = 0, \quad 0 < x_1 < \infty.$$

Consider a second elastodynamic field u_i^* for this body that also satisfies the mode I symmetry conditions, so that this field is subject to the boundary conditions $(5.7.22)_2$ and $(5.7.22)_3$ as well. For the time being, it is also assumed that $u_i^* = 0$ for $t < 0$, and that a crack face pressure distribution $p^*(x_1, t)$ begins to act at time $t = 0$. This distribution p^* has no relationship to p, except that its range of definition is the same, and no restrictions are placed on the asymptotic crack tip behavior of u_i^*.

The elastodynamic reciprocal theorem is stated in (1.2.59). In the present case, the region of the body is the half plane $x_2 \geq 0$ except for a cylindrical cut-out of arbitrarily small radius r centered on the crack edge, and the boundary of the region is made up of the intervals $-\infty < x_1 < -r$ and $r < x_1 < \infty$ of the line $x_2 = 0$, plus the cylindrical surface $x_1^2 + x_2^2 = r^2$, $x_2 \geq 0$. In the present instance, the reciprocal theorem implies that

$$\int_0^t \int_{-\infty}^{0^-} p(x_1, \tau)\, u_2^*(x_1, 0^+, t - \tau)\, dx_1\, d\tau$$

$$+ \int_0^t \lim_{r \to 0} \int_0^\pi \sigma_{ij}(r, \theta, \tau) n_j(\theta) u_i^*(r, \theta, t - \tau)\, r d\theta\, d\tau$$

$$= \int_0^t \int_{-\infty}^{0^-} p^*(x_1, \tau)\, u_2(x_1, 0^+, t - \tau)\, dx_1\, d\tau$$

$$+ \int_0^t \lim_{r \to 0} \int_0^\pi \sigma_{ij}^\star(r, \theta, \tau) n_j(\theta) u_i(r, \theta, t - \tau) \, r d\theta \, d\tau . \quad (5.7.23)$$

This relationship incorporates the boundary conditions on both elastodynamic fields.

Guidance on a direction for further development is provided by (5.7.23). The first term on the left side has the form of (5.7.16), which suggests that $u_i^\star$ itself may be the influence function. The first term on the right side, on the other hand, is the integral of an arbitrary function $p^\star$ times a displacement distribution that is not known a priori. This term can be eliminated by taking advantage of the arbitrariness of $p^\star$ to require that $p^\star = 0$. At first sight, this step might seem to be self-defeating because it suggests that $u_i^\star$ is an elastodynamic field satisfying homogeneous boundary conditions. The uniqueness theorem of elastodynamics states that the only elastodynamic field satisfying homogeneous boundary conditions is a deformation-free motion. In the present case, this would imply that $u_2^\star = 0$ everywhere. However, the uniqueness theorem applies for fields for which total mechanical energy is bounded. It is advantageous to consider $u_i^\star$ to be outside this class of fields, for reasons to become evident, and thus $p^\star = 0$ is assumed.

Evaluation of the remaining two terms in (5.7.23) involves integration of stress and deformation fields along a semicircular arc at an arbitrarily small radial distance from the crack tip. The crack tip stress field for the problem of interest σ_{ij} varies with r as $r^{-1/2}$. The integral over θ appearing as the second term on the left side of (5.7.23) will have a finite nonzero value only if $u_i^\star$ also varies as $r^{-1/2}$ for arbitrarily small r. This suggests that the elastodynamic "displacement" field should be square root singular at the crack tip. It is not necessary for $u_i^\star$ to represent a physically realizable field, so there can be no fundamental objection to this singularity. The crack tip displacement field for the problem of interest u_i varies with r as $r^{1/2}$. Thus, the remaining integral on the right side of (5.7.23) will have a finite nonzero value only if $\sigma_{ij}^\star$ varies as $r^{-3/2}$ for arbitrarily small r. This suggests that the "stress" field $\sigma_{ij}^\star$ which is derived from the "displacement" field $u_i^\star$ is singular in r to the power $-3/2$, which is consistent with the singularity in $u_i^\star$. Thus, this singular behavior is adopted as a requirement on the field $u_i^\star$.

Next, the time dependence of the terms in the integrands of these same two integrals in (5.7.23) is considered. The time dependence of the stress σ_{ij} is reflected in the stress intensity factor history

$K_I(t)$ which is to be determined. The integral on the left side is a convolution of $K_I(\tau)$ with u_i^*. Thus, if the time dependence of u_i^* is required to be an impulse, or a delta function in time, then the sifting property of the delta function will render the second term on the left side of (5.7.23) proportional to $K_I(t)$. Likewise, for the second term on the right side, the time dependence of u_i is $K_I(t - \tau)$, hence if σ_{ij}^* has time dependence $\delta(\tau)$ then this term will also be proportional to $K_I(t)$. Thus, it is required that the asymptotic crack tip field of u_i^* must have delta function time dependence. If a function u_i^* can indeed be found which satisfies the requirements identified, then (5.7.23) will have precisely the form (5.7.1) with u_i^* being proportional to the influence function.

The crack tip singular field of u_i^* is presented next. The spatial structure of the singular field is dictated by the highest order spatial derivatives in the governing differential equations, so that the structure for a stationary crack tip is identical to that for an equilibrium field. If the tensile "stress" on the crack plane directly ahead of the crack tip has the form

$$\sigma_{22}^* \sim \frac{\delta(t)}{r^{3/2}} \qquad (5.7.24)$$

for arbitrarily small r, then the angular variations of the "stress" and "displacement" components near the crack tip are

$$\begin{pmatrix} \sigma_{11}^* \\ \sigma_{12}^* \\ \sigma_{22}^* \end{pmatrix} \sim \frac{\delta(t)}{4r^{3/2}} \begin{pmatrix} \cos \frac{3}{2}\theta + 3\cos \frac{7}{2}\theta \\ 3\sin \frac{7}{2}\theta - 3\sin \frac{3}{2}\theta \\ 7\cos \frac{3}{2}\theta - 3\cos \frac{7}{2}\theta \end{pmatrix},$$

$$\begin{pmatrix} u_1^* \\ u_2^* \end{pmatrix} \sim -\frac{\delta(t)}{4\mu r^{1/2}} \begin{pmatrix} (3 - 8\nu)\cos \frac{1}{2}\theta + \cos \frac{5}{2}\theta \\ (8\nu - 9)\sin \frac{1}{2}\theta + \sin \frac{5}{2}\theta \end{pmatrix}. \qquad (5.7.25)$$

If these expressions are substituted into (5.7.23) and the pertinent integrals are evaluated, it is found that, as $r \to 0$,

$$\int_0^t \int_0^\pi \sigma_{ij}(r,\theta,\tau)n_j(\theta)u_i^*(r,\theta,t-\tau)\,r d\theta\, d\tau \to -\frac{(1-\nu)}{\mu}\sqrt{\frac{\pi}{2}}\,K_I(t),$$

$$\int_0^t \int_0^\pi \sigma_{ij}^*(r,\theta,\tau)n_j(\theta)u_i(r,\theta,t-\tau)\,r d\theta\, d\tau \to \frac{(1-\nu)}{\mu}\sqrt{\frac{\pi}{2}}\,K_I(t).$$

$$(5.7.26)$$

In a similar analysis, Burridge (1976) circumvented this integration step by adopting a result from the theory of generalized functions due to Fisher (1971). If the reciprocal theorem is applied in a direct manner on the boundary $x_2 = 0$ without taking special precaution to deflect the integration path around the crack tip then the second term on the right side of (5.7.23) is

$$\int_0^t \lim_{\epsilon \to 0} \int_{-\epsilon}^\epsilon \sigma_{22}(x_1, 0^+, \tau) u_2^*(x_1, 0^+, t - \tau) \, dx_1 \, d\tau$$

$$= \frac{(1 - \nu)}{\mu} \sqrt{\frac{2}{\pi}} K_I(t) \lim_{\epsilon \to 0} \int_{-\epsilon}^\epsilon x_+^{-1/2} x_-^{-1/2} \, dx, \quad (5.7.27)$$

where, for any real λ,

$$x_+^\lambda = \begin{cases} |x|^\lambda & \text{for } x > 0, \\ 0 & \text{for } x < 0, \end{cases} \qquad x_-^\lambda = \begin{cases} 0 & \text{for } x > 0, \\ |x|^\lambda & \text{for } x < 0. \end{cases} \quad (5.7.28)$$

Fisher (1971) showed that

$$x_-^\lambda \, x_+^{-1-\lambda} = -\frac{\pi \delta(x)}{2 \sin(\pi \lambda)} \quad (5.7.29)$$

within the framework of generalized functions. This approach leads to precisely the same results as in (5.7.26). By either approach, the conditions imposed on u_i^* reduce (5.7.23) to

$$K_I(t) = \frac{\mu}{(1 - \nu)\sqrt{2\pi}} \int_0^t \int_{-\infty}^0 p(x_1, \tau) \, u_2^*(x_1, 0^+, t - \tau) \, dx_1 \, d\tau \,.$$

$$(5.7.30)$$

The remaining step is to actually determine $u_2^*(x_1, 0^+, t)$, and this process is greatly facilitated by the analysis in Section 2.5. There, the case of a suddenly applied crack face pressure of magnitude σ^* for the same geometrical configuration is considered in some detail, and the solution of the boundary value problem is reduced to the solution of the functional equation (2.5.11) by means of the Wiener-Hopf technique. The corresponding functional equation in the present case must be identical, except that $\sigma^* = 0$. Therefore, in the present case,

$$F_+(\zeta)\Sigma_+(\zeta) = -\frac{\mu\kappa}{b^2}\frac{U_-(\zeta)}{F_-(\zeta)}, \qquad (5.7.31)$$

where all quantities are defined as in Section 2.5. The functions $U_-(\zeta)$ and $\Sigma_+(\zeta)$ are proportional to the double transforms of the unknown crack face normal displacement for $x_1 < 0$ and the unknown crack line normal stress for $x_1 > 0$, respectively. In the present context,

$$U_-(\zeta) = s^\gamma \int_{-\infty}^{0} \widehat{u}_2^*(x_1, 0^+, s)\, e^{-s\zeta x_1}\, dx_1 . \qquad (5.7.32)$$

This expression differs from the corresponding expression in (2.5.8) where the exponent on s is fixed by the nature of the applied loading. Here, a factor as a power of s remains in anticipation of a homogeneous solution, but the exponent γ is left arbitrary until it can be fixed by one of the conditions to be imposed on the field u_i^*. Likewise, in the present context,

$$\zeta^{-1}\Sigma_+(\zeta) = -s^\gamma \int_{0}^{\infty} \int_{x_1}^{\infty} \widehat{\sigma}_{22}^*(x', 0, s)\, dx'\, e^{-s\zeta x_1}\, dx_1 , \qquad (5.7.33)$$

where the special form of this expression is dictated by the presence of a nonintegrable singularity in $\widehat{\sigma}_{22}^*$ at $x_1 = 0$.

Attention is now focused on (5.7.31) with the interpretations (5.7.32) and (5.7.33) in mind. Each side is the analytic continuation of the other into its complementary half plane, so each side equals one and the same entire function. The Abel relationships between asymptotic properties of functions and their transforms (2.5.34) imply that both $U_-(\zeta)$ and $\zeta^{-1}\Sigma_+(\zeta)$ in (5.7.31) are $O(\zeta^{-\frac{1}{2}})$ as $|\zeta| \to \infty$ and, consequently, the entire function is at most a constant, say E_o, according to Liouville's theorem. If the steps involved in inverting the transformed solution that are outlined in Section 2.5 are followed here, then it is found that

$$u_2^*(x_1, 0^+, t) = \frac{E_o b^2}{\mu\kappa\pi\Gamma(\gamma)} \frac{\partial}{\partial t} \int_{a}^{-t/x_1} \frac{\mathrm{Im}\left\{F_-(\zeta)\right\}\, d\zeta}{(t + \zeta x_1)^{1-\gamma}} \qquad (5.7.34)$$

for $t > 0$ and $x_1 < 0$, where the limiting value of the integrand as $\mathrm{Im}(\zeta) \to 0^+$ is assumed. $\Gamma(\gamma)$ is the gamma function, and it has been assumed that γ is in the range $0 \leq \gamma < 1$. The inversion is similar for values of γ outside this range but, based on experimenting with different values of γ, it is concluded that this is the pertinent range for present purposes.

Indeed, suppose that the choice $\gamma = \frac{1}{2}$ is made and that (5.7.34) is rewritten as

$$u_2^*(x_1, 0^+, t) = -\frac{E_o b^2}{\mu \kappa \sqrt{-\pi x_1}} \frac{\partial}{\partial t} \left\{ \frac{1}{2\pi i} \int_{C_1} \frac{F_-(\zeta)}{(-\zeta - t/x_1)^{\frac{1}{2}}} \, d\zeta \right\},$$

(5.7.35)

where C_1 is an integration path that begins at $\zeta = -t/x_1 - i0$, and then runs along the lower side of the real axis to the branch point at $\zeta = a$, around the branch point, and along the upper side of the real axis to $\zeta = -t/x_1 + i0$. If $-t/x_1 > c$ then the integration path is augmented by small semicircular loops around $\zeta = c$ on both sides of the branch cut along the real axis. These loops make no net contribution to the value of the integral. The function $(-\zeta - t/x_1)^{1/2}$ is made single-valued by cutting the ζ-plane along the real axis from $\zeta = -t/x_1$ to $\zeta \to \infty$, and the branch with a positive real value at $\zeta = 0$ is assumed.

Observe that if $-t/x_1 > c$ then the path of integration can be closed around the branch point at $\zeta = -t/x_1$ and that all singularities of the integrand are inside this path. Consequently, according to Cauchy's integral theorem, the value of the integral along C_1 is equal to the value along a clockwise circular path of indefinitely large radius centered at the origin. The latter integral can be determined from the asymptotic properties of the integrand to be

$$u_2^*(x_1, 0^+, t) = -\frac{E_o b^2}{\mu \kappa \sqrt{-\pi x_1}} \frac{\partial}{\partial t} \left\{ -\frac{1}{2\pi i} \oint \frac{d\zeta}{\zeta} H(t) \right\}$$

(5.7.36)

$$= -\frac{E_o b^2}{\mu \kappa \sqrt{-\pi x_1}} \delta(t)$$

for $0^- < t < -c_R x_1$. If this result is compared with (5.7.25) then it is evident that $E_o = -2\sqrt{\pi}$. Thus, in view of (5.7.1),

$$K_I(t) = \int_0^t \int_{-\infty}^0 p(x_1, \tau) \, h(x_1, t - \tau) \, dx_1 \, d\tau,$$

(5.7.37)

where

$$h(x_1, t) = \sqrt{\frac{-2}{\pi x_1}} \frac{\partial}{\partial t} \left\{ \frac{1}{2\pi i} \int_{C_1} \frac{F_-(\zeta)}{(-\zeta - t/x_1)^{\frac{1}{2}}} \, d\zeta \right\}. \tag{5.7.38}$$

This is the problem-independent weight function for transient mode I crack problems.

Two simple illustrations of the application of (5.7.37) to situations considered previously are noted. First, if $p(x_1, t) = p^* \delta(x_1 + l)H(t)$ for some positive constant distance l, then the steps leading to the stress intensity factor are very simple. A second example considered in Section 2.5 is that when $p(x_1, t) = \sigma^* H(t)$. In this case, $p(x_1, t)$ is independent of x_1 and it is advantageous to write (5.7.38) in the form

$$h(x_1, t) = \sqrt{\frac{2}{\pi}} \frac{\partial}{\partial t} \left\{ \frac{1}{2\pi i} \int_{C_\infty} \frac{F_-(\zeta)H(\zeta x_1 + t)}{(\zeta x_1 + t)^{\frac{1}{2}}} \, d\zeta \right\}, \tag{5.7.39}$$

where C_∞ is the limiting case of C_1 as $x_1 \to 0^-$. Then, from (5.7.37),

$$K_I(t) = \frac{-\sigma^*}{2\pi i} \sqrt{\frac{2}{\pi}} \int_0^t \frac{\partial}{\partial \tau} \left\{ \int_{C_\infty} F_-(\zeta) \int_{\frac{(\tau - t)}{\zeta}}^0 \frac{dx_1}{(\zeta x_1 + t)^{\frac{1}{2}}} \, d\zeta \right\} d\tau$$

$$= 2\sigma^* \sqrt{\frac{2t}{\pi}} \frac{1}{2\pi i} \int_{C_\infty} \zeta^{-1} F_-(\zeta) \, d\zeta$$

$$= 2\sigma^* F_-(0) \sqrt{2t/\pi}. \tag{5.7.40}$$

This result is identical to (2.5.44).

Finally, a special property of $h(x_1, t)$ is noted. From the foregoing development, it is evident that, for any time $t > 0$, the function $h(x_1, t)$ is nonzero on the crack faces only for $-c_d t < x_1 < -c_R t$. In other words, the impulse singularity acts at time $t = 0$ and the displacement of any material point on the crack face, in the direction of the normal to the face, is zero for all time after the Rayleigh wave emitted from the tip at time $t = 0$ has passed that point.

The corresponding normal stress σ_{22}^* ahead of the crack tip is, following inversion of (5.7.33) for $\gamma = \frac{1}{2}$,

$$\sigma_{22}^*(x_1, 0, t) = -\frac{\partial^2}{\partial t \partial x_1} \left\{ \frac{1}{2\pi i} \int_{C_1} \frac{d\zeta}{\zeta F_+(\zeta)(t + \zeta x_1)^{\frac{1}{2}}} \right\} \tag{5.7.41}$$

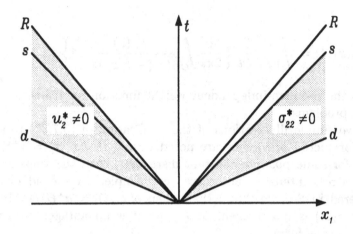

Figure 5.11. The parts of the x_1, t-plane over which u_2^* and σ_{22}^* are nonzero, shown as the shaded regions. The lines labeled d, s, and R represent trajectories of the dilatational, shear, and Rayleigh waves, respectively, emitted from $x_1 = 0$ at $t = 0$.

for $t > 0$ and $x_1 > 0$, where the path C_1 now begins at $\zeta = -t/x_1 - i0$, and then extends along the lower side of the real axis to the branch point at $\zeta = -a$, around this point, and along the upper side of the real axis to $\zeta = -t/x_1 + i0$. It is readily verified by means of calculations parallel to those performed above in studying u_2^* that

$$\sigma_{22}(x_1, 0, t) \sim \frac{\delta(t)}{x_1^{3/2}} \qquad (5.7.42)$$

for $t > 0$ and arbitrarily small values of $x_1 > 0$. In view of the fact that $\sigma_{22}^*(x_1, 0, t) = 0$ on $x_1 < 0$, it can be shown from (5.7.41) that this same condition holds for $0 < x_1 < c_s t$ for all $t > 0$. The regions of the x_1, t–plane over which u_2^* and σ_{22}^* are nonzero are shown in Figure 5.11. The remarkable feature is that both of these fields are *identically zero* for $-c_R t < x_1 < c_s t$ for all time $t > 0$. Thus, even though the present situation presumes that the crack tip is stationary at $x_1 = 0$, it could equally well meander according to $x_1 = l(t)$, with $l(0) = 0$ for $t > 0$, and the fields denoted by a superimposed star would still apply. This feature was exploited to great advantage by Burridge (1976) in his study of dynamic crack growth.

5.8 Energy radiation from an expanding crack

In this section, expressions for the energy radiated from an expanding planar crack in an elastic solid are considered. The discussion is limited to bodies that are unbounded except for the crack surface, and the material is assumed to be an isotropic and homogeneous linear elastic material. The body may be subjected to some initial equilibrium stress distribution, but any transient stress field is assumed to arise from expansion of the crack in the body.

The energy radiated from an expanding crack is of particular interest in the field of seismology, where the energy radiated from a dynamically expanding fault in the earth, inferred from transient displacement measurements at remote points, is used to infer a value for the energy released in the faulting process. Alternatively, this energy is estimated from changes in the equilibrium deformation of the earth's surface at points relatively near the epicenter of the earthquake. These energies are equal only under certain conditions, and the present model provides a framework for considering this equivalence and other issues associated with radiated seismic energy (Rudnicki and Freund 1981).

Consider an expanding crack that has a characteristic dimension l. The total energy radiated from this crack is defined as

$$E_R = \int_{-\infty}^{\infty} \int_S \sigma_{ij} n_j \dot{u}_i \, dS \, dt \,, \qquad (5.8.1)$$

where u_i is the change in displacement from the initial equilibrium distribution and σ_{ij} is the corresponding change in stress. The surface S is a spherical surface of radius r centered at the expanding crack, with $r \gg l$, and n_i is the unit normal to S pointing toward the crack. The integrand of (5.8.1) is the rate of work done on the material outside S by the tractions on S. If the stress change σ_{ij} and the particle velocity $\dot{u}_i$ are each proportional to r^{-1} as r becomes very large, as is the case in some point source representations for far-field motion, then E_R is expected to have a finite nonzero value.

Kostrov (1973) has shown that the energy radiated from the crack can be expressed in terms of the crack surface traction and particle velocity. Let S_f be the surface coinciding with the crack faces, excluding the edge, let S_t be a tube enclosing the crack edge, and let n_i be a unit vector normal to S, S_f, and S_t and *pointing into* the region bounded by these surfaces. A work rate balance for this region implies that

$$-\int_S \overline{\sigma}_{ij} n_j \dot{u}_i \, dS - \dot{T} - \dot{\overline{U}} = \overline{F}(t) + \int_{S_f} \overline{\sigma}_{ij} n_j \dot{u}_i \, dS, \qquad (5.8.2)$$

where $\dot{T}$ and $\dot{\overline{U}}$ are the rates of increase of kinetic energy and strain energy, respectively, and $\overline{F}(t)$ is the rate of energy flow to the extending crack edge through S_t. The total stress is

$$\overline{\sigma}_{ij} = \sigma_{ij} + \overline{\sigma}_{ij}^o \qquad (5.8.3)$$

and the total particle displacement is

$$\overline{u}_i = u_i + \overline{u}_i^o, \qquad (5.8.4)$$

where $\overline{u}_i^o$ is the initial equilibrium displacement and $\overline{\sigma}_{ij}^o$ is the corresponding stress. The left side of (5.8.2) is the excess of the total power input over the increase in internal energy, and the right side is the energy dissipation at the crack edge and over the crack surfaces. Adapting the general energy flux expression (5.2.11), the rate of energy flow into the crack edge is

$$\overline{F}(t) = \lim_{S_t \to 0} \int_{S_t} \left\{ \overline{\sigma}_{ij} n_j \dot{u}_i + \tfrac{1}{2} \left(\overline{\sigma}_{ij} \overline{u}_{i,j} + \rho \dot{u}_i \dot{u}_i \right) v n_k m_k \right\} dS, \quad (5.8.5)$$

where v is the local normal speed of the crack edge in the plane of the crack and m_k is the local normal direction of crack front motion.

    In view of (5.8.3) and the definition of radiated energy, (5.8.2) becomes

$$-\dot{E}_R - \int_S \overline{\sigma}_{ij}^o n_j \dot{u}_i \, dS - \dot{T} - \dot{\overline{U}} = \overline{F}(t) + \int_{S_f} \overline{\sigma}_{ij} n_j \dot{u}_i \, dS. \quad (5.8.6)$$

If this expression is integrated over all time then

$$E_R + \int_S \overline{\sigma}_{ij} n_j u_i^{final} \, dS + \left(\overline{U}_{final} - \overline{U}_{initial} \right)$$

$$= -\int_{-\infty}^{\infty} \overline{F}(t) \, dt - \int_{-\infty}^{\infty} \int_{S_f} \overline{\sigma}_{ij} n_j \dot{u}_i \, dS \, dt, \qquad (5.8.7)$$

where $T_{final} = T_{initial} = 0$ has been assumed. For crack expansion under equilibrium conditions, $E_R = 0$ and (5.8.7) reduces to a formula for the change of strain energy in the body, that is,

$$\overline{U}_{final} - \overline{U}_{initial} + \int_S \overline{\sigma}_{ij}^0 n_j u_i^{final} \, dS$$

$$= -\int_{-\infty}^{\infty} \overline{F}_0(t) \, dt - \int_{-\infty}^{\infty} \int_{S_f} [\overline{\sigma}_{ij} n_j \dot{u}_i]_0 \, dS \, dt, \quad (5.8.8)$$

where the subscript zero denotes values during an equilibrium expansion. If (5.8.8) is subtracted from (5.8.7), then

$$E_R = \int_{-\infty}^{\infty} \int_{S_f} \left\{ [n_j \overline{\sigma}_{ij} \dot{u}_i]_0 - [n_j \overline{\sigma}_{ij} \dot{u}_i] \right\} \, dS \, dt$$

$$+ \int_{-\infty}^{\infty} \left\{ \overline{F}_0(t) - \overline{F}(t) \right\} \, dt, \quad (5.8.9)$$

where the superposed bar may be omitted in this equation. The integrand of the second term is the difference between values of energy flux to the crack edge for quasi-static and dynamic propagation of the crack edge, and the integrand of the first term is the difference between values of the work rate of the crack surface traction away from the edge for quasi-static and dynamic expansion. For bounded crack face traction, the first term in (5.8.9) can be integrated by parts with the result that

$$E_R = \int_{-\infty}^{\infty} \int_{S_f} \left\{ [n_j \dot{\overline{\sigma}}_{ij} u_i] - [n_j \dot{\overline{\sigma}}_{ij} u_i]_0 \right\} \, dS \, dt$$

$$+ \int_{-\infty}^{\infty} \left\{ \overline{F}_0(t) - \overline{F}(t) \right\}^{\cdot} dt, \quad (5.8.10)$$

where the integrated terms are zero because both the initial and final states are time-independent. This form of E_R makes clear the role of the time-dependent surface tractions in radiating energy. It shows that it is impossible to obtain an unambiguous expression for radiated energy in terms of only changes from the initial equilibrium state of the medium to the final equilibrium. Furthermore, the first term of

(5.8.10) makes clear the important role for the radiated energy of transients in fault plane traction due to variable interaction of the surfaces undergoing relative sliding (Kostrov 1973).

An alternate form for the radiated energy can be obtained by using (5.8.3) in the first term on the right side of (5.8.8), with the result that

$$\int_{-\infty}^{\infty} \int_{S_f} [\bar{\sigma}_{ij} n_j \dot{u}_i]_0 \, dS \, dt =$$

$$\int_{-\infty}^{\infty} \int_{S_f} [\sigma_{ij} n_j \dot{u}_i]_0 \, dS \, dt + \int_{S_f} \sigma_{ij}^0 n_j u_i^{final} \, dS . \quad (5.8.11)$$

If the stress σ_{ij} depends on time only through its dependence on u_i, then

$$\int_{-\infty}^{\infty} \int_{S_f} [\sigma_{ij} n_j \dot{u}_i]_0 \, dS \, dt = \int_{S_f} \left[\langle \sigma_{ij} n_j \rangle_{avg} u_i^{final} \right] dS , \quad (5.8.12)$$

where this equation defines the average crack face traction $\langle \sigma_{ij} n_j \rangle_{avg}$. In the special case where $\sigma_{ij} n_j$ is proportional to u_i

$$\langle \sigma_{ij} n_j \rangle_{avg} = \tfrac{1}{2} \left\{ [\sigma_{ij} n_j]_{final} + [\sigma_{ij} n_j]_{initial} \right\} . \quad (5.8.13)$$

Note that $[\sigma_{ij} n_j]_{initial}$ is not the traction existing ahead of the crack on the prospective fracture plane, but rather that which exists on the crack surface just after passage of the edge. If (5.8.12) is substituted into (5.8.11) and if the relationship

$$\langle \sigma_{ij} n_j \rangle_{avg} = \langle \bar{\sigma}_{ij} n_j \rangle_{avg} - \sigma_{ij}^0 n_j \quad (5.8.14)$$

is incorporated, then

$$\int_{-\infty}^{\infty} \int_{S_f} [\bar{\sigma}_{ij} n_j \dot{u}_i]_0 \, dS \, dt = \int_{-\infty}^{\infty} \int_{S_f} [\langle \bar{\sigma}_{ij} n_j \rangle_{avg} \dot{u}_i]_0 \, dS . \quad (5.8.15)$$

Consequently, (5.8.8) can be written as

$$\bar{U}_{final} - \bar{U}_{initial} + \int_{S} \sigma_{ij}^0 n_j u_i^{final} \, dS$$

$$= -\int_{-\infty}^{\infty} \bar{F}_0(t) \, dt + \int_{-\infty}^{\infty} \int_{S_f} [\langle \bar{\sigma}_{ij} n_j \rangle_{avg} \dot{u}_i]_0 \, dS \, dt . \quad (5.8.16)$$

If $\overline{F}_0$ and the integral over S are assumed to be negligible in the particular application of dynamic earth faulting, then this equation reduces to the formula for strain energy change associated with faulting as discussed by Aki and Richards (1980). Yet another expression for the radiated energy is obtained by subtracting (5.8.16) from (5.8.7) with the result that

$$E_R = \int_{-\infty}^{\infty} \int_{S_f} \left\{ [\langle \overline{\sigma}_{ij} n_j \rangle_{avg} \dot{u}_i]_0 - [\overline{\sigma}_{ij} n_j \dot{u}_i] \right\} dS \, dt$$

$$+ \int_{-\infty}^{\infty} \left\{ \overline{F}_0(t) - \overline{F}(t) \right\} dt, \quad (5.8.17)$$

where, as before, the superposed bars can be omitted.

If the crack surface traction is time-independent, except possibly in the small region very near the crack edge, then the first term in (5.8.9), (5.8.10), or (5.8.17) vanishes and the radiated energy is

$$E_R = \int_{-\infty}^{\infty} \left[\overline{F}_0(t) - \overline{F}(t) \right] dt. \quad (5.8.18)$$

A special case of the expression (5.8.18) was used by Husseini and Randall (1976) in a study of the efficiency of a seismic source in converting strain energy released to radiated wave energy.

The concept of the seismic moment of an earthquake source was introduced by Aki (1966) to provide a quantitative connection between source properties and the far-field, long wavelength energy radiated by the source. Under the simplest circumstances of relative slip on a planar surface in a homogeneous and isotropic elastic body, the seismic moment is defined to be

$$M = \mu[u]A_{slip}, \quad (5.8.19)$$

where μ is the shear modulus of the material, A_{slip} is the total area of slip, and $[u]$ is the spatial average of the relative slip or displacement discontinuity, say $[u_k]$, across the fault plane

$$[u] = \frac{1}{A_{slip}} \int_{A_{slip}} \sqrt{[u_k][u_k]} \, dS. \quad (5.8.20)$$

The radiated energy E_R is also a parameter of fundamental significance in the study of earthquake sources. It can be measured remotely, essentially according to (5.8.1), and its value is representative of physical characteristics of the fault region. The foregoing analysis provides the basis for observations on the possible connection between M and E_R.

The qualitative notion of seismic efficiency has been used in the past to establish such a relationship. In terms of quantities already introduced above, seismic efficiency is defined as the ratio $\eta = E_R / |\overline{U}_{final} - \overline{U}_{initial}|$. The strain energy change of the material in the volume bounded by the surfaces $S + S_t + S_f$ which appears in (5.8.7) can be expressed in terms of the elastic compliance of the material M_{ijkl} by

$$
\overline{U}_{final} - \overline{U}_{initial} = \int_V \tfrac{1}{2} M_{ijkl} \left[\overline{\sigma}_{ij}\overline{\sigma}_{kl} - \sigma_{ij}^0 \sigma_{kl}^0 \right] dV
$$

$$
= \int_V M_{ijkl}\sigma_{ij} \left[\tfrac{1}{2}\sigma_{kl} + \overline{\sigma}_{kl}^0 \right] dV
$$

$$
= \int_{S+S_t+S_f} -\tfrac{1}{2} \left[\overline{\sigma}_{ij}^{final} + \overline{\sigma}_{ij}^0 \right] n_j u_i^{final} \, dS. \qquad (5.8.21)
$$

The last step in the calculation follows from the relation $2M_{ijkl}\sigma_{ij} = u_{k,l} + u_{l,k}$ and the divergence theorem. Suppose that the contributions from the remote surface S and the crack edge surface S_t are assumed to be negligibly small compared to the fault surface contribution. If $\overline{\tau}^{final}$ and $\overline{\tau}^0$ denote the average values of the traction vectors $\overline{\sigma}_{ij}^{final} n_j$ and $\overline{\sigma}_{ij}^0 n_j$ over S_f projected onto S_f, and if $[\![u]\!]$ is the spatial average of the discontinuity in displacement across the slip surface of area A_{slip}, then

$$
|\overline{U}_{final} - \overline{U}_{initial}| \approx \tfrac{1}{2} \left(\overline{\tau}^{final} + \overline{\tau}^0 \right) [\![u]\!] A_{slip}. \qquad (5.8.22)
$$

From the definition of seismic efficiency,

$$
E_R = \eta \, |\overline{U}_{final} - \overline{U}_{initial}|
$$
$$
= \tfrac{1}{2} \left(\overline{\tau}^{final} - \overline{\tau}^0 \right) [\![u]\!] = \tfrac{1}{2}\eta \left(\overline{\tau}^{final} + \overline{\tau}^0 \right) M/\mu. \qquad (5.8.23)
$$

The quantity $\eta(\overline{\tau}^{final} + \overline{\tau}^o)/2$ was termed the "apparent stress" by Wyss and Brune (1968). This idea is discussed more completely by Kostrov and Das (1989).

Another approximate relationship between M and E_R that does not rely on the idea of radiation efficiency can be derived directly from (5.8.7). Again, suppose that the contributions to this relationship from the remote surface S and the crack edge surface S_t are negligibly small in comparison to the fault plane contributions, so that

$$E_R \approx -\int_{-\infty}^{\infty} \overline{\sigma}_{ij} n_j \dot{u}_i \, dS \, dt - \left(\overline{U}_{final} - \overline{U}_{initial}\right) . \qquad (5.8.24)$$

If the fault plane traction is time-independent then the time integration in the first term on the right side of (5.8.24) is readily evaluated. This equation becomes

$$E_R = -\int_{S_f} \overline{\sigma}_{ij}^{final} n_j u_i^{final} \, dS + \int_{S_f} \tfrac{1}{2} \left(\overline{\sigma}_{ij}^{final} + \overline{\sigma}_{ij}^o\right) n_j u_i^{final} \, dS \,,$$

$$(5.8.25)$$

where the second term follows from (5.8.21). If the various fields are replaced by their spatial averages, then

$$E_R = \tfrac{1}{2} \left(\overline{\tau}^{final} - \overline{\tau}^o\right) [\![u]\!] = \tfrac{1}{2} \left(\overline{\tau}^{final} - \overline{\tau}^o\right) M/\mu . \qquad (5.8.26)$$

Thus, the ratio of radiated energy to the seismic moment is equal to the ratio of one-half the average fault plane stress drop $(\overline{\tau}^{final} - \overline{\tau}^o)$ to the shear modulus. Comparison of (5.8.23) with (5.8.26) suggests that the seismic efficiency is the ratio of the stress drop to twice the fault plane effective stress $(\overline{\tau}_{final} + \overline{\tau}^o)/2$, that is,

$$\eta = \frac{\overline{\tau}^{final} - \overline{\tau}^o}{\overline{\tau}^{final} + \overline{\tau}^o} , \qquad (5.8.27)$$

as noted by Kostrov (1973). Though these results on seismic moment and its connection to radiated energy are commonly used in characterizations of earthquake sources, it must be recognized that they do not account for time dependence in the fault plane stress distribution, for any transient aspects of the dynamic faulting process, or for inhomogeneities in material properties on the fault plane.

6

ELASTIC CRACK GROWTH
AT CONSTANT SPEED

6.1 Introduction

The focus in this chapter, as well as in the following chapter, is on analytical models of crack growth phenomena based on nominally elastic material response. The analysis of Chapter 4 provides information on the nature of crack tip fields during rapid crack growth for several categories of material response, and parameters that characterize the strength or intensity of these fields are also identified. A main purpose in formulating and solving boundary value problems concerned with crack propagation is to determine the dependence of the crack tip field characterizing parameters on the applied loading and on the configuration of the body.

For any particular crack growth process, it is often the case that its analytical modeling can be pursued at several scales of observation. For example, consider the case of crack growth in an engineering structure. The full structure can be analyzed to determine the stress resultants that are imposed at the fracturing section. Under suitable circumstances, these stress resultants can, in turn, be viewed as the applied loads in a two-dimensional plane stress or plane strain problem from which the dynamic stress intensity factor can be determined. At a finer scale of observation, the elastic stress intensity factor represents the applied loading in an analysis of nonlinear processes very close to the crack edge. By focusing on dominant effects at adjacent levels of observation in this way, various aspects of dynamic fracture phenomena can be analyzed.

In this chapter, attention is limited to crack growth at constant speed in a nominally elastic material. The imposed condition of constant speed results in significant simplification of the boundary value problems to be solved for the governing partial differential equations. Under certain circumstances, the restriction to constant speed crack growth permits a reduction of the number of independent variables by one. Thus, problems involving two space dimensions and time can be reduced to boundary value problems in two independent variables. It is shown in this chapter that solutions of the equations of elastodynamics for certain crack growth problems can be expressed in terms of one or more general solutions of elementary partial differential equations in two independent variables. The specific solutions are determined by the boundary conditions.

The first problem class considered is steady state crack growth. Here, the crack tip has been moving at constant speed for all time and the mechanical fields are invariant with respect to an observer moving with the crack tip. The prototype problem in this category is the two-dimensional Yoffe problem of a crack of fixed length propagating in a body subjected to uniform remote tensile loading. The second class of problems considered is self-similar crack growth. In this case, the crack tip has been moving at constant speed since some initial instant, and certain mechanical fields (to be identified in the context of a particular problem) are invariant with respect to an observer moving steadily away from the process being observed. The prototype problem in this category is the two-dimensional Broberg problem of a crack that suddenly expands symmetrically from zero initial length at constant speed in a body subjected to a uniform remote tension.

Next, the situation of the sudden constant speed extension of a preexisting crack in a body initially at rest and subjected to time-independent loading is considered. The particular cases analyzed are also self-similar, but they are solved by means of integral transform methods rather than by direct appeal to similarity arguments. This problem is significant because the indirect analytical approach developed to analyze it, based on superposition over a fundamental elastodynamic solution and inverse arguments, opens the way for analysis of certain problems of crack growth at nonuniform speed, which are described in Chapter 7.

As already noted, the objective of the analysis of crack growth is to understand the connection between crack tip motion in a body of given material and the loading applied to that body. Obviously, the potential for improved understanding is severely diminished if the motion of the crack is assumed to be uniform from the outset. However, the analysis of constant speed crack growth has been important for two main reasons. In some cases, the prevailing conditions are such that crack growth in a real material is indeed steady for a long enough time so that these constant speed results are useful in the interpretation of the phenomenon. In addition, the progress in the study of crack growth at constant speed has resulted in the analytical tools that have been successful in the analysis of crack growth at nonuniform rates.

6.2 Steady dynamic crack growth

Consider a planar crack extending in an elastic solid under conditions of two-dimensional plane strain deformation. Suppose that the crack tip of interest moves in the plane of deformation at constant speed v. Furthermore, suppose that both the configuration of the body and the traction distribution are time invariant in a reference frame translating with the crack tip, and that this situation has prevailed for a "long time." In the typical configuration assumed in steady crack growth analysis, the body is either unbounded or it has straight boundaries parallel to the crack plane. (Other configurations can be adopted for special purposes, say to develop numerical procedures for dynamic crack growth analysis, but the boundary conditions must then be chosen to simulate the property of translational invariance of the configuration in the direction of crack growth.) Under these conditions a *steady state solution* of the boundary value problem is sought. The assumption that a steady state has been achieved permits the reduction of the number of independent variables in the problem from three to two, and the analysis is simplified considerably. It must always be kept in mind, of course, that some features of the crack growth process that are physically significant are overlooked in this approach, and the results must be assessed accordingly.

A rectangular coordinate system labeled x, y, z is introduced in the body in such a way that the plane of deformation is the x, y-plane, and the crack is in the plane $y = 0$. The crack tip of interest advances in the x-direction at constant speed and, without loss of generality, it is assumed that the crack tip passes $x = 0$ at time $t = 0$. If other crack edges are present, each must be at a fixed distance from the edge of primary interest. Let $f(x, y, t)$ denote any field over the body.

If this field is an element of a steady state solution then

$$f(x, y, t) = f(\xi, y), \qquad (6.2.1)$$

where $\xi = x - vt$. The functions on opposite sides of the equation in (6.2.1) are different, of course, but the values of the functions represent one and the same physical quantity, so there is little danger of confusing the two functions. The ξ, y-coordinate system translates at speed v in the x- or ξ-direction with the crack tip of primary interest, and this crack tip is at the origin of coordinates in this system.

6.2.1 General solution procedure

The main properties of the mechanical fields which result from the steady state condition (6.2.1) are that, for any field f,

$$\frac{\partial f}{\partial x}(x, y, t) = \frac{\partial f}{\partial \xi}(\xi, y), \qquad \frac{\partial f}{\partial t}(x, y, t) = -v\frac{\partial f}{\partial \xi}(\xi, y), \qquad (6.2.2)$$

where the interpretation of the function is made clear by including the arguments explicitly. Following Sneddon (1952), the basic field quantities are chosen to be the dilatational and shear displacement potentials $\phi(\xi, y)$ and $\psi(\xi, y)$, respectively. An alternative approach based on stress potential functions was introduced independently by Radok (1956), who considered applications to moving punch and dislocation problems, as well as crack problems. The equivalence of the two methods for the problem at hand was noted by Sneddon (1958). A representation in terms of stress functions that has been useful in the solution of some steady state problems for anisotropic elastic materials was introduced by Stroh (1962).

The displacement potentials satisfy their respective scalar wave equations (1.2.61) which, in the case of steady state motion, take the form

$$\alpha_d^2 \frac{\partial^2 \phi}{\partial \xi^2} + \frac{\partial^2 \phi}{\partial y^2} = 0, \qquad \alpha_s^2 \frac{\partial^2 \psi}{\partial \xi^2} + \frac{\partial^2 \psi}{\partial y^2} = 0 \qquad (6.2.3)$$

over the region of the ξ, y-plane occupied by the body, where $\alpha_d = \sqrt{1 - v^2/c_d^2}$ and $\alpha_s = \sqrt{1 - v^2/c_s^2}$. As noted in Section 1.3, the equations have the general solutions

$$\phi(\xi, y) = \text{Re}\{F(\zeta_d)\}, \qquad \psi(\xi, y) = \text{Im}\{G(\zeta_s)\}, \qquad (6.2.4)$$

where $\zeta_d = \xi + i\alpha_d y$ and $\zeta_s = \xi + i\alpha_s y$, and the functions F and G are analytic functions of their complex arguments in the interior region of the body. These functions must be determined from boundary conditions. For the particular case of mode I deformation, the potentials have the symmetries $\phi(\xi, -y) = \phi(\xi, y)$ and $\psi(\xi, -y) = -\psi(\xi, y)$, whereas for mode II deformation $\phi(\xi, -y) = -\phi(\xi, y)$ and $\psi(\xi, -y) = \psi(\xi, y)$. These symmetries imply that

$$F(\bar{\zeta}) = \pm\overline{F(\zeta)} \quad \text{and} \quad G(\bar{\zeta}) = \pm\overline{G(\zeta)} \tag{6.2.5}$$

for mode I (upper sign) and mode II (lower sign), where the bar denotes the complex conjugate.

Displacement and stress components are readily expressed in terms of the unknown functions F and G by means of (1.2.68) and (1.2.69). For example,

$$u_y(\xi, y) = -\text{Im}\{\alpha_d F'(\zeta_d) + G'(\zeta_s)\},$$

$$\sigma_{yy}(\xi, y) = -\mu\text{Re}\{(1 + \alpha_s^2)F''(\zeta_d) + 2\alpha_s G''(\zeta_s)\}, \tag{6.2.6}$$

$$\sigma_{xy}(\xi, y) = -\mu\text{Im}\{2\alpha_d F''(\zeta_d) + (1 + \alpha_s^2)G''(\zeta_s)\},$$

where the prime denotes differentiation with respect to the argument.

The foregoing mathematical results are now applied to obtain the solution of representative steady crack growth problems. The first boundary value problem considered is that formulated by Yoffe (1951) in a pioneering continuum analysis of fracture dynamics. The second case studied is the mode II propagation of a half plane crack under the action of crack face traction. The mode I counterpart of the second problem was first considered by Craggs (1960). The section concludes with a brief discussion of the concept of crack tip cohesive zone, within the context of steady crack growth, as it has been applied to simulate crack tip plasticity in ductile materials, interatomic interaction in pure cleavage, and various frictional interactions between the sliding faces of a shear crack. The essential ideas of the cohesive zone models were introduced in Section 5.3.

6.2.2 The Yoffe problem

The plane strain problem analyzed by Yoffe (1951) is that of a crack of fixed length $2a$ gliding through an otherwise unbounded solid at

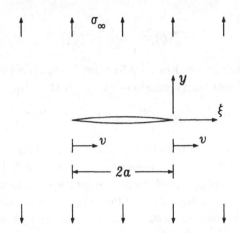

Figure 6.1. Diagram of the plane strain problem considered by Yoffe (1951), showing a crack of fixed length moving steadily at speed v through a body subjected to uniform remote tension of magnitude σ_∞.

speed v, with the crack opening at the leading crack tip at $\xi = 0$, $y = 0$ and closing at the trailing crack tip at $\xi = -2a$, $y = 0$; see Figure 6.1. The body is loaded by means of a uniform remote tension of magnitude σ_∞ in the y-direction. The crack faces are traction free. It is convenient (but not essential) to superimpose a uniform compressive stress of magnitude σ_∞ in the y-direction. Thus, the crack faces are pushed apart by a uniform normal traction, $T_x(\xi) = 0$ and $T_y(\xi) = \pm\sigma_\infty$ on $y = \pm 0$, $-2a < \xi < 0$. All stress components vanish as $\xi^2 + y^2 \to \infty$. The boundary conditions are

$$\sigma_{yy}(\xi, 0^\pm) = -\sigma_\infty, \qquad \sigma_{xy}(\xi, 0^\pm) = 0 \qquad (6.2.7)$$

for $-2a < \xi < 0$. The problem possesses the symmetry characteristic of mode I deformation, and all fields are continuous across $y = 0$ for $\xi > 0$ and $\xi < -2a$.

In terms of the unknown functions F and G, the boundary conditions take the form

$$(1 + \alpha_s^2) \left[F_+''(\xi) + F_-''(\xi) \right] + 2\alpha_s \left[G_+''(\xi) + G_-''(\xi) \right] = 2\sigma_\infty/\mu,$$

$$2\alpha_d \left[F_+''(\xi) - F_-''(\xi) \right] + (1 + \alpha_s^2) \left[G_+''(\xi) - G_-''(\xi) \right] = 0$$

$$(6.2.8)$$

for $-2a < \xi < 0$, where the prime denotes differentiation with respect to the argument,

$$F''_\pm(\xi) = \lim_{y \to 0^\pm} F''(\zeta_d), \tag{6.2.9}$$

and the symmetries noted in (6.2.5) have been incorporated.

The second boundary condition implies that the function

$$E(\zeta) = 2\alpha_d F''(\zeta) + (1 + \alpha_s^2) G''(\zeta), \tag{6.2.10}$$

which is analytic everywhere in the plane of the body except possibly along the crack line, is continuous across the crack line. The only singularities that $E(\zeta)$ could have are poles on the crack line, but this possibility is precluded by the boundary conditions and the tacit assumption that concentrated forces do not act at the crack tips. Consequently, $E(\zeta)$ is an entire function. Furthermore, the vanishing of stress at remote points implies that $E(\zeta) \to 0$ as $|\zeta| \to 0$. Liouville's theorem on bounded entire functions then implies that $E(\zeta) = 0$, or that

$$G''(\zeta) = -\frac{2\alpha_d}{1 + \alpha_s^2} F''(\zeta) \tag{6.2.11}$$

throughout the region of the body.

In light of (6.2.11) the first boundary condition (6.2.8) takes the form

$$F''_+(\xi) + F''_-(\xi) = -\frac{2(1 + \alpha_s^2)}{D} \frac{\sigma_\infty}{\mu}, \qquad -2a < \xi < 0, \tag{6.2.12}$$

where D is given in (4.3.8) as $4\alpha_d\alpha_s - (1 + \alpha_s^2)^2$. This equation has the homogeneous solution

$$F''_H(\zeta) = \frac{P(\zeta)}{Q(\zeta)\sqrt{\zeta(\zeta + 2a)}}, \tag{6.2.13}$$

where P and Q are polynomial functions of ζ. The ζ-plane is cut along the crack line, and the branch of $\sqrt{\zeta(\zeta + 2a)}$ that is positive and real on the positive real ζ-axis is assumed. The polynomial $Q(\zeta)$ may have zeros only at the ends of the crack line, but this possibility is precluded by the condition that there can be no concentrated forces or dislocations there. Thus, $Q(\zeta) = 1$ without loss of generality. The equation (6.2.12) has the particular solution

$$F''_P(\zeta) = -\frac{(1 + \alpha_s^2)}{D} \frac{\sigma_\infty}{\mu}. \tag{6.2.14}$$

Then, to ensure that the stress vanishes at remote points, $P(\zeta)$ must be a polynomial of degree one, that is, $P(\zeta) = P_0 + P_1\zeta$ with the constant P_1 having the value

$$P_1 = \frac{1 + \alpha_s^2}{D} \frac{\sigma_\infty}{\mu}. \tag{6.2.15}$$

To ensure continuity of displacement across the crack plane behind the advancing crack,

$$P_0 = aP_1. \tag{6.2.16}$$

This condition renders $F''(\zeta)$ symmetric with respect to the center of the crack. The function $F''(\zeta)$ is completely determined as the sum of the homogeneous and particular solutions,

$$F''(\zeta) = \frac{1 + \alpha_s^2}{D} \frac{\sigma_\infty}{\mu} \left[\frac{\zeta + a}{\sqrt{\zeta(\zeta + 2a)}} - 1 \right], \tag{6.2.17}$$

which implies from (6.2.11) that

$$G''(\zeta) = -\frac{2\alpha_d}{D} \frac{\sigma_\infty}{\mu} \left[\frac{\zeta + a}{\sqrt{\zeta(\zeta + 2a)}} - 1 \right]. \tag{6.2.18}$$

With the functions $F''(\zeta)$ and $G''(\zeta)$ determined, features of the solution of physical interest can be extracted. For example, from $(6.2.6)_2$, the normal stress distribution $\sigma_{yy}(\xi, 0)$ on the crack line ahead of the advancing crack tip is

$$\sigma_{yy}(\xi, 0) = \sigma_\infty \left[\frac{\xi + a}{\sqrt{\xi(\xi + 2a)}} - 1 \right]. \tag{6.2.19}$$

For the original problem of uniform remote tension in the direction perpendicular to the crack plane and traction-free crack faces, the second term in brackets is absent. It is noteworthy that the distribution does not depend on the crack speed v. From (4.3.9), the dynamic stress intensity factor is

$$K_I = \lim_{\xi \to 0^+} \sqrt{2\pi\xi}\, \sigma_{yy}(\xi, 0) = \sigma_\infty \sqrt{\pi a}, \tag{6.2.20}$$

which is identical to the corresponding equilibrium result for any v. By integrating (6.2.17) and (6.2.18), the crack opening displacement is found from (6.2.6)$_1$ to be

$$u_y(\xi, 0^+) = \frac{v^2 \alpha_d}{c_s^2 D} \frac{\sigma_\infty}{\mu} \sqrt{-\xi(\xi + 2a)}, \quad -2a < \xi < 0. \qquad (6.2.21)$$

The crack opening profile is elliptic and it depends on crack speed v. In the limit as $v/c_s \to 0$, the coefficient $v^2 \alpha_d/c_s^2 D \to (1 - \nu)$, so the expression (6.2.21) reduces to the corresponding result for the classical Griffith crack, and the magnitude of displacement increases monotonically with increasing v for fixed σ_∞.

Finally, it must be recognized that the mathematical problem considered in this section is not a realistic model of a physical situation because of the feature that the crack closes at one end at the same rate at which it opens at the other end. In her study of dynamic fracture phenomena, on the basis of this analysis Yoffe (1951) drew conclusions only from those features of the solution that are *independent* of the fictitious crack length $2a$. In particular, she was interested in the angular variation of the near tip stress field. As was mentioned in Section 4.3, this variation was established some years later to be universal for a given material. Features of the solution of this problem or of similar problems that do depend on the fictitious crack length must be viewed as having limited significance. Another unrealistic feature of the system emerges if it is considered from the energy flow point of view. For $0 \le v < c_R$, the leading edge of the advancing crack of fixed length absorbs a certain amount of energy from the body according to the result in Section 5.3. The system is in a steady state, however, so the total mechanical energy cannot change and, indeed, it does not. If the energy flux at the trailing crack edge is examined, it is found that mechanical energy is *added* to the body there at exactly the same rate that it is lost at the leading crack edge. Because the crack opening is symmetric with respect to the center of the crack, the tractions acting on the crack faces do no net work. For $c_R < v < c_s$, the roles of the two crack tips as source and sink of energy are reversed. For this case, if the crack is indeed open then the normal stress on the crack plane ahead of the leading tip is compressive and the local energy flow is out of the crack tip. Similarly, the trailing crack tip absorbs energy at the rate at which it is radiated at the leading tip.

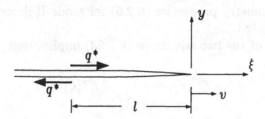

Figure 6.2. Diagram of steady crack growth in mode II due to the action of an opposed pair of concentrated forces acting on the crack faces at a fixed distance l behind the crack tip.

6.2.3 Concentrated shear traction on the crack faces

As a second illustration of the analysis of a steady crack growth problem, consider the plane strain situation of a semi-infinite crack along $y = 0$, $\xi < 0$ with the crack edge at $\xi = 0$ advancing in the ξ-direction at speed v in an otherwise unbounded body. For the time being suppose that there are no remote tractions on the body but that the crack faces are subjected to concentrated loads tangential to the crack faces at a distance l from the crack edge, that is, $T_x(\xi) = \pm q^* \delta(\xi + l)$, $T_y(\xi) = 0$ on $y = \pm 0$, $\xi < 0$. The magnitude of each of the concentrated forces is q^* and the point of application of each moves along the faces with the same speed v as the crack edge. The problem possesses the symmetries of a mode II deformation field. The problem is depicted in Figure 6.2 and a steady state solution is sought.

In terms of stress components, the boundary conditions are

$$\sigma_{xy}(\xi, \pm 0) = -q^* \delta(\xi + l), \qquad \sigma_{yy}(\xi, \pm 0) = 0 \qquad (6.2.22)$$

for $-\infty < \xi < 0$. Stress vanishes at points far from the crack tip compared to l. A general solution exists in the form of two analytic functions of a complex variable through (6.2.4). These two functions are again denoted by F and G. The conditions on these functions imposed by the boundary conditions (6.2.22) are

$$(1 + \alpha_s^2) \left[F_+''(\xi) - F_-''(\xi) \right] + 2\alpha_s \left[G_+''(\xi) - G_-''(\xi) \right] = 0,$$

$$2\alpha_d \left[F_+''(\xi) + F_-''(\xi) \right] + (1 + \alpha_s^2) \left[G_+''(\xi) + G_-''(\xi) \right] = 2iq^* \delta(\xi + l)/\mu,$$

$$(6.2.23)$$

where the symmetry properties (6.2.5) for mode II deformation have been incorporated.

The first of the two equations (6.2.23) implies that

$$G''(\zeta) = -\frac{1+\alpha_s^2}{2\alpha_s} F''(\zeta) \qquad (6.2.24)$$

throughout the region of the body. The arguments leading to this conclusion are similar to those presented in the preceding subsection. The function G can be eliminated from $(6.2.23)_2$ by means of (6.2.24), with the result that

$$F_+''(\xi) + F_-''(\xi) = \frac{4i\alpha_s}{D} \frac{q^*}{\mu} \delta(\xi + l) \qquad (6.2.25)$$

on $-\infty < \xi < 0$. A solution of this equation that is consistent with the constraints of the physical problem is

$$F''(\zeta) = -\frac{2i\alpha_s\sqrt{l}}{\pi D \sqrt{\zeta}(\zeta + l)} \frac{q^*}{\mu}. \qquad (6.2.26)$$

With $F''(\zeta)$ determined throughout the region of the body, and consequently $G''(\zeta)$ determined, any feature of the solution can be examined. For example, the shear stress distribution $\sigma_{xy}(\xi, 0)$ on the crack line ahead of the tip $\xi > 0$ is found to be

$$\sigma_{xy}(\xi, 0) = \frac{q^*\sqrt{l}}{\pi\sqrt{\xi}(\xi + l)} \qquad (6.2.27)$$

and the mode II stress intensity factor is

$$K_{II} = \lim_{\xi \to 0^+} \sqrt{2\pi\xi}\, \sigma_{xy}(\xi, 0) = q^*\sqrt{\frac{2}{\pi l}}. \qquad (6.2.28)$$

Again, the feature that both the stress distribution on the crack line and the stress intensity factor are independent of the crack speed v is observed.

6.2.4 Superposition and cohesive zone models

As noted in Section 4.3, the study of the advance of a sharp crack on the basis of an elastodynamic model leads naturally to the result

that the stress is unbounded at the crack edge. At a certain level
of observation, the result is accepted on the basis that the size of
the region over which the material behavior deviates from linear elasticity is small enough to admit the use of the stress intensity factor
concept. The cohesive zone idea provides a simple yet useful device
for examining crack tip phenomena at a level of observation for which
deviations from linearity in material response are included. On the
basis of energy concepts, it was determined in Section 5.3.2 that stress
intensity factor in a mode I deformation field can be related to the
constitutive characteristics of the cohesive zone through (5.3.18), for
example. However, it was not possible to determine the size of the
cohesive zone in terms of the remote loading and the constitutive
properties of the zone by means of the global energy arguments.
Instead, the determination of the size Λ must follow from a study
of the stress distribution.

The main ideas of cohesive zone models are simpler to convey
for mode I deformation than for mode II deformation. Although the
concentrated force problem analyzed in the preceding subsection was
a case of mode II, the mode I results corresponding to (6.2.27) and
(6.2.28) are obvious. The first is

$$\sigma_{yy}(\xi,0) = \frac{p^*\sqrt{l}}{\pi\sqrt{\xi}(\xi+l)} \tag{6.2.29}$$

for normal stress on the crack line $\xi > 0$, where p^* is the magnitude
of each concentrated force acting at $\xi = -l$ to open the crack. The
corresponding stress intensity factor is

$$K_I = p^*\sqrt{\frac{2}{\pi l}} . \tag{6.2.30}$$

The essential idea of the cohesive zone model for plane strain
mode I crack growth under the simplest circumstances is outlined
in Section 5.3.2, and the notation of that section is followed here.
The stress intensity factor due to the cohesive stress is determined by
superposition. The concentrated force p^* in (6.2.29) is first replaced
by a traction $-\sigma(l)$ distributed over the infinitesimal interval between
$\xi = -l$ and $\xi = -(l+dl)$, and the result is then integrated over the
range $0 < l < \Lambda$ to yield

$$K_{Icoh} = -\sqrt{\frac{2}{\pi}} \int_0^\Lambda \frac{\sigma(l)}{\sqrt{l}}\, dl . \tag{6.2.31}$$

The total stress is nonsingular or $K_I = 0$ if Λ is chosen to satisfy

$$K_{Iappl} = \sqrt{\frac{2}{\pi}} \int_0^\Lambda \frac{\sigma(l)}{\sqrt{l}} \, dl \qquad (6.2.32)$$

for given $\sigma(l)$. The weighted integral of the cohesive stress on the right side of (6.2.32) is sometimes called the *modulus of cohesion*, a terminology apparently due to Barenblatt (1959a, 1959b).

To make further progress with the cohesive zone idea, it is necessary to be more specific about the nature of the cohesive stress $\sigma(l)$. The simplest case of a uniform cohesive stress was introduced by Dugdale (1960) to represent a plastic zone near a crack tip in a thin sheet of an elastic-plastic nonhardening material under tension. He assumed the cohesive stress to be the tensile flow stress σ_o of an ideally plastic material. Then, from (6.2.32), the traction on the plane $y = 0$ will be less than or equal to σ_o everywhere provided that Λ is

$$\Lambda = \frac{\pi}{8} \left(\frac{K_{Iappl}}{\sigma_o} \right)^2 . \qquad (6.2.33)$$

The size of the crack tip inelastic zone scales with the length parameter $(K_I/\sigma_o)^2$ which could have been anticipated on the basis of dimensional considerations. With Λ determined, the crack line loading is completely specified and any feature of the solution can be extracted by following the steps outlined in the foregoing subsections. It should be noted that the physical crack tip is at $\xi = -\Lambda$ and not at $\xi = 0$. The fact that the crack tip is initially taken to be at $\xi = 0$ is simply a mathematical artifice of the approach.

Although the result (6.2.33) is based on an assumption that the crack growth process is steady, it is reasonable to expect that it will be accurate for nonsteady growth provided that cohesive zone length is small compared to overall crack size and that the crack speed does not change appreciably over growth distances equal to several times the zone length. As an application for nonsteady crack growth, consider the result (6.2.32) for the case when the cohesive stress σ depends on position l only through a dependence on the local crack opening displacement $\delta(l)$. Another relationship between the applied dynamic stress intensity factor and the cohesive stress is given in (5.3.17) on the basis of energy considerations. If K_{Iappl} is eliminated from the two equations, it follows that

$$\int_0^{\delta_t} \sigma(\delta)\,d\delta = \frac{2(1-\nu^2)}{\pi E} A_I(v) \left[\int_0^\Lambda \frac{\sigma[\delta(l)]}{\sqrt{l}}\,dl \right]^2 , \qquad (6.2.34)$$

where $A_I(v)$ is given in (5.3.11). If the crack is assumed to grow with a fixed value of crack opening displacement δ_t then the left side of (6.2.34) is essentially a material constant. Suppose now that the crack grows with an ever increasing remote driving force so that v continues to increase. The function $A_I(v)$ increases monotonically with speed v, becoming unbounded as $v \to c_R$, so that the integral in square brackets in (6.2.34) must decrease as the crack speed increases. This suggests that Λ must decrease as v increases such that $\Lambda \to 0$ as $v \to c_R$. Because the magnitude of particle displacement must increase from zero to δ_t over a distance Λ, large displacement gradients and thus potentially enormous stress components must be generated at points near the cohesive zone. This observation was developed by Rice (1980) within the context of steady crack growth analysis applied to earthquake rupture mechanics.

To pursue the matter a bit further, suppose that the interaction stress has the uniform value σ_0 everywhere within the cohesive zone (or slip zone for the case of shear modes). The relation (6.2.34) can be rewritten as

$$\Lambda/\Lambda_0 = A(v)^{-1} , \qquad (6.2.35)$$

where Λ_0 is the length of the cohesive zone or slip zone when the crack growth condition of fixed crack tip opening displacement is satisfied under equilibrium conditions. The function $A(v)$ in (6.2.35) is written without a subscript because the relationship is valid for each of the three modes of crack tip opening. The form of $A(v)$ for each mode is given in (5.3.11). For the present case, it is evident that Λ does indeed become smaller as v increases. For mode II deformation and the coordinate system established in Figure 6.2, Rice (1980) noted that the average value of stress σ_{xx}, say $\langle\sigma_{xx}\rangle$, along the stretched side of the slip zone is

$$\langle\sigma_{xx}\rangle = \frac{2\mu}{1-\nu}\langle\epsilon_{xx}\rangle = \frac{2\mu}{1-\nu}\frac{\delta_t}{2\Lambda} = \frac{4}{\pi}A_{II}(v)\tau_0 , \qquad (6.2.36)$$

where τ_0 is the uniform shear traction in the zone, $\Lambda = \pi K_{Iappl}^2/8\tau_0^2$, and $(1-\nu^2)A_{II}(v)K_{Iappl}^2 = E\tau_0\delta_t$ for mode II. If $\langle\sigma_{xx}\rangle$ becomes large enough, local tensile fracturing off the main crack plane may occur. This effect can contribute to making shear fracture of the kind thought to model seismic faulting unstable at high crack speed. Though stresses at points off the main crack plane may become large during rapid crack growth for a number of reasons, the influence of off-plane stresses has received only limited consideration in dynamic fracture studies.

Finally, it is observed that some specific results obtained are identical to the corresponding equilibrium results for the same configurations, but that they are nonetheless exact for steady dynamic crack growth. Once again, it is noted that the cohesive zone idea was introduced above with the feature that this zone was completely embedded within a stress intensity factor field. This is the small-scale yielding situation introduced in Section 1.1.2. It is not essential that this be the case for the idea to be useful, and an example for which the situation is relaxed is given in Section 7.6. For a crack of finite length with cohesive zones in a body subjected to uniform remote tension of magnitude σ_∞ in a direction transverse to the crack plane, Rice (1968) showed that the cohesive zones may be viewed as being embedded within a stress intensity factor field provided that the ratio of applied stress to uniform cohesive zone stress σ_∞/σ_0 is less than about 0.2. Because the stress distribution on the crack plane is independent of crack speed for steady elastodynamic crack growth, presumably the same restriction applies for this case as well.

6.2.5 Approach to the steady state

A steady state situation for a physical system should always be viewed as a state that is approached in some limiting sense. In considering steady state crack propagation solutions, it is important to have some understanding of the rate of approach of a transient field to its steady state limit, assuming that such a limit does indeed exist. It is a difficult issue to resolve unambiguously. If transient solutions could be found in cases of interest to permit examination of their long-time behaviors, then there would be little motivation for considering the corresponding steady state problems. Fortunately, a limited number of transient solutions that have steady state limits can be found, thus providing some guidance in the matter.

It appears that configurations that do not have a characteristic length will not approach a long time limit of interest. This is suggested by the dimensional argument leading up to (2.4.1). In the absence of a characteristic length, the limit as $t \to \infty$ is indistinguishable from the limit as the observation point moves toward the crack tip. Thus, as time becomes very large, a feature of the solution is that stress and particle velocity increase without bound everywhere in the finite part of the plane. For this reason, attention in this section is limited to configurations with a characteristic length in considering the question of approach to the steady state.

For a configuration with a single characteristic length, say l_o, and no characteristic time, an important time period is the time for a wave to travel the distance l_o at speed c, where $c = c_d$, c_s, or c_R. Suppose that a transient process having these features begins at time $t = 0$. A rule of thumb in wave propagation analysis is that the solution is approximated by its long-time limit for time t greater than two or three times such a wave transit time, that is, for $t > nl_o/c$, where $n = 2$ or 3. For example, suppose that the Yoffe crack in Section 6.2 is initially at rest in an unbounded body subjected to uniform remote tension. The characteristic length in this case is the length of the crack $2a$. At time $t = 0$, the crack tips both begin to move at speed $v < c_R$ in the x-direction. On the basis of the qualitative argument above, it is expected that the steady state stress intensity factor (6.2.20) is a good approximation to the actual transient stress intensity factor for time $t > nc_R/2a$ for $n = 2$ or 3. The speed c_R is chosen for this case because the main interaction between the crack edges occurs through Rayleigh waves propagating on the crack faces, as suggested by the analysis of Section 3.3.

A simple but instructive mode III crack propagation problem which sheds further light on the matter of approach to the steady state and which can be solved completely is the following. A half plane crack exists in a body that is initially at rest and stress free. In terms of rectangular coordinates, the plane of the crack is $y = 0$, and the crack occupies $x < 0$. At time $t = 0$ the crack begins to move at speed $v < c = c_s$. Simultaneously, a traction distribution given by

$$\tau_y(x, 0^\pm, t) = \begin{cases} -\tau^* & \text{if } -l_o < x - vt < 0, \\ 0 & \text{if } x < l_o \end{cases} \tag{6.2.37}$$

begins to act on the crack faces. The traction on the crack faces is uniform over an interval of length l_o behind the moving crack tip, and it is zero elsewhere. The steady state stress intensity factor for this configuration is the equilibrium stress intensity factor $K_{III}(\infty) = \tau^*\sqrt{2l_o/\pi}$.

This transient crack propagation problem can be solved exactly by means of Green's method introduced in Chapter 2. The complete traction distribution on the plane $y = 0$ between the leading wavefront at $x = ct$ and the crack tip at $x = vt$ is found to be

$$\tau_y(x, 0, t) = \frac{\tau^*}{\pi} \left\{ -\frac{\pi}{2} + \sin^{-1} \left[\frac{1 - q(x, t)}{1 + q(x, t)} \right] + \sqrt{q(x, t)} \right\}, \quad (6.2.38)$$

where

$$q(x, t) = \begin{cases} \dfrac{ct - x}{x - vt} & \text{if } ct - l_o < x < ct, \\[2ex] \dfrac{l_o}{x - vt} & \text{if } vt < x < ct - l_o. \end{cases} \quad (6.2.39)$$

The corresponding dynamic stress intensity factor is

$$K_{III}(t) = \tau^* \sqrt{\frac{2}{\pi}} \begin{cases} \sqrt{ct(1 - v/c)} & \text{if } 0 < t < l_o/(c - v), \\[1ex] \sqrt{l_o} & \text{if } l_o/(c - v) < t < \infty. \end{cases} \quad (6.2.40)$$

In this case, the transient stress intensity factor assumes its long-time, steady state limit at $t = l_o/(c - v)$, and it maintains this value thereafter. The time $t = l_o/(c - v)$ is the time at which the wavefront emitted at $x = -l_o$ and $t = 0$ overtakes the moving crack tip. This is a case where the rule of thumb argument put forward above actually under estimates the time required for the transient stress intensity factor to approach its steady state limit.

Experience with various two-dimensional transient elastodynamic crack solutions leads to the following conjecture for this class of problems. Consider the wavefronts generated through sudden application of boundary loading or through scattering of waves at crack edges and other geometrical features. For a particular situation, if all wavefronts are generated within a finite time interval, then the transient stress intensity factor assumes its long-time limit immediately upon generation of the last wavefront. The mode III crack propagation problem discussed in this section provides an illustration of this behavior. On the other hand, if new wavefronts continue to appear as time goes on, then the transient stress intensity factor gradually approaches its long time limit in a oscillatory manner. The case of transient loading of a crack of finite length exhibits this kind of behavior. This conjecture is based only on the few solutions that have been found, and no general mathematical argument is available to support it.

6.3 Self-similar dynamic crack growth

Once again, consider a planar crack growing dynamically in an elastic solid under two-dimensional plane strain conditions. To introduce the notion of self-similarity in concrete terms, attention is focused on the symmetric expansion of a crack at constant rate from zero initial length. The crack grows in an otherwise unbounded body which is subjected to uniform tensile stress at remote points. The magnitude of the remote stress is σ_∞ and the tensile direction is normal to the crack plane. Due to the linearity of the problem, it can be assumed that the crack grows in a body that is stress free and at rest prior to the onset of crack growth, and to grow under the action of compressive normal traction of magnitude σ_∞ on the crack faces.

A rectangular coordinate system is introduced so that the plane of deformation is the x, y-plane, and the crack begins to grow at time $t = 0$. The system of differential equations, boundary conditions, and initial conditions is a linear system. Consequently, any stress component has the representation

$$\sigma_{ij}(x,y,t) = \sigma_\infty f_{ij}(x,y,t) , \qquad (6.3.1)$$

where f_{ij} is a *dimensionless* function of its arguments.

Next, it is observed that the physical system involves neither a characteristic length nor a characteristic time. The crack length scales with time t, so it is not a characteristic length. It follows that if f_{ij} is dimensionless and if there is no characteristic length or time against which the independent variables can be scaled then f can only depend on dimensionless combinations of x, y, and t. For example, it can be assumed that

$$f_{ij}(x,y,t) = f_{ij}\left(ct/x, ct/y\right) , \qquad (6.3.2)$$

where c is a wave speed. Thus, the function f_{ij} is a homogeneous function of its argument of degree zero. Solutions of boundary value problems having this property are commonly called *homogeneous solutions* and the fields that they describe are called *self-similar fields*. The latter term derives from the fact that when such fields are compared at two different times they are found to be identical except for their spatial scale. Evidently, the property of self-similarity reduces the number of independent variables from three to two in this case and the mathematical problem becomes more manageable. Some early examples of the use of homogeneous solutions of the scalar wave equation are reviewed by Bateman (1955). Applications to elastodynamic problems involving stationary cracks were presented by Miles (1960) and Papadopoulos (1963). A thorough discussion of the analysis of self-similar mixed boundary value problems in elastodynamics was presented by Willis (1973). He considered problems in both two and three space dimensions, and his general results embody virtually all known self-similar elastodynamic solutions. The approach exploited here has evolved from the study of external flows in steady supersonic aerodynamics, and a general discussion from this point of view is presented by Ward (1955) under the heading of Busemann's method of conical fields in supersonic flow. In that case, the axial coordinate of the conical region is a spatial coordinate, whereas in the present approach the axial coordinate is time.

The idea of self-similar fields was motivated by the introduction of a particular crack growth problem that will be solved subsequently. It is an idea of broad applicability that can be used to great advantage. As a second elementary illustration within plane strain elastodynamics, suppose that two parallel edge dislocations that are initially coincident begin to move away from each other at a uniform rate. If the magnitude of the Burgers vector of each dislocation is Δ, then each component of displacement has the form

$$u_i(x, y, t) = \Delta\, g_i\left(ct/x, ct/y\right), \qquad (6.3.3)$$

where g_i is a homogeneous function of x, y, and t of degree zero. Furthermore, the derivative with respect to any of the original independent variables of a function that is homogeneous of degree m is itself homogeneous of degree $m-1$. Consequently, if the displacement field is homogeneous of degree zero, then the velocity field and the stress field are homogeneous of degree -1, the acceleration field and the stress rate field are homogeneous of degree -2, the displacement potential fields are homogeneous of degree $+1$, and so on.

6.3.1 General solution procedure

Suppose that $f(x, y, t)$ is a homogeneous function of its arguments of degree zero of the type introduced above. Further, suppose that f is a wave function with characteristic wave speed c, that is,

$$\frac{\partial^2 f}{\partial x^2} + \frac{\partial^2 f}{\partial y^2} - \frac{1}{c^2}\frac{\partial^2 f}{\partial t^2} = 0. \qquad (6.3.4)$$

As already noted, the property of homogeneity permits reduction of the number of independent variables from three to two. The choice of the two reduced independent variables is not unique, and for present purposes the particular choice is

$$f(x, y, t) = f(\xi, \theta),$$ (6.3.5)

where

$$\xi = ct/r, \qquad \theta = \tan^{-1}(y/x), \qquad r = \sqrt{x^2 + y^2}.$$ (6.3.6)

Again, the functions on opposite sides of the equal sign in (6.3.5) are different but they are represented by the same variable name, with little likelihood of confusion.

If the derivatives appearing in (6.3.4) are formed by differentiating both sides of (6.3.5), taking into account the implicit dependence of ξ and θ on x, y, and t, then the partial differential equation takes the form

$$(\xi^2 - 1)\frac{\partial^2 f}{\partial \xi^2} + \xi \frac{\partial f}{\partial \xi} + \frac{\partial^2 f}{\partial \theta^2} = 0.$$ (6.3.7)

This second-order equation in two independent variables is elliptic if $\xi > 1$ and hyperbolic if $0 < \xi < 1$. Note that the discriminating condition $\xi = 1$ is the equation of a cone in x, y, t-space. An additional transformation of independent variables is sought that will transform the differential equation to Laplace's equation in the former case and to the elementary wave equation in the latter case. To this end, introduce a new variable s that is related to ξ through a function $s = \omega(\xi)$ that is to be determined. Then, $f(\xi, \theta) = f(\omega(\xi), \theta) = f(s, \theta)$ and the partial differential equation takes the form

$$(\xi^2 - 1)\omega'(\xi)^2 \frac{\partial^2 f}{\partial s^2} + \left[(\xi^2 - 1)\omega''(\xi) + \xi\omega'(\xi) \right] \frac{\partial f}{\partial s} + \frac{\partial^2 f}{\partial \theta^2} = 0.$$ (6.3.8)

If (6.3.8) is to reduce to Laplace's equation for $\xi > 1$ then the coefficient of $\partial^2 f/\partial s^2$ must be unity, or

$$\omega'(\xi) = \pm(\xi^2 - 1)^{-1/2}.$$ (6.3.9)

Fortuitously, for this choice of $\omega'(\xi)$, the coefficient of $\partial f/\partial s$ in (6.3.8) vanishes identically and the differential equation is indeed reduced to

$$\frac{\partial^2 f}{\partial \theta^2} + \frac{\partial^2 f}{\partial s^2} = 0.$$ (6.3.10)

The function $\omega(\xi)$ itself is determined by integration of (6.3.9) with the result that

$$\omega(\xi) = \pm \cosh^{-1} \xi, \qquad \xi > 1, \tag{6.3.11}$$

where an inconsequential constant of integration has been taken as zero. The transformation (6.3.11) is commonly known as the Chaplygin transformation. Likewise, the differential equation is reduced to the wave equation

$$\frac{\partial^2 f}{\partial \theta^2} - \frac{\partial^2 f}{\partial s^2} = 0 \tag{6.3.12}$$

if $\omega(\xi) = \sin^{-1} \xi$ or $\cos^{-1} \xi$ for $0 < \xi < 1$.

A major advantage is gained by means of the transformation, in the sense that general solutions are known for both Laplace's equation and the wave equation in two independent variables. For example, a solution of (6.3.10) is

$$f = \operatorname{Re}\{F_1(\theta + is)\}, \tag{6.3.13}$$

where F_1 is an analytic function of its argument in a semi-infinite strip of the complex $\theta + is$-plane. If the plus sign is chosen in (6.3.11) then the strip extends over $0 < s < \infty$ and an appropriate range of θ. If the minus sign is chosen in (6.3.11) then the semi-infinite strip extends over $-\infty < s < 0$ and the same range of θ. The latter case is chosen here for a reason to be clear in the subsequent development. Furthermore, θ is assumed to be in the interval $-\pi < \theta < \pi$. Other cases, such as $0 < \theta < \pi$, can be handled with minor modifications. Next, it is observed that the strip $-\infty < s < 0$, $-\pi < \theta < \pi$ of the complex $\theta + is$-plane is simply a rectangle with one edge at infinity. Thus, a degenerate Schwarz-Christoffel transformation can be used to conformally map the part of the semi-infinite strip with $\theta < 0$ into the lower half of the complex ζ-plane and the part with $\theta > 0$ into the upper half of the ζ-plane. The transformation is

$$\zeta = \frac{1}{c} \cos(\theta + is) = \frac{1}{c} \cos(\theta - i \cosh^{-1} \xi), \tag{6.3.14}$$

where the factor $1/c$ appears only for convenience. If the cosine expression is expanded, and θ and ξ are written in terms of physical coordinates, then

$$\zeta = \frac{x}{r^2} t + i \frac{y}{r^2} \sqrt{t^2 - r^2/c^2}, \qquad \frac{ct}{r} > 1. \tag{6.3.15}$$

Thus, the portion of the physical plane $y > 0$ ($y < 0$) is mapped into the upper (lower) half of the complex ζ-plane, a correspondence assured by the choice of the negative sign in (6.3.11). It is noted that the wavefront $r = ct - 0$ for $y > 0$ ($y < 0$) maps into the portion of the real axis with $\zeta = x/cr + i0$ ($\zeta = x/cr - i0$), where $-1 < x/r < 1$. Also, the portion of the plane $y = 0^{\pm}$ bounded by the wavefront maps into the portion of the real axis with $\zeta = t/x \pm i0$, where $-1/c < t/x < 1/c$. The corners of the rectangle in the $\theta + is$-plane are singular points of the Schwarz-Christoffel transformation, and these points map into $\mathrm{Re}(\zeta) = \pm c^{-1}$, $\mathrm{Im}(\zeta) = 0$, which are branch points in the complex ζ-plane. A branch cut between these two points is introduced, a step that could have been anticipated when it was noted that two points on the wavefront in the physical plane map into the same point on the real axis of the ζ-plane. The points on the wavefront in the upper (lower) half of the physical plane are mapped into the points on the upper (lower) side of this branch cut. For future reference, it is noted that the expression (6.3.15) for ζ is a root of the equation

$$x\zeta + y\sqrt{c^{-2} - \zeta^2} = t, \qquad (6.3.16)$$

where the branch of $\sqrt{c^{-2} - \zeta^2}$ with positive (negative) real values on the upper (lower) side of the cut is assumed.

Because the mapping (6.3.14) is conformal, the wave function f has the representation

$$f = \mathrm{Re}\{F_2(\zeta)\}, \qquad (6.3.17)$$

where F_2 is an analytic function of ζ for $y > 0$ and $y < 0$. The function F_2 may have singularities on the real axis other than the branch points at $\zeta = \pm c^{-1}$. Thus, the determination of the wave function satisfying (6.3.4) has been reduced by the property of self-similarity to determination of an analytic function of a complex variable in either a half plane or a whole plane. The analytic function must be determined from its boundary values, a process greatly facilitated by the fact that the complex variable is related to the physical coordinates in the simple way shown in (6.3.15). Next, the solution for a particular crack growth problem is constructed by taking advantage of self-similarity. The general approach is as follows. First, the field quantities that are homogeneous of degree zero are identified by anticipating certain qualitative features of the solution. Then these fields are represented as sums of wave functions, each of which can be expressed in terms of an analytic function. The geometrical symmetries, boundary conditions, and other constraints on the solution are then invoked to determine the analytic functions, and thereby the complete solution.

6.3.2 The Broberg problem

The scope of mathematical solutions available for interpretation of dynamic fracture phenomena was extended significantly with the solution of the problem of the self-similar expansion of a crack from zero initial length in a uniform tension field by Broberg (1960). This is the problem described in the opening remarks of this section which provided the motivation for pursuing the implications of self-similarity in analysis of certain boundary value problems. Broberg used solutions of a sequence of particular problems to produce a complete solution to the crack expansion problem of interest. Though the property of self-similarity played a role in his procedure, the approach to be followed here is more direct and, in the course of describing it, a general approach for solving a wide range of problems having the property of self-similarity emerges. A number of illustrations are presented by Cherepanov and Afanasev (1973), who developed a general framework for analysis of self-similar fields.

A rectangular x, y-coordinate system is introduced in the plane of deformation, oriented so that crack growth occurs in the plane $y = 0$. The crack begins to expand symmetrically from zero initial length at time $t = 0$ with each tip moving at a constant speed v that is less than the Rayleigh wave speed of the material. At any later time, the crack occupies the interval $-vt < x < vt$. Both crack faces are subjected to compressive normal traction of magnitude σ_∞. The shear traction is zero on the crack faces, and no other loads act on the body. The material is stress free and at rest for $t \le 0$. Thus, displacement potentials ϕ and ψ that satisfy the wave equations

$$\frac{\partial^2 \phi}{\partial x^2} + \frac{\partial^2 \phi}{\partial y^2} - \frac{1}{c_d^2}\frac{\partial^2 \phi}{\partial t^2} = 0, \quad \frac{\partial^2 \psi}{\partial x^2} + \frac{\partial^2 \psi}{\partial y^2} - \frac{1}{c_s^2}\frac{\partial^2 \psi}{\partial t^2} = 0 \quad (6.3.18)$$

are sought subject to the boundary conditions

$$\sigma_{yy}(x, 0^\pm, t) = -\sigma_\infty, \quad \sigma_{xy}(x, 0^\pm, t) = 0 \quad \text{for} \quad |x| < vt. \quad (6.3.19)$$

The solution is expected to have a symmetry with respect to reflection in the x-axis typical of mode I fields, namely,

$$u_x(x, -y, t) = u_x(x, y, t), \quad u_y(x, -y, t) = -u_y(x, y, t). \quad (6.3.20)$$

As anticipated in (6.3.2), the stress and particle velocity components are homogeneous functions of x, y, and t of degree zero, and a solution is sought on this basis.

The in-plane stress components are given in terms of the displacement potentials by

$$\frac{\sigma_{xx}}{\mu} = \frac{c_d^2}{c_s^2}\frac{\partial^2 \phi}{\partial x^2} + \left(\frac{c_d^2}{c_s^2} - 2\right)\frac{\partial^2 \phi}{\partial y^2} + 2\frac{\partial^2 \psi}{\partial x \partial y},$$

$$\frac{\sigma_{yy}}{\mu} = \left(\frac{c_d^2}{c_s^2} - 2\right)\frac{\partial^2 \phi}{\partial x^2} + \frac{c_d^2}{c_s^2}\frac{\partial^2 \phi}{\partial y^2} - 2\frac{\partial^2 \psi}{\partial x \partial y}, \qquad (6.3.21)$$

$$\frac{\sigma_{xy}}{\mu} = 2\frac{\partial^2 \phi}{\partial x \partial y} + \frac{\partial^2 \psi}{\partial y^2} - \frac{\partial^2 \psi}{\partial x^2},$$

where μ is the elastic shear modulus. Likewise, the particle velocity components are

$$\dot{u}_x = \frac{\partial^2 \phi}{\partial x \partial t} + \frac{\partial^2 \psi}{\partial y \partial t}, \qquad \dot{u}_y = \frac{\partial^2 \phi}{\partial y \partial t} - \frac{\partial^2 \psi}{\partial x \partial t}. \qquad (6.3.22)$$

The left side of each equation in (6.3.21) is a homogeneous function of degree zero which has certain symmetry properties with respect to reflection in the plane $y = 0$. These properties will be assured if solutions of the wave equations (6.3.18) are sought such that each term on the right side of each equation in (6.3.21) has the same properties as the function on the left side. Thus, in light of the discussion in Section 6.3.1, the function $\partial^2 \phi/\partial x^2$ can be represented as the real part of a function, say F_{xx}, of the complex variable

$$\zeta_d = \frac{x}{r^2}t + i\frac{y}{r^2}\sqrt{t^2 - r^2/c_d^2}, \quad \frac{c_d t}{r} > 1, \qquad (6.3.23)$$

that is analytic in the complex ζ_d-plane except at certain singular points on the real axis. In a similar way,

$$\frac{\partial^2\phi}{\partial x^2} = \text{Re}\{F_{xx}(\zeta_d)\}, \quad \frac{\partial^2\phi}{\partial x\partial y} = \text{Im}\{F_{xy}(\zeta_d)\},$$

$$\frac{\partial^2\phi}{\partial y^2} = \text{Re}\{F_{yy}(\zeta_d)\}, \qquad\qquad (6.3.24)$$

$$\frac{\partial^2\phi}{\partial x\partial t} = \text{Re}\{F_{xt}(\zeta_d)\}, \quad \frac{\partial^2\phi}{\partial y\partial t} = \text{Im}\{F_{yt}(\zeta_d)\}$$

and, likewise,

$$\frac{\partial^2\psi}{\partial x^2} = \text{Im}\{G_{xx}(\zeta_s)\}, \quad \frac{\partial^2\psi}{\partial x\partial y} = \text{Re}\{G_{xy}(\zeta_s)\},$$

$$\frac{\partial^2\psi}{\partial y^2} = \text{Im}\{G_{yy}(\zeta_s)\}, \qquad\qquad (6.3.25)$$

$$\frac{\partial^2\psi}{\partial x\partial t} = \text{Im}\{G_{xt}(\zeta_s)\}, \quad \frac{\partial^2\psi}{\partial y\partial t} = \text{Re}\{G_{yt}(\zeta_s)\},$$

where

$$\zeta_s = \frac{x}{r^2}t + i\frac{y}{r^2}\sqrt{t^2 - r^2/c_s^2}, \quad \frac{c_s t}{r} > 1. \qquad (6.3.26)$$

The required symmetries of the stress field with respect to the crack plane are assured if

$$F_{\alpha\beta}(\overline{\zeta_d}) = \overline{F_{\alpha\beta}(\zeta_d)}, \quad G_{\alpha\beta}(\overline{\zeta_s}) = \overline{G_{\alpha\beta}(\zeta_s)} \qquad (6.3.27)$$

for α, β as any two of x, y, or t, where the overbar denotes complex conjugate.

The functions F_{xx}, F_{xy}, F_{yy}, and F_{yt} are all derivable from the *same* function ϕ according to (6.3.24), so they are not independent. Indeed, it is evident from (6.3.24) that the functions F_{xx} and F_{xy} are compatible only if

$$\frac{\partial}{\partial y}\text{Re}\{F_{xx}(\zeta_d)\} = \frac{\partial}{\partial x}\text{Im}\{F_{xy}(\zeta_d)\} \qquad (6.3.28)$$

which is assured if

$$\sqrt{a^2 - \zeta^2}\, F'_{xx}(\zeta) = -i\zeta F'_{xy}(\zeta) \qquad (6.3.29)$$

everywhere in the complex plane, where $a = c_d^{-1}$ and the prime denotes differentiation with respect to the argument. In establishing (6.3.29), it is necessary to evaluate the partial derivatives of $\zeta_d(x, y, t)$ with respect to x and y. This is done most conveniently by differentiating the relevant special case of (6.3.16) with respect to x and y to find that

$$\frac{\partial \zeta_d}{\partial x} = -\frac{\zeta_d \sqrt{a^2 - \zeta_d^2}}{x\sqrt{a^2 - \zeta_d^2} - y\zeta_d}, \qquad \frac{\partial \zeta_d}{\partial y} = -\frac{a^2 - \zeta_d^2}{x\sqrt{a^2 - \zeta_d^2} - y\zeta_d}. \qquad (6.3.30)$$

Likewise, it is found that

$$-i\sqrt{a^2 - \zeta^2}\, F'_{xy}(\zeta) = \zeta F'_{yy}(\zeta) = i\zeta \sqrt{a^2 - \zeta^2}\, F'_{yt}(\zeta) \qquad (6.3.31)$$

identically in ζ, and

$$(b - \zeta^2) G'_{xx}(\zeta) = i\zeta \sqrt{b^2 - \zeta^2}\, G'_{xy}(\zeta)$$

$$= \zeta^2 G'_{yy}(\zeta) = -\zeta(b^2 - \zeta^2) G'_{xt}(\zeta). \qquad (6.3.32)$$

As a first step in the determination of the various analytic functions in terms of which the physical fields are represented, consider the dilatational contribution to $\dot{u}_y$, namely, $\partial^2 \phi / \partial y \partial t$ in (6.3.22). Due to symmetry of the fields, this function vanishes on $y = 0$ for $vt < |x| < c_d t$. This interval maps into the portion of the real axis $a < |\mathrm{Re}(\zeta_d)| < h$ in the complex ζ_d-plane, where $h = v^{-1}$.

Consider next the behavior of $\partial^2 \phi / \partial y \partial t$ at the wavefront $r = c_d t$. It is evident that this quantity is zero everywhere ahead of the wavefront, that is, for $r > c_d t$. It is well known from the theory of linear plane strain elastodynamics that sudden application of a force of constant magnitude in the elastic plane produces a dilatational wavefront that carries a square root singularity in particle velocity and stress. Sudden application of a force with a magnitude that increases linearly in time from zero initial magnitude produces a dilatational wavefront across which the particle velocity and stress are continuous. The applied loading in the present problem is even less singular than the second of these two cases because it is essentially a self-equilibrating pair of linearly increasing forces. Consequently, it is anticipated that stress and particle velocity will be continuous across the wavefront $r = c_d t$ in the present problem. An immediate implication of the initial conditions is that $\partial^2 \phi / \partial y \partial t = 0$ on $r = c_d t - 0$ for $-\pi < \theta < \pi$. This wavefront is mapped by (6.3.23) into the upper and lower sides of the real axis in the ζ_d-plane in the interval $-a < \mathrm{Re}(\zeta_d) < a$.

The implication of the foregoing arguments based on symmetry and the initial conditions is that

$$\frac{\partial^2 \phi}{\partial y \partial t} = \mathrm{Im}\{F_{yt}(\xi)\} = 0 \quad \text{for} \quad -h < \xi < h \qquad (6.3.33)$$

on the real axis of the complex ζ_d-plane. An analytic function $F_{yt}(\zeta_d)$ that satisfies this condition and that results in a particle velocity that is square root singular at the crack tips is

$$F_{yt}(\zeta) = \frac{P_F(\zeta)}{Q_F(\zeta)\sqrt{h^2 - \zeta^2}}, \qquad (6.3.34)$$

where $P_F(\zeta)$ and $Q_F(\zeta)$ are polynomial functions with real coefficients. The square root function in the denominator is single valued in the plane cut along $h < |\mathrm{Re}(\zeta)| < \infty$, $\mathrm{Im}(\zeta) = 0$ and the branch with positive real value at the origin is selected. Similar arguments applied to $\partial^2 \psi / \partial x \partial t$ lead to a representation for $G_{xt}(\zeta)$ in the complex ζ-plane with the same general features,

$$G_{xt}(\zeta) = \frac{P_G(\zeta)}{Q_G(\zeta)\sqrt{h^2 - \zeta^2}}. \qquad (6.3.35)$$

Because neither of these functions represents a physical quantity by itself, anticipated features of the solution cannot be used to determine the nature of the polynomial functions. However, the *difference* of these functions represents the particle velocity component $\dot{u}_y$. The polynomial functions $Q_F(\zeta)$ and $Q_G(\zeta)$ can have zeros only on the real axis. But if such zeros exist, they can correspond only to concentrated forces or dislocations in the physical plane. Thus, the polynomials are at most constants and, without loss of generality, $Q_F(\zeta) = Q_G(\zeta) = 1$. The symmetry property that $\dot{u}_y(-x, -y, t) = -\dot{u}_y(x, y, t)$ implies that the polynomials $P_F(\zeta)$ and $P_G(\zeta)$ must be odd functions of ζ. Finally, if the particle velocity is to have a finite limiting value at any point in the physical plane as $t \to \infty$, then the polynomials can be at most linear in ζ. Consequently, the difference of the two analytic functions that comprise the velocity component $\dot{u}_y$ has the form

$$F_{yt}(\zeta) - G_{xt}(\zeta) = \frac{\alpha\zeta}{h\sqrt{h^2 - \zeta^2}}, \qquad (6.3.36)$$

where α is a real dimensionless constant to be determined from the boundary conditions. Although α is constant with respect to ζ, it may depend on crack speed or other system parameters. As $|\zeta| \to \infty$, the right side of (6.3.36) has the form of a constant plus terms that are $O(\zeta^{-2})$. The presence of terms that are $O(\zeta^{-1})$ would suggest a logarithmic singularity in the long time behavior of stress in the finite plane.

The condition that the shear stress vanishes on the plane $y = 0$ along with continuity of stress across the wavefronts implies that

$$\mathrm{Im}\,\{2F_{xy}(\xi) + G_{yy}(\xi) - G_{xx}(\xi)\} = 0 \quad \text{for} \quad -\infty < \xi < \infty \quad (6.3.37)$$

for $\mathrm{Im}(\zeta) \to 0^\pm$. Thus, in view of the property of analyticity of the functions involved, this condition is assured if

$$2F_{xy}(\zeta) + G_{yy}(\zeta) - G_{xx}(\zeta) = 0 \qquad (6.3.38)$$

identically in ζ. Actually, the right side of (6.3.38) could be some suitable rational function with poles on the real axis, but other arguments can be used to reduce it to the result shown in (6.3.38). If this equation is differentiated with respect to ζ, and the resulting derivatives are eliminated in favor of either $F'_{yt}(\zeta)$ or $G'_{xt}(\zeta)$ by means of (6.3.31) or (6.3.32), respectively, then

$$2\zeta^2 F'_{yt}(\zeta) + (b^2 - 2\zeta^2)G'_{xt}(\zeta) = 0. \qquad (6.3.39)$$

The two relations (6.3.39) and (6.3.36), following differentiation of the latter, provide two linear equations for $F'_{yt}(\zeta)$ and $G'_{xt}(\zeta)$. The solution is

$$F'_{yt}(\zeta) = \frac{\alpha(b^2 - 2\zeta^2)h}{b^2(h^2 - \zeta^2)^{3/2}}, \quad G'_{xt}(\zeta) = -\frac{2\alpha\zeta^2 h}{b^2(h^2 - \zeta^2)^{3/2}}. \qquad (6.3.40)$$

Except for evaluating the constant α, the solution of the boundary value problem is complete.

The normal stress $\sigma_{yy}(x, 0, t)$ on the plane $y = 0$ is given by

$$\sigma_{yy} = \text{Re} \left\{ \frac{b^2}{a^2} F_{yy}(\xi) + \left(\frac{b^2}{a^2} - 2 \right) F_{xx}(\xi) - 2G_{xy}(\xi) \right\}_{\xi = t/x} \quad (6.3.41)$$

for $a < |t/x| < \infty$. If this expression is differentiated with respect to ξ, each term can be expressed in terms of one or the other function in (6.3.40). The result is

$$\frac{\partial}{\partial \xi} \left(\frac{\sigma_{yy}}{\mu} \right) = -\alpha \frac{h}{b^2} \text{Im} \left\{ \frac{R(\xi)}{(h^2 - \xi^2)^{3/2} \sqrt{a^2 - \xi^2}} \right\}, \quad (6.3.42)$$

where $R(\xi) = (b^2 - 2\xi^2)^2 + 4\xi^2 \sqrt{a^2 - \xi^2} \sqrt{b^2 - \xi^2}$ is the Rayleigh wave function introduced in Section 2.5. Continuity of stress at the wave front $r = c_d t$ implies that $\sigma_{yy}(x, 0, t) = 0$ for $x = t/a = c_d t$. Consequently, the normal stress $\sigma_{yy}(x, 0, t)$ at any point in the interval $a < t/x < h$ is given by the definite integral

$$\sigma_{yy} = -\alpha \mu \frac{h}{b^2} \text{Im} \left\{ \int_a^{t/x} \frac{R(\xi)}{(h^2 - \xi^2)^{3/2} \sqrt{a^2 - \xi^2}} \, d\xi \right\}. \quad (6.3.43)$$

Finally, the constant α is determined by applying the boundary condition $(6.3.19)_1$ to this integral expression. Because of the term $(h^2 - \xi^2)^{3/2}$ in the denominator of the integrand, the integral cannot be evaluated as a line integral along the real axis of the ζ-plane for $t/x > h$. The path of integration is readily distorted into the complex plane by means of Cauchy's integral theorem in order to provide an alternate representation for $\sigma_{yy}(x, 0, t)$, avoiding the strong singularity at $\zeta = h$. Thus, the value of α can be determined from

$$\frac{\sigma_\infty}{\mu} = \alpha \frac{h}{b^2} \int_0^\infty \frac{R(i\eta)}{(h^2 + \eta^2)^{3/2} \sqrt{a^2 + \eta^2}} \, d\eta. \quad (6.3.44)$$

This expression is rewritten as

$$\alpha = \frac{\sigma_\infty}{\mu} I(b/h), \quad (6.3.45)$$

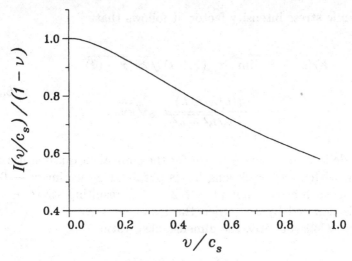

Figure 6.3. Graph of the function $I(v/c_s)$ introduced in (6.3.45) versus the normalized crack speed v/c_s for $\nu = 0.3$.

where the value of $I(b/h) = I(v/c_s)$ is determined by numerical evaluation of the integral in (6.3.44). The result is shown graphically in Figure 6.3. From the property that $R(i\eta) \to 2\eta^2(b^2-a^2)$ as $\eta \to \infty$, it can be shown that $I(b/h) \to b^2/2(b^2 - a^2) = (1 - \nu)$ as $h/b \to \infty$ or as $v/c_s \to 0$.

For points close to the crack tip at $x = vt$ or $\xi = h$, the factor $R(\xi)/(h+\xi)^{3/2}\sqrt{a^2 - \xi^2}$ in (6.3.42) is analytic and it can be approximated by the first term of its Taylor series expansion. Thus, for small but positive $x - vt$ or $h - \xi$,

$$\frac{\partial}{\partial \xi}\left(\frac{\sigma_{yy}}{\mu}\right) \sim \frac{\alpha R(h)}{2b^2(h - \xi)^{3/2}\sqrt{2h(h^2 - a^2)}}. \qquad (6.3.46)$$

For points close to the crack tip, the stress component σ_{yy} varies as

$$\frac{\sigma_{yy}}{\mu} \sim \frac{\alpha R(h)}{b^2\sqrt{2h(h^2 - a^2)(h - \xi)}} \qquad (6.3.47)$$

plus contributions that are bounded as $\xi \to h^-$. From the definition of dynamic stress intensity factor, it follows that

$$K_I(t, v) = \lim_{x \to vt^+} \sigma_{yy}(x, 0, t)\sqrt{2\pi(x - vt)}$$

$$= -\frac{I(b/h)R(h)}{b^2 h\sqrt{h^2 - a^2}}\sigma_\infty\sqrt{\pi vt}.$$

(6.3.48)

The mode I stress intensity factor for the equivalent quasi-static crack problem with total crack length l is obtained by letting $v \to 0$ and $t \to \infty$ in such a way that $vt \to l/2$. If the resulting stress intensity factor is denoted by K_{Io}, then the result that $K_{Io} = \sigma_\infty\sqrt{\pi l/2}$ is obtained. Consequently, the dimensionless ratio

$$\frac{K_I(t, v)}{K_{Io}} = -\frac{I(b/h)R(h)}{b^2 h\sqrt{h^2 - a^2}}$$

(6.3.49)

is of interest because it indicates the influence of inertial effects in the process. It is evident from the asymptotic behavior of α or I as $h/b \to \infty$ or $v/c_s \to 0$ that $K_I(t, v)/K_{Io} \to 1$ as $v/c_s \to 0$, which is consistent with the corresponding quasi-static result if vt is identified with the half crack length $l/2$. This ratio is plotted as a function of v/c_s in Figure 6.4.

The crack tip singular field is completely determined by (6.3.49) once the stress intensity factor is determined. The next most significant contributions to the crack tip stress field are those terms in the local expansion that are bounded and nonzero in the limit as the crack tip is approached. These terms are denoted by $\sigma_{ij}^{(1)}$ in the development in Section 4.3. For the Broberg problem, $\sigma_{yy}^{(1)} = -\sigma_\infty$ for a crack face pressure of magnitude σ_∞ or $\sigma_{yy}^{(1)} = 0$ for remote tension of magnitude σ_∞ with traction-free crack faces, and $\sigma_{xy}^{(1)} = 0$ due to symmetry of the fields. The remaining term of order unity $\sigma_{xx}^{(1)}$ must be worked out from the complete solution. The field $\sigma_{xx}(x, 0, t)$ can be determined from (6.3.40) in precisely the same way that the expression for σ_{yy} in (6.3.43) was obtained. The result is

$$\sigma_{xx} = -\alpha\mu\frac{h}{b^2}\mathrm{Im}\left\{\int_a^{t/x}\frac{2(b^2 - a^2)(b^2 - 2\xi^2) - R(\xi)}{(h^2 - \xi^2)^{3/2}\sqrt{a^2 - \xi^2}}\,d\xi\right\}.$$

(6.3.50)

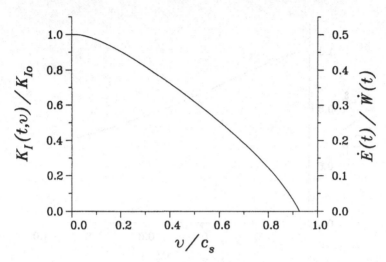

Figure 6.4. The stress intensity factor for the Broberg problem, normalized by the corresponding equilibrium stress intensity factor for a crack of length $2vt$, versus normalized crack speed v/c_s from (6.3.49). Also shown is the ratio of the rate of energy flux into the moving crack tips to the rate of work done on the body by the applied tractions versus v/c_s for $\nu = 0.3$.

From the observation that the integrand is zero for $t/x > h$, it is clear that σ_{xx} is a constant on the crack faces. Thus, this constant must be the value of $\sigma_{xx}^{(1)}$. By following the arguments leading to (6.3.44), it can be shown that the value of $\sigma_{xx}^{(1)}$ is given by

$$\frac{\sigma_{xx}^{(1)}}{\mu} = -\alpha \frac{h}{b^2} \int_0^\infty \frac{2(b^2 - a^2)(b^2 + 2\eta^2) - R(i\eta)}{(h^2 + \eta^2)^{3/2}\sqrt{a^2 + \eta^2}} \, d\eta . \qquad (6.3.51)$$

For the equilibrium limit of a finite length crack subjected to uniform crack face pressure of magnitude σ_∞, the stress component parallel to the crack line that is of order unity is $-\sigma_\infty$. A plot of the ratio $-\sigma_{xx}^{(1)}/\sigma_\infty$ versus crack speed is shown in Figure 6.5. It is emphasized that this result applies for the case of pressure loading on the crack faces.

In light of (6.3.22) and (6.3.40), the distribution of the particle velocity of points on the crack faces at any time t is

$$\dot{u}_y(x, 0^\pm, t) = \mathrm{Im}\left\{ \frac{\alpha\zeta}{h\sqrt{h^2 - \zeta^2}} \right\}_{\zeta = t/x \pm i0} = \frac{\pm \alpha v}{\sqrt{1 - (x/vt)^2}} \qquad (6.3.52)$$

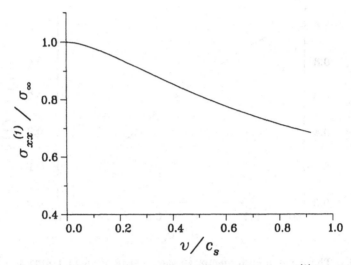

Figure 6.5. Graph of the nonsingular tensile stress component $\sigma_{xx}^{(1)}$ normalized by σ_∞ versus dimensionless crack speed v/c_s for $\nu = 0.3$.

for $x^2 < (vt)^2$. The normal displacement of the crack faces is obtained by integration with respect to time from zero initial displacement, with the result that

$$u_y(x, 0^\pm, t) = \pm \alpha v t \sqrt{1 - (x/vt)^2}, \quad x^2 < (vt)^2. \qquad (6.3.53)$$

Thus, the crack opening profile is elliptical with the axis in the plane $y = 0$ having length $2vt$ and the transverse axis having length $2\alpha vt$. For the corresponding quasi-static crack problem with $vt \to l/2$ that was discussed above, the total opening displacement at the midpoint of the crack is $u_{yo} = \sigma_\infty l(1-\nu)/2\mu$. The ratio of $u_y(0, 0^+, t) = \alpha vt$ to this corresponding quasi-static result is $I(b/h)/(1-\nu)$, which indicates the influence of inertia on the magnitude of crack opening at a given crack length. This ratio is the quantity plotted versus crack speed in Figure 6.3 for $\nu = 0.3$.

Finally, some global energy rates for this crack growth process are considered. Because the crack face traction is constant and the crack face velocity in the direction of the applied traction is known explicitly, the rate of work being done on the solid (per unit thickness in the z-direction) is

$$\dot{W}(t) = 2 \int_{-vt}^{vt} \sigma_\infty \dot{u}_y(x, 0^+, t)\, dx, \qquad (6.3.54)$$

where the fact that the rates of work on the upper and lower crack faces are identical has been incorporated. Substitution of (6.3.52) into (6.3.54) and evaluation of the integral yields

$$\dot{W}(t) = 2\pi\sigma_\infty^2 I(v/c_s)v^2 t/\mu, \tag{6.3.55}$$

where $I(v/c_s)$ is defined in (6.3.45).

The rate at which energy is absorbed into the moving crack edges (per unit thickness in the z-direction), each moving at speed v, was found by Broberg (1967). Here, the result is obtained by substituting the explicit expression for stress intensity factor (6.3.48) for this problem into the general energy rate expression (5.3.9). The result is

$$\dot{E}(t) = \frac{2\pi\sigma_\infty^2 I(v/c_s)v^2 t}{\mu} \left[\frac{I(v/c_s)D(v)c_s^2}{2v^2\sqrt{1-v^2/c_d^2}} \right], \tag{6.3.56}$$

where, as before, $D(v) = 4\sqrt{1-v^2/c_d^2}\sqrt{1-v^2/c_s^2} - (2-v^2/c_s^2)^2$. The expression has been written so that $\dot{W}(t)$ is an obvious factor, and the ratio of the energy flux rate into the crack edges to the rate of work on the body is

$$\frac{\dot{E}(t)}{\dot{W}(t)} = \frac{I(v/c_s)D(v)c_s^2}{2v^2\sqrt{1-v^2/c_d^2}} = -\frac{I(b/h)R(h)}{2b^2 h\sqrt{h^2-a^2}} \tag{6.3.57}$$

for any crack speed $v = 1/h$. For the limiting case of $v/c_s \to 0$, this ratio approaches one-half, a feature of the corresponding quasi-static solution that can be verified independently. As the crack speed v/c_s increases from zero, this ratio decreases monotonically, approaching zero as the crack speed approaches the Rayleigh wave speed. It is noted that this ratio is exactly one-half of the ratio $K_I(t,v)/K_{Io}$ given in (6.3.49), so it is also shown in Figure 6.4 with an adjusted scale.

The physical interpretation of the ratio (6.3.57) is that it is the fraction of the energy being added to the body that is dissipated in the fracture process at a fixed crack speed. For example, if $v/c_s = 0.5$ then about one-third of the work being done on the body is being consumed in the fracture process. The remaining two-thirds of the work done is deposited as a distribution of strain energy and kinetic energy radiating out behind the wavefronts. The question of whether or not this constant speed crack growth process is physically realizable depends on whether or not a crack growth criterion is satisfied, a matter that will be addressed in Chapter 7.

6.3.3 Symmetric expansion of a shear crack

The self-similar expansion of a finite length crack under in-plane shear loading may be analyzed in the same way as the corresponding mode I crack treated in the preceding subsection. Only a few of the details are included here. As before, the material is assumed to be stress free and at rest for time $t \leq 0$. A rectangular x, y-coordinate system is introduced in the plane of deformation, oriented so that crack growth occurs in the plane $y = 0$. The crack begins to expand symmetrically from zero initial length at time $t = 0$ with each tip moving at a constant speed v that is less than the Rayleigh wave speed of the material. At any later time, the crack occupies the interval $-vt < x < vt$. The crack faces are subjected to opposed shear traction of magnitude τ_∞. The normal traction is zero on the crack faces, and no other loads act on the body. Thus, displacement potentials ϕ and ψ that satisfy the wave equations (6.3.18) are sought subject to the boundary conditions

$$\sigma_{yy}(x, 0^\pm, t) = 0 , \quad \sigma_{xy}(x, 0^\pm, t) = -\tau_\infty \quad \text{for} \quad |x| < vt . \quad (6.3.58)$$

The solution is expected to have a symmetry with respect to reflection in the x-axis typical of mode II fields, namely,

$$u_x(x, -y, t) = -u_x(x, y, t), \quad u_y(x, -y, t) = u_y(x, y, t) . \quad (6.3.59)$$

Following the reasoning of the preceding section, the stress and particle velocity components are anticipated to be homogeneous functions of x, y and t of degree zero, and a solution is sought on this basis. The solution of the corresponding problem of symmetric growth of a crack with traction-free faces under the action of a remotely applied shear traction is obtained by adding a uniform shear stress field to the present solution.

In light of the symmetries in the mode II case, the second derivatives of displacement potentials, in terms of which the stress components are represented in (6.3.21), are expressed in terms of analytic functions as

$$\frac{\partial^2 \phi}{\partial x^2} = \text{Im}\{F_{xx}(\zeta_d)\}, \quad \frac{\partial^2 \phi}{\partial x \partial y} = \text{Re}\{F_{xy}(\zeta_d)\},$$

$$\frac{\partial^2 \phi}{\partial y^2} = \text{Im}\{F_{yy}(\zeta_d)\}, \tag{6.3.60}$$

$$\frac{\partial^2 \phi}{\partial x \partial t} = \text{Im}\{F_{xt}(\zeta_d)\}, \quad \frac{\partial^2 \phi}{\partial y \partial t} = \text{Re}\{F_{yt}(\zeta_d)\}$$

and, likewise,

$$\frac{\partial^2 \psi}{\partial x^2} = \text{Re}\{G_{xx}(\zeta_s)\}, \quad \frac{\partial^2 \psi}{\partial x \partial y} = \text{Im}\{G_{xy}(\zeta_s)\},$$

$$\frac{\partial^2 \psi}{\partial y^2} = \text{Re}\{G_{yy}(\zeta_s)\}, \tag{6.3.61}$$

$$\frac{\partial^2 \psi}{\partial x \partial t} = \text{Re}\{G_{xt}(\zeta_s)\}, \quad \frac{\partial^2 \psi}{\partial y \partial t} = \text{Im}\{G_{yt}(\zeta_s)\},$$

where the complex variables ζ_d and ζ_s are defined in terms of physical coordinates in (6.3.23) and (6.3.26), respectively. Only two of the various functions labeled $F_{\alpha\beta}$ and $G_{\alpha\beta}$ above are independent. By following the development of the preceding section, it can be shown that

$$(a^2 - \zeta^2)F'_{xx}(\zeta) = i\zeta\sqrt{a^2 - \zeta^2}\, F'_{xy}(\zeta)$$

$$= \zeta^2 F'_{yy}(\zeta) = -\zeta(a^2 - \zeta^2)F'_{xt}(\zeta). \tag{6.3.62}$$

A similar compatibility condition can be written for the functions designated $G_{\alpha\beta}$.

The complete solution, except for evaluating a real constant β from the boundary condition (6.3.58)$_2$ is embodied in the two functions

$$G'_{yt}(\zeta) = \frac{\beta(b^2 - 2\zeta^2)h}{b^2(h^2 - \zeta^2)^{3/2}}, \quad F'_{xt}(\zeta) = \frac{2\beta\zeta^2 h}{b^2(h^2 - \zeta^2)^{3/2}}. \tag{6.3.63}$$

The shear stress $\sigma_{xy}(x, 0, t)$ at any point in the interval $a < t/x < h$ is given by the definite integral

$$\sigma_{xy} = -\beta\mu\frac{h}{b^2}\text{Im}\left\{\int_a^{t/x} \frac{R(\xi)}{(h^2 - \xi^2)^{3/2}\sqrt{b^2 - \xi^2}}\,d\xi\right\}. \qquad (6.3.64)$$

The value of the constant β is determined from the condition that

$$\frac{\tau_\infty}{\mu} = \beta\frac{h}{b^2}\int_0^\infty \frac{R(i\eta)}{(h^2 + \eta^2)^{3/2}\sqrt{b^2 + \eta^2}}\,d\eta. \qquad (6.3.65)$$

The similarities between these results and the corresponding expressions (6.3.43) and (6.3.44) for mode I are obvious. The expression (6.3.65) is rewritten as

$$\beta = \frac{\tau_\infty}{\mu}I_{II}(b/h), \qquad (6.3.66)$$

where the value of the quantity I_{II} is determined by numerical evaluation of the integral in (6.3.65).

From the definition of dynamic stress intensity factor, it follows that

$$K_{II}(t, v) = \lim_{x \to vt^+} \sigma_{xy}(x, 0, t)\sqrt{2\pi(x - vt)}$$

$$= -\frac{I_{II}(b/h)R(h)}{b^2h\sqrt{h^2 - b^2}}\tau_\infty\sqrt{\pi vt}. \qquad (6.3.67)$$

The mode II stress intensity factor for the equivalent quasi-static crack problem with total crack length l is obtained by letting $v \to 0$ and $t \to \infty$ in such a way that $vt \to l/2$. If the resulting stress intensity factor is denoted by K_{IIo}, then the result that $K_{IIo} = \tau_\infty\sqrt{\pi l/2}$ is obtained. Consequently, the dimensionless ratio

$$\frac{K_{II}(t, v)}{K_{IIo}} = -\frac{I_{II}(b/h)R(h)}{b^2h\sqrt{h^2 - a^2}} \qquad (6.3.68)$$

is of interest because it indicates the influence of inertial effects in the process. It is evident from the asymptotic behavior of β or I_{II} as $h/b \to \infty$ or $v/c_s \to 0$ that $K_{II}(t, v)/K_{IIo} \to 1$ as $v/c_s \to 0$, which is

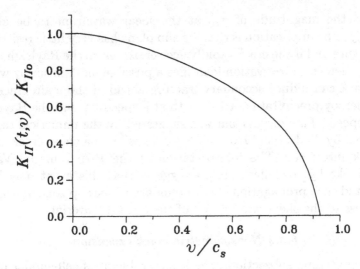

Figure 6.6. Graph of the mode II stress intensity factor for symmetric expansion of a shear crack, normalized by the corresponding equilibrium stress intensity factor for a crack of length $2vt$, versus dimensionless crack speed v/c_s for $\nu = 0.3$ from (6.3.68).

consistent with the corresponding quasi-static result if vt is identified with the half crack length $l/2$. This ratio is plotted as a function of v/c_s in Figure 6.6.

Burridge (1973) made an interesting observation on this solution for the special case when the crack speed is the Rayleigh wave speed, that is, when $v = c_R$ or $h = c$. In this case the shear stress on the crack plane is nonsingular at the crack tip, as can be seen from the result that the stress intensity factor vanishes as $v \rightarrow c_R$. The shear stress σ_{xy} has some finite value at the crack tip which must be $-\tau_\infty$. By direct calculation, he showed that the distribution of σ_{xy} has a positive peak in amplitude at the shear wavefront $x = c_s t$. This feature is indeed suggested by the form of (6.3.64).

Suppose that (6.3.64) is differentiated with respect to $\xi = t/x$. Then it can be shown by substitution that $\partial \sigma_{xy}/\partial \xi$ is square root singular with a negative coefficient at $\xi = b + 0$, and that it is also square root singular with a negative coefficient at $\xi = c - 0$. Demonstration of the latter result depends on the observation that $dR(c)/d\xi < 0$. Furthermore, $\partial \sigma_{xy}/\partial \xi$ has no apparent zeros in the interval between these two values. It is therefore concluded that σ_{xy} increases very rapidly from $x = c_R t$ to $x = c_s t$. Burridge (1973) reports the magnitude of σ_{xy} at the shear wavefront to be about $1.63\,\tau_\infty$. The implication is that if a slip plane has very little resistance to fracture and if the crack rapidly accelerates up to the Rayleigh wave speed, then this observation identifies a possible mechanism by which the crack can induce secondary fracture ahead of the main crack tip and thereby precipitate crack growth at a speed beyond the Rayleigh wave speed. This mechanism was suggested by the numerical results reported by Andrews (1976b) on the expansion of a shear crack along a weak interface. The interpretation of the 1979 Imperial Valley earthquake by Archuleta (1982) suggests that this event may have exhibited such propagation of the boundary of the slip zone at a speed in excess of the shear wave speed of the crustal material.

6.3.4 Nonsymmetric crack expansion

In the preceding subsections, the basic problems of self-similar crack expansion in modes I and II were considered. In both cases, it was assumed that the expansion is symmetrical, that is, that the two crack edges move with the *same* crack tip speed. This is not an essential feature for the fields to be self-similar. To result in self-similar fields, it is only necessary that the crack edges each move at *constant* speed as the crack expands from zero initial length. The constant speeds need not be the same, nor is it essential that the crack tips move in opposite directions. Such a situation of nonsymmetric growth is considered in this subsection, but only briefly.

Consider once again the growth of a crack in a uniform remote tensile field of intensity σ_∞, as in Section 6.3.2. The equivalent situation of crack growth with crack face pressure is discussed here. A rectangular x, y-coordinate system is introduced in the plane of deformation, oriented so that crack growth occurs in the plane $y = 0$. The crack begins to expand from zero initial length at time $t = 0$ with each tip moving at a constant speed which is less than the Rayleigh wave speed of the material. The tip advancing in the positive x-direction has speed $v_+ > 0$ and the tip advancing in the negative x-direction has speed $v_- \geq 0$; see the inset in Figure 6.7. The case of $v_- < 0$ can be handled in a similar manner, but this case is not considered here. At any time $t > 0$, the crack occupies the interval $-v_- t < x < v_+ t$. Both crack faces are subjected to compressive normal traction of magnitude σ_∞. The shear traction is zero on the crack faces, and no other loads act on the body. The material is stress free and at rest for $t \leq 0$.

The governing equations are developed in Section 6.3.2, and these steps are not repeated here. However, arguments similar to those presented there lead to the conclusion that the difference of the two analytic functions that comprise the velocity component $\dot{u}_y$ has the form

$$F_{yt}(\zeta) - G_{xt}(\zeta) = \frac{\alpha(2\zeta - h_+ + h_-)}{2\sqrt{h_+ h_-}(h_+ - \zeta)(h_- + \zeta)} , \qquad (6.3.69)$$

where $h_\pm = 1/v_\pm$ and α is a real dimensionless function of v_+/c_s and v_+/v_- which must be determined from the boundary conditions.

If the steps leading up to (6.3.43) are followed, then it is found that the normal stress $\sigma_{yy}(x, 0, t)$ on the plane $y = 0$ ahead of the crack tip advancing in the direction of increasing x is

$$\sigma_{yy} = -A\,\text{Im}\left\{\int_a^{t/x} \frac{R(\xi)}{(h_+ - \xi)^{3/2}(h_- + \xi)^{3/2}\sqrt{a^2 - \xi^2}}\,d\xi\right\} \qquad (6.3.70)$$

for $a < t/x < h_+$, where $A = \alpha\mu(h_+ + h_-)^2/4b^2\sqrt{h_+ h_-}$. The value of α is determined for any values of v_+/v_- and v_+/c_s by evaluating the integral in the equation

$$\frac{\sigma_\infty}{\mu} = A\,\text{Im}\left\{\int_{\Gamma_{ap}} \frac{R(\zeta)}{(h_+ - \zeta)^{3/2}(h_- + \zeta)^{3/2}\sqrt{a^2 - \zeta^2}}\,d\zeta\right\} , \qquad (6.3.71)$$

where $p > h_+$ is a point on the real axis and Γ_{ap} is any simple path in the complex ζ-plane from $\zeta = a$ to $\zeta = p$ along which the integrand is analytic.

The stress intensity factor is

$$\frac{K_I}{K_{Io}} = \alpha\frac{(b^2 - 2h_+^2)^2 - 4h_+^2\sqrt{h_+^2 - a^2}\sqrt{h_+^2 - b^2}}{b^2 h_+ \sqrt{h_+^2 - a^2}} , \qquad (6.3.72)$$

where $K_{Io} = \sigma_\infty\sqrt{\pi t(v_+ + v_-)/2}$ is the equilibrium stress intensity factor for any crack length $t(v_+ + v_-)$. Graphs of stress intensity factor versus v_+/c_s are shown in Figure 6.7 for $v_-/v_+ < 1$, which includes all possible cases through interchange of v_+ and v_-.

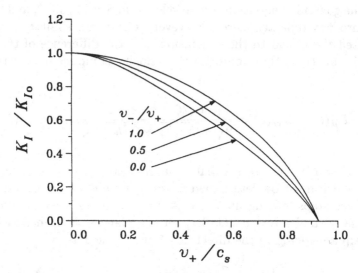

Figure 6.7. Graph of the dynamic stress intensity factor for nonsymmetrical self-similar expansion of a crack normalized by the corresponding equilibrium stress intensity factor for a crack of the same length versus dimensionless speed of the rightmost crack tip v_+/c_s for $\nu = 0.3$ from (6.3.72). The speed of the left crack tip is v_-.

6.3.5 Expansion of circular and elliptical cracks

Suppose that an unbounded, homogeneous, and isotropic elastic solid is subjected to a uniform remote tensile stress of magnitude σ_∞ in a particular direction, and that the material is initially at rest. A single component of stress is nonzero and the stress field is initially uniform. At a certain instant of time, say $t = 0$, a circular crack begins to grow from nominally zero initial radius in a plane transverse to the direction of the initial tension. The faces of the crack are free of traction, and the crack speed v is assumed to be constant and less than the Rayleigh wave speed of the material.

Due to the linearity of the boundary value problem, the situation can be analyzed by considering the same configuration with the uniform normal pressure σ_∞ acting on the faces of the expanding crack and with the material initially stress free and at rest. As in the case of the Broberg problem, the complete solution is then obtained by superimposing a uniform tensile stress field on the result. Consider cylindrical coordinates r, θ, z with the crack occupying the circular region $r < vt$ of the plane $z = 0$ for $t \geq 0$. The configuration and load distribution are invariant with respect to rotation about the z-axis, so the component of displacement in the θ-direction is zero at every point and all fields are independent of θ. The nonzero components of stress with reference to cylindrical coordinates are σ_{rr}, $\sigma_{\theta\theta}$, σ_{zz}, and σ_{rz}. The boundary conditions on the crack faces are

$$\sigma_{zz}(r, 0^\pm, t) = -\sigma_\infty, \quad \sigma_{rz}(r, 0^\pm, t) = 0 \quad \text{for} \quad r \leq vt. \quad (6.3.73)$$

In view of the symmetry with respect to reflection in the plane $z = 0$,

$$u_r(r, -z, t) = u_r(r, z, t), \quad u_z(r, -z, t) = -u_z(r, z, t) \quad (6.3.74)$$

everywhere, and

$$u_z(r, 0, t) = 0, \quad \sigma_{rz}(r, 0, t) = 0 \quad \text{for} \quad r \geq vt \quad (6.3.75)$$

on the crack plane.

The anticipated solution of this problem is a function of the three coordinates r, z, t. The geometrical configuration of the body at any time is obtained by self-similar expansion of the configuration at any earlier time and, furthermore, the boundary values of the specified stress components are time-independent. Under these conditions, a solution for which the stress and particle velocity components are homogeneous functions of r, z, t of degree zero may be anticipated. A solution of the problem was obtained on this basis by Kostrov (1964a) and Craggs (1966). The steps in the solution procedure are not included here, but the principal results are summarized.

The normal velocity of a point on the crack surface is found to be

$$\dot{u}_z(r, 0^+, t) = \frac{\sigma_\infty}{\mu} \frac{v^2 \gamma}{\sqrt{v^2 - r^2/t^2}}, \quad r < vt, . \quad (6.3.76)$$

where γ is a parameter depending on v/c_s and c_d/c_s which is to be determined from the boundary conditions. The function on the right side of (6.3.76) is indeed a homogeneous function of degree zero and it exhibits the characteristic square root singularity as $r \to vt^-$. The crack opening displacement can be obtained from (6.3.76) by integration with the result that

$$u_z(r, 0^+, t) = \sigma_\infty \gamma \sqrt{v^2 t^2 - r^2} / \mu. \quad (6.3.77)$$

According to Kostrov (1964a), the value of γ is determined from the relation

$$\gamma \int_0^\infty \frac{(1+2\eta)^2 - 4\eta\sqrt{(\eta + c_s^2/c_d^2)(\eta + 1)}}{(\eta + c_s^2/v^2)^2\sqrt{\eta + c_s^2/c_d^2}}\, d\eta = \frac{v}{c_s}. \qquad (6.3.78)$$

The mode of crack opening near the crack edge is mode I, and the stress intensity factor is

$$K_I(t, v) = \lim_{r \to vt^+} \sigma_{zz}(r, 0, t) 2\pi\sqrt{r - vt}, \qquad (6.3.79)$$

which is uniform around the crack edge. In terms of parameters already defined, the stress intensity factor for the expanding circular crack is

$$K_I(t, v) = \frac{\sigma_\infty c_s^2 \gamma \sqrt{\pi vt}\, D(v)}{v^2 \alpha_d}, \qquad (6.3.80)$$

where $D(v) = 4\sqrt{1 - v^2/c_d^2}\sqrt{1 - v^2/c_s^2} - (2 - v^2/c_s^2)^2$ is defined in (4.3.8). The equilibrium stress intensity factor for a uniform pressure σ_∞ acting on the faces of a circular crack of radius vt in an unbounded elastic solid is (Tada, Paris, and Irwin 1985)

$$K_{Io} = \frac{2\sigma_\infty\sqrt{\pi vt}}{\pi}. \qquad (6.3.81)$$

Thus, the ratio of the dynamic stress intensity factor to the equilibrium stress intensity factor for the same loading and crack size is

$$\frac{K_I(t, v)}{K_{Io}} = \frac{\pi c_s^2 \gamma D(v)}{2v^2 \alpha_d}. \qquad (6.3.82)$$

This ratio is shown for Poisson's ratio $\nu = 0.3$ in Figure 6.8. By standard asymptotic methods, it can be established that $\gamma \to 2(1 - \nu)/\pi$ as $v/c_s \to 0$, so the ratio (6.3.82) also approaches unity as $v/c_s \to 0$.

Much of the discussion following the analysis of the self-similar expansion of a crack under plane strain conditions in Section 6.3.2 applies in the present case as well. The dependence of stress intensity factor on speed v is similar, and the stress intensity factor vanishes

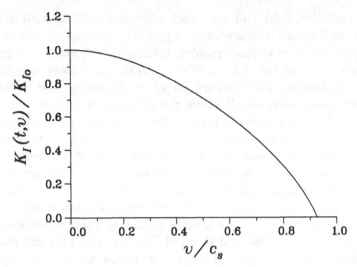

Figure 6.8. Ratio of the dynamic stress intensity factor for self-similar expansion of a circular crack with uniform crack face pressure to the equilibrium stress intensity for the same loading and crack size versus crack speed, according to (6.3.82).

as the speed approaches the Rayleigh wave speed. Energy variations with time may also be considered for the circular crack. One difference between the two-dimensional and three-dimensional cases is that the length of the crack edge increases in the latter case (it is the circumference of the circle) but it is constant in the former case. In both cases, the rate at which energy is absorbed from the material per unit length of crack edge through advance of the crack is proportional to time t; see (6.3.56), for example. In the case of a circular crack, however, the length of the crack also increases in proportion to time t, so that the *total* rate of energy dissipation is proportional to t^2.

Although few exact analytical solutions are available for three-dimensional crack growth, several other studies are noted. The self-similar expansion of a circular crack in an unbounded body subjected to remote homogeneous shear stress was also studied by Kostrov (1964b). Burridge and Willis (1969) obtained the solution for self-similar expansion of an elliptical crack in an unbounded anisotropic body subjected to general remote homogeneous stress. They essentially anticipated the shape of the crack opening and then verified that all conditions of the boundary value problem could be satisfied exactly. The particular case of an expanding elliptical shear crack was studied by Richards (1973), who presented a thorough discussion of the radiation field and the crack edge singularities with a view toward applications in seismology. A general, systematic study of self-similar problems in elastodynamics, including both crack and punch problems, was carried out by Willis (1973). This work provides a unified mathematical framework for all of the self-similar problems discussed previously, and it offers the prospect of finding new solutions, particularly within the realm of three-dimensional crack growth.

The particular situation of planar crack growth in an unbounded homogeneous elastic solid is ideal for reduction of the linear field equations and boundary conditions to an integral equation on the crack plane, analogous to the procedure in Section 2.3 for a relatively simple two-dimensional problem. An efficient numerical solution technique of the type commonly known as boundary integral methods was developed by Das (1980) and Das and Kostrov (1987) for this class of problems. This technique and related techniques are reviewed by Das (1985), and the area of study is fully developed in the monograph by Kostrov and Das (1989).

6.4 Crack growth due to general time-independent loading

Suppose that a crack exists in a body of elastic material, and that the material is subjected to time-independent loads in the form of surface tractions or body forces. The mechanical fields prior to crack growth are equilibrium fields. If the loading is increased to a sufficiently large magnitude, then the crack will begin to extend. It is a purpose of fracture mechanics to predict the way in which the crack will grow on the basis of the laws of continuum mechanics and a crack growth criterion. The special case in which the crack suddenly begins to grow dynamically at a fixed speed is considered in this section; the more general case of nonuniform crack tip motion is studied in the next chapter. The general approach to the problem of the sudden extension of a preexisting crack under the action of arbitrary time-independent loading is illustrated for the case of the extension of a planar mode I crack under plane strain conditions. The analysis is carried out for a semi-infinite crack in an otherwise unbounded body, so it is strictly applicable to crack growth in a bounded body only until waves reflected from the boundaries perturb the fields in the crack tip region. The method of analysis is based on integral transform techniques and the Wiener-Hopf technique. Important steps in the solution of dynamic crack problems by means of this approach were introduced by Baker (1962), who studied a particular situation of crack growth under stress wave loading. The problem studied by Baker will be discussed in Section 6.5. The application of integral transform methods and the Wiener-Hopf technique in the analysis of stationary crack problems was discussed in Chapter 2, which provides a framework for the present approach.

Consider crack growth under two-dimensional plane strain conditions. A rectangular x, y, z-coordinate system is introduced in the body in such a way that the crack is in the plane $y = 0$, the edge of the crack is parallel to the z-axis, and the crack edge moves in the positive x-direction. Thus, the plane of deformation is the x, y-plane. The crack tip is initially at $x = 0$ and it begins to move at constant speed v at time $t = 0$. The position of the crack tip at any time $t > 0$ is then $x = vt$.

Suppose that the configuration of the body and the applied loading system have reflective symmetry in the plane $y = 0$, so that the deformation field prior to crack growth has the symmetries representative of mode I crack fields. The applied loads induce a traction distribution on the crack plane ahead of the crack tip, and the process of crack growth is essentially the negation of this traction distribution. This idea is exploited to obtain a complete solution for general loading by means of superposition. Denote the tensile traction distribution on $y = 0$, $x > 0$ prior to crack growth by $\sigma_{yy}(x, 0) = p(x)$. The shear traction on the crack plane is zero due to symmetry. If a solution can be found for elastodynamic crack growth which satisfies the conditions that the initial stress and deformation are zero, the crack edge moves with position $x = vt$ for $t > 0$, and the crack faces are subject to a compressive traction of magnitude $p(x)$ over $0 < x < vt$, then the superposition of this result and the initial equilibrium field provides the complete solution to the problem of sudden growth of the crack under general time-independent loading. The initial equilibrium solution is assumed to be known, and the requisite elastodynamic crack growth fields are determined next. This is accomplished in two steps (Freund 1972a). First, the situation of crack growth with a pair of opposed concentrated forces acting on fixed material points on the crack faces is analyzed, giving rise to a very useful result called the *fundamental solution* for the problem. Then, the corresponding field quantities for any distribution of tractions on the crack faces can be determined directly by superposition over this fundamental solution. The fundamental solution is derived in the next subsection.

6.4.1 The fundamental solution

For time $t \leq 0$, the crack tip is at $x = 0$ and the material is stress free and at rest everywhere. At time $t = 0$, the crack tip begins to move in the positive x-direction at speed v. As the tip moves away from the origin of the coordinate system, it leaves behind a pair of concentrated forces, each of magnitude p^* (force per unit length in the z-direction) and tending to separate the crack faces. The crack faces are traction free at points other than $x = 0$. As the crack advances, the crack faces separate, the forces do work, and an elastodynamic field is generated. This field is the fundamental solution. In view of the symmetry of the problem, the solution is required only in the half plane $-\infty < x < \infty$, $0 < y < \infty$. Thus, displacement potentials ϕ and ψ that satisfy the wave equations

$$\frac{\partial^2 \phi}{\partial x^2} + \frac{\partial^2 \phi}{\partial y^2} - \frac{1}{c_d^2}\frac{\partial^2 \phi}{\partial t^2} = 0, \quad \frac{\partial^2 \psi}{\partial x^2} + \frac{\partial^2 \psi}{\partial y^2} - \frac{1}{c_s^2}\frac{\partial^2 \psi}{\partial t^2} = 0 \quad (6.4.1)$$

are sought subject to the boundary conditions

$$\sigma_{yy}(x,0,t) = -p^*\delta(x)H(t) \quad \text{for} \quad -\infty < x < vt,$$

$$\sigma_{xy}(x,0,t) = 0 \quad \text{for} \quad -\infty < x < \infty, \quad\quad\quad (6.4.2)$$

$$u_y(x,0,t) = 0 \quad \text{for} \quad vt < x < \infty,$$

where $\delta(\cdot)$ and $H(\cdot)$ represent the Dirac delta function and the unit step function, respectively. The last condition in (6.4.2) follows from the symmetry properties of displacement fields for mode I deformation, namely,

$$u_x(x,-y,t) = u_x(x,y,t), \quad u_y(x,-y,t) = -u_y(x,y,t). \quad (6.4.3)$$

It is advantageous to change variables from material coordinates x, y in the plane of deformation to crack tip coordinates ξ, y, where

$$\xi = x - vt, \quad\quad\quad\quad\quad\quad (6.4.4)$$

as shown in Figure 6.9. The displacement potentials are then sought as functions of position in the crack tip coordinate system. It should

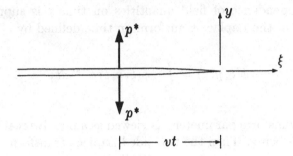

Figure 6.9. Diagram of the boundary value problem leading to the fundamental solution described in Section 6.4.1.

be noted that, at any material point, $\phi(\xi, y, t)$ depends on time explicitly through its last argument and implicitly through ξ. The wave equation governing ϕ in crack tip coordinates is

$$\left(1 - \frac{a^2}{h^2}\right)\frac{\partial^2 \phi}{\partial \xi^2} + \frac{\partial^2 \phi}{\partial y^2} + \frac{2a^2}{h}\frac{\partial^2 \phi}{\partial \xi \partial t} - a^2 \frac{\partial^2 \phi}{\partial t^2} = 0, \qquad (6.4.5)$$

where $a = 1/c_d$ and $h = 1/v$. The shear wave potential $\psi(\xi, y, t)$ is governed by the same partial differential equation except that a is replaced by $b = 1/c_s$.

In the moving coordinate system, the boundary conditions (6.4.2) take the form

$$\sigma_{yy}(\xi, 0, t) = \sigma_+(\xi, t) - p^* \delta(\xi + vt) H(t) H(-\xi),$$

$$\sigma_{xy}(\xi, 0, t) = 0, \qquad (6.4.6)$$

$$u_y(\xi, 0, t) = u_-(\xi, t)$$

for $-\infty < \xi < \infty$, where σ_+ and u_- have the same roles as the corresponding functions in (2.5.1). In particular, they are unknown functions which may take on nonzero values only for positive or negative values of ξ, as indicated by the subscript $+$ or $-$. In this form the boundary conditions have certain important characteristics in common with the corresponding boundary conditions (2.5.1), which suggests the Wiener-Hopf technique as a method of analysis. Indeed, from this point onward, the analysis parallels that in Section 2.5 although the details differ somewhat. Only the main steps are included here.

The dependence of field quantities on time t is suppressed by application of the Laplace transform on time defined by

$$\widehat{\phi}(\xi, y, s) = \int_0^\infty \phi(\xi, y, t)\, e^{-st}\, dt\,, \qquad (6.4.7)$$

where the transform parameter s is viewed as a positive real parameter for the time being. Then the two-sided Laplace transform

$$\Phi(\zeta, y, s) = \int_{-\infty}^\infty \widehat{\phi}(\xi, y, s)\, e^{-s\zeta\xi}\, d\xi \qquad (6.4.8)$$

is applied to the governing equations and boundary conditions, where the transform parameter is $s\zeta$, the factor s being simply a scaling parameter. The notation for transforms follows that established in Sections 2.4 and 2.5. Furthermore, based on the anticipated far-field asymptotic properties of $\widehat{\phi}(\xi, y, s)$, it is anticipated that the transform (6.4.8) will converge in the strip $-a_- < \mathrm{Re}(\zeta) < a_+$, where $a_\pm = a/(1 \pm a/h)$.

Application of the transforms to the differential equations (6.4.1) yields two ordinary differential equations whose solutions, bounded as $y \to \infty$, are

$$\Phi(\zeta, y, s) = s^{-3} P(\zeta)\, e^{-s\alpha y}\,, \qquad \Psi(\zeta, y, s) = s^{-3} Q(\zeta)\, e^{-s\beta y}\,, \quad (6.4.9)$$

where

$$\alpha(\zeta) = (a^2 - \zeta^2 + a^2\zeta^2/h^2 - 2a^2\zeta/h)^{1/2} \qquad (6.4.10)$$

and $\beta(\zeta)$ is the same function of ζ except that a is replaced by b. Furthermore, for (6.4.9) to be admissible for any ζ, the branch of α or β with $\mathrm{Re}(\alpha) \geq 0$ or $\mathrm{Re}(\beta) \geq 0$ must be selected. This is assured if branch cuts are introduced in the complex ζ-plane, running from the branch points $-a_-$, a_+, $-b_-$, and b_+ outward along the real axis (away from the origin) to $\pm\infty$, and if the branches with positive real values at the origin are selected. The functions $P(\zeta)$ and $Q(\zeta)$ are unknown at this point, and the coefficients of these functions in (6.4.9) have been selected in anticipation of a solution with stress and particle velocity fields that are homogeneous functions of degree negative one.

The transformed boundary conditions (6.4.6) are

$$\mu\left[\left(\frac{b^2}{a^2}-2\right)s^2\zeta^2\Phi + \frac{b^2}{a^2}\frac{d^2\Phi}{dy^2} - 2s\zeta\frac{d\Psi}{dy}\right]_{y=0+} = \frac{\Sigma_+(\zeta)}{s} - \frac{hp^*}{s(h-\zeta)},$$

$$\mu\left[2s\zeta\frac{d\Phi}{dy} + \frac{d^2\Psi}{dy^2} - s^2\zeta^2\Psi\right]_{y=0+} = 0,$$

$$\left[\frac{d\Phi}{dy} - s\zeta\Psi\right]_{y=0+} = \frac{U_-(\zeta)}{s^2}.$$

$$(6.4.11)$$

The special form of the terms on the right side of (6.4.11) with powers of s as factors has been selected for convenience. At this point, it is still possible that the transforms

$$\Sigma_+(\zeta) = s\int_0^\infty \hat{\sigma}_+(\xi,s)\,e^{-s\zeta\xi}d\xi,$$

$$(6.4.12)$$

$$U_-(\zeta) = s^2\int_{-\infty}^0 \hat{u}_-(\xi,s)\,e^{-s\zeta\xi}d\xi$$

will depend on s as well as on ζ. As suggested by the special form of (6.4.11), however, these functions will be found to depend only on ζ. Based on a consideration of the far-field behavior of the transformed fields, Σ_+ is expected to be analytic in $\text{Re}(\zeta) > -a_-$ and U_- is expected to be analytic in the overlapping half plane $\text{Re}(\zeta) < a_+$. In particular, both functions are expected to be analytic in the strip $-a_- < \text{Re}(\zeta) < a_+$.

Next, the transformed potentials (6.4.9) are substituted into the transformed boundary conditions (6.4.11). The result is a system of three linear algebraic equations for the four unknown functions $P(\zeta)$, $Q(\zeta)$, $\Sigma_+(\zeta)$, and $U_-(\zeta)$. The system of equations is valid for all values of ζ in the common strip. As was anticipated, the system of equations does *not* involve the parameter s. If two of the equations are used to eliminate P and Q in favor of Σ_+ and U_-, then the remaining equation is

$$\frac{hp^*}{(\zeta-h)} + \Sigma_+(\zeta) = -\frac{\mu h^2}{b^2}\frac{R(\zeta)}{\alpha(\zeta)(\zeta-h)^2}U_-(\zeta) \qquad (6.4.13)$$

which is valid in the common strip of analyticity $-a_- < \mathrm{Re}(\zeta) < a_+$, where

$$R(\zeta) = 4\zeta^2 \alpha(\zeta)\beta(\zeta) + (2\zeta^2 - b^2 - b^2\zeta^2/h^2 + 2b^2\zeta/h)^2 \quad (6.4.14)$$

is a modified form of the Rayleigh wave function. The function R is analytic everywhere in the complex plane except at the branch points $\zeta = \pm a_\pm$ and $\zeta = \pm b_\pm$. For the branches of α and β that were selected above, R is single valued in the ζ-plane cut along $a_\pm \leq |\mathrm{Re}(\zeta)| \leq b_\pm$, $\mathrm{Im}(\zeta) = 0$. The similarity between the Wiener-Hopf equations (6.4.13) and (2.5.11) is obvious, but it must be emphasized that the functions α, β, and R here are different from their counterparts introduced in connection with (2.5.11). The solution proceeds according to the outline in Section 2.5.

The factorization of the function $\alpha(\zeta)$ into a product of sectionally analytic functions $\alpha_+\alpha_-$ can be carried out by inspection with the result that

$$\alpha_\pm = \{a \pm \zeta(1 \mp a/h)\}^{1/2} . \quad (6.4.15)$$

The function R must also be factored and, to do so, the behavior of this function at infinity, as well as the locations of its zeros, must be known. Repeated application of the binomial expansion theorem for large values of $|\zeta|$ yields the result that $R(\zeta) \sim -\kappa\zeta^4$ as $|\zeta| \to \infty$, where

$$\kappa = 4\sqrt{(1 - a^2/h^2)(1 - b^2/h^2)} - (2 - b^2/h^2)^2 \quad (6.4.16)$$

is real and positive for all crack speeds between zero and the shear wave speed. Furthermore, it can be shown by direct calculation that $R(\zeta)$ vanishes for $\zeta = h, c_+, -c_-$, where $c_\pm = c/(1 \pm c/h)$ and $c = c_R^{-1}$ is the inverse Rayleigh wave speed. The zero at $\zeta = h$ has multiplicity two, whereas the other zeros are simple zeros. The zero counting formula of complex integration theory (2.5.26) can be applied to prove that the function R has no other zeros in the entire complex ζ-plane. With the step in (2.5.28) as a guide, it is convenient here to introduce an auxiliary function $S(\zeta)$ defined by

$$S(\zeta) = \frac{R(\zeta)}{\kappa(c_+ - \zeta)(c_- + \zeta)(\zeta - h)^2} . \quad (6.4.17)$$

This function has the important properties that $S(\zeta) \to 1$ as $|\zeta| \to \infty$ and that it has neither zeros nor poles in the complex ζ-plane. In light of these properties, the factorization theorem (2.5.23) can be applied directly with the result that

$$S_\pm(\zeta) = \exp\left\{-\frac{1}{\pi}\int_{a_\mp}^{b_\mp} \arctan\left[V(\eta)\right]\frac{d\eta}{\eta \pm \zeta}\right\},$$

$$(6.4.18)$$

$$V(\eta) = \left[\frac{4\eta^2\beta(\eta)|\alpha(\eta)|}{(2\eta^2 - b^2 - b^2\eta^2/h^2 \mp 2b^2\eta/h)^2}\right].$$

The factors $S_\pm(\zeta)$ have a simple relationship to the corresponding factors for $v = 0$ as obtained in (2.5.29). Suppose that the latter are denoted here and subsequently by $S_\pm^0(\zeta)$ to reflect the fact that

$$S_\pm^0(\zeta) = \lim_{v\to 0} S_\pm(\zeta) \qquad (6.4.19)$$

identically in ζ. If the change of variable $\eta \to -\eta/(1 + \eta/h)$ is introduced into (6.4.18) then it can be shown that

$$S_\pm(\zeta) = \frac{S_\mp^0\left(\frac{\zeta h}{\zeta - h}\right)}{S_\mp^0(h)}. \qquad (6.4.20)$$

With the factors (6.4.15) and (6.4.20) in hand, the Wiener-Hopf equation (6.4.13) is conveniently rewritten in terms of the composite function

$$F_\pm(\zeta) = \frac{\alpha_\pm(\zeta)}{S_\pm(\zeta)(c_\mp \pm \zeta)}. \qquad (6.4.21)$$

Then (6.4.13) becomes

$$\frac{hp^*}{(\zeta - h)}\left[F_+(\zeta) - F_+(h)\right] + F_+(\zeta)\Sigma_+(\zeta)$$

$$= -\frac{hp^*F_+(h)}{(\zeta - h)} - \frac{\mu\kappa h^2}{b^2}\frac{U_-(\zeta)}{F_-(\zeta)}, \qquad (6.4.22)$$

which is valid in the strip $-a_- < \text{Re}(\zeta) < a_+$ of the complex plane. The left side of (6.4.22) is analytic in the half plane $\text{Re}(\zeta) > -a_-$

and the right side is analytic in the overlapping half plane $\text{Re}(\zeta) <$ a_+. Therefore, by the identity theorem of analytic functions noted in Section 1.3.1, each side provides the analytic continuation of the other side into the complementary half plane. Together the two sides represent one and the same entire function. Just as in Section 2.5, the conditions that the displacement must be continuous at $x = 0$ and that the stress must be only square root singular there imply that the value of this entire function is zero. It follows that

$$\Sigma_+(\zeta) = \frac{hp^*}{(\zeta - h)}\left[\frac{F_+(h)}{F_+(\zeta)} - 1\right],$$

$$U_-(\zeta) = -\frac{b^2}{\mu\kappa h^2}\frac{hp^*}{(\zeta - h)}F_-(\zeta)F_+(h).$$

(6.4.23)

A number of features of the solution will be examined in the remainder of this chapter and in the next chapter, but the crack tip stress intensity factor will be extracted first. It is evident from (6.4.21) that $F_\pm(\zeta) = O\left(\zeta^{-1/2}\right)$ as $|\zeta| \to \infty$, hence $\Sigma_+(\zeta) = O\left(\zeta^{-1/2}\right)$ as $|\zeta| \to \infty$. This implies that the stress will indeed be square root singular at the crack edge. The strength of the singularity can be determined from the transformed solution by direct application of the Abel theorem (1.3.18) governing asymptotic properties of functions and their transforms,

$$\lim_{\zeta\to\infty}(s\zeta)^{1/2}\frac{1}{s}\Sigma_+(\zeta) = \lim_{\xi\to0^+}(\pi\xi)^{1/2}\hat{\sigma}_+(\xi, s) = \frac{1}{\sqrt{2}}\hat{K}_I(s, v),\quad(6.4.24)$$

where $\hat{K}_I(s, v)$ is the Laplace transform on time of the stress intensity factor as viewed by the crack tip observer moving at speed v. Direct application of this general result to the solution of the Wiener-Hopf equation (6.4.23) yields

$$\hat{K}_I(s, v) = p^*\sqrt{\frac{2}{vs}}\,k(h),$$

(6.4.25)

where the function of crack speed $k(h)$ is given by

$$k(h) = \frac{(1 - c/h)}{S_+(h)\sqrt{1 - a/h}}.$$

(6.4.26)

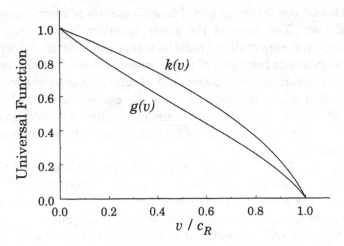

Figure 6.10. Graph of the universal function $k(v)$ versus crack tip speed v/c_R for $\nu = 0.3$ from (6.4.26). Also shown is the universal function $g(v)$ from (6.4.35).

This function of crack speed recurs throughout the study of elasto-dynamic crack tip fields. The factor $S_+(h) = 1/S_-^0(h)$ is not too different from unity over the full range of its argument, and a useful approximation to $k(h)$ for most practical purposes is $k(h) \approx (1 - c/h)/\sqrt{1-a/h}$ or $k(v) \approx (1 - v/c_R)/\sqrt{1 - v/c_d}$. The function depends on the material properties through the elastic wave speeds, but it is independent of the loading on the body. In this sense, it is a *universal function of crack tip speed*. It will be written as a function of inverse crack speed $h = 1/v$, or, alternatively, as a function of crack speed, that is, as $k(v)$, depending on the context. However, the function defined in (6.4.26) is understood in all cases. A graph of the dimensionless function k versus crack speed normalized by the Rayleigh wave speed v/c_R is shown in Figure 6.10. Its general features are evident from its definition or from the graph. In particular, it has the value $k = 1$ when $v/c_R = 0$, has the value $k = 0$ when $v/c_R = 1$ and is monotonic between these values.

The Laplace transform of the mode I stress intensity factor in (6.4.25) is readily inverted. The result is

$$K_I(vt, v) = p^* \sqrt{\frac{2}{\pi v t}} \, k(v) \,, \qquad (6.4.27)$$

where the arguments of the function have been shown as the amount of

crack growth and the crack tip speed in anticipation of more general
results to follow. The form of the stress intensity factor result is
consistent with the expectation that the stress and particle velocity
fields are homogeneous functions of degree negative one. Furthermore,
the result has the correct form in the limit of vanishing inertial effects.
If vt is identified with the distance from the concentrated loads of
magnitude p^* to the crack tip, say l, under equilibrium conditions,
then (6.4.27) implies that $K_I = p^* \sqrt{2/\pi l}$ for $v/c_R \to 0$, which is the
correct equilibrium result.

Although the focus here has been on the determination of the
dynamic elastic stress intensity factor, the full elastodynamic field
can be determined from the transformed solution. Once $\Sigma_+(\zeta)$ and
$U_-(\zeta)$ have been found by means of the Wiener-Hopf procedure, the
functions $P(\zeta)$ and $Q(\zeta)$ introduced in (6.4.9) are also known. For
the problem at hand, these functions are found to be

$$P(\zeta) = \frac{hp^*(b^2 - 2\zeta^2 + b^2\zeta^2/h^2 - 2b^2\zeta/h)F_+(h)}{\mu(\zeta - h)R(\zeta)F_+(\zeta)},$$

$$Q(\zeta) = \frac{2hp^*\alpha(\zeta)\zeta F_+(h)}{\mu(\zeta - h)R(\zeta)F_+(\zeta)}.$$

$$(6.4.28)$$

The function $\phi(\xi, y, t)$ itself is given by the double Laplace inversion
integral

$$\phi(\xi, y, t) = \frac{1}{2\pi i} \int_{s_0 - i\infty}^{s_0 + i\infty} \frac{1}{s^2} \frac{1}{2\pi i} \int_{\zeta_0 - i\infty}^{\zeta_0 + i\infty} P(\zeta) e^{s(t + \zeta\xi - \alpha y)} d\zeta \, ds,$$

$$(6.4.29)$$

where s_0 is a positive real number and ζ_0 is a positive real number
between $-a_-$ and a_+. Such a transform inversion integral is most
effectively evaluated by means of the Cagniard-de Hoop technique
(de Hoop 1961).

6.4.2 Arbitrary initial equilibrium field

Attention now is returned to the case of an arbitrary initial equilibrium stress field. Under the action of an applied loading on the
cracked solid which results in mode I deformation, a normal traction

$p(x)$ is induced on the prospective fracture plane ahead of the crack tip, where the x-direction is the crack growth direction. The process of crack growth is essentially the negation of this traction distribution, as described in the introduction of this section. An elastodynamic solution is sought for time $t > 0$ satisfying the condition of zero initial stress and particle velocity in the body and the condition that the normal *compressive* traction on the newly created crack faces, say over $0 < x < vt$, is $p(x)$. Such a solution, when superimposed on the initial equilibrium solution, is the desired complete solution of the problem. Superposition of solutions is not valid, in general, for moving boundary problems even though they are governed by linear equations. However, if two solutions can be found for the same motion of the boundary, then a linear sum of the two solutions will also be a solution. The solution for arbitrary $p(x)$ is now constructed from the fundamental solution obtained in the preceding subsection.

Consider the same problem as in Section 6.4.1, except that the pair of concentrated loads appears on the crack faces at $x = x' > 0$ instead of at $x = 0$. The crack tip still begins to move at constant speed v from the position $x = 0$ at time $t = 0$. The tip therefore passes the point x' at time $t = x'/v$. Let $p^* f(x, y, t)$ denote a scalar component of any field quantity in the fundamental solution. Then, the solution of the modified problem for the same physical quantity is $p^* f(x - x', y, t - x'/v)$. Furthermore, suppose that the load that appears on the crack faces behind the crack tip as it passes the point $x = x'$ is not a concentrated load of magnitude p^* but that it is a load of intensity $p(x')$ distributed over the infinitesimal interval from x' to $x' + dx'$. The solution for the same physical quantity for this problem is $f(x - x', y, t - x'/v) p(x') dx'$ to first order in dx'. Finally, the solution for the case of a distributed traction appearing through the crack tip and giving rise to the normal traction $p(x)$ on $0 < x < vt$ is given by the superposition integral

$$\int_0^{vt} f(x - x', y, t - x'/v) p(x') \, dx' . \qquad (6.4.30)$$

It is obvious by direct substitution that the integral in (6.4.30) is a wave function if $f(x, y, t)$ is a wave function, and that the integral satisfies the correct initial conditions. Furthermore, by choosing the function f to be any of the particular fields that are subject to the correct boundary conditions, the superposition integral itself satisfies the correct boundary conditions. This superposition provides a solution to the problem of interest for the entire physical plane. Of special interest for the time being is the stress intensity factor for general crack face loading. This result is extracted next.

If the stress intensity factor for general loading is sought, then from (6.4.27) it is evident that the function $f(x, y, t)$ in the superposition integral (6.4.30) should be selected to be $k(v)\sqrt{2/\pi vt}$. Thus, the stress intensity factor for general loading is

$$K_I(vt, v) = k(v)\sqrt{\frac{2}{\pi}} \int_0^{vt} \frac{p(x)}{\sqrt{vt - x}}\, dx \qquad (6.4.31)$$

for $t > 0$. Note that the result has the form of the universal function $k(v)$ times a function that depends on v and t only through the combination vt, that is, through the amount of crack growth since $t = 0$. Indeed, the factor of $k(v)$ in (6.4.31) is exactly the equilibrium stress intensity factor for a crack tip at $x = l = vt$ and crack face pressure $p(x)$ over $0 < x < l$. Thus, the dynamic stress intensity factor (6.4.31) is simply the universal function $k(v)$ times the equilibrium stress intensity factor for the given applied loading and the instantaneous amount of crack growth,

$$K_I(vt, v) = k(v)K_I(vt, 0). \qquad (6.4.32)$$

With the stress intensity factor in hand for general loading $p(x)$, the energy release rate can be calculated by means of the generalized Irwin relationship (5.3.9), with the result that

$$G(vt, v) = \frac{1 - \nu^2}{E} A_I(v)\, K_I(vt, v)^2. \qquad (6.4.33)$$

In view of the special form of the stress intensity factor result, the energy release rate expression can be rewritten as

$$G(vt, v) = \frac{1 - \nu^2}{E} K_I(vt, 0)^2\, A_I(v)k(v)^2. \qquad (6.4.34)$$

This form is immediately recognized as the corresponding equilibrium energy release rate times a universal function of crack tip speed $g(v)$, so that

$$G(vt, v) = g(v)\, G(vt, 0), \qquad g(v) = A_I(v)k(v)^2. \qquad (6.4.35)$$

Thus, the dynamic energy release rate also has the form of a universal function of crack tip speed times the equilibrium energy release rate for the specified crack face loading and a crack tip at the position corresponding to the instantaneous amount of dynamic growth. The dimensionless function $g(v)$ is also shown in Figure 6.10 as a function of crack speed normalized by the Rayleigh wave speed of the material v/c_R. It has the features that $g = 1$ for $v/c_R = 0$ and $g = 0$ for $v/c_R = 1$. It varies monotonically between these limiting values, and it can be approximated by the linear function $g(v) \approx 1 - v/c_R$ for virtually all applications.

The stress wave fields radiated by a crack once it begins to grow dynamically have not been considered in any detail here. However, these fields are of central importance in methods used in seismology and acoustic emission techniques of materials testing. It was noted by Freund (1972e) that a mode I crack radiates a finite discontinuity in stress and particle velocity fields if it suddenly begins to grow at a constant speed. The issue of stress wave radiation from a growing crack was pursued in much greater depth by Rose (1981), who considered radiation from a semi-infinite mode I crack due to suddenly starting or stopping crack growth, and from a three-dimensional crack. He also considered the influence of multiple scattering of the kind studied in Section 3.3 on the radiation from a growing crack of finite length.

6.4.3 Some illustrative cases

As a simple illustration of the general stress intensity factor result, consider the special case in which the advancing crack relieves an equilibrium stress intensity factor field as it begins to grow from rest at speed v. Thus, it is assumed that

$$p(x) = \frac{K_0}{\sqrt{2\pi x}} \quad \text{for} \quad 0 < x < vt, \tag{6.4.36}$$

where K_0 is the initial equilibrium stress intensity factor. The dynamic stress intensity factor for this case is

$$K_I(vt, v) = k(v)\frac{K_0}{\pi} \int_0^{vt} \frac{dx}{\sqrt{x(vt - x)}} = k(v)\,K_0\,. \tag{6.4.37}$$

Thus, at the onset of crack growth at constant speed v, the stress intensity factor changes *discontinuously* from its initial value of K_0

to its dynamic value $k(v) K_0$. Because $k(v) < 1$ for $v/c_R > 0$, the stress intensity factor decreases discontinuously. Likewise, the energy release rate experiences a discontinuous drop at the instant crack growth begins. For the case of time-independent loading, this suggests that crack growth under equilibrium or slowly rising applied loading probably does not begin abruptly with finite crack speed but, instead, that the crack tip will accelerate from zero initial speed under dynamic conditions. For example, suppose that a crack is assumed to grow with a fixed, material specific value of the stress intensity factor. Then the applied loading must be increased to a sufficient level in order to bring the stress intensity factor to the critical level. If the crack then begins to grow with a constant nonzero speed, however, the dynamic crack tip stress intensity factor will jump to a value *below* the critical value. On the other hand, if the process of fracture initiation requires that a sharp crack must be induced to grow from an initial blunt crack, a sudden drop in stress intensity factor cannot be ruled out. Indeed, this situation has been exploited in certain crack propagation and arrest experiments. A more thorough discussion of this situation must be postponed until the solution for *nonuniform* motion of the crack tip is developed in the next chapter.

As a second illustration, consider the case when

$$p(x) = p^*\delta(x) - \frac{p^*}{\pi x}\sqrt{\frac{l}{x-l}}\, H(x-l),\qquad (6.4.38)$$

where l is a positive parameter with the dimension of length and $H(\cdot)$ is the unit step function. The loading function (6.4.38) represents a concentrated force p^* acting at $x = 0$ to open the crack plus a square root singular normal traction distribution acting over $l < x < vt$ and tending to close the crack. Again, the crack begins to grow from $x = 0$ at time $t = 0$ with constant speed v, and the normal compressive traction (6.4.38) appears on the crack faces behind the crack tip. For $vt < l$, the solution for this special case is simply the fundamental solution. For $vt > l$, on the other hand, the dynamic stress intensity factor $K_I(vt, v)$ is

$$p^* k(v)\sqrt{\frac{2}{\pi}}\left[\frac{1}{\sqrt{vt}} - \frac{\sqrt{l}}{\pi}\int_l^{vt}\frac{dx}{x\sqrt{(x-l)(vt-x)}}\right] = 0.\qquad (6.4.39)$$

Thus, the dynamic stress intensity factor corresponding to the loading (6.4.38) is *identically zero* for all time $t > l/v$. This remarkable result arises in a natural way in the discussion of nonuniform crack growth in Section 7.2.

Finally, the special case when the traction $p(x)$ is a spatially uniform pressure of magnitude σ^* is considered. In this case, the dynamic stress intensity factor is

$$K_I(vt, v) = k(v)\sigma^* \sqrt{\frac{2}{\pi}} \int_0^{vt} \frac{dx}{\sqrt{vt - x}} = 2\sigma^* k(v)\sqrt{2vt/\pi}\,. \quad (6.4.40)$$

The stress intensity factor therefore increases indefinitely with time as the crack tip advances at constant speed v.

6.4.4 The in-plane shear mode of crack growth

The situation of dynamic growth of a semi-infinite crack at constant speed under the action of an applied loading that produces a state of mode II deformation can be handled in the same way as the foregoing mode I situation. The analysis has been summarized by Fossum and Freund (1975). Of particular note is the way in which the dynamic stress intensity factor depends on the initial loading. Suppose that the distribution of shear traction on the prospective fracture plane is $-q(x)$ prior to crack growth. Then the significant step in the analysis is the solution for the case of the shear stress $q(x)$ appearing on each crack face as the crack advances at speed v. It is found that the dynamic stress intensity factor for this situation has the same general properties that were established above for the case of mode I deformation. In particular, it is found that

$$K_{II}(vt, v) = k_{II}(v)\, K_{II}(vt, 0)\,, \quad (6.4.41)$$

where

$$k_{II}(v) = \frac{1 - c/h}{S_+(h)\sqrt{1 - b/h}} = \frac{1 - v/c_R}{S_+(v^{-1})\sqrt{1 - v/c_s}}\,. \quad (6.4.42)$$

As before, $h = 1/v$ and $S_+(h)$ is given in (6.4.18). That is, the dynamic stress intensity factor has the form of a universal function of crack tip speed for mode II deformation times the equilibrium stress intensity factor for the given initial load distribution represented by the function $q(x)$ and the instantaneous amount of growth vt, namely,

$$K_{II}(vt, 0) = \sqrt{\frac{2}{\pi}} \int_0^{vt} \frac{q(x)}{\sqrt{vt - x}}\, dx\,. \quad (6.4.43)$$

For purposes of calculation, the function in (6.4.42) can be approximated by $k_{II}(v) \approx (1 - v/c_R)/\sqrt{1 - v/c_s}$ for most practical purposes.

6.4.5 The antiplane shear mode of crack growth

The corresponding situation of dynamic growth of a semi-infinite crack at constant speed under the action of time-independent loading that results in a state of mode III deformation can also be analyzed by means of the integral transform methods outlined above. Suppose that the distribution of shear traction on the prospective fracture plane ahead of the crack tip prior to crack advance is given as a function of position x by $-q(x)$. At time $t = 0$, the crack begins to grow at the constant speed v. It can be shown that

$$K_{III}(vt, v) = k_{III}(v)K_{III}(vt, 0), \qquad (6.4.44)$$

where K_{III} is also given by the right side of (6.4.43) and

$$k_{III}(v) = \sqrt{1 - v/c_s} = \sqrt{1 - b/h}. \qquad (6.4.45)$$

This result is derived for arbitrary crack motion in Section 7.2, following an approach introduced by Kostrov (1966). The result for the special case of constant speed crack growth is included here for the sake of completeness.

6.5 Crack growth due to time-dependent loading

In the preceding section, the situation of a semi-infinite crack in an otherwise unbounded body subjected to general time-independent loading was considered. Here, the same physical system is again considered but the material in the region of the crack is assumed to be stress free and at rest initially. At a certain instant of time, say $t = 0$, the crack tip region is stressed due to the passage of a plane stress wave or to the sudden application of uniform crack face traction. The crack will begin to extend at some later time, and the purpose is to examine the mechanical fields associated with such a crack extension process. The development of this section is less general than that of the preceding section, in the sense that only a uniform crack face pressure or a normally incident plane stress pulse is considered. The time-independent loading of the preceding section, by comparison, is quite general in its spatial variation. The restriction to relatively simple spatial distributions of loading is not essential, but its adoption permits an examination of the structure of the elastodynamic crack fields that is not possible when a completely general time-dependent loading is considered. The more general case is discussed briefly in Section 7.5.2.

The material is stress free and at rest for time $t < 0$. At time $t = 0$, a uniform compressive normal traction of magnitude σ^* begins to act on each face of the crack [problem (b) in Figure 6.11]. The connection between this situation and loading by an incident stress pulse is discussed in a subsequent subsection. A rectangular x, y, z-coordinate system is introduced in the body in such a way that the crack is in the plane $y = 0$ for $x \leq 0$, the edge of the crack is parallel to the z-axis, and the crack edge moves in the positive x-direction once it begins to advance. Thus, the plane of deformation is the x, y-plane and the mode of deformation is mode I. The crack tip is initially at $x = 0$. Prior to crack growth, this is exactly the situation studied in Section 2.5, where it was observed that the crack edge stress intensity factor is proportional to σ^* due to linearity and that it must increase in proportion to $\sqrt{c_d t}$ on the basis of dimensional considerations. If the material is of limited strength, then the crack will begin to extend at some later time, say $t = \tau$, where τ is the *delay time* for the process. It is the purpose of this section to study the situation of crack growth at constant speed v beginning at time $t = \tau$ due to the sudden application of crack face pressure at time $t = 0$. The position of the crack tip at any time $t > \tau$ is then $x = v(t - \tau)$. The special case of constant speed crack growth with $\tau = 0$ was first analyzed by Baker (1962). The possibility of crack growth at a nonuniform rate after an arbitrary delay τ will be considered in Chapter 7.

The suddenly applied crack face pressure induces a transient traction distribution on the crack plane ahead of the crack tip, and the process of crack growth is essentially the negation of this traction distribution. As in the preceding section, this idea is exploited to obtain a complete solution by means of superposition. Denote the tensile traction distribution on $y = 0$, $x > 0$ prior to crack growth by

$$\sigma_{yy}(x, 0) = p(x/t) = \sigma_+(x, t). \tag{6.5.1}$$

This stress distribution was determined in Section 2.5, where it was denoted by $\sigma_+(x, t)$. The particular form $p(x/t)$ in (6.5.1) is selected to make clear the important property that this function is a homogeneous function of x and t of degree zero. The shear traction on the crack plane is zero due to symmetry. If a solution for elastodynamic crack growth can be found which satisfies the conditions that the initial stress and deformation are zero everywhere, the crack edge moves with position $x = v(t-\tau)$ for $t > \tau$, and the crack faces are subject to compressive traction of magnitude $p(x/t)$ over $0 < x < v(t-\tau)$, then the superposition of this result [problem (c) in Figure 6.11] and the initial transient field of problem (b) provides the complete solution to the problem of the growth of the crack under suddenly applied crack face pressure loading over $-\infty < x < 0$. The requisite elastodynamic crack growth fields are determined next. This is accomplished by first deriving a very useful result for a special distribution of traction on the crack faces called the *fundamental solution* for the problem (Freund 1973a). Then, the corresponding field quantities for the distribution of traction on the crack faces given in (6.5.1) can be determined by superposition over this fundamental solution. The fundamental solution is derived in the next subsection.

6.5.1 The fundamental solution

As a result of the sudden application of the crack face pressure, the normal stress distribution (6.5.1) is induced on the prospective fracture plane $x > 0$, $y = 0$. When the crack begins to grow at time $t = \tau$, it must in effect negate this traction distribution. Stated in another way, if a solution can be found for zero initial stress and particle velocity everywhere in the body and for crack growth at speed v beginning at time $t = \tau$ with compressive normal crack face traction of magnitude $p(x/t)$ over $0 < x < v(t - \tau)$, then the sum of this solution and the solution of Section 2.5 yields the desired result. These two problems whose solutions are to be superimposed will be called simply the stationary crack problem (b) and the moving crack problem (c) in this section.

The mathematical property of homogeneity of the distribution (6.5.1) has an important implication for this analysis. It implies that any fixed stress level in the scattered field radiates out along the x-axis at constant speed, that is, the stress level $p(u)$ travels from the crack tip at $x = 0$ at speed u for time $t > 0$. The speed u lies in the range from zero to the longitudinal wave speed c_d. The x-coordinate at time t of the stress levels moving with speeds u and $u + du$ are ut and $(u + du)t$, respectively, where du is understood to be an infinitesimal

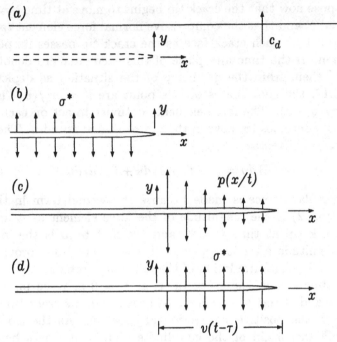

Figure 6.11. Schematic representation of the various boundary value problems considered in constructing the solution for constant-speed crack growth under stress pulse loading.

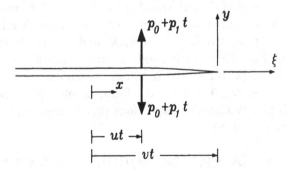

Figure 6.12. Diagram of the boundary value problem leading to the fundamental solution described in Section 6.5.1.

increment in speed. Thus, to within first order in the infinitesimal quantity du, the total force (per unit length in the z-direction) due to all stress levels with speeds between u and $u + du$ is $tp(u)\,du$, and this force acts at $x = ut$.

Suppose now that the crack tip begins to move at time $t = \tau > 0$ with speed v, and that the compressive normal force element $tp(u) \, du$ begins to act on each crack face as the crack tip passes its point of application. If the time and place of this encounter are denoted by t' and x' then, from the geometry of the situation as depicted in Figure 6.12, the coordinates of this point are $t' = v\tau/(v - u)$ and $x' = uv\tau/(v - u)$. The force element continues to act on each crack face and it continues to move in the x-direction at speed u. The force element is rewritten as

$$tp(u) \, du = (t - t')p(u) \, du + t'p(u) \, du. \qquad (6.5.2)$$

This particular choice is made because the second term in the expression (6.5.2) is the magnitude of the force element as it appears at the crack tip at time $t = t'$, and the first term is the increase in the magnitude after it appears. These observations provide the motivation for the formulation of the following problem.

For time $t \leq 0$, the crack tip is at $x = 0$ and the material is stress free and at rest everywhere. At time $t = 0$, the crack tip begins to move in the positive x-direction at speed v. As the tip moves away from the origin of the coordinate system, it leaves behind a pair of concentrated forces tending to separate the crack faces, each of magnitude p_0 (force per unit length in the z-direction). Each of these forces increases in magnitude at rate p_1, and the forces move in the positive x-direction at speed u for $t > 0$. The crack faces are traction free at points other than $x = ut$. As the crack advances, the crack faces separate, the forces do work, and an elastodynamic field is generated. This field is the fundamental solution. In view of the symmetry of the problem, the solution is required only in the half plane $-\infty < x < \infty$, $0 < y < \infty$. Thus, displacement potentials ϕ and ψ which satisfy the wave equations (6.4.1) are sought subject to the boundary conditions

$$\sigma_{yy}(x, 0, t) = -(p_0 + p_1 t)\delta(x - ut)H(t), \quad -\infty < x < vt,$$

$$\sigma_{xy}(x, 0, t) = 0, \quad -\infty < x < \infty, \qquad (6.5.3)$$

$$u_y(x, 0, t) = 0, \quad vt < x < \infty,$$

where $\delta(\cdot)$ and $H(\cdot)$ represent the Dirac delta function and the unit step function, respectively. The last condition follows from the symmetry properties of displacement fields for mode I deformation (6.4.3).

It is again advantageous to change variables from material coordinates x, y in the plane of deformation to crack tip coordinates ξ, y, where $\xi = x - vt$. The solution procedure for this particular boundary value problem follows that in the preceding section quite closely, and the details are not included here. The solution procedure is simplified if two separate problems are treated, one with $p_0 \neq 0$ and $p_1 = 0$ and the other with $p_0 = 0$ and $p_1 \neq 0$. A solution for the full physical plane can be found, but the only elements of the solution included here are the normal stress on the prospective fracture plane ahead of the crack tip and the normal component of displacement of the crack face at $y = 0^+$. The double transforms of this stress component and displacement component, $\Sigma_{yy}(\zeta, 0, s)$ and $U_y(\zeta, 0^+, s)$, are

$$
\Sigma_{yy} = p_0 \frac{wF_+(w)}{s(\zeta - w)F_+(\zeta)} - p_1 \frac{w^2}{s^2 F_+(\zeta)} \left[\frac{F_+(w)}{(\zeta - w)} \right]',
$$

$$
U_y = -p_0 \frac{wb^2 F_-(\zeta)F_+(w)}{s^2 \mu \kappa h^2 (\zeta - w)} + p_1 \frac{w^2 b^2 F_-(\zeta)}{s^3 \mu \kappa h^2} \left[\frac{F_+(w)}{(\zeta - w)} \right]',
$$

(6.5.4)

where $w = 1/(v - u)$ is the inverse relative speed and the other parameters are defined in the preceding section. In particular, the function $F_\pm(\zeta)$ is given in (6.4.21).

Of special interest is the stress intensity factor for the fundamental solution. The behavior of the Laplace transform of the stress intensity factor $\widehat{K}_I(s)$ can be determined by examining the asymptotic behavior of $\Sigma_{yy}(\zeta, 0, s)$ as $\zeta \to \infty$. The result can be inverted by inspection with the result that

$$
K_I(t) = p_0 \sqrt{\frac{2}{\pi}} \frac{wF_+(w)}{\sqrt{1 - a/h}} t^{-1/2} - 2p_1 \sqrt{\frac{2}{\pi}} \frac{w^2 F_+'(w)}{\sqrt{1 - a/h}} t^{1/2}. \quad (6.5.5)
$$

If the time dependence of the concentrated load in $(6.5.3)_1$ is generalized to $p_0 + p_1 t + p_2 t^2 + \cdots$ then the stress intensity factor can be determined to correspondingly higher fractional powers in t, with the coefficients depending on higher order derivatives of $F_+(w)$. The fundamental solution as derived, however, is adequate to determine the dynamic stress intensity factor for the situation of suddenly applied crack face pressure described above and, if desired, to determine the full elastodynamic field. This is demonstrated next.

6.5.2 Arbitrary delay time with crack face pressure

Let $p_0 f_0(\xi, y, t; u) + p_1 t f_1(\xi, y, t; u)$ represent any element of the fundamental solution, such as a stress component or a displacement component. It is evident from the linearity of the problem that one part of the solution will always be proportional to p_0 and another part to p_1. The dependence of the solution on the load point speed u is made explicit for this development. Consider now the same problem as that studied in Section 6.5.1 except that the crack begins to move at time $t = \tau$ instead of time $t = 0$. The concentrated loads still appear on the crack faces at $x = 0$ at the instant that the crack begins to extend. The solution of this modified problem is then $p_0 f_0(\xi, y, t - \tau; u) + p_1 (t - \tau) f_1(\xi, y, t - \tau; u)$, where ξ is now understood to be $\xi = x - v(t - \tau)$. Furthermore, suppose that the concentrated loads appear on the crack faces as the crack tip passes the point $x = x'$ instead of at $x = 0$. The crack tip still begins to move at time $t = \tau$ at the point $x = 0$. Then, the solution is $p_0 f_0(\xi, y, t - \tau - x'/v; u) + p_1 (t - \tau - x'/v) f_1(\xi, y, t - \tau - x'/v; u)$.

With a view toward constructing the solution for the case of a compressive pressure $p(x/t)$ on the crack faces, x'/u is chosen as $t' = v\tau/(v - u) = w\tau/h$. Furthermore, with (6.5.2) in mind, p_0 is set equal to $t' p(u)\, du$ and p_1 is set equal to $p(u)\, du$. The solution for the moving crack face load (6.5.2) is then

$$f_0(\xi, y, t - t'; u)\, p(u)\, du + (t - t') f_1(\xi, y, t - t'; u)\, p(u)\, du. \quad (6.5.6)$$

The complete field representation for the element of the solution under consideration can then be found by superposition over the appropriate range of u. In view of the interpretation of u as the speed at which the stress level $p(u)$ moves from the crack tip in the stationary crack problem, the range of u is the range of x/t between $x = 0$, $t = \tau$ and $x = v(t - \tau)$, $t = t$, that is, the range of u is $0 \leq u \leq v(t - \tau)/t$. Thus, the element of the solution is given by

$$\int_0^{v(t-\tau)/t} [f_0(\xi, y, t - t'; u)\, p(u) + (t - t') f_1(\xi, y, t - t'; u)\, p(u)]\, du.$$

$$(6.5.7)$$

The superposition integral (6.5.7) can be applied to determine the stress intensity factor for the moving crack problem with the crack tip at $x = v(t - \tau)$ and the compressive normal traction $p(x/t)$ acting on the crack faces. In view of the fact that the various functions in (6.5.7) depend on u mainly through $w = 1/(v - u)$, it is more convenient to integrate with respect to w. Thus, the change of variable of integration from u to w is effected with $du = dw/w^2$. The range of integration over w is then $h < w < h^*$, where $h^* = th/\tau$. Following some manipulation of the resulting expression, it is found that

$$K_I(t) = -2\sqrt{\frac{2\tau}{\pi(h - a)}} \int_h^{h^*} \left[F_+(w)(h^* - w)^{1/2}\right]' p(w)\, dw, \quad (6.5.8)$$

where it should be noted that $p(u)$ has been replaced by $p(w)$. Although p is a different function of u than it is of w, the notation should not cause any confusion because the function always represents the same traction distribution. The integral in (6.5.8) can be evaluated, but it requires special care because the term $p(w)$ is singular at $w = h$ ($u = 0$) and its factor in the integrand is singular at $w = h^*$.

The explicit form of $p(w)$ is required to carry out the evaluation. From (2.5.40), this function is given by

$$p(w) = \frac{\sigma^*}{\pi} \int_a^{wh/(w-h)} \mathrm{Im}\left\{\frac{F_+^0(0)}{\eta F_+^0(-\eta)}\right\} d\eta, \quad (6.5.9)$$

where F_+^0 is the function defined in (2.5.31). The superscript 0 has been added to make clear the distinction between functions defined in Section 2.5 for the stationary crack problem and functions represented by the same letter in the present section. The connection between these functions is simply

$$\lim_{v \to 0} F_+(\zeta) = F_+^0(\zeta), \quad \lim_{v \to 0} \alpha_+(\zeta) = \alpha_+^0(\zeta), \quad (6.5.10)$$

and so on. These various functions satisfy some identity relationships that are important for extracting the principal result. These identities are not obvious, but they can be established by direct manipulation of the quantities involved. For example, it can be shown that

$$S_+(w) = \frac{S_-^0(u^{-1})}{S_-^0(v^{-1})}, \quad w = \frac{1}{v - u}. \quad (6.5.11)$$

This follows directly from the identity (6.4.20). The relationship (6.5.11) is important, in turn, in establishing that

$$F_+(w) = ik(h)\sqrt{\frac{h-a}{w-h}}\, F_+^0\left(-\frac{wh}{w-h}\right), \qquad (6.5.12)$$

where $k(h)$ is the universal function of crack speed defined in (6.4.26).

The integral in (6.5.7) is evidently convergent and, at first sight, it appears to be ideally suited for evaluation by integration by parts. This is not possible, however, as was already noted previously. To circumvent the difficulty, the lower limit of integration h is replaced by $h+\epsilon$, where ϵ is a positive number that is arbitrarily small compared to h. The resulting integral can be integrated by parts to yield

$$K_I(t) = 2\sigma^* k(h)\sqrt{2/\pi}\left\{\frac{\sqrt{v\tau}}{\pi}F_+(0)\left[2\sqrt{\frac{h(h^*-h)}{\epsilon}}\right.\right.$$

$$\left.\left. -\int_{h+\epsilon}^{h^*}\frac{h^{3/2}(h^*-w)^{1/2}}{w(w-h)^{3/2}}\,dw\right] - \sqrt{vt-v\tau}\right\} \qquad (6.5.13)$$

plus terms that are $o(1)$ as $\epsilon \to 0$. The surprisingly simple form of the expression in the integrand of (6.5.13) is a result of applying the identity (6.5.11). The remaining integral is easily evaluated in terms of algebraic functions with the result that

$$\int_{h+\epsilon}^{h^*}\frac{h^{3/2}(h^*-w)^{1/2}}{w(w-h)^{3/2}}\,dw = 2\sqrt{\frac{h(h^*-h)}{\epsilon}} - \pi\sqrt{h^*} \qquad (6.5.14)$$

plus terms that are bounded as $\epsilon \to 0$. Thus, the divergent part of the integral (6.5.13) exactly cancels the divergent term resulting from the integration by parts. The remainder of the expression is unaffected by letting $\epsilon \to 0$. The stress intensity factor for the moving crack problem is therefore

$$K_I(t,v) = 2\sigma^* k(v)\sqrt{\frac{2}{\pi}}\left[\frac{\sqrt{c_d t(1-2v)/2}}{(1-\nu)} - \sqrt{v(t-\tau)}\right] \qquad (6.5.15)$$

for $t \geq \tau$. The universal function $k(v)$ is shown in the graph in Figure 6.10. For $0 < t < \tau$, the stress intensity factor is given by (2.5.44).

6.5.3 Incident plane stress pulse

Attention is turned once again to the case of a plane pulse incident on the crack edge. Suppose that a plane pulse parallel to the crack plane propagates through the body so that the wavefront reaches the crack plane at time $t = 0$. Suppose, further, that this pulse carries a jump in tensile stress σ_{yy} from its zero initial value to σ^*. If the material was not cracked on the plane $y = 0$ then the stress pulse would induce a tensile traction of magnitude σ^* on this plane, as already discussed in Section 2.5; see problem (a) in Figure 6.11. The complete elastodynamic field for an incident stress pulse can be obtained by superimposing the elementary solution for the uncracked solid and the solution for a suddenly applied crack face pressure of magnitude σ^*.

This superposition scheme carries over without modification to the case of crack growth. The stress intensity factor (6.5.15) already includes the influence of the crack face pressure σ^* over the original surface $x < 0$ [problem (b)], and it includes the influence of the diffracted wave field [problem (c)]. To extract the stress intensity factor for the case of an incident step stress pulse, the only additional effect that must be taken into account is the uniform crack face pressure σ^* on the newly created crack surfaces $0 < x < v(t - \tau)$ [problem (d) in Figure 6.11]. From the expression (6.4.40), the stress intensity due to this contribution is $2\sigma^* k(v) \sqrt{2v(t - \tau)/\pi}$. If this contribution is added to (6.5.15), then it is found that the stress intensity factor for an incident step stress pulse of magnitude σ^* and delay time τ is

$$K_I(t, v) = 2\sigma^* k(v) \frac{\sqrt{c_d t(1 - 2\nu)/\pi}}{(1 - \nu)}. \qquad (6.5.16)$$

This result has a couple of noteworthy features. First, it was derived for arbitrary delay time τ but it is found to be independent of τ, except for the condition that it applies only for $t > \tau$. Second, the stress intensity factor has the form of the stress intensity factor for a stationary crack subjected to the same stress wave loading multiplied by the universal function of crack speed $k(v)$. That is, it has the form

$$K_I(t, v) = k(v) K_I(t, 0), \qquad (6.5.17)$$

which is a remarkable property shared with the corresponding result for time-independent loading obtained in Section 6.4.

The foregoing analysis has been carried out for step stress loading of the crack plane at time $t = 0$. The stress intensity factor result is easily generalized to the case of an incident stress pulse of general profile, following the procedure in Section 2.7. Suppose that the wavefront of a plane stress pulse parallel to the plane of the crack reaches $y = 0$ at time $t = 0$ and that the profile of the pulse is given by

$$\sigma_{yy}^{inc}(x,0,t) = \begin{cases} 0 & \text{if } t \le 0, \\ p(t) & \text{if } t > 0. \end{cases} \quad (6.5.18)$$

Then the transient stress intensity factor for the onset of crack growth at speed v at any time $t = \tau$ is

$$K_I(t,v) = 2k(v)\frac{\sqrt{(1-2\nu)/\pi}}{(1-\nu)} \int_{0-}^{t} \sqrt{c_d(t-t_o)}\,\frac{dp(t_o)}{dt_o}\,dt_o. \quad (6.5.19)$$

Numerous examples of $K_I(t,0)$ are given in Section 2.7 for various profiles $p(t)$, and the corresponding result for growth is obtained from (6.5.19) by multiplication by $k(v)$. As in the case of time-independent loading, the sudden onset of crack growth at some speed v always implies a discontinuous drop in the level of the stress intensity factor. As in that case, if the level of stress intensity is thought to control the process of crack growth and a critical level is achieved just prior to the onset of growth, then it is unlikely that growth will commence at a large crack speed. Instead, the more likely behavior is for crack growth to begin gradually and the speed to increase with the amount of crack growth. The actual response depends on the details of the growth criterion and the applied loading, and several cases are examined in Chapter 7.

7

ELASTIC CRACK GROWTH
AT NONUNIFORM SPEED

7.1 Introduction

The restriction to constant crack speed in the early analytical work on
dynamic crack propagation, as reflected in the contents of Chapter 6,
was not motivated by physical considerations. Instead, the restriction
was imposed in order to render the mathematical models tractable.
In this chapter, progress toward relaxing this restriction for rapid
crack growth in nominally elastic materials is described. Results are
limited, but sufficient to provide a relatively complete conceptual basis
for analysis of crack propagation under two-dimensional conditions or
in simple structural elements.

The study of a problem involving crack growth at nonuniform
rate proceeds in two steps. First, the underlying boundary value
problem is considered for arbitrary motion of the crack tip, with
a view toward obtaining a full description of the mechanical fields
near the crack edge during growth. Then, an additional physical
postulate in the form of a crack propagation criterion is imposed on
the mechanical fields in order to determine an equation of motion for
the crack tip. This approach is illustrated for antiplane shear crack
growth, for plane strain crack growth under quite general loading
conditions, and for one-dimensional string and beam models. The
technologically important issue of crack arrest (Bluhm 1969) is an
essential feature in this development; arrest is identified through the
crack tip equation of motion as the point beyond which crack growth
cannot be sustained.

The difficulty in extending the analytical approaches that were so successful for stationary cracks and cracks growing at a constant rate to the case of nonuniform crack growth are evident from the analysis in Chapters 2, 3, and 6. The boundary value problems considered are strictly linear and, for the most part, the mechanical fields obtained are steady state or self-similar. In such cases, solution procedures based on the properties of analytic functions of a complex variable and on integral transform techniques are very effective. For the case of crack growth at a nonuniform rate, however, the solution depends on the crack tip motion in a way that precludes the use of these techniques. The mechanical fields are neither steady nor self-similar. Consequently, these same analytical methods cannot be applied directly. A solution procedure that is an exception to this observation is Green's method, as described in Section 2.3, which relies on the linearity of the governing differential equations but not on a particular form of the boundary conditions. Indeed, the first exact stress intensity factor solution for two-dimensional transient crack growth under conditions of antiplane shear deformation, or mode III crack propagation, was obtained by Kostrov (1966) through application of this method. The approach and principal conclusions are presented in Section 7.2. Direct approaches to the corresponding plane strain crack growth problem have been elusive. However, certain features of the solution obtained by Kostrov (1966), as well as a similar solution obtained by Eshelby (1969a), suggested an indirect approach based on the notion of fundamental solutions (Freund 1972a, 1972b). This approach is described in Section 7.3 for the case of mode I deformation and general time dependent loading. The corresponding mode II analysis was reported by Fossum and Freund (1975). The approach was generalized by Freund (1973a), Kostrov (1975), and Burridge (1976) to other cases of loading.

The chapter concludes with two sections devoted to ancillary results. In Section 7.6, it is shown that the mathematical solution for growth of a mode I crack at nonuniform speed can be applied to determine the speed at which a strip yield zone grows from a crack tip in a body subjected to dynamic loading. Finally, in the last section, the question of uniqueness of elastodynamic crack growth solutions is addressed. The proof of the classical uniqueness theorem for elastodynamics due to Neumann is based on consideration of global energy variations associated with a difference solution of the governing equations. Because a moving crack tip is a sink of mechanical energy, it is essential that possible implications for the conditions of the uniqueness theorem are considered.

7.2 Antiplane shear crack growth

Consider an elastic body which contains a half plane crack but which is otherwise unbounded. Introduce a rectangular x, y, z coordinate system so that the z-axis lies along the crack edge. The crack is in the plane $y = 0$ and occupies the half plane $x \leq 0$. Suppose that the material is initially at rest, and that *time-independent* loading is applied which produces a state of antiplane shear deformation in the body. Thus, the only nonzero component of displacement is the z-component $u_z(x, y, t) = w(x, y, t)$. The displacement field is independent of z and it has the symmetry property that $w(x, -y, t) = -w(x, y, t)$. Prior to crack advance, the stress distribution on the prospective fracture plane ahead of the crack tip is $\sigma_{yz}(x, 0) = \tau_y(x, 0) = q(x)$.

In Section 6.4, the process of crack advance in such a situation was viewed as the process of negating the traction $q(x)$, which suggested a superposition scheme for construction of solutions to crack growth problems. The same general point of view is adopted here. Suppose that at time $t = 0$ the crack begins to advance in the x-direction so that the position of the crack edge is $x = l(t)$ for $t > 0$. The function $l(t)$ specifies the amount of crack advance at any instant of time, and the instantaneous crack tip speed is $\dot{l}(t)$. The function $l(t)$ is continuous and its rate of change is restricted by

$$0 \leq \dot{l}(t) < c \quad \text{for} \quad 0 < t < \infty, \qquad (7.2.1)$$

where c denotes the elastic shear wave speed c_s in this development; otherwise, the function is arbitrary. Following the reasoning in Section 6.4, it is assumed that the material is stress free and at rest for $t \leq 0$. As the crack advances for $t > 0$, it does so under the action of a traction distribution on the crack faces that is equal and opposite to the initial traction, that is,

$$\sigma_{yz}(x, 0^\pm, t) = \tau_y(x, 0^\pm, t) = -q(x), \quad 0 \leq x < l(t). \qquad (7.2.2)$$

No other loads act on the body. The solution of this problem, when superimposed on the original equilibrium state, provides a complete solution for the situation of crack growth in a body subjected to some initial time-independent loading. Also, as demonstrated in Section 6.4, the principal features of the solution are exhibited by a relatively simple canonical problem, or fundamental solution. Thus, the procedure is developed for the particular case of opposed concentrated forces on the crack faces of magnitude q^*, that is,

$$q(x) = q^* \, \delta(x). \tag{7.2.3}$$

More general loadings are then handled by superposition.

The general framework for a solution procedure based on Green's method is outlined in Section 2.3. The component of displacement $w(x, y, t)$ satisfies the wave equation in two space dimensions and time,

$$\nabla^2 w - \frac{1}{c^2} \frac{\partial^2 w}{\partial t^2} = 0. \tag{7.2.4}$$

If attention is restricted to the half plane $-\infty < x < \infty$, $0 < y < \infty$ then Green's formula (2.3.1) leads to a representation for the value of w at any point in the body and any time in terms of the Green's function and the values of w and $\partial w / \partial y$ on the boundary $y = 0^+$. The representation formula appears as (2.3.5) and the corresponding Green's function is given in (2.3.4). This representation formula was applied to solve a stationary crack problem in Section 2.3. That problem is identical to the one being considered here except that $l(t) = 0$ in the former case. It was observed in Section 2.3 that if the field point in the representation theorem approaches the boundary $y = 0$ ahead of the crack tip, then the condition that the displacement vanishes on $y = 0$, $x > 0$ due to symmetry yields an integral equation for the unknown traction on the crack plane ahead of the crack tip. For a stationary crack tip, this integral equation is (2.3.6). In the present situation involving crack growth according to $x = l(t)$, the representation theorem (2.3.5) is likewise applicable. Furthermore, the symmetry condition that

$$w(x, 0, t) = 0 \quad \text{for} \quad x > l(t) \tag{7.2.5}$$

implies that

$$\int_0^{t_o} \int_{x_o - c(t_o - t)}^{x_o + c(t_o - t)} \frac{\tau_y(x, 0, t)}{\sqrt{c^2(t_o - t)^2 - (x_o - x)^2}} \, dx \, dt = 0 \tag{7.2.6}$$

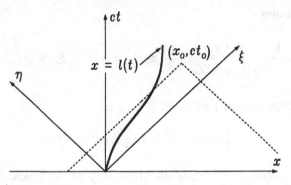

Figure 7.1. A portion of the x, t-plane showing the region of integration in (7.2.6). The time axis has been rescaled by wave speed c for convenience. The crack tip path in this plane is $x = l(t)$.

for $x_o > l(t_o)$. The integral equation (7.2.6) is identical to (2.3.6), of course, except that the location of the field point is restricted by $x_o > l(t_o)$ rather than $x_o > 0$. A solution of this integral equation can be found by following the procedure outlined in Section 2.3, as adapted from the work of Kostrov (1966) on antiplane shear crack growth at a nonuniform rate. With reference to (7.2.6), it is noted that $\tau_y(x, 0^\pm, t) = -q^*\delta(x)$ for $x < l(t)$ but that $\tau_y(x, 0, t)$ is unknown for $x > l(t)$. Thus, the first step is to determine this stress distribution ahead of the crack tip.

Consider the portion of the x, t-plane shown in Figure 7.1, where the time coordinate has been scaled by c for convenience. This is the plane $y = 0^+$ in the three-dimensional x, y, t-space of the solution $w(x, y, t)$. A generic crack path $x = l(t)$ is shown, and a field point (x_o, t_o) ahead of the crack tip is chosen where condition (7.2.6) applies. The lines $x + ct = x_o + ct_o$ and $x - ct = x_o - ct_o$ running backward in time from (x_o, t_o) are shown as dashed in Figure 7.1. The solution at (x_o, t_o) depends on boundary values and initial values only within the domain of dependence enclosed by the lines for $t < t_o$. The solution procedure is facilitated by a change of integration variables in (7.2.6) to

$$\xi = ct + x, \qquad \eta = ct - x. \qquad (7.2.7)$$

The ξ, η-coordinate axes are also shown in Figure 7.1. This transformation carries the point (x_o, t_o) into (ξ_o, η_o) and the integral equation

(7.2.6) becomes

$$\int_0^{\eta_o} \frac{1}{\sqrt{\eta_o - \eta}} \int_{-\eta}^{\xi_o} \frac{\tau_y(\xi, 0, \eta)}{\sqrt{\xi_o - \xi}} \, d\xi \, d\eta = 0, \qquad \xi_o > g(\eta_o), \qquad (7.2.8)$$

where the function $g(\cdot)$ satisfies

$$\tfrac{1}{2}\left[g(\eta) - \eta\right] = l\left(\tfrac{1}{2}\left[g(\eta) + \eta\right]\right) \qquad (7.2.9)$$

identically. Thus, $\xi = g(\eta)$ is the trajectory of the crack tip in terms of the ξ, η-coordinates. The variable representing the y-component of stress has been preserved as τ_y in writing (7.2.8) even though it is obviously a different function of its arguments than it was in the original equation (7.2.6). The partial solution of this integral (2.3.7) implied by the initial conditions has been incorporated into (7.2.8).

The integral equation (7.2.8) will be satisfied if the inner integral vanishes for all η in the interval $0 < \eta < \eta_o$. This condition can be written as the Volterra integral equation of the first kind

$$\int_{g(\eta_o)}^{\xi_o} \frac{\tau_y(\xi, \eta_o)}{\sqrt{\xi_o - \xi}} \, d\xi = \int_{-\eta_o}^{g(\eta_o)} \frac{q\left(\tfrac{1}{2}[\xi - \eta_o]\right)}{\sqrt{\xi_o - \xi}} \, d\xi \qquad (7.2.10)$$

for $\xi_o > g(\eta_o)$. In writing (7.2.10), the boundary condition on traction (7.2.2) has been incorporated. Finally, for the special case in which the boundary traction is the pair of opposed concentrated forces at $x = 0$, namely, $q(x) = q^* \delta(x)$, the integral equation (7.2.10) becomes

$$\int_{g(\eta_o)}^{\xi_o} \frac{\tau_y(\xi, 0, \eta_o)}{\sqrt{\xi_o - \xi}} \, d\xi = \frac{2q^*}{\sqrt{\xi_o - \eta_o}}, \qquad \xi_o > g(\eta_o), \qquad (7.2.11)$$

where the scaling property of the delta function $\delta(x/2) = 2\delta(x)$ has been used. With a change of integration variable to $\zeta = \xi - g(\eta_o)$ this is again an integral equation with the standard form of an Abel integral equation with a square root singular kernel as given in (2.3.12). The general solution is given in (2.3.15). In the present case, the solution for any point with $\xi_o > g(\eta_o)$ is

$$\tau_y(\xi_o, 0, \eta_o) = \frac{2q^* \sqrt{g(\eta_o) - \eta_o}}{\pi(\xi_o - \eta_o)\sqrt{\xi_o - g(\eta_o)}}. \qquad (7.2.12)$$

To transform this result from the coordinates ξ, η to the coordinates x, t, let the physical coordinates of the point $\xi = g(\eta_0)$, $\eta = \eta_0$ be (x^*, t^*). The special significance of this point is that if the crack tip emits an elementary wave signal at time t^* when it is at the place $x^* = l(t^*)$ and if the signal radiates with speed c, then the signal will arrive at position x at time t. From the geometry of Figure 7.1, it is evident that

$$g(\eta_0) - \eta_0 = 2l(t^*),$$

$$\xi_0 - g(\eta_0) = 2[x_0 - l(t^*)], \qquad (7.2.13)$$

$$\xi_0 - \eta_0 = 2x_0.$$

Therefore, in terms of the physical coordinates, the stress distribution ahead of the crack tip (7.2.12) is

$$\tau_y(x, 0, t) = \frac{q^* \sqrt{l(t^*)}}{\pi x \sqrt{x - l(t^*)}}, \qquad x > l(t), \qquad (7.2.14)$$

where the subscript "o" has been dropped because its use is no longer essential. For any fixed x, t and any crack motion, the parameter t^* is a solution of

$$c = \frac{x - l(t^*)}{t - t^*}, \qquad (7.2.15)$$

which is simply a statement of the fact that a signal traveling at speed c that is emitted at the position $l(t^*)$ at time t^* will reach position x at time t.

The solution (7.2.14) has a number of noteworthy features. First, it is remarkably simple. Indeed, the mathematical expression is exactly the same as the corresponding expression for the equilibrium stress distribution ahead of a crack tip at $x = l_0$ due to a pair of opposed concentrated loads q^* applied at $x = 0$, except that the quantity $l(t^*)$ replaces the constant l_0 in the equilibrium case. Another surprising feature emerges for the situation where $\dot{l}(t) \neq 0$ for $0 < t < t_a$ but $\dot{l}(t) = 0$ for $t > t_a$, that is, if crack growth stops at the *arrest time* $t = t_a$. The result (7.2.14) implies that the correct equilibrium field for the given loading and crack length $l(t_a)$ radiates out from the crack tip following its arrest, and that the equilibrium field is *fully established* behind a point moving with the wave speed c, that is, for $l(t_a) < x < c(t - t_a)$. This is an extraordinary feature for a two-dimensional elastodynamic field. Such fields are normally characterized by long transient "tails" after any abrupt change in loading. This feature was observed and discussed in some detail by Eshelby (1969b), who noted that the corresponding stress distribution for a nonuniformly moving dislocation does not share this remarkable feature. Some broad implications of this feature will play a central role in the discussions in subsequent sections in this chapter.

The dynamic elastic stress intensity factor is readily extracted from the complete stress distribution (7.2.14). If $x - l(t)$ is vanishingly small, then $(t - t^*)$ is also small. Thus, for points close to the crack edge in this sense,

$$x - l(t^*) = x - l(t) + \dot{l}(t)(t - t^*) + o(t - t^*) \qquad (7.2.16)$$

as $t^* \to t$. In light of (7.2.15), this can be rewritten as

$$x - l(t^*) = \frac{x - l(t)}{1 - \dot{l}(t)/c} \qquad (7.2.17)$$

for points x near $l(t)$. Therefore, the dynamic stress intensity factor is

$$K_{III}(l, \dot{l}) = \lim_{x \to l(t)^+} \tau_y(x, 0, t)\sqrt{2\pi(x - l(t))}$$

$$= q^*\sqrt{\frac{2\left[1 - \dot{l}(t)/c\right]}{\pi l(t)}} . \qquad (7.2.18)$$

The equilibrium stress intensity factor for this same loading applied to the faces of a stationary crack at a fixed distance l_o behind the crack tip is

$$K_{III}(l, 0) = q^*\sqrt{\frac{2}{\pi l_o}} . \qquad (7.2.19)$$

Therefore, the dynamic stress intensity factor for *arbitrary nonuniform motion* of the crack edge has the form of a function of instantaneous crack speed times the corresponding equilibrium stress intensity factor for the given applied loading and an amount of crack growth equal to the instantaneous value of $l(t)$,

$$K_{III}(l, \dot{l}) = \sqrt{1 - \dot{l}/c}\, K_{III}(l, 0) . \qquad (7.2.20)$$

Although the specific results obtained so far in this section apply only for a very special loading situation, generalization to a more complex loading situation can be accomplished by means of the superposition arguments presented in Section 6.4. For the case of general crack face traction $-q(x)$ behind the advancing crack edge, the result for the dynamic stress intensity factor is

$$K_{III}(l, \dot{l}) = \sqrt{1 - \dot{l}/c} \sqrt{\frac{2}{\pi}} \int_0^{l(t)} \frac{q(x)}{\sqrt{l(t) - x}} \, dx \,. \qquad (7.2.21)$$

Again, the expression has the form of the universal function of crack speed times the corresponding equilibrium stress intensity factor for the specified loading and a crack length equal to the instantaneous length $l(t)$. In effect, the result (7.2.21) provides the dynamic stress intensity factor for the most general time-independent loading conditions. For example, the function $q(x)$ could be the initial stress distribution on the prospective crack plane ahead of a finite length crack. The result (7.2.21) then provides the stress intensity factor for dynamic advance of a finite length crack up until the time at which waves generated at the advancing crack edge are reflected back onto the moving crack tip from the other, still stationary, edge of the crack. The matter of interaction between the ends of a crack of finite length when both are moving was considered by Rose (1976c).

The discussion of antiplane shear crack growth at nonuniform rate has been developed under the assumption of time-independent loading, the situation for which the features of the analysis and the interpretation of the results are most transparent. This is not an essential restriction, and cases of time-dependent loading of the crack faces can be handled in much the same way. For example, the situation in which the crack faces are subjected to suddenly applied, spatially uniform shear traction $\tau_y(x, 0^\pm, t) = -\tau^* H(t)$ was studied in Sections 2.2 and 2.3 for the case of a stationary crack tip. The dynamic stress intensity factor in that case was found to be $K_{III}(t, 0) = 2\tau^* \sqrt{2ct/\pi}$. If the crack tip actually moves according to $x = l(t)$, then an analysis similar to that just presented yields the result that

$$K_{III}(t, \dot{l}) = \sqrt{1 - \dot{l}/c} \, K_{III}(t, 0) \,. \qquad (7.2.22)$$

In other words, the dynamic stress intensity factor for arbitrary motion of the crack tip is again the same universal function of crack tip speed that appeared in (7.2.20) times the time-dependent stress intensity factor that the crack tip would have if it were not moving. Solutions for more general cases of time-dependent loading and oblique stress wave loading have been given by Achenbach (1970a, 1970b).

The displacement representation theorem takes the form

$$w(x_o, y_o, t_o) = \frac{c}{2\pi\mu} \int_0^{t_o} \int_{x_o-c(t_o-t)}^{x_o+c(t_o-t)} \frac{\tau_y(x,0,t)\,dx\,dt}{\sqrt{c^2(t_o-t)^2-R^2}} \tag{7.2.23}$$

for the problem at hand, where $R^2 = (x_o - x)^2 + (y_o - y)^2$. The stress component $\tau_y(x,0,t)$ is known on the entire boundary $y = 0^+$, either through the boundary condition (7.2.2) or the result (7.2.14) for the stress distribution ahead of the crack tip, so the displacement can be evaluated according to (7.2.23) anywhere in the body. The particular case of $w(x_o, 0, t_o)$ for the pair of opposed concentrated forces is now developed in some detail for purposes of illustration.

For the situation that has been under consideration in this section, namely, crack growth due to crack face traction that is operative for $t \geq 0$, the change of integration variables (7.2.7) yields

$$w(\xi_o, 0, \eta_o) = \frac{1}{2\pi\mu} \int_0^{\eta_o} \frac{1}{\sqrt{\eta_o-\eta}} \int_{-\eta}^{\xi_o} \frac{\tau_y(\xi,0,\eta)}{\sqrt{\xi_o-\xi}}\,d\xi\,d\eta. \tag{7.2.24}$$

The point (x_o, t_o) is assumed here to be on the crack face behind the tip that is advancing according to $x = l(t)$. In view of (7.2.8), the inner integral vanishes for all $\eta \leq \eta^* = ct^* - x^*$, where the point (x^*, t^*) is defined as above. The integral is now evaluated for the case when $\tau_y(x, 0, t) = -q^*\delta(x)$ for $x < l(t)$ and for an observation point for which $0 < x_o < l(t_o)$. In this case, the integral becomes

$$w(\xi_o, 0, \eta_o) = \frac{q^*}{\pi\mu} \int_{\eta^*}^{\eta_o} \frac{1}{\sqrt{\eta_o-\eta}\sqrt{\xi_o-\xi}}\,d\eta. \tag{7.2.25}$$

This integral is readily evaluated in terms of elementary functions. The result can be expressed in terms of physical coordinates by noting that $\eta^* = \eta_o - 2(x^* - x_o)$ and $x^* = l(t^*)$, so that

$$w(\xi_o, 0, \eta_o) = \frac{q^*}{\pi\mu} \ln\left[\frac{l(t^*) - \sqrt{l(t^*)\,[l(t^*) - x_o]}}{l(t^*) + \sqrt{l(t^*)\,[l(t^*) - x_o]}}\right] \tag{7.2.26}$$

for $0 < x_o < l(t_o)$. The displacement field on the crack face behind the moving crack tip has the same property exhibited by the stress distribution ahead of the tip, that is, the expression for the displacement is the same as the corresponding equilibrium expression except that the quantity $l(t^*)$ appears in place of the crack length parameter. An immediate consequence is that, if the crack tip stops at some time t_a, then the equilibrium crack face displacement appropriate for the given crack face loading and the crack length parameter $l(t_a) = l_a$ radiates out along the crack face from the crack tip behind a point traveling with the speed c. Although the result is established for a simple case of crack face loading, the solution for any other case of crack face loading can be constructed by linear superposition over this solution, so that all solutions for time-independent loading share this remarkable property.

The features of the elastodynamic fields for growth of a semi-infinite crack in the antiplane shear mode at a nonuniform rate can be summarized as follows, whether the crack face loading is time-independent or time-dependent. The stress intensity factor has the form of a universal function of crack tip speed times the stress intensity factor that would be found for the same loading but with a stationary crack tip. Furthermore, if the crack tip stops at a certain instant of time, then the complete field for the given loading and the amount of crack growth at that instant radiates out from the crack tip behind a wavefront traveling at speed c, and it is *fully established* everywhere behind this wavefront. Finally, the results apply equally well for the case of extension of a crack of finite length or extension of a crack in a body of finite extent, but only for times up until waves reflected from the remote boundary, reflected from the other stationary crack tip or emitted from the other crack tip reach the crack tip of interest. Thus, in any practical situation, the solutions are strictly valid only for a short time, roughly twice the travel time for a wave to propagate from the crack tip to the nearest boundary or to the other crack tip, if the other tip is stationary.

An interesting application of Kostrov's solution was developed by Nilsson (1977a). He considered a mode III crack that was propagating under steady state conditions up until a certain instant of time. The class of steady state elastodynamic crack problems was considered in Section 6.2. At a certain instant of time, the motion of the crack tip was assumed to become nonuniform. Nilsson (1977a) determined the mode III stress intensity factor for the nonuniformly moving crack tip for both acceleration and deceleration from the steady state speed. The resulting stress intensity factor was shown to depend on the features of the steady state problem and on the difference in distance traveled by the crack tip between the steady state and nonuniform crack tip motions. The solution provides a framework for discussing issues connected with the arrest of a rapidly growing crack.

The results of this section were obtained by following the analysis of Kostrov (1966) for mode III crack growth based on Green's method. For the particular case of general time-independent loading, these same results were obtained by Eshelby (1969a) who followed a completely different approach. In a study of electromagnetic radiation due to moving sources in two space dimensions and time, Bateman (1955) established that if a function of position $w(x, y)$ satisfies the Laplace equation $\nabla^2 w = 0$ and is homogeneous of degree one-half in x, y, then $w(x - l(\tau), y)$ satisfies the wave equation (7.2.4) with

$$c(t - \tau) = \sqrt{[x - l(\tau)]^2 + y^2} \,. \tag{7.2.27}$$

The "time" $\tau(x, y, t)$ satisfying (7.2.27) has the following interpretation. A point moves in the x-direction so that its position at any time t is $x = l(t)$. As it moves, the point continuously radiates cylindrical wavelets which expand at speed c. The particular wavelet sensed at location x, y and time t was emitted by the moving source at time τ when the source was located at $x = l(\tau)$, $y = 0$. For any given motion $l(t)$ and any particular x, y, t, this time τ is specified by (7.2.27). By means of this result, Eshelby (1969a) showed that the essential structure of the near tip fields for the stationary mode III crack are preserved for an arbitrarily moving crack. The solution for complete crack fields for general time-independent loading was obtained by building up general crack motions from a sequence of infinitesimal constant speed segments. This constructive approach to the problem provided some motivation for pursuing the analysis of mode I crack growth at arbitrary rate presented in the following sections.

7.3 Plane strain crack growth

In Section 6.4, the extension of a tensile crack in a body subjected to general time-independent loading was considered. It was assumed that the crack tip is stationary up until the instant of onset of crack growth, and that the speed of propagation of the crack tip is constant for all later time. The process of crack extension was viewed as the negation of traction on the prospective fracture plane, and the problem was reduced to examination of a loading situation involving crack face traction but no other loading. The case of crack extension at constant rate was solved in two steps. First, the relatively simple case in which a pair of opposed concentrated normal forces of constant magnitude acting on the crack faces are left behind the crack tip as it advances at speed v through an otherwise unloaded body was studied. The forces continue to act on the same material particles as the crack tip moves away. The elastodynamic field generated in this way is called the fundamental solution of the problem. Then, the solution for general time-independent loading was obtained from the fundamental solution by means of linear superposition.

The discussion of this same physical problem is resumed here, since the features of this problem indicate a path to determining the stress intensity factor history for the same physical system when the crack tip propagates at a *nonuniform* rate. For nonuniform propagation under the assumed plane strain conditions, the crack edge is at $x = l(t)$ for $t > 0$, where $l(t)$ is a general continuous function of time. Motivated by the results of Section 7.2 on the corresponding antiplane shear problem of crack growth at a nonuniform rate, an inverse method is developed that leads to the desired stress intensity factor result. The development proceeds in several steps. First, the situation when the constant speed crack of Section 6.4 suddenly stops is examined. The solution for this case exhibits the remarkable feature that the stress intensity takes on the correct equilibrium value for the given applied loading and the new crack length immediately upon the stopping of the crack. Furthermore, the corresponding equilibrium stress distribution radiates out ahead of the crack tip behind the wavefront traveling with the shear wave speed. These features of the response provide a basis for construction of the stress intensity factor for general motion, again by linear superposition. The notation established in Section 6.4 is followed here, except that variables and functions in the fundamental solution are denoted by a superposed star $\star$ to distinguish them from corresponding variables and functions representing the same physical quantities in the present case.

7.3.1 Suddenly stopping crack

Recall the general features of the constant speed analysis in which crack growth under two-dimensional plane strain conditions is considered. A rectangular coordinate system x, y, z is introduced in the body in such a way that the crack is in the plane $y = 0$, the edge of the crack is parallel to the z-axis, and the crack edge moves in the positive x-direction. Thus, the plane of deformation is the x, y-plane. The crack tip is initially at $x = 0$ and it begins to move at constant speed v at time $t = 0$. The position of the crack tip at any time $t > 0$ is then $x = vt$.

For time $t \le 0$, the crack tip is at $x = 0$ and the material is stress free and at rest everywhere. At time $t = 0$, the crack tip begins to move in the positive x-direction at speed v. As the tip moves away from the origin of the coordinate system, it leaves behind a pair of concentrated forces, each of magnitude p^* (force per unit length in the z-direction) and tending to separate the crack faces. The crack faces are traction free at points other than $x = 0$. The field generated is the fundamental solution.

Now suppose that the crack tip advances a distance l_a at the constant speed v and then suddenly stops, that is, the speed of the crack tip changes abruptly from v to zero. The fundamental solution still solves the problem for $vt < l_a$ but some modification is required for $vt > l_a$. For $vt > l_a$ the field equations (6.4.1) remain unchanged, of course, but the boundary conditions (6.4.2) are replaced by

$$\sigma_{yy}(x, 0, t) = -p^* \delta(x) H(t) \quad \text{for} \quad -\infty < x < l_a,$$

$$\sigma_{xy}(x, 0, t) = 0 \quad \text{for} \quad -\infty < x < \infty, \qquad (7.3.1)$$

$$u_y(x, 0, t) = 0 \quad \text{for} \quad l_a < x < \infty.$$

As before, only the region $y \ge 0$ must be considered due to symmetry of the fields.

Suppose for the moment that the crack tip does not actually stop at $x = l_a$ but that it continues to advance at speed v. Furthermore, suppose that a time-independent distribution of normal traction $\sigma_{yy}(x, 0, t) = p(x)$ appears on the crack faces behind the advancing crack tip for $x > l_a$. If the solution for a stopping crack is sought, then the stress intensity factor at the moving crack tip must be zero for $x > l_a$, and the crack face normal displacement must be zero for $l_a < x < vt$. What must this distribution $p(x)$ be so that the stress intensity factor for $x > l_a$ is identically zero? This is one feature that the solution must have if it is to actually represent the field for the suddenly stopping crack. The choice of a time-independent traction distribution is motivated by the observation that the corresponding traction distribution for the mode III problem analyzed in Section 7.2 is time-independent.

The answer to the question on $p(x)$ is provided by the stress intensity factor expression (6.4.31) for general crack face loading. In the present case, the crack face traction is $-p^*\delta(x) + p(x)$ for $-\infty < x < vt$, where $p(x) = 0$ for $vt < l_a$. If this expression is substituted for $p(x)$ in (6.4.31) and the condition that the resulting stress intensity factor is identically zero is imposed, then the integral equation

$$\int_{l_a}^{vt} \frac{p(x)}{\sqrt{vt - x}}\, dx = \frac{p^*}{\sqrt{vt}}, \qquad vt > l_a, \qquad (7.3.2)$$

is the result. With a change of integration variable to $x' = x - l_a$, this equation takes the standard form of an Abel integral equation as given in (2.3.12). The unique solution based on (2.3.15) is

$$p(x) = \frac{p^*}{\pi x}\sqrt{\frac{l_a}{x - l_a}}, \qquad x > l_a. \qquad (7.3.3)$$

This distribution is recognized as the *equilibrium* normal traction distribution on $y = 0$, $x > l_a$ in a body containing a half plane crack occupying $y = 0$, $x < l_a$ with the crack faces subjected to concentrated normal forces of magnitude p^* at $x = 0$. The stress intensity factor for this equilibrium distribution is $p^*\sqrt{2/\pi l_a}$ at $x = l_a$. It is therefore established that if the traction distribution (7.3.3) acts on the crack faces behind the crack tip as it advances at constant speed then the stress intensity factor is zero for $x > l_a$. It cannot be concluded that the crack tip has actually stopped at $x = l_a$, however, because it is not yet known if the boundary condition (7.3.1)$_3$ is satisfied. The possibility that $u_y(x, 0, t) = 0$ for $x > l_a$ is examined next.

To address the matter, it is more convenient to consider the gradient of normal crack face displacement $\partial u_y/\partial x$ rather than the displacement itself. If this displacement gradient for the fundamental solution is the wave function $f(x, y, t)$ in the general superposition integral (6.4.30) and if the particular traction distribution (6.4.38) is assumed, then the complete crack face displacement gradient on

$y = 0^+$ is

$$\frac{\partial u_y}{\partial x}(x,0,t) = \frac{\partial u_y^*}{\partial x}(x,0,t) - \int_{l_a}^{vt} \frac{\partial u_y^*}{\partial x}(x - x', 0, t - x'/v)\, p(x')\, dx'.$$

(7.3.4)

From (6.4.12) the crack face displacement gradient is

$$\frac{\partial u_y^*}{\partial x}(x, 0^+, t) = \frac{1}{\pi\xi} \text{Im}\Big\{ \zeta U_-(\zeta) \Big\}_{\zeta = -t/\xi + i0}$$

(7.3.5)

for $0 < -a_+\xi < t$ or $-c_d t < x < vt$, where the function $U_-(\zeta)$ is given in (6.4.23).

It is important to recall some of the properties of the function $U_-(\zeta)$. The factor $S_-(\zeta)$ has branch points at $\zeta = a_+$ and $\zeta = b_+$, and it is single valued in the ζ-plane cut along the real axis between these points. It is analytic everywhere except at these two branch points. Furthermore, $S_-(\zeta) \to 1$ as $|\zeta| \to \infty$, so $U_-(\zeta) = O(\zeta^{3/2})$ as $|\zeta| \to \infty$. The function $U_-(\zeta)$ also has simple poles at $\zeta = c_+$ and $\zeta = h$. The relative magnitudes of the singular points are such that $a_+ < b_+ < c_+ < h$.

If the expression (7.3.5) is substituted into the integral term in (7.3.4) and the change of variable of integration $x' = v\xi(\zeta + t/\xi)$ is made, the resulting integral is

$$\frac{1}{\pi^2}\sqrt{\frac{-l_a}{v\xi^3}} \int_{a_+}^{a^*} \text{Im}\left\{ \frac{\zeta U_-(\zeta)}{(a^* - \zeta)^{1/2}(\zeta + t/\xi)} \right\} d\zeta,$$

(7.3.6)

where $a^* = -(t + l_a h)/\xi$. For $l_a < x < vt$, the upper limit of integration is in the range $h < a^* < -t/\xi$. Because $U_-(\zeta)$ has poles on the path of integration, the integral in (7.3.6) is interpreted in the principal value sense.

The integral (7.3.6) is a real integral. By viewing it as a line integral in the complex ζ-plane, however, its evaluation becomes a simple matter. With this objective in mind, the definition of $(a^* - \zeta)^{1/2}$ is extended to the complex plane and the branch with nonnegative real part everywhere in the ζ-plane is chosen. The function $\zeta U_-(\zeta)/(a^* - \zeta)^{1/2}$ is single valued in the plane cut along $\text{Im}(\zeta) = 0$,

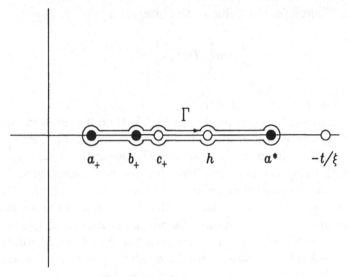

Figure 7.2. The path of integration Γ in the complex ζ-plane used to evaluate (7.3.7) encloses the branch cut running from a_+ to a^*.

$a_+ \leq \operatorname{Re}(\zeta) \leq a^*$. The integral (7.3.6) is now viewed as a complex line integral along the upper side of this cut, as shown in Figure 7.2. When evaluated at $\bar{\zeta}$, the integrand is the complex conjugate of its value at ζ, so the path of integration can be converted to the closed contour Γ embracing the branch cut, as shown in Figure 7.2. Note that each of the small semicircular arcs added to the contour near a pole of the integrand makes a contribution to the value of the integral equal in magnitude to one-half the residue of the pole. Because of the branch cut passing through the pole in each case, however, the algebraic signs of the contributions differ on opposite sides of the cut so that the net contribution is zero. The integral (7.3.6) becomes

$$\frac{1}{\pi}\sqrt{\frac{-l_a}{v\xi^3}}\,\frac{1}{2\pi i}\int_\Gamma \frac{\zeta U_-(\zeta)}{(a^* - \zeta)^{1/2}(\zeta + t/\xi)}\,d\zeta. \qquad (7.3.7)$$

The integral can be evaluated by direct application of Cauchy's theorem (1.3.4). The only singularity of the integrand exterior to the contour Γ is a simple pole at $\zeta = -t/\xi$. The value of the integrand taken along Γ is equal to the value of the residue of this pole plus the value of the integral along a circle of indefinitely large radius in the ζ-plane. The integrand is $O(\zeta^{-2})$ as $|\zeta| \to \infty$, so the latter contribution is zero. Therefore, the value of the integral is

$$\frac{1}{\pi\xi}\mathrm{Im}\left\{\zeta U_-(\zeta)\right\}_{\zeta=-t/\xi+i0} \tag{7.3.8}$$

for $0 < -a_+\xi < t$ or $-c_d t < x < vt$.

If this result is compared with (7.3.5) then it is seen immediately that the displacement gradient due to the action of the traction distribution $p(x)$ on the crack faces behind the advancing crack tip exactly cancels the displacement gradient due to the concentrated loads in the fundamental solution for x in the range $l_a < x < vt$. The total displacement is thus a constant in this interval, and the value of this constant must be zero because the normal crack face displacement is zero at the crack tip $x = vt$. Therefore, the fundamental solution has been extended to the case where the crack tip grows at constant speed v up to the point $x = l_a$ and then suddenly stops.

The foregoing analysis leads to the result that if the constant speed crack suddenly stops, then the stress intensity factor instantaneously takes on the value appropriate for the given crack length and applied loading *under equilibrium conditions*. Furthermore, the equilibrium stress distribution is fully established on the extension of the crack line ahead of the crack tip behind the shear wavefront, a feature that has been verified experimentally by Vu and Kinra (1981). This is a remarkable result for a two-dimensional elastodynamic field, and it indicates that the stress intensity factor result for general motion may have the same simple features that were observed for the case of antiplane shear crack growth in the preceding section. If this is to be the case, however, other special properties will have to be demonstrated. In particular, it must be shown that the appropriate equilibrium stress distribution radiates out along the crack line behind a front traveling with an elastic wave speed when the crack tip suddenly stops, that is, at a speed beyond any possible crack tip speed in the present analysis. This point is considered next, along with the nature of the crack face displacement that is radiated out from the crack edge upon sudden stopping of the crack tip at $x = l_a$.

Consider the displacement of the crack faces first. Suppose that the crack tip suddenly stops at $x = l_a$ and that a Rayleigh pulse is radiated out along the crack faces with speed $c_R = 1/c$. Attention is now focused on points on the crack faces between the crack tip and this Rayleigh pulse. Thus, at any time t, the value of x is restricted to be in the range $l_a - (t - l_a/v)c_R < x < l_a$. Thus, the parameter a^* introduced as the limit of integration in (7.3.6) is in the range $c_+ < a^* < h$. The steps outlined in going from (7.3.6) to (7.3.8) can again be followed in extracting the surface displacement gradient in this case. The only difference here is that the integrand has two singularities exterior to the contour Γ, namely, simple poles at $\zeta = -t/\xi$ and at $\zeta = h$. The value of $\partial u_y/\partial x$ in the range of interest is therefore the residue of the pole at $\zeta = h$, which is

$$\frac{\partial u_y}{\partial x}(x, 0^+, t) = -\frac{p^*(1 - \nu)}{\mu\pi x}\sqrt{\frac{l_a}{l_a - x}}\,, \qquad (7.3.9)$$

where ν is Poisson's ratio and μ is the shear modulus of the material.

   This displacement is the *equilibrium* displacement gradient of the crack face $y = 0^+$ of a semi-infinite crack on $y = 0$, $x \le l_a$ due to an opposed pair of concentrated loads of intensity p^* acting at $x = 0$ and tending to open the crack. Because $u_y(l_a, 0, t) = 0$, the displacement itself is the appropriate equilibrium displacement. If the same procedure is followed for the case when x is farther from the crack tip than the Rayleigh pulse that is emitted when the tip suddenly stops, then it is found that the surface displacement has a complicated transient variation. Thus, it is concluded that if the constant-speed crack propagation described by the fundamental solution suddenly stops, then the equilibrium displacement appropriate for the specified load and new crack length radiates out along the crack faces behind a front traveling with the Rayleigh wave speed of the material. Though this has been established here only for the case of the fundamental solution, the generalization to general time-independent loading conditions is straightforward and is discussed in what follows.

   Next, the distribution of normal stress σ_{yy} on the crack line ahead of the crack tip is considered for times after the crack tip suddenly stops. Based on the discussion leading to (7.3.4), the expression for this stress distribution is

$$\sigma_{yy}(x, 0, t) = \sigma_{yy}^*(x, 0, t) - \int_{l_a}^{vt} \sigma_{yy}^*(x - x', 0, t - x'/v)\, p(x')\, dx'\,,$$

$$(7.3.10)$$

where σ_{yy}^* is the stress component indicated from the fundamental solution. By means of an analysis paralleling that presented above in the discussion of crack face displacement, it can be shown that

$$\sigma_{yy}(x, 0, t) = \frac{p^*}{\pi x} \sqrt{\frac{l_a}{x - l_a}} \qquad (7.3.11)$$

for $vt < x < l_a + (t - l_a/v)c_d$. The expression (7.3.11) is recognized as the equilibrium stress distribution on the plane $y = 0$, $x > vt$ for a semi-infinite crack occupying $y = 0$, $x \le l_a$ with the crack faces subjected to an opposed pair of concentrated forces of magnitude p^* at $x = 0$. It was shown in deriving (7.3.3) that precisely the same stress distribution prevails over the region $l_a < x < vt$. Thus, the important conclusion is reached that if the constant speed crack represented by the fundamental solution suddenly stops, then the equilibrium normal stress distribution appropriate for the given loads and the new crack length radiates out from the tip along the prospective fracture plane and it is fully established behind a point traveling with the shear wave speed of the material c_s. Again, the result can be extended to any case with time-independent mode I loading. This strong result makes it possible to construct the stress intensity factor for a crack extending at a nonuniform rate.

The foregoing analysis applies only in the case of concentrated loads of magnitude p^* left behind the crack tip as it grow from $x = 0$. Suppose now that a normal traction distribution $-p(x; 0)$ acts on the crack faces behind the crack tip as it advances at constant speed from $x = 0$ to $x = vt$. This traction distribution might result from the application of general time-independent loading on the cracked body, resulting in the normal stress distribution $p(x; 0)$ ahead of the crack tip for $t < 0$. After the moving crack tip passes the point $x = l_a$, an additional traction distribution $p(x; l_a)$ is left behind the advancing crack tip on $l_a < x < vt$. If the net stress intensity factor is to be zero for $x > l_a$ or $t > l_a/v$ then a linear superposition over (7.3.3) yields the relationship

$$p(x; l_a) = p(x; 0) - \frac{1}{\pi \sqrt{x - l_a}} \int_0^{l_a} \frac{p(x; 0)\sqrt{l_a - x'}}{x' - x} \, dx' \qquad (7.3.12)$$

for $x > l_a$. The important observations made above concerning the radiation of equilibrium fields from the crack tip upon sudden stopping of the advancing crack represented by the fundamental solution can be extended to the case of general loading by means of (7.3.12). Of special interest is the stress intensity factor for the case of a crack tip that advances at constant speed under general time-independent loading and then suddenly stops. If the traction left behind the crack tip as it advances from $x = 0$ is $p(x; 0)$ and if the normal stress distribution that develops ahead of the tip following stopping is $p(x; l_a)$ then the stress intensity factor following stopping is

$$K_I(l_a, 0) = \lim_{x \to l_a^-} \sqrt{2\pi(x - l_a)}\, p(x; l_a)$$

$$= \sqrt{\frac{2}{\pi}} \int_0^{l_a} \frac{p(x'; 0)}{\sqrt{l_a - x'}}\, dx'\,.$$

(7.3.13)

This is precisely the equilibrium stress intensity factor for the given applied loads when the crack tip is at $x = l_a$.

The stress intensity factor history for the constant speed growth segment from $x = 0$ to $x = l_a$ can be summarized as follows. Initially, say at time $t = 0$, the stress intensity factor has some time-independent value $K_I(0,0)$. Then, when the crack begins to advance at speed v, negating the traction distribution $p(x; 0)$, the stress intensity factor changes discontinuously to $k(v)K_I(0,0)$ and it has the value $k(v)K_I(vt, 0)$ thereafter, as demonstrated in Section 6.4. Then, upon stopping at $x = l_a$, the crack tip stress intensity factor changes discontinuously from $k(v)K_I(l_a, 0)$ to $K_I(l_a, 0)$, and this value is maintained indefinitely. Thus, for the entire growth segment, the stress intensity factor has the form of the universal function of crack tip speed $k(v)$ times the equilibrium stress intensity factor for the given time-independent loading and the instantaneous crack length. The stress intensity factor history for growth of a crack at general nonuniform rate is next constructed as a sequence of constant speed segments.

7.3.2 Arbitrary crack tip motion

Consider once again the elastic body containing a half plane crack but that is otherwise unbounded, as described in the foregoing section. The body is under the action of time-independent loads that produce a state of mode I plane strain deformation. The coordinate system introduced earlier in this section is used once again. Initially the crack tip is at $x = 0$ on the plane $y = 0$. The normal stress distribution on

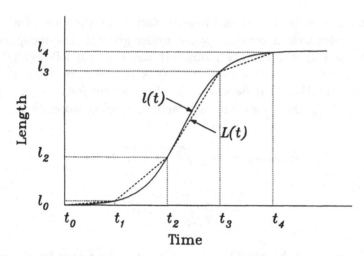

Figure 7.3. The piecewise linear curve $L(t)$ with vertices at $t_k = k\Delta t$, $k = 0, 1, 2, \ldots$, approximates the actual crack tip trajectory $l(t)$ ever more closely, in terms of both position and speed, as $\Delta t \to 0$.

the prospective fracture plane $x > 0$, $y = 0$ is given by $\sigma_{yy} = p(x; 0)$ for time $t < 0$. If the same time independent loading acts on the body but the crack tip is at $x = l$, then the normal stress distribution on the plane ahead of the tip is $p(x; l)$. At time $t = 0$ the crack begins to extend, and for $t > 0$ the crack tip position is given by $x = l(t)$. The function $l(t)$ is continuous, nondecreasing, and restricted by the condition that the crack speed is always less than the Rayleigh wave speed of the material, $0 \leq \dot{l}(t) < c_R$. Otherwise, the crack tip motion is arbitrary.

The function $l(t)$ is now approximated by a polygonal curve $L(t)$ which has its vertices lying on the original curve, as shown in Figure 7.3. Successive times $t_k = k\Delta t$, $k = 0, 1, 2, \ldots$ are marked off, and the vertices of $L(t)$ are located at $(0, 0)$, (t_1, l_1), and so on, where $l_k = l(t_k) = L(t_k)$. According to this approximation, the crack tip moves with constant speed $v_0 = (l_1 - 0)/(t_1 - 0)$ during the time interval $0 < t < t_1$, with speed $v_1 = (l_2 - l_1)/(t_2 - t_1)$, during the time interval $t_1 < t < t_2$, and so on.

If the crack tip motion is actually described by the polygonal curve, then the analysis of the preceding subsection provides the exact stress intensity factor as a functional of $L(t)$. Initially, the crack extends by negating the normal stress $p(x; 0)$, or, in other words, the crack advances under the action of crack face normal traction $-p(x; 0)$.

From (6.4.31), the stress intensity factor is then given by

$$K_I(L(t), v_0) = k(v_0)\sqrt{\frac{2}{\pi}} \int_0^{v_0 t} \frac{p(x; 0)}{\sqrt{v_0 t - x}}\, dx \qquad (7.3.14)$$

for $0 < t < t_1$.

Suppose that the crack tip suddenly stops at time $t = t_1 - 0$. According to the analysis in Section 7.3.1, the normal stress distribution $p(x; l_1)$ is radiated out along $y = 0$, $x > l_1$ behind a point traveling with the shear wave speed of the material c_s. In order to extend further the crack must negate the normal stress $p(x; l_1)$ on the prospective fracture plane. In view of the remarkable fact that this distribution is also time-independent, the original solution can be applied once again. Thus, at time $t = t_1 + 0$ the crack tip begins to move with speed v_1 and the stress intensity factor is

$$K_I(L(t), v_1) = k(v_1)\sqrt{\frac{2}{\pi}} \int_{l_1}^{L(t)} \frac{p(x; l_1)}{\sqrt{L(t) - x}}\, dx \qquad (7.3.15)$$

for $t_1 < t < t_2$.

If the relationship (7.3.12) is multiplied by $(L(t) - x)^{-1/2}$ and integrated with respect to x then it follows that

$$\int_{l_1}^{L(t)} \frac{p(x; l_1)}{\sqrt{L(t) - x}}\, dx = \int_0^{L(t)} \frac{p(x; 0)}{\sqrt{L(t) - x}}\, dx\,. \qquad (7.3.16)$$

In light of this result, (7.3.15) can also be written in the form

$$K_I(L(t), v_1) = k(v_1)K_I(L(t), 0)\,. \qquad (7.3.17)$$

This procedure can be continued indefinitely with the result that the stress intensity factor has the exact form

$$K_I(L(t), v_k) = k(v_k)K_I(L(t), 0) \qquad (7.3.18)$$

in *any* time interval $t_k < t < t_{k+1}$. It is clear that the polygonal curve $L(t)$ approaches the actual crack tip trajectory $l(t)$ as $\Delta t \to 0$. Furthermore, the approach is not only in magnitude but also in slope, that is, $\dot{L}(t)$ also approaches $\dot{l}(t)$ pointwise in time as $\Delta t \to 0$. In view of the fact that the change in the stress intensity factor at a discontinuity in crack speed is determined completely by the crack tip speed before the discontinuity, the discontinuity in speed itself, and the crack length at the instant of the jump, it is anticipated that the stress intensity factor for crack tip motion $L(t)$ approaches the stress intensity factor for crack tip motion $l(t)$ as $\Delta t \to 0$.

Thus, the conclusion is reached that the stress intensity factor for arbitrary motion of the crack tip depends on the crack tip motion only through the instantaneous amount of crack advance $l(t)$ and the instantaneous crack tip speed $\dot{l}(t)$. Furthermore, the dynamic stress intensity factor is given as a functional of crack tip motion by

$$K_I(l, \dot{l}) = k(\dot{l}) K_I(l, 0). \qquad (7.3.19)$$

In other words, the dynamic stress intensity factor for arbitrary motion of the crack tip has the form of the universal function of instantaneous crack tip speed defined in (6.4.26) and shown in Figure 6.10 times the equilibrium stress intensity factor for the specified time-independent loading and an amount of crack growth equal to the instantaneous crack growth. The result (7.3.19) implies that if the crack tip speed changes abruptly from $\dot{l}^-$ to $\dot{l}^+$ for some particular crack length l, then the ratio of the stress intensity factor immediately after the jump to the value immediately preceding the jump is independent of the applied loading and is given by

$$\frac{K_I(l, \dot{l}^+)}{K_I(l, \dot{l}^-)} = \frac{k(\dot{l}^-)}{k(\dot{l}^+)}. \qquad (7.3.20)$$

This relationship carries over to all modes of crack growth for both time-independent and time-dependent loading conditions.

In deriving the general result (7.3.19), the process of crack advance was viewed as the negation of a stress distribution on the prospective fracture path. There is another class of elastodynamic crack problems which can be analyzed by considering a displacement distribution, rather than a stress distribution, to be negated, but which leads to exactly the same result for a suddenly stopping crack. The approach is based on the use of continuously distributed moving dislocations, and illustrations were presented in Sections 3.2 and 3.3 where stress intensity factor solutions were obtained by implementing this idea. The same method can be used to determine the transient stress intensity factor history for particular cases of sudden stopping of a constant speed crack tip. This was demonstrated by Freund (1976a) for the sudden stopping of a symmetrically expanding crack, and by Nilsson (1977b) for the sudden stopping of a crack growing steadily in a uniform strip. In each case, the results provided not only the value of the stress intensity factor immediately after the stopping of the crack according to (7.3.20), but also the transient stress intensity factor for some time after stopping.

The result (7.3.19) shows that the main property noted for antiplane shear crack growth at a nonuniform rate in Section 7.2 is shared by the corresponding mode I crack growth process. The property is unusual for two reasons. The dynamically advancing crack creates a stress wave field due to a sudden release of traction that radiates out from the crack edge. The crack tip propagates through this stress wave field, presumably sensing waves that were radiated when the crack tip was at relatively long past positions. However, the crack tip field shows no sensitivity to this wave field in its dominant singular term, in the sense that the instantaneous stress intensity factor depends on the instantaneous crack tip position and speed but not on the history of crack tip motion. This is unlike the corresponding situation in dislocation dynamics when a dislocation moves rapidly and passes through the stress wave field created by its motion. In this case, there is a stronger sensitivity of the singular field to the past positions of the dislocation. A second reason for surprise at this property becomes evident when the two-dimensional process is viewed as a three-dimensional situation. If the dynamically advancing crack edge suddenly stops, a material particle near a particular point on the crack edge receives information concerning the stopping of the crack edge from ever more distant points along the edge. A gradual decay toward the long-time limiting field near the crack edge could be anticipated, and this is indeed the common situation in two-dimensional elastodynamics. In this case, however, the field near the crack edge takes on its long time value instantaneously as the crack edge stops.

With this particular solution in hand, the way is opened for obtaining a number of other results on crack tip fields for nonuniform crack tip motion. Some of these are discussed subsequently. However, the great value of such a solution is that it makes possible the development of a crack tip equation of motion and this idea is pursued after a brief summary of corresponding results for mode II crack growth.

7.3.3 In-plane shear crack growth

The analysis of growth of a mode II in-plane shear crack parallels that of mode I crack growth, and only the main results are summarized here. This analysis is based on the growth of a half plane crack in an otherwise unbounded elastic solid. The mode II crack has particular significance for the area of earthquake source modeling. In seismology, more attention has been devoted to the study of dynamic faults of limited extent than to half plane crack models. The obvious reason for this emphasis is that actual faults have finite dimensions. If primary interest is on the details of seismic radiation due to fault motion, for example, then the fault dimensions are of fundamental importance. On the other hand, if the primary interest is on the fracture process then the actual fault dimensions are of lesser importance and the semi-infinite crack models appear to be suitable. In general, for a specific characterization of fracture resistance the main influence on crack tip motion is the increased area over which the stress drop acts as the crack increases in size, and this effect can be included in the semi-infinite plane crack models. The influence on crack tip motion at some point on the fault edge due to stress waves generated at some other points on the fault edge seems to be minor by comparison, and this effect is precluded in the half plane crack models. With regard to three-dimensional shear cracks, the edge deformation can be resolved into a combination of mode II and mode III deformations. With this point of view, the results obtained from analysis of plane models have bearing on three-dimensional plane crack propagation as well.

The specific mode II crack problem considered in this section is similar to the mode I problem analyzed in detail above. At the initial time $t = 0$ the crack begins to grow in the plane $y = 0$ from some initial position $l(0) = l_0 \geq 0$, and at time t the edge has advanced to the position $x = l(t)$. Equal but opposite shear tractions corresponding to $\sigma_{xy}(x, 0, t) = -q(x)$ act on the crack faces over $0 < x < l(t)$ as shown. The traction may be viewed as the difference between a remotely applied shear stress and a frictional stress on the crack faces due to relative sliding of the faces. In this sense, $q(x)$ may be viewed as a stress drop associated with crack advance, and it acts over an ever increasing area as the crack advances.

The stress intensity factor for crack growth at the constant speed

$\dot{l} = v$ for $t > 0$ is given in Section 6.4.4. The stress intensity factor solution for growth at nonuniform speed in the range $0 < \dot{l}(t) < c_R$ was determined by Fossum and Freund (1975). The result is

$$K_{II}(l, \dot{l}) = k_{II}(\dot{l}) K_{II}(l, 0),\qquad (7.3.21)$$

where k_{II} is the universal function of crack speed defined in (6.4.42) and $K_{II}(l, 0)$ is the equilibrium stress intensity factor for the given applied loading and crack length. The equilibrium stress intensity factor for this case is given in (6.4.43). This property is identical in form to the corresponding mode I result in (7.3.19), and the discussion following that result applies in this case as well. The generalization to time-dependent loading also parallels that for mode I deformation in Section 6.5.

7.4 Crack tip equation of motion

In the study of the propagation of a crack through a solid within the context of continuum mechanics, the field equations can be solved for any motion of the crack edge, in principle. That is, if the motion of the crack edge is specified, along with the configuration of the body and the details of the loading, then the resulting mechanical fields can be determined. In the development of a theoretical model of a dynamic crack growth process, however, the motion of the crack edge should not be specified a priori but, instead, it should follow from the analysis. Unless the constitutive equation for bulk response already includes the possibility of material separation in some sense, the ingredient that must be added to the governing equations is a mathematical statement of a *crack growth criterion*. Such a criterion must be stated as a physical postulate on material behavior, separate from the kinematical theorems governing deformation and the other physical postulates governing momentum balance and material response at a point in the continuous medium. The most common form for such a criterion is the requirement that the crack must grow in such a way that some parameter defined as part of the crack edge mechanical field maintains a value that is specific to the material. This value represents the resistance of the material to advance of the crack, and it must be a material property.

In the study of crack growth processes in materials which fail in a purely brittle manner, or in which any inelastic crack tip zone is contained within a surrounding elastic crack tip zone, the most common crack growth criteria are the generalizations of Irwin's critical stress intensity factor criterion and Griffith's critical energy release rate criterion. According to the generalized Griffith criterion, for example, the crack must grow in such a way that the crack tip energy release rate is always equal to the dynamic fracture energy of the material. The energy release rate represents the effect of the applied loading, the geometrical configuration of the body, and the bulk material parameters on the material in the crack tip region for any motion of the crack tip; it is a property of the local mechanical fields. The dynamic crack propagation energy or fracture energy, on the other hand, represents the resistance of the material to crack advance; it is assumed to be a property of the material and its values can be determined only through laboratory measurement. In the simplest case, the amount of energy that must be supplied per unit crack advance and per unit length along the crack edge to sustain growth is a constant, say Γ. The parameter Γ is termed the specific fracture energy. It is a material property with physical dimensions of energy/area or force/length, and its values can be determined only by experiment.

Under the simplest circumstances of two-dimensional deformation, the local energy release rate depends on the instantaneous crack length $l(t)$, the instantaneous crack tip speed $\dot{l}(t)$, the applied loading, the geometrical configuration, and the bulk elastic moduli. Then, the statement of the growth criterion is

$$G(l, \dot{l}, t, \text{loading, configuration, moduli}, \ldots) = \Gamma, \qquad (7.4.1)$$

which is in a deceptively simple form. This relation is called an *equation of motion* for the crack tip because it has the form of an ordinary differential equation for the crack tip position $l(t)$ as a function of time, analogous to the equation of motion for a particle in elementary dynamics. Similar statements can be made concerning the generalization of Irwin's stress intensity factor criterion to the case of dynamic crack propagation. In this criterion, the material specific resistance to crack advance is called the dynamic fracture toughness. At any instantaneous crack speed v, the dynamic energy release rate and the dynamic stress intensity factor are related through (5.3.10). Consequently, if either the energy criterion or the stress intensity criterion is considered to provide an acceptable postulate for describing crack growth then either the specific fracture energy or the dynamic fracture toughness characterizes the resistance of the material to crack growth, while the other material function is derived from (5.3.10). Specific examples of the equation of motion are available for several cases of rapid crack advance, and some of these are discussed in the remainder of this section. The nature of the crack growth resistance will be assumed to be very simple in these illustrations. Qualitative observations on the dynamic fracture response of real materials will be discussed in parallel with the development of equations of motion, and again in Chapter 8.

The way in which a crack tip equation of motion is used depends on the particular approach to a crack growth problem. For example, if a problem is approached analytically as a boundary value problem modeling some process, then the objective is to determine the crack tip characterizing parameter as a functional of crack tip motion for all possible motions. The role of the growth criterion is then to select the actual crack tip motion, according to this criterion, from the class of all possible motions. On the other hand, if a crack growth problem is approached by numerical methods, then the objective is to incrementally integrate the field equations while holding the value of the crack tip characterizing parameter at its critical value. The equation of motion is not solved directly but, instead, it is applied incrementally. And if a crack growth phenomenon is approached experimentally, then the applied loads and crack motion are measured or inferred from measurements. In this case the equation of motion itself provides a means for determining the crack growth resistance from the measurements.

7.4.1 Tensile crack growth

For crack growth in the plane strain tensile opening mode, or mode I, the dynamic energy release rate G is given in terms of the dynamic stress intensity factor K_I for arbitrary nonuniform motion of the crack tip in (5.3.9). If the energy balance crack growth criterion is adopted and isothermal conditions are assumed, then $G = \Gamma$, where, as before, Γ is the specific fracture energy. The specific fracture energy may depend on crack length or crack speed, and it is viewed as a property that characterizes the resistance of the material to crack growth. Suppose that the stress intensity factor is known for arbitrary motion $l(t)$ of the crack tip. For example, it might be expressible as a function of time t, instantaneous crack length $l(t)$, and instantaneous crack tip speed $\dot{l}(t)$. The equation of motion of the crack tip according to the energy balance growth criterion is then

$$\frac{1-\nu^2}{E} A_I(\dot{l}) K_I(t,l,\dot{l})^2 = \Gamma, \qquad (7.4.2)$$

where ν and E are the elastic constants and $A_I(\dot{l})$ is a universal function of crack speed defined in (5.3.11). If Γ is a function of l and $\dot{l}$, for example, then (7.4.2) provides an ordinary differential equation for $l(t)$ that can be identified as an equation of motion for the crack tip under the stated conditions. The relationship (7.4.2) is quite general in the sense that it applies for any case for which the conditions evident in the statement are satisfied. To summarize, the relationship is based on the assumptions of nominally elastic response of a homogeneous and isotropic material, plane strain opening mode of deformation, a sharp crack tip, a region around the crack tip in which the mechanical field is dominated by a stress intensity factor field, and applicability of the energy balance idea. Validity of (7.4.2) does not hinge on the crack plane being semi-infinite or the body being otherwise unbounded; it applies without modification for a bounded body containing an edge crack or an internal crack.

To draw conclusions from (7.4.2) on the motion of a crack tip, some knowledge of the way in which K_I depends on its arguments is required. For the particular case of a semi-infinite crack growing at an arbitrary rate in an unbounded body subjected to time-independent loading, it was found in Section 7.3 that K_I depends on time t only through l and $\dot{l}$, and that K_I has the separable form of a universal function of crack tip speed $k(\dot{l})$ times an equivalent stationary crack stress intensity factor. This property is embodied in (7.3.19). It will be shown subsequently in this chapter that this special property exists for virtually any transient mode I crack growth situation. That is, any dynamic stress intensity factor for general crack face loading which results in mode I deformation has the form

$$K_I(t,l,\dot{l}) = k(\dot{l}) K_I(t,l,0), \qquad (7.4.3)$$

where $K_I(t,l,0)$ is the time-independent stress intensity factor that would have resulted from the applied loading if the crack tip had always been at its instantaneous position represented by l. Virtually any crack growth problem, including one with transient boundary tractions, body forces, and a complex geometrical configuration, can be reduced through superposition to a situation involving general crack face tractions on a semi-infinite crack in an otherwise unbounded body. Consequently, the equation is relatively unrestricted within the class of mode I crack growth problems.

In light of (7.4.3), the crack tip equation of motion (7.4.2) can be rewritten as

$$\frac{E\Gamma}{(1-\nu^2)K_I(t,l,0)^2} = A_I(\dot{l})k(\dot{l})^2 . \qquad (7.4.4)$$

The effects of the bulk properties of the material are included in the elastic constants, the effects of loading and geometrical configuration are represented by $K_I(t,l,0)$, and the resistance of the material to crack growth is represented by Γ. The equation of motion requires that the crack must move in such a way that the combination of these effects on the left side of (7.4.4) always equals the universal function of crack speed

$$g(\dot{l}) = A_I(\dot{l})k(\dot{l})^2 \approx 1 - \dot{l}/c_R \qquad (7.4.5)$$

on the right side, where c_R is the Rayleigh wave speed. A graph of $g(\dot{l})$ versus $\dot{l}/c_R$ is shown in Figure 6.9 for $\nu = 0.3$. It is evident that the linear approximation to g in (7.4.5) is very accurate over the whole range of crack speed $0 < \dot{l} < c_R$. This approximation is used whenever the function $g(\dot{l})$ is required in subsequent analysis. Of course, if the left side of (7.4.4) is greater than unity then crack growth will not occur.

Some general conclusions can be drawn from the equation of motion in the form (7.4.4). For example, consider the case in which Γ is a constant and crack growth is unstable in the equilibrium sense, that is, $K_I(t,l,0)$ is an increasing function of l. These are features of the original crack tip equation of motion model developed by Mott (1948). If the loading is increased to a level sufficiently great to initiate crack growth, then the left side of (7.4.4) decreases as l increases and, according to the equation of motion, $g(\dot{l})$ must also decrease. As the crack length becomes large, the crack tip speed approaches the Rayleigh wave speed asymptotically. Because the analysis leading to the result that c_R is the upper limit on crack speed is exact, it must be concluded that inertial effects are not responsible for the fact that the greatest attainable crack speed for any known material appears to be considerably less than c_R. Thus, although Mott's (1948) qualitative description of crack tip motion on the basis of an energy criterion has been important to the development of the subject, the approximations invoked are valid only for very small crack speed and the attempts to quantify this description (for example, Roberts and Wells 1954) are inconsistent with the above exact results. The agreement between predicted terminal velocities much less than c_R and the experimental results is fortuitous. The fact that crack speeds are usually substantially less than c_R must be attributed to influences that are not taken into account in the model at this level. This topic will be discussed further in Chapter 8.

As a second general conclusion, consider the form of the equation of motion (7.4.4) when Γ depends on crack speed $\dot{l}$, and possibly on crack length l. It is noteworthy that the ordinary differential equation for $l(t)$ is a first-order equation, that is, it does not involve second and higher derivatives of $l(t)$. If the analogy between the moving crack tip and an elementary mass particle is recalled, this observation implies that the crack tip moves like a massless particle. In other words, the crack tip speed changes in phase with the applied driving force, whereas for a mass particle the acceleration changes in phase with the driving force. This does not imply that inertia of the material is not important in dynamic crack propagation; material inertia can have a strong influence on momentum transfer in the material. Instead, it is an observation on the relationship between the crack driving force and the crack tip motion. This observation is perhaps relevant in considering phenomena such as crack arrest, crack propagation through periodic media or other materials with nonuniform fracture resistance such as weldments, or in situations where the crack direction changes abruptly such as in bifurcation. The feature of no apparent mass of a crack tip moving rapidly through a brittle material was anticipated in the observations made by Kuppers (1967) on crack growth in glass using the ultrasonic wave fractographic technique developed by Kerkhof (1970, 1973). By using modulating ultrasonic waves of two different frequencies during crack growth in a glass plate subjected to a large quasi-static load, he was able to modify the local stress amplitude with the lower frequency wave and to discern the response to the stress amplitude change by means of the higher frequency wave. He found that the crack speed changed in phase with the local driving force.

A third general observation on (7.4.4) concerns the separation of a cohesionless interface. Suppose that flat surfaces of identical elastic bodies are pressed against each other and then, by means of some local action, the interface is forced to separate very rapidly under plane strain conditions. If the leading edge of the separation region is regarded as a crack edge with zero specific fracture energy, then substitution of $\Gamma = 0$ into (7.4.4) implies that $\dot{l} = c_R$ for the edge. In other words, interface separation with no energy absorption occurs at the Rayleigh wave speed of the material. This idea was discussed by Freund (1974b) in an analysis of the problem described in Section 3.2, and it was incorporated into a very sophisticated analysis of diffraction of a stress pulse by a crack by Papadopoulos (1963), who took into account the possibility that the crack faces could be in contact over portions of the boundary.

As a specific illustration of the equation of motion (7.4.4), consider the following plane strain situation. A half plane crack exists in an otherwise unbounded isotropic elastic solid on the half plane $y = 0$, $x \leq l_0$. The crack faces are subjected to an opposed pair of concentrated forces of magnitude p^* at $x = 0$ that tend to separate the crack faces, and no other loads are applied to the body. The specific fracture energy is assumed to be a constant Γ_c, but the initial crack tip is slightly blunted so that an energy release rate of $n\Gamma_c$ is required to initiate crack growth, where $n > 1$. The parameter n used in this way may be called a bluntness parameter. The force magnitude p^* is gradually increased until the crack tip energy release rate is $n\Gamma_c$, and the crack then grows with $G = \Gamma_c$ and with no further increase in p^*. The total distance from the crack tip to the load point is denoted by $l(t)$.

The equilibrium stress intensity factor for this loading case is $p^*\sqrt{2/\pi l}$. At the onset of crack growth

$$G = \frac{1 - \nu^2}{E} K_I^2 = \frac{1 - \nu^2}{E} \frac{2p^{*2}}{\pi l_0} = n\Gamma_c, \qquad (7.4.6)$$

hence

$$p^* = \sqrt{\frac{n\Gamma_c E \pi l_0}{2(1 - \nu^2)}} \qquad (7.4.7)$$

during growth. The equation of motion (7.4.4) during growth is

$$l/nl_0 \approx 1 - \dot{l}/c_R, \qquad (7.4.8)$$

subject to the initial condition that $l(0) = l_0$. The initial speed of the crack tip is therefore $c_R(n-1)/n$. As the crack grows, it continuously decelerates from this speed, and it comes to rest when $l = nl_0$. This is also the equilibrium length for a sharp crack subjected to the given loading. The complete crack tip trajectory, that is, the integral of (7.4.8), is

$$l(t) = l_0 \left[n - (n-1) \exp(-c_R t / n l_0) \right] . \qquad (7.4.9)$$

Note that $n l_0 / c_R$ is the characteristic time for the transient response.

As a second illustration of the equation of motion (7.4.4), consider again a crack occupying the half plane $y = 0$, $x \leq l_0$ in an otherwise unbounded elastic solid. The crack faces are subjected to pressure of magnitude σ^* over $0 \leq x \leq l_0$. At time $t = 0$, the crack begins to grow and the crack tip position is $x = l(t)$ for $t > 0$. The pressure σ^* acts over $0 \leq x \leq l(t)$ for $t > 0$ so that the equilibrium stress intensity factor due to the applied loading increases with crack length. This equilibrium stress intensity factor for any amount of crack growth is given in (6.4.40) as $K_I(l, 0) = 2\sigma^* \sqrt{2l/\pi}$. Suppose that the fracture resistance varies with crack speed according to

$$\Gamma = \Gamma_c f(\dot{l}), \quad f(\dot{l}) = 1 + \beta \dot{l}/c_R , \qquad (7.4.10)$$

where β is a constant. The fracture resistance of the material increases (decreases) with speed if $\beta > 0$ ($\beta < 0$), and $\beta = 0$ is the case considered in the preceding illustration. In the present case, the equation of motion of the crack tip is

$$\frac{l_c}{l} \left[1 + \beta \frac{\dot{l}}{c_R} \right] = 1 - \frac{\dot{l}}{c_R} , \qquad (7.4.11)$$

where $l_c = \pi E \Gamma_c / 8 \sigma^{*2} (1 - \nu^2)$ and the approximation (7.4.5) has been used. It is evident from (7.4.11) that crack growth will begin only if $l_0 > l_c$ and the initial speed will be $\dot{l} = c_R (l_0 - l_c)/(l_0 + \beta l_c)$. The integration of the differential equation subject to the initial condition $l(0) = l_0$ yields

$$\frac{c_R t}{l_c} = (1 + \beta) \ln \frac{l - l_c}{l_0 - l_c} + \frac{l - l_0}{l_c} . \qquad (7.4.12)$$

The crack speed approaches c_R asymptotically as $t \to \infty$ for any β, but the rate of approach increases as β decreases. The acceleration

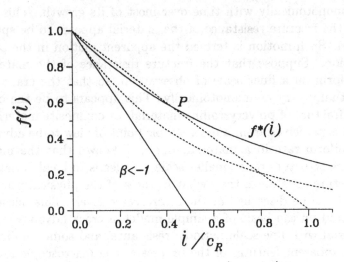

Figure 7.4. Schematic diagram of the functions $g(\dot{l})$ and $f(\dot{l})$ for $\beta < -1$, plus a smooth function $f^*(\dot{l})$ that is assumed to approach $f(\dot{l})$. For any particular $f^*(\dot{l})$ the equation of motion can be solved only for speeds $\dot{l} > \dot{l}_P$.

to speed c_R is instantaneous for $\beta = -1$, and it appears that no physically acceptable solution to the differential equation exists for $\beta < -1$.

This peculiar result on the nature of the solution for $\beta < -1$ can be resolved in the following way. Note that $f(\dot{l})$ in (7.4.10) is negative for $-c_R/\beta < \dot{l} < c_R$. Suppose that the definition of $f(\dot{l})$ is extended for $\beta < -1$ so that it is given by (7.4.10) for $0 < \dot{l} < -c_R/\beta$ and that $f(\dot{l}) = 0$ for $-c_R/\beta < \dot{l} < c_R$. Now, consider the case when $\Gamma = \Gamma_c f^*(\dot{l})$, where f^* is the smooth function shown schematically as the solid curve in Figure 7.4. For this case, the equation of motion is satisfied by having the crack begin to grow with initial speed $\dot{l}_P$ defined by the condition $g(\dot{l}_P)/f^*(\dot{l}_P) = 1$. The crack tip then accelerates from this speed to c_R. As the function f^* becomes closer to f, as suggested by the dashed curve in Figure 7.4, the speed $\dot{l}_P$ approaches c_R. Thus, it appears that the case with $\beta < -1$ is indistinguishable from the case of a cohesionless interface discussed above.

7.4.2 Fine-scale periodic fracture resistance

In the analysis of dynamic fracture phenomena or in the interpretation of experiments, it is commonly assumed that the crack tip speed varies monotonically with time over most of its growth. This is the case if the fracture resistance of the material appears to be spatially uniform. Such motion is termed the apparent motion in the present discussion. Suppose that the fracture resistance of the material is nonuniform on a finer scale of observation, so that the crack speed will actually vary nonmonotonically. This appears to be the case in brittle fracture of polycrystalline materials or composite materials. If a crack is perceived from a macroscopic point of view to be advancing at a uniform rate in a material, but it is known that the material has some structure on a smaller scale that leads to local variation of the fracture resistance, then what features of the physical process of fracture can be discerned at the macroscopic level? This question is considered on the basis of a simple model of crack advance through a material with fine-scale periodic resistance, and some implications for the apparent features of the process at the macroscopic level are examined (Freund 1987b).

Consider rapid growth of a planar crack through a nominally elastic isotropic material under plane strain conditions. The tensile opening mode of deformation, or mode I, is assumed. The elastic modulus and Poisson's ratio of the material are E and ν, respectively. Attention is focused on points close to the crack tip compared to the in-plane dimensions of the body in which the crack grows, that is, the case of a semi-infinite crack in an otherwise unbounded body is considered. At any instant, the mechanical field surrounding the crack tip is characterized by the stress intensity factor and the speed of the crack tip. The instantaneous crack length measured with respect to some arbitrary reference point is denoted by $l(t)$ and the instantaneous crack tip speed is $\dot{l}(t)$. It is assumed that the crack grows according to an overall energy balance criterion. Thus, the crack tip equation of motion (7.4.4) based on this criterion applies.

On a macroscopic scale of observation, the material appears to have a constant spatially homogeneous fracture resistance Γ_o. When observed on a fine scale, however, the material actually has a periodic (not necessarily sinusoidal) resistance to crack growth, say

$$\Gamma = \Gamma_m \gamma(l), \tag{7.4.13}$$

where Γ_m is a constant and $\gamma(l)$ is a periodic function of crack tip position with spatial period λ, defined for all l. Suppose $\gamma(l)$ has the properties that

$$\max_{0<l\le\lambda} \gamma(l) = 1, \qquad \min_{0<l\le\lambda} \gamma(l) \ge 0 . \tag{7.4.14}$$

Thus, the constant Γ_m is the maximum value of the nonnegative periodic specific fracture energy Γ.

The *actual crack speed* as observed at the finer level of observation is determined by examining the crack tip equation of motion (7.4.4) for the periodic specific fracture energy (7.4.13),

$$\frac{\dot{l}}{c_R} = 1 - \frac{E}{1 - \nu^2}\frac{\Gamma_m}{K_o^2}\gamma(l) . \tag{7.4.15}$$

The applied stress intensity factor $K_I(l, 0) = K_o$ is assumed to have little variation over distances equal to many times λ or during times equal to many times t_λ, so it is taken to be a constant for purposes of this discussion. The quantity t_λ is the time required for the crack tip to travel a distance equal to one wave length λ, and its dependence on loading level is obtained by integration as

$$c_R t_\lambda = \int_0^\lambda \frac{dl}{1 - B\gamma(l)} , \tag{7.4.16}$$

where B is the dimensionless combination of parameters $E\Gamma_m/(1 - \nu^2)K_o^2$ which has the range $0 < B < 1$. The extreme values of $B \to 1^-$ and $B \to 0^+$ correspond, respectively, to the applied stress intensity factor at a level just large enough to push the crack tip past the peaks in the fracture resistance and at a level many times greater than this minimum level.

Consider now the same process at a macroscopic level at which the crack tip appears to move along steadily at constant speed, say v_o, under the action of the uniform applied stress intensity factor K_o. This crack speed must be the average speed in the periodic fine-scale fracture resistance, that is,

$$v_o(B) = \frac{\lambda}{t_\lambda} = c_R \left[\int_0^1 \frac{d\xi}{1 - B\gamma(\lambda\xi)}\right]^{-1} . \tag{7.4.17}$$

If the crack tip equation of motion is applied at the macroscopic level, then

$$\frac{E}{1-\nu^2}\frac{\Gamma_o}{K_o^2} = 1 - \frac{v_o(B)}{c_R}, \tag{7.4.18}$$

where Γ_o is the macroscopically uniform specific fracture energy. The parameters on the left side of (7.4.18) are conveniently expressed in terms of B, so the ratio of the apparent macroscopic fracture energy to the maximum in the fine scale periodic variation is

$$\frac{\Gamma_o}{\Gamma_m} = \frac{1 - v_o(B)/c_R}{B} = \frac{1 - v_o/c_R}{B(v_o)}. \tag{7.4.19}$$

This is the main result for the apparent fracture energy in terms of the details of the periodic variation of the fine-scale resistance and the applied load level. It is written in two ways to make clear that the ratio can be expressed in terms of the load parameter B or the average crack speed v_o, these two quantities being related through (7.4.17). A relationship similar to (7.4.53) was derived by Dahlberg and Nilsson (1977) in considering fluctuations in velocity measurements for rapid crack propagation in metals.

The limiting values of the ratio (7.4.19) for very slow and very fast crack growth are easily deduced for arbitrary $\gamma(l)$. Recall that for very slow crack growth, $B \to 1^-$. In view of the property (7.4.14)$_1$ it is clear that $t_\lambda \to \infty$ as $B \to 1^-$. This, in turn, implies that $v_o(B) \to 0$, hence

$$\Gamma_o/\Gamma_m \to 1 \qquad \text{as} \qquad v_o/c_R \to 0 \qquad \text{or} \qquad B \to 1. \tag{7.4.20}$$

On the other hand, for severely overdriven crack growth, $B \ll 1$. Thus, to first order in small values of B,

$$c_R t_\lambda \approx \int_0^\lambda \{1 + B\gamma(l)\}\,dl = \lambda(1 + B\overline{\gamma}), \tag{7.4.21}$$

where $\overline{\gamma} < 1$ is the average value of $\gamma(l)$ over one wave length λ. The uniform macroscopic speed is then approximately $(1 - B\overline{\gamma})c_R$, so

$$\Gamma_o/\Gamma_m \to \overline{\gamma} \qquad \text{as} \qquad v_o/c_R \to 1 \qquad \text{or} \qquad B \to 0. \tag{7.4.22}$$

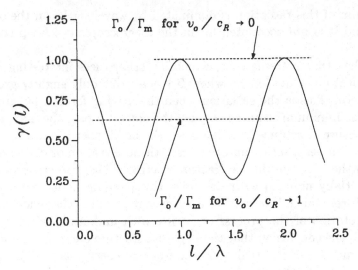

Figure 7.5. Periodic variation of the specific fracture energy of a material with position along the fracture path. The horizontal dashed lines represent the apparent fracture energy for the two extreme cases of very slow and very fast crack growth represented by (7.4.20) and (7.4.22), respectively.

These extreme cases (7.4.20) and (7.4.22) can be interpreted as follows. Consider first the case when B is only slightly less than unity or $v_o/c_R \ll 1$. Suppose that the variation in fracture energy has the qualitative features shown in Figure 7.5. The crack tip driving force is barely large enough to push the crack tip past the maxima in the fracture energy and, according to the equation of motion, the crack tip moves slowly as it passes a maximum in fracture energy. Between the maxima, the crack tip rapidly accelerates to a relatively high speed and then decelerates as it approaches the next maximum in $\gamma(l)$. Of the total time required for the crack tip to traverse a wavelength λ in fracture energy variation, the crack tip spends most of its time in regions where the fracture energy is Γ_m. If this result is now interpreted in terms of a uniform crack tip speed v_o, then the apparent time rate of energy flux into the crack tip is $\Gamma_m v_o$ for most of the time. This is the conclusion reached in (7.4.20). Of course, the actual energy absorbed per wavelength λ in the fracture process differs from the apparent energy absorbed by the amount represented as a shaded area in Figure 7.5. This amount of energy, which is included in the macroscopic energy flow into the crack tip region, is radiated outward from the crack tip as it accelerates and decelerates. The spectrum of this radiation is dominated by wavelengths on the order of λ and it is *not* accounted for in the macroscopic crack tip energy flux.

The other extreme case, which is perhaps less interesting from a physical point of view, is when B has a value only slightly greater than zero. This is the situation when the driving force is far greater than the minimum required to push the crack tip past the maxima in the fracture resistance. In this case, the crack speed is always near the ideal upper limit c_R and there is little acceleration or deceleration due to the variation in the fracture energy. Thus, the crack motion is essentially uniform with the crack tip spending about equal time at each resistance level within a wavelength. Thus, the macroscopic interpretation of the process occurring with uniform crack speed v_o leads to the conclusion that the apparent fracture energy is the wavelength average of fine-scale periodic fracture energy, as in (7.4.22).

In general, the difference between the apparent rate of energy flow into the crack tip and the actual energy consumed in the fracture process is radiated as high frequency wave motion. The amount of energy radiated per unit crack advance is

$$E_R = \frac{1}{\lambda} \int_0^\lambda \{\Gamma_o - \Gamma_m \gamma(l)\}\, dl\,, \qquad (7.4.23)$$

which is the difference in area under the graphs of the apparent fracture energy and the actual fracture energy versus distance.

Although the main results are evident in (7.4.20) and (7.4.22) for arbitrary periodic variation of the specific fracture energy, it is instructive to consider some special cases of $\gamma(l)$. For example, suppose that $\gamma(l) = \cos^2(\pi l/\lambda)$. This variation obviously satisfies the conditions in (7.4.14). The integral in (7.4.17) can be evaluated with the result that

$$\frac{\Gamma_o}{\Gamma_m} = \frac{1 - \sqrt{1 - B}}{B} = \frac{1}{1 + v_o/c_R}\,. \qquad (7.4.24)$$

It is evident that this special case has the general properties in (7.4.20) and (7.4.22). A second instructive example is the piecewise constant function $\gamma(l) = 1$ for $0 < l < n\lambda$ and $\gamma(l) = q$ for $n\lambda < l < \lambda$, where $0 < n, q < 1$. Evaluation of the integral in (7.4.17) for the particular case with $n = q = \frac{1}{2}$ yields

$$\frac{\Gamma_o}{\Gamma_m} = \frac{6(v_o/c_R) + 2\sqrt{4 - 4(v_o/c_R) + 9(v_o/c_R)^2}}{9(v_o/c_R) - 2 + 3\sqrt{4 - 4(v_o/c_R) + 9(v_o/c_R)^2}} \, . \qquad (7.4.25)$$

Again, the general features already discussed are evident in the result. A piecewise constant fracture resistance was used by Das and Aki (1977b) in their study of strength inhomogeneities on a crustal fault in the earth during seismic slip. In such an application, it is anticipated that three-dimensional aspects of the fracture resistance would significantly influence the results.

It is noted that the general features of the analysis here would be the same if a stress intensity factor criterion for crack growth had been adopted, instead of the energy criterion. One issue raised by this simple calculation is whether or not the macroscopic fracture energy of a material perceived through brittle crack propagation experiments has a direct relationship with the true surface energy of the material. If the fracture energy of the material is inhomogeneous on a fine scale, then the answer would appear to be negative. Instead, the macroscopically perceived fracture energy represents a maximum, rather than an average, of the fine-scale fracture resistance. That is, if the fracture resistance Γ has a periodic variation in the material, then the *maximum* of this variation governs crack growth behavior for speeds near $v = 0$ but the *average* of the variation governs behavior for crack speeds v approaching the Rayleigh wave speed of the material c_R.

7.4.3 Propagation and arrest of a mode II crack

In this section, the results of Section 7.3.3 on mode II crack growth at nonuniform speed are coupled with a growth criterion to consider propagation and arrest of a shear crack due to a spatially varying stress drop, defined by $q(x)$ in Section 7.3.3, and varying fracture resistance in the growth condition (Husseini et al. 1975; Freund 1979a). The discussion of the growth of a tensile mode I crack in Sections 7.4.1 and 7.4.2 was based on the assumption that crack growth is governed by an energy growth condition, which was stated symbolically in (7.4.1). It was noted in the introduction to Section 7.4 that similar results are obtained if a critical stress intensity factor growth condition is assumed, and the discussion of this section is based on the latter postulate. Furthermore, in discussing mode I crack growth, it was assumed that the specific fracture energy was either constant or that it varied with crack tip speed. In this case, the parameter characterizing fracture resistance of the material is not speed dependent. Instead, it is assumed to depend on *position* along the fracture path. With a view toward modeling the earthquake faulting process, this is one way to account for variability of slip resistance along the fault surface as the edge of a shear crack advances. Thus, it is assumed that the crack tip equation of motion follows from the condition that

$$K_{II}(l, \dot{l}) = k_{II}(\dot{l})K_{II}(l, 0) = C(l), \qquad (7.4.26)$$

where $C(l)$ is a function which represents the resistance of the material to mode II fracture at the position $x = l$ on the fracture plane (refer to Figure 7.6).

Under the conditions described, a running fracture can be arrested because either the driving force is reduced to a subcritical level or the fracture resistance is increased to a supercritical level as the crack advances. These two cases are considered qualitatively for the equation of motion based on the critical stress intensity factor criterion. For mode II crack growth, this equation of motion is

$$k_{II}(\dot{l})\sqrt{\frac{2}{\pi}} \int_0^{l(t)} \frac{q(x)}{\sqrt{l-x}}\, dx = C(l), \qquad (7.4.27)$$

subject to the initial condition $l(0) = l_0$. It is assumed that the magnitudes of the various parameters involved are appropriate for fracture initiation at crack position $x = l_0$.

Consider first the case in which the critical stress intensity factor is a constant, independent of position along the fracture path, but in which the applied stress distribution $q(x)$ (the stress drop) decreases to negative values on the crack faces as the crack advances. This situation is represented schematically in Figure 7.6a. The linear decrease in $q(x)$ implies that the integral in (7.4.27) will first increase with l to a local maximum, and it will then decrease. The equation of motion requires that the product of $k_{II}(\dot{l})$ and $K_{II}(l, 0)$ is a constant. Thus, so long as $K_{II}(l, 0)$ increases, $k_{II}(\dot{l})$ must decrease and the crack tip accelerates to ever larger crack speeds, as shown in Figure 7.6a. Likewise, as $K_{II}(l, 0)$ decreases beyond $x = l_c$, the crack tip decelerates.

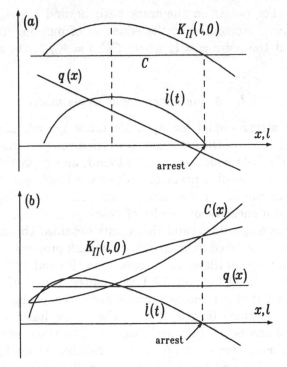

Figure 7.6. A representation of the variation along the fracture path of the physical quantities appearing in the crack tip equation of motion (7.4.27) for a mode II crack. (a) The critical stress intensity factor is constant C, but the crack face stress distribution $q(x)$ decreases with distance along the fracture path. (b) The critical stress intensity factor increases along the fracture path, but the crack face stress distribution is uniform.

Growth is arrested at $x = l_a$, where the equilibrium stress intensity factor has been reduced to the level satisfying $K_{II}(l_a, 0) = C$.

A similar illustration of the crack tip equation of motion is represented schematically in Figure 7.6b for the case when the stress drop $q(x) = q_o$ is uniform over the faces of the expanding crack but the crack tip encounters ever increasing fracture resistance $C(l)$ as it advances. Under these conditions, the equilibrium stress intensity factor $K_{II}(l, 0)$ increases parabolically with increasing crack length. With reference to the equation of motion (7.4.27), the factor $k_{II}(\dot{l})$ will decrease and the crack will accelerate so long as the ratio $C(l)/K_{II}(l, 0)$ decreases. Thus, the crack tip will accelerate to the position $x = l_c$ on the fracture plane at which $C'(l_c)/C(l_c) = K'_{II}(l_c, 0)/K_{II}(l_c, 0)$, where the prime denotes the derivative with respect to the crack tip position coordinate. For points on the crack path beyond $x = l_c$, the crack tip decelerates. According to the equation of motion, crack growth is arrested at the point $x = l_a$ where $C(l_a) = K_{II}(l_a, 0)$, as shown in Figure 7.6b.

7.4.4 A one-dimensional string model

The simple string model for dynamic crack growth introduced in Section 5.5.1 is one of the few cases in which a complete mathematical solution of the field equations can be found, and it provides the rare opportunity to see all aspects of a dynamic crack growth event in a simple framework. It was noted there that the string model is equivalent to a shear beam model of crack propagation in a double cantilever beam specimen, and the results obtained through analysis of the string model shed light on observed crack propagation phenomena in the double cantilever beam specimen (Freund 1977, 1979b).

The notation of Section 5.5.1 is adopted here without modification. Recall that a stretched elastic string lies along the positive x-axis, that is, along $0 < x < \infty$. The string has mass per unit length ρ and characteristic wave speed c. The transverse deflection is $w(x,t)$, the elastic strain from the undeflected configuration is $\gamma(x,t) = \partial w / \partial x$, and the transverse particle velocity is $\eta(x,t) = \partial w / \partial t$. Initially, the string is free to move in the interval $0 < x < l_0$ and it is bonded to a rigid, flat surface for $x > l_0$. A particular boundary condition on $w(0,t)$ is introduced here, namely, that

$$w(0, t) = w_0, \qquad (7.4.28)$$

where w_0 is a constant. With this boundary condition, there is no energy exchange between the deformable body and its surroundings. Consistent with this boundary condition of fixed end deflection, the initial deflected shape is $w(x, 0) = w_0(1 - x/l_0)$. If the string is initially at rest, then the initial conditions in terms of strain and particle velocity are $\eta(x, 0) = 0$ and $\gamma(x, 0) = \gamma_0 = -w_0/l_0$. At time $t = 0$, the string begins to peel away from the surface, so at some later time $t > 0$ the free length is $0 < x < l(t)$. The field equations and the condition to be satisfied at the moving crack tip are given as (5.5.1) and (5.5.2), and the crack tip energy release rate is found in (5.5.5) to be

$$G = \tfrac{1}{2}\rho c^2 (1 - \dot{l}^2/c^2)\gamma(l^-, t)^2. \qquad (7.4.29)$$

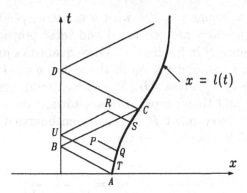

Figure 7.7. The x, t-plane for the string problem showing the path of the crack tip $x = l(t)$. At any instant of time, the string is free over $0 < x < l(t)$ and it is bonded to a rigid foundation for $x \geq l(t)$.

If a symmetrical structure, such as a double cantilever beam, is considered then both halves would contribute the same energy supply and the factor of one-half in the energy release rate expression would be absent. With reference to the foregoing discussion concerning the ingredients essential to deriving an equation of motion for the crack tip, this energy release rate expression provides a crack tip characterizing parameter as a functional of crack motion for all possible motions provided that $\gamma(l^-, t)$ can be found for arbitrary motion of the crack tip from the field equations. Thus, $\gamma(l^-, t)$ is determined next.

The governing partial differential equations (5.5.1) form a first-order hyperbolic system of equations. The characteristic curves are straight lines in the x, t-plane with slope c^{-1} or $-c^{-1}$, and the characteristic relations are

$$d\eta \mp c \, d\gamma = 0 \quad \text{along} \quad dx/dt = \pm c. \tag{7.4.30}$$

Local momentum balance requires that discontinuities in strain and particle velocity, denoted by $[\![\gamma]\!]$ and $[\![\eta]\!]$, satisfy

$$[\![\eta]\!] \pm c[\![\gamma]\!] = 0 \quad \text{along} \quad dx/dt = \pm c. \tag{7.4.31}$$

A sketch of the x, t-plane is shown in Figure 7.7 and values of strain and particle velocity can be determined at any time and any point on the string by a straightforward application of the characteristic relations and jump conditions for arbitrary crack tip motion with speed in the range $0 \leq \dot{l} < c$. Only certain pertinent results are included here.

If the crack begins to grow with a nonzero speed $\dot{l}_A$, then discontinuities in η and γ are generated and these propagate along the characteristic line AB in Figure 7.7. These quantities jump from their initial values of $\eta = 0$ and $\gamma = \gamma_0$ to the values $\eta = -\dot{l}_A \gamma_0/[1 + \dot{l}_A/c]$ and $\gamma = \gamma_0/[1 + \dot{l}_A/c]$ across AB. The discontinuities reflect from the fixed end at B and then propagate back toward the crack tip along BC. At an arbitrary point P in the region bounded by ABC, the strain and particle velocity have the values

$$\eta^P = -\frac{\dot{l}_Q \gamma_0}{[1 + \dot{l}_Q/c]}, \qquad \gamma^P = \frac{\gamma_0}{[1 + \dot{l}_Q/c]}, \qquad (7.4.32)$$

where the subscripts indicate the points in the x, t-plane, according to Figure 7.7, at which the various quantities are evaluated and PQ is a segment of a characteristic line.

The discontinuities propagating along BC reflect from the moving crack tip at C and propagate back toward the fixed end along CD. If R is any point in the region BCD then

$$\eta^R = \frac{\gamma_0[\dot{l}_T - \dot{l}_S]}{[1 + \dot{l}_T/c][1 + \dot{l}_S/c]}, \qquad \gamma^R = \frac{\gamma_0[1 - \dot{l}_S \dot{l}_T/c^2]}{[1 + \dot{l}_T/c][1 + \dot{l}_S/c]}, \qquad (7.4.33)$$

where UR, RS and TU are segments of characteristic lines. The solution procedure can be continued to indefinitely large times, but the results obtained so far are sufficient for present purposes. Next, an equation of motion for $l(t)$ is derived.

The crack propagation criterion that is adopted here is the energy criterion according to which the crack must propagate in such a way that the energy release rate G is always equal to a material specific fracture energy Γ. At first, this fracture energy is assumed to be constant, but the possibility that it depends on crack tip speed is considered subsequently. Furthermore, suppose that the crack tip is initially blunt, that is, that the quasi-static energy release rate required to cause the onset of growth, which is $G_0 = c^2 \rho \gamma_0^2/2$, exceeds the level of G required to sustain growth of the sharp crack, which is Γ. The relationship between the initiation level and the propagation level of the fracture energy is conveniently expressed through the value of the bluntness parameter $n = G_0/\Gamma > 1$. Given a value of n, the relation $n = G_0/\Gamma$ actually specifies the end displacement w_0 which must be imposed to initiate the process at $t = 0$. For subsequent time $t > 0$ the crack must propagate with $G = \Gamma$, or else $\dot{l} = 0$ with $G < \Gamma$.

By letting the observation point P in (7.4.32) approach the crack edge, it is evident that the crack tip strain appearing in the energy release rate expression (7.4.29) is given by

$$\gamma(l^-, t) = \frac{\gamma_0}{[1 + \dot{l}(t)/c]} \qquad (7.4.34)$$

at any point in the segment AC in Figure 7.7. The energy growth criterion then implies that

$$\frac{\rho c^2 \gamma_0^2 (1 - \dot{l}^2/c^2)}{2(1 + \dot{l}/c)^2} = \Gamma \qquad (7.4.35)$$

for all time until the reflected wave overtakes the crack tip. If the parameter γ_0 is eliminated in favor of the equivalent parameter n, then the simple result that

$$\frac{\dot{l}}{c} = \frac{n-1}{n+1} \qquad (7.4.36)$$

follows immediately. Thus, the energy criterion implies that the crack speed is constant in the interval AC and its value depends only on the initial bluntness parameter. It is noted in passing that if a growth criterion based on a critical value of the force Q in (5.5.6) were used instead of the energy criterion, then the crack tip speed would also be found to be constant. The force criterion is the counterpart of the stress intensity factor criterion within the context of this simple one-dimensional model.

What happens after the reflected wave overtakes the crack tip at point C in Figure 7.7? The two crack speeds $\dot{l}_S$ and $\dot{l}_T$ that appear in (7.4.33) are both equal to the constant value (7.4.36), hence in the region BCD,

$$\eta = 0, \qquad \gamma = \sqrt{2\Gamma/n\rho c^2} \qquad (7.4.37)$$

for the actual motion. At time $t = t_C$, the body is therefore entirely at rest and the crack tip strain is below the level required to meet the crack growth criterion (7.4.35) for a positive crack speed. Because the fields (7.4.37) represent an equilibrium solution, it is concluded that the crack growth is arrested at point C. From the geometry of the x, t-plane the arrest length of the crack is seen to be $l_C = n l_0$, which is significantly larger than the equilibrium value of the length $\sqrt{n}$ for end displacement w_0. From the fact that the instant of crack arrest corresponds to the instant at which the reflected wave BC overtakes the crack tip, it is evident that $\dot{l} t_C = l_C - l_0$ and that $c t_C = l_C + l_0$. If t_C is eliminated then a relationship between the crack speed up to arrest and the arrest length is obtained in the form

$$\frac{\dot{l}}{c} = \frac{l_C - l_0}{l_C + l_0}. \tag{7.4.38}$$

It is instructive to compare the actual dynamic energy release rate as a function of crack length with the corresponding equilibrium energy release rate for fixed end displacement w_0. The equilibrium rate is simply $\rho c^2 \gamma^2 / 2 = n \Gamma (l_0 / l)^2$. The dynamic energy release rate, on the other hand, has the value $n \Gamma$ for $l = l_0$. It then drops abruptly to Γ as the crack begins to grow, and this value is maintained until arrest of the growth process. Upon arrest, the energy release rate again drops abruptly to Γ / n when $l = l_0 n$, the crack length at arrest. The two energy release rate variations have three points in common. Because the initial and final fields are equilibrium fields, the two energy release rates must coincide for the corresponding crack lengths, as indicated in Figure 7.8. The intermediate intersection point also has some significance. It occurs at $l = l_0 \sqrt{n}$ and the energy release rate there is the equilibrium value for the end displacement w_0. If the crack tip were initially sharp and the end displacement were increased very slowly to its final value of w_0, then the final energy release rate would be Γ at crack length $l = l_0 \sqrt{n}$. These features seem to capture the essence of data on rapid crack propagation and arrest in double cantilever beam specimens of a brittle plastic obtained by Kalthoff, Beinert, and Winkler (1976, 1977), although the quantitative details are beyond the capabilities of the model.

The reason why the crack tip propagates beyond this equilibrium length in the dynamic growth situation can be found in the wave propagation character of the mechanical fields. Initially, the crack grows as though the material that is strained to the level γ_0 extends indefinitely far behind the tip. If this were indeed the case, then there would be a large reservoir of energy available to drive the crack growth process, and the crack advances accordingly. The wave generated at

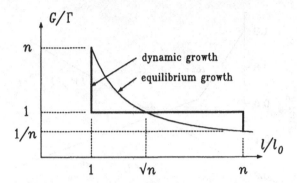

Figure 7.8. A graph of the normalized energy release rate G/Γ versus amount of crack growth l/l_0 for the string problem. The darker line is the actual energy release rate variation for crack growth governed by a constant critical energy release rate criterion, whereas the lighter curve is the equilibrium energy release rate for the imposed end displacement w_0.

the onset of growth (AB in Figure 7.7) which is ultimately reflected back toward the crack tip (BC in Figure 7.7) carries the information that the body is bounded and that the energy supply is limited. By the time that this information reaches the crack tip, however, the tip has already advanced beyond the equilibrium position for the available energy supply. The crack tip mechanical field is subcritical thereafter, and the crack growth process arrests abruptly with a corresponding drop in energy release rate, as shown in Figure 7.8.

Because of the simplicity of the solution in this case, both the total strain energy U_{tot} and the total kinetic energy T_{tot} of the specimen can be calculated as explicit functions of crack length or, equivalently, as functions of time. Expressions for the energy densities are given in (5.5.3) and the result of this calculation is

$$\frac{U_{tot}}{U_0} = 1 + \frac{n+1}{2n}\left(1 - \frac{l}{l_0}\right), \quad \frac{T_{tot}}{U_0} = -\frac{n-1}{2n}\left(1 - \frac{l}{l_0}\right) \quad (7.4.39)$$

for $0 \le t < l_0/c$, and

$$\frac{U_{tot}}{U_0} = \frac{n^2 + 2n - 1}{2n^2} + \frac{n+1}{2n^2}\left(1 - \frac{l}{l_0}\right),$$

$$\frac{T_{tot}}{U_0} = \frac{(n-1)^2}{2n^2} + \frac{n-1}{2n^2}\left(1 - \frac{l}{l_0}\right) \quad (7.4.40)$$

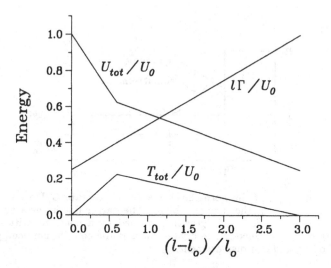

Figure 7.9. The variation of potential energy, kinetic energy and fracture energy with amount of crack growth for the string problem, as given by (7.4.39) and (4.4.40), for the case of $n = 4$.

for $l_0/c \leq t \leq (n+1)l_0/c$, where $U_0 = \frac{1}{2}\rho c^2 \gamma_0^2 l_0$ is the strain energy stored in the body at crack growth initiation. The time l_0/c corresponds to point B in Figure 7.7 and the time $(n+1)l_0/c$ is the time of crack arrest.

The variation of energy with crack length is shown in Figure 7.9 for the case of $n = 4$. For this case, the crack speed up to arrest is $3c/5$, the arrest length is $4l_0$, and the crack length at time l_0/c is $8l_0/5$. The interpretation of the energy variations in Figure 7.9 in terms of the wave propagation process is straightforward. At the instant of fracture initiation, the strain γ is reduced at the crack tip from γ_0 to the value appropriate to satisfy the crack growth criterion, and this reduction in strain implies a reduction in strain energy. Because of the rate of this reduction in γ, however, the inertia of the material comes into play. The value of γ is reduced behind the wavefront AB traveling toward the fixed end of the string. The strain energy decreases and the kinetic energy increases as the wave engulfs more and more of the free length of the body. Because energy is drawn from the body at the crack tip, the decrease in U_{tot} is not balanced by the increase in T_{tot}. When the stress wave reflects from the fixed end $x = 0$, the fixed grip condition requires that it must do so in just the right way so as to cancel the particle velocity. Thus, as the wave reflects back onto itself, the particle velocity η is reduced to zero and the strain γ is further reduced behind the wavefront BC traveling back toward the crack tip. The kinetic energy decreases after wave reflection, and the strain energy also continues to decrease but at a rate slower than before. Until the reflected wave overtakes the crack tip, the tip moves as though it was advancing in an unbounded body. The strain level carried by the reflected wave brings the crack tip strain to a value below that required to sustain crack growth, and the motion is abruptly arrested. It should be noted that the fact that the kinetic energy is exactly zero at the instant of arrest is a special feature of this model. For more realistic models of the double cantilever beam specimen, the kinetic energy typically increases and then decreases with time, but not necessarily to zero.

Up to this point in the analysis of the one-dimensional model, it has been assumed that the specific fracture energy is independent of crack speed. The extension of the main ideas to the case of a fracture energy that is speed dependent is straightforward, and some results are included here. A particular speed dependence that models the behavior of some materials is

$$\Gamma = \Gamma_c \left(\frac{1 + \dot{l}/c}{1 - \dot{l}/c} \right)^{m-1} , \qquad (7.4.41)$$

where Γ_c is a constant and $m \geq 1$. The fracture energy is constant if $m = 1$.

Suppose that the same crack growth situation that was examined above is considered again for the speed-dependent fracture energy (7.4.41). If the energy growth criterion is applied then it is found that the crack will grow with the constant speed

$$\frac{\dot{l}}{c} = \frac{n^{\alpha_1} - 1}{n^{\alpha_1} + 1} , \qquad (7.4.42)$$

where $\alpha_k = (1 - 2/m)^{k-1}/m$, $k = 1, 2, 3, \ldots$, from time $t = 0$ until the time at which the wave generated at the initiation site reflects from the end $x = 0$ and overtakes the moving crack tip. The parameter n is a measure of bluntness of the initial crack tip, as before. At the instant when the reflected wave overtakes the crack tip, the body is completely at rest and the energy release rate takes on the value

$n^{1-2/m}\Gamma_c$. Thus, if $1 \leq m \leq 2$ then this value of energy release rate is less than or equal to the minimum level necessary to sustain crack growth. The process is arrested with a crack length of $n^{1/m}l_0$. As before, the energy release rate decreases abruptly at arrest except for the special case $m = 2$.

If $m > 2$, then the crack propagation criterion implies that the crack grows with the constant speed

$$\frac{\dot{l}}{c} = \begin{cases} \dfrac{n^{\alpha_2} - 1}{n^{\alpha_2} + 1} & \text{for } n^{\alpha_1}l_0 < l < n^{(\alpha_1+\alpha_2)}l_0\,, \\[3mm] \dfrac{n^{\alpha_3} - 1}{n^{\alpha_3} + 1} & \text{for } n^{(\alpha_1+\alpha_2)}l_0 < l < n^{(\alpha_1+\alpha_2+\alpha_3)}l_0\,, \end{cases} \qquad (7.4.43)$$

and so on. The crack speed is piecewise constant in time, and it decreases gradually with each wave reflection to zero as time becomes very large. The crack length itself approaches a finite limit, however, that is given by

$$\lim_{t \to \infty} l(t) = l_0 n^{(\alpha_1+\alpha_2+\cdots)} = l_0\sqrt{n}\,, \qquad (7.4.44)$$

which does not depend on m. It is interesting to note that the crack length (7.4.44) is exactly the length that would result from slow crack growth under equilibrium conditions to the final state.

As a final point on this problem, the condition that the end $x = 0$ of the specimen is held rigidly fixed is relaxed to introduce the possibility of flexibility in the loading apparatus. A common manner of loading the double cantilever beam specimen, which was used to motivate this simple model, is to gradually force a pair of wedges between loading pins that pass through the specimen arms near the free ends. If these wedges are extremely stiff compared to the flexibility of the arms of the specimen then the assumption of fixed grip loading is quite realistic (unless the pins lose contact with the wedges at some time in the process). This might be the case for hard steel wedges and pins used to load a brittle plastic specimen. If the specimen is also very stiff, then the flexibility of the loading apparatus can influence the crack growth process. Thus, the boundary condition of constant displacement at the end $x = 0$ is replaced by the condition that the transverse shear force at the end $-\rho c^2\gamma(0,t)$ varies linearly with the transverse deflection at the end of the specimen, for example,

$$-\rho c^2\gamma(0,t) = k[w_0 - w(0,t)] - \rho c^2\gamma_0\,, \qquad (7.4.45)$$

where k is the stiffness of an elastic spring. This boundary condition corresponds to the case of a linear elastic spring of stiffness k that is compressed to a length w_0 by a force of magnitude $\rho c^2 \gamma_0$ that maintains the initial equilibrium deflection of the string and that acts according to (7.4.45) in a direction transverse to the string at $x = 0$.

The condition (7.4.45) is a nonstandard boundary condition for the application of the method of characteristics to determine the solution for arbitrary motion of the crack tip, but it can be converted into a standard form in the following way. The result of time differentiation of (7.4.45) is

$$\frac{\partial \gamma}{\partial t}(0, t) = \frac{k}{\rho c^2} \eta(0, t). \qquad (7.4.46)$$

With reference to Figure 7.10, for any point R on the boundary such that $t_B < t_R < t_D$ the method of characteristics yields

$$\eta^R = c\gamma_0/n - c\gamma^R, \qquad (7.4.47)$$

where γ_0 and n are related as before. Substitution of η^R into (7.4.46) yields a differential equation for γ^R that has the solution

$$\gamma^R = \left[(n-1) \exp\left\{ -(t - l_0/c)/\tau \right\} + 1 \right] \gamma_0/n \qquad (7.4.48)$$

which satisfies the initial condition $\gamma(0, l_0/c) = \gamma_0$, where τ is the characteristic time $\rho c/k$. The condition (7.4.48) is a standard boundary condition, and the method of characteristics can be applied to determine the values of γ and η at point P in terms of the unknown crack length and speed there. Imposition of the constant energy release crack growth criterion then yields the differential equation for $l(t)$

$$\frac{\dot{l}(t)}{c} = \frac{1 - m(t)}{1 + m(t)}, \qquad (7.4.49)$$

$$m(t) = n\left[1 + 2(n-1) \exp\left\{ [l(t) + l_0 - ct]/c\tau \right\} \right]^{-2}$$

for $t_C < t < t_E$, subject to the initial condition that $l = nl_0$ when $t = (n+1)l_0/c$. This equation has been integrated numerically for $n = 4$,

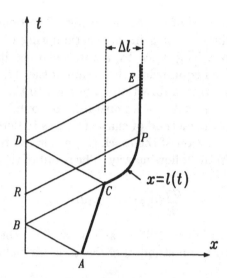

Figure 7.10. The x, t-plane for the string problem, showing the path of the crack tip moving according to the critical energy release criterion when loading compliance is taken into account.

and the qualitative features of the result are shown in Figure 7.10. Of particular interest is the *additional* amount of crack growth due to loading apparatus flexibility. It was found that this amount of crack growth is $0.31l_0$, $0.06l_0$, and $0.01l_0$ for $c\tau/l_0 = 0.5$, 0.25, and 0.10, respectively.

With reference to Figure 7.10, the observation that the crack speed increases at C before decaying to zero can be explained in the following way. When the unloading wave AB reaches the boundary $x = 0$, the particle velocity of the load point can change instantaneously without affecting the applied force. This is a feature of dead weight loading with an indefinitely large energy supply, and the crack tip responds to the waves carrying this information by accelerating. After some time has elapsed, however, the load point actually displaces and the load level decays with characteristic time τ. This information is also communicated to the crack tip by means of reflected waves, and the crack tip responds by decelerating. If the loading is very stiff compared to the specimen, then τ is very small and the reduction of the crack speed to zero is very rapid.

The differences between the crack tip motion predicted here and the motion found in Section 7.4.1 point out the importance of the shape of the elastic body in which a crack is growing on the nature of the crack tip motion. For example, in the case of the string model, the crack grows at constant speed and then is arrested with a jump in speed. In the illustration presented near the end of Section 7.4.1, on the other hand, the crack speed decreases gradually to zero from its initial value. Furthermore, the final length of the crack in the string model is found to be much greater than the equilibrium length for the imposed end displacement, whereas in the plane strain case the final length is found to be identical to the equilibrium length. These differences are apparently due to the way in which waves are radiated from, and subsequently reflected back onto, the moving crack tip. In one-dimensional structures, loads can be transferred over long distances without dilution of their effect and waves are guided along the length. In the plane strain case, on the other hand, the influence of an applied load becomes increasingly diffuse with increasing distance from its point of application and waves are continuously scattered rather than guided.

7.4.5 Double cantilever beam: approximate equation of motion

The crack tip energy release rate for rapid crack growth along the symmetry line of a body in the double cantilever beam configuration was determined in Section 5.5.2 and an application to steady state crack growth was considered in Section 5.5.3. Here, a transient crack growth problem for the double cantilever beam configuration is analyzed. The analysis is approximate; no exact solution is available for crack growth at an arbitrary rate in this configuration. With reference to the notation established in Section 5.5, a solution is required for the equation governing the transverse deflection $w(x,t)$ of a uniform beam

$$EI\frac{\partial^4 w}{\partial x^4} = \rho A \frac{\partial^2 w}{\partial t^2} \qquad (7.4.50)$$

over the interval $0 < x < l(t)$, where E is the elastic modulus, I is the area moment of inertia of the beam cross section, A is the area of the beam cross section, and ρ is the mass density of the material. The solution of this differential equation is subject to the boundary conditions

$$w(l,t) = 0, \qquad \frac{\partial w}{\partial x}(l,t) = 0 \qquad (7.4.51)$$

at the crack tip. Appropriate end conditions must be specified at

$x = 0$, say the conditions of fixed displacement with zero moment,

$$w(0, t) = w_0, \qquad \frac{\partial^2 w}{\partial x^2}(0, t) = 0 \qquad (7.4.52)$$

for $0 < t < \infty$. Initial conditions on deflection and particle velocity are also required; for example,

$$w(x, 0) = \frac{w_0}{2}\left(2 - 3\frac{x}{l_0} + \frac{x^3}{l_0^3}\right), \qquad \frac{\partial w}{\partial t}(x, 0) = 0 \qquad (7.4.53)$$

for $0 < x < l(t)$. The function $l(t)$ is a continuous, monotonically increasing function that specifies the position of the crack tip as a function of time.

The crack tip dynamic energy release rate for this crack growth process is, from (5.5.14),

$$G = \frac{M(l^-, t)^2}{EI}, \qquad (7.4.54)$$

where $M(l^-, t)$ is the total internal bending moment in the beam at the crack tip end. This form takes into account energy contributions from both halves of the specimen, and it is the form of the energy release rate that is strictly consistent with Bernoulli-Euler beam theory. As given by (7.4.54), G is the mechanical energy released from the body per unit crack advance, taking full account of the width of the specimen, and it has the physical dimension of force. The difficulty in writing an exact equation of motion for this case is that the quantity $M(l^-, t)$ which appears in (7.4.54) cannot be determined for arbitrary $l(t)$. Thus, an approximate approach based on Hamilton's principle is considered. The problem represented by the equation (7.4.50), the boundary conditions (7.4.51) and (7.4.52), and the initial conditions (7.4.53) is examined to make the development specific, but the same approach can be applied to a range of problems.

The kinetic energy and strain energy of the symmetrically deformed double cantilever beam specimen at any time t are

$$T_{tot} = \int_0^l \rho A\left(w_{,t}\right)^2 dx, \qquad U_{tot} = \int_0^l EI\left(w_{,xx}\right)^2 dx \qquad (7.4.55)$$

in terms of quantities already defined. The familiar factor one-half is absent to account for both halves of the symmetric body. For any given crack tip motion $l(t)$, Hamilton's principle of stationary action (Gelfand and Fomin 1963) requires that the differential equation (7.4.50) and the boundary conditions (7.4.51) and (7.4.52) must be satisfied if the space of admissible deflections includes all continuous and sufficiently differentiable functions $w(x, t)$ which satisfy the kinematic or essential boundary conditions. If the crack length $l(t)$ is also viewed as a generalized coordinate, then the additional condition that

$$\frac{d}{dt}\left(\frac{\partial(T_{tot} - U_{tot})}{\partial \dot{l}}\right) - \frac{\partial(T_{tot} - U_{tot})}{\partial l} = -G \qquad (7.4.56)$$

is obtained, where the dynamic energy release rate is the "dissipative" generalized force that is work-conjugate to $l(t)$. If $w(x, t)$ is a solution of the complete boundary value problem represented by the partial differential equation (7.4.50) plus its initial and boundary conditions then (7.4.56) reduces to (7.4.50). As already noted, however, such an exact solution is not available for arbitrary $l(t)$, so the following approximate approach based on variational methods is included here.

The approach is developed for the particular crack growth problem introduced above, and extensions to other cases are evident. The Bernoulli-Euler idealization of a beam does not admit propagation of sharp-fronted stress waves, so a modal approach appears to be reasonable for some purposes. If the tracking of discrete wave pulses is important in the analysis, then the Bernoulli-Euler idealization itself is deficient and a higher order beam idealization is required. Suppose that at any time t the deflection of the beam is given by

$$w(x, t) = \frac{w_0}{2}\left(2 - 3\frac{x}{l(t)} + \frac{x^3}{l(t)^3}\right) + \sum_{n=1}^{\infty} a_n(t)\, \phi_n\big(x/l(t)\big), \quad (7.4.57)$$

where $l(t)$ is the instantaneous crack length, $a_n(t)$ is an unknown amplitude, and $\phi_n(x/l)$ is the nth normal mode shape for free vibration of a beam subjected to *homogeneous* boundary conditions. In the present case, these homogeneous boundary conditions are that $\phi_n(0) = \phi_n''(0) = \phi_n(1) = \phi_n'(1) = 0$ for any n, where the prime denotes differentiation with respect to the argument. This form is suggested because it satisfies all boundary conditions, it satisfies the initial conditions if $a_n(0) = 0$ for all n, and the set of functions $\phi_n(\cdot)$, $n = 1, 2, \ldots$, is a complete orthogonal set. Thus, it appears that any solution can be represented by (7.4.57), where the functions $a_n(\cdot)$ are determined from

$$\frac{d}{dt}\left(\frac{\partial(T_{tot} - U_{tot})}{\partial \dot{a}_n}\right) - \frac{\partial(T_{tot} - U_{tot})}{\partial a_n} = 0, \quad n = 1, 2, \ldots, \quad (7.4.58)$$

according to Hamilton's principle.

Although this approach offers some promise for analytical treatment of problems in this class, it has not been extensively developed. Some preliminary steps were taken by Benbow and Roessler (1957) and Gilman, Knudsen, and Walsh (1958) for crack growth under equilibrium conditions, and by Burns and Webb (1970a, 1970b) for a case of dynamic crack growth somewhat different from that being considered here. In all three cases, the analysis was carried out in order to interpret fracture propagation experiments in which a wedge was driven into the end of a relatively long single crystal to induce cleavage crack growth. Other perspectives on this class of problems are provided by Berry (1960a, 1960b) and Sih (1970). The present analysis is restricted to the case where $a_n = 0$ for all n in (7.4.57) and, in this sense, it is similar to the analysis of Burns and Webb (1970a, 1970b).

If the deflection is assumed to be given by the first term in (7.4.57) then the kinetic energy and the potential energy expressions in (7.4.55) can be evaluated to yield

$$T_{tot} = \frac{6w_0^2\rho A l^2}{35l}, \quad U_{tot} = \frac{3w_0^2 EI}{l^3}. \quad (7.4.59)$$

If the energy balance crack growth criterion $G = \Gamma$ is assumed, where Γ is the constant specific fracture energy, then substitution of $T_{tot} - U_{tot}$ into (7.4.56) yields the differential equation for $l(t)$

$$\ddot{l} = \frac{1}{2}\frac{\dot{l}^2}{l} + \frac{105}{4}\frac{EI}{\rho A}\frac{1}{l^3} - \frac{35}{12}\frac{\Gamma l}{w_0^2 \rho A}. \quad (7.4.60)$$

Just prior to the onset of crack growth, $T_{tot} = 0$, $U_{tot} = U_0$, and the energy release rate is $9w_0^2 EI/l_0^4$, and it is assumed that this is equal to $n\Gamma$ to account for some initial bluntness of the crack tip. In this sense,

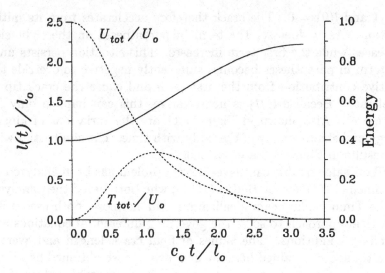

Figure 7.11. The crack tip motion predicted by the approximate equation of motion (7.4.61) for the double cantilever beam configuration when $n = 4$ and $l_0 = 1.45\,h$. The kinetic energy and strain energy of the body versus time are shown as the dashed curves.

n is essentially the same bluntness parameter that was introduced in Section 7.4.1. The differential equation then becomes

$$\ddot{l} = \frac{1}{2}\frac{\dot{l}^2}{l} + \frac{105}{4}\frac{EI}{\rho A}\left(\frac{1}{l^3} - \frac{l}{nl_0^4}\right). \qquad (7.4.61)$$

This is the equation of motion for the crack tip, subject to the initial conditions that $l(0) = l_0$ and $\dot{l}(0) = 0$. The latter condition follows from the assumption that the total kinetic energy of the system is initially zero. The equilibrium length of the crack is evident from the equation of motion when $\dot{l} = \ddot{l} = 0$, namely, $l = l_0 n^{1/4}$. Based on the results of the previous section, crack growth beyond this length due to inertial effects might be anticipated.

The differential equation (7.4.61) has been integrated numerically for $n = 4$ and $l_0 = 1.45\,h$. The result is shown in Figure 7.11 in the dimensionless form of $l(t)/l_0$ versus $c_o t/l_0$, where $c_o = \sqrt{E/\rho}$. The equilibrium length of the crack in this case is $1.414\,l_0$, but it is evident that it has grown to a final length of about $2.2\,l_0$. The reason for this dynamic overshoot can be seen from the equation of motion (7.4.61). Initially, the term in parentheses on the right side is positive for any

$n > 1$ and $\dot{l}(0) = 0$. The crack therefore accelerates from its initial position. As it does so, the term in parentheses on the right side decreases while the first term increases. This situation persists until the term in parentheses becomes sufficiently negative to override the positive contribution from the first term and cause the crack tip to decelerate. Because l^2/l_0 is nonnegative, this can happen only for $l > l_0 n^{1/4}$. Also shown in Figure 7.11 are the variations of kinetic energy and strain energy of the body with time. The similarities with the results in Figure 7.9 are obvious.

This same double cantilever beam problem has been analyzed by Kanninen (1974) and by Hellan (1981), who both based their analyses on the Timoshenko beam idealization. Furthermore, both used a full numerical approach to solve the governing differential equations and boundary conditions. The values of final crack length and average crack tip speed obtained here are similar to those obtained by means of the more complete numerical analyses for comparable values of the system parameters n and h/l_0. Finally, it is noted that generalization of the analysis of this section to cases for which the specific fracture energy Γ depends on crack position or crack speed is straightforward.

7.5 Tensile crack growth under transient loading

In this section, the matter of calculating the transient stress intensity factor history as a feature of the stress distribution is discussed for the case of a crack subjected to time-dependent loading. For background, the situation that is equivalent to the present case except for the crack edge being stationary was analyzed in Sections 2.4 and 2.5, and the case of constant speed crack growth with an arbitrary delay time was analyzed in Section 6.5. The notation established in these sections is adopted here without modification. The plane wave loading case discussed next is the simplest example of nonuniform plane strain crack growth under transient loading conditions (Freund 1973a). An analysis which generalized this result to include most of the results of this chapter as particular cases was outlined by Kostrov (1975) who considered all three modes of crack deformation fields and crack face loading varying in both position and time.

7.5.1 Incident plane stress pulse

Consider once again a body of elastic material which contains a half plane crack but that is otherwise unbounded. In the present instance, the material is stress free and at rest everywhere for $t < 0$, and the crack faces $y = \pm 0$, $x \leq 0$ are subjected to uniform normal pressure of magnitude σ^* for $t \geq 0$. For points near to the crack face compared to distance to the crack edge, the transient field consists only of a plane wave parallel to the crack face and traveling away from it at speed c_d. Near the crack edge, on the other hand, the deformation field is more complex. It was shown in Section 2.5 that the stress intensity factor for this dynamic loading situation is

$$K_I(t) = \lim_{x \to 0^+} \sqrt{2\pi x}\ \sigma_{yy}(x, 0, t) = C_I \sigma^* \sqrt{2\pi c_d t}\ , \qquad (7.5.1)$$

where $C_I = \sqrt{2(1 - 2\nu)}/\pi(1 - \nu)$ in terms of Poisson's ratio ν. Although this stress intensity factor was obtained for the case of crack face loading, it was given an interpretation in terms of an incident plane stress wave in Section 2.5.

Next, suppose that the crack edge remains stationary at $x = 0$ for some time after the load begins to act, but at some later time $t = \tau > 0$ the crack edge begins to advance in the x-direction at constant speed $v < c_R$. The time τ is called the *delay time*. An exact stress intensity factor solution for this situation has been provided in Section 6.5 with the result that

$$K_I = \lim_{x \to vt^+} \sqrt{2\pi(x - vt)}\ \sigma_{yy}(x, 0, t) = k(v) C_I \sigma^* \sqrt{2\pi c_d t}\ . \qquad (7.5.2)$$

This result, which appears as (6.5.16), differs from the corresponding result for a stationary crack only through the dimensionless factor $k(v)$, which is the universal function of instantaneous crack tip speed introduced in (6.4.26).

The stress intensity factor expression equivalent to (7.5.2) for the case of arbitrary crack tip motion $x = l(t)$ was obtained by solving a sequence of special boundary value problems (Freund 1973a). The key step was showing that if the constant velocity crack described in Section 6.5 suddenly stops, say at position $x = l_a$, then a normal stress distribution is radiated out along the crack plane ahead of the crack tip, behind a point traveling with the shear wave speed of the material, that is precisely the normal stress distribution that would have resulted if the crack tip had *always* been at its instantaneous position $x = l_a$ and if the crack had been struck by the incident stress pulse at time $t = 0$. This stress distribution is denoted by $p[(x - l_a)/t]$

in Section 6.5, indicating the fact that it is a homogeneous function of distance ahead of the crack tip $x - l_a$ and elapsed time t after the stress pulse reaches the crack plane. In other words, it is the stress distribution appropriate for the given incident wave and the new crack length, and it is independent of the fact that the crack tip had been in motion in the past. It follows that, upon sudden stopping of the crack tip, the stress intensity factor jumps immediately from the value given by (7.5.2) to the stationary crack value given by (7.5.1).

With this strong result in hand, it becomes a simple matter to construct the stress intensity factor history for any crack motion by approximating the actual crack speed history by a piecewise constant crack speed history, as was done for the case of time-independent loading in Section 7.3. A piecewise constant crack speed history can be found that is arbitrarily close to the actual history, in terms of both position and speed. It is concluded that the instantaneous value of the stress intensity factor for any motion of the crack tip depends on crack motion only through time and the instantaneous value of crack speed $\dot{l}(t)$. This dependence has the separable form

$$K_I(t, \dot{l}) = \lim_{x \to l^+} \sqrt{2\pi(x - l)}\, \sigma_{yy}(x, 0, t) = k(\dot{l})C_I\sigma^*\sqrt{2\pi c_d t}\,, \quad (7.5.3)$$

where $k(\dot{l})$ is again the universal function of crack tip speed given in (6.4.26) and shown graphically in Figure 6.10.

This result thus broadens the class of situations for which the transient stress intensity factor has this separable form, as has already been suggested in (7.4.3). In Section 6.5, the stress intensity factor result for an incident stress pulse carrying a jump in tensile stress is generalized to the case of an incident tensile pulse of arbitrary stress profile. This generalization extends without modification to the case of arbitrary crack tip motion. Consequently, for arbitrary crack tip motion described by $l(t)$ and for an incident plane stress pulse with general tensile stress variation, the stress intensity factor history is

$$K_I(t, \dot{l}) = k(\dot{l})K_I(t, 0)\,. \quad (7.5.4)$$

That is, at any instant of time, the stress intensity factor has the value given by the universal function of crack tip speed $k(\dot{l})$ times the value the stress intensity factor would have if the crack tip had been at the instantaneous position on the fracture plane $x = l(t)$ for all time. A

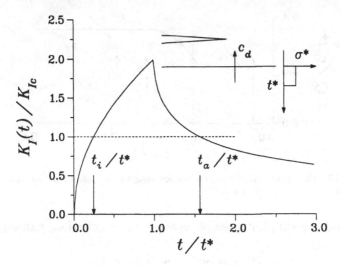

Figure 7.12. The solid line shows the stress intensity factor versus time for the case of a rectangular tensile stress pulse of magnitude σ^* and duration t^* striking a long crack in a body initially at rest, as shown in the inset, assuming that crack growth does not occur. If the crack grows according to a critical stress intensity factor criterion $K_I = K_{Ic}$, then t_i and t_a indicate the times of initiation and arrest of crack advance.

number of specific examples of stress intensity history $K_I(t, 0)$ for a stationary crack tip have been introduced in Section 2.7.

To conclude this section, a specific example of stress wave loading is considered. In particular, the equation of motion for the crack tip for the case of loading by a rectangular stress pulse is obtained and its implications for dynamic crack growth according to a stress intensity factor growth criterion under these conditions are studied. This configuration has been realized experimentally by Ravichandran and Clifton (1989). Consider a tensile rectangular stress pulse of magnitude σ^* and duration t^* normally incident on a traction-free crack. If the crack does *not* extend, then it is clear from (7.5.1) that the crack tip stress intensity factor will increase in proportion to $\sqrt{t}$ for $0 < t < t^*$, and it will decrease in proportion to $\sqrt{t} - \sqrt{t - t^*}$ for $t^* < t < \infty$. The time dependence of K_I without extension is shown in Figure 7.12 as the solid line. The largest value of stress intensity factor without extension is $C_I \sigma^* \sqrt{2\pi c_d t^*}$, so the crack will grow only if $C_I \sigma^* \sqrt{2\pi c_d t^*} > K_{Ic}$, the critical value of stress intensity factor for the material being studied. It is assumed that the crack does indeed

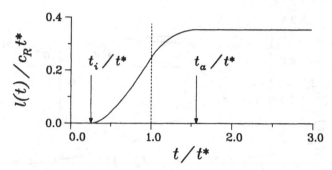

Figure 7.13. The crack tip motion corresponding to the case of stress wave loading of a crack as shown in Figure 7.12.

grow, and the simplest possible crack growth criterion, namely,

$$\dot{l}(t) = 0 \quad \text{with} \quad K_I(t,0) < K_{Ic}$$

$$\text{or} \qquad\qquad\qquad\qquad (7.5.5)$$

$$\dot{l}(t) \geq 0 \quad \text{with} \quad K_I(t,l) = K_{Ic},$$

is adopted. It is observed that the growth criterion is first satisfied at time $t = t_i$, where

$$t_i = \frac{1}{2\pi c_d}\left(\frac{K_{Ic}}{C_I \sigma^*}\right)^2. \qquad (7.5.6)$$

This time is the *initiation time* for the process, and the crack will grow for $t > t_i$. The growth criterion imposes the condition that

$$K_{Ic} = k(\dot{l})C_I\sigma^*\sqrt{2\pi c_d}\begin{cases} \sqrt{t} & \text{if } t_i < t \leq t^*, \\ \sqrt{t} - \sqrt{t - t^*} & \text{if } t^* < t \leq t_a, \end{cases} \qquad (7.5.7)$$

where the function k was introduced in (7.5.4) and

$$t_a = (t_i + t^*)^2/4t_i \qquad (7.5.8)$$

is the arrest time, that is, the time at which the decreasing $K_I(t,0)$ passes the value K_{Ic}; see Figure 7.12. The relationship (7.5.7) is an ordinary differential equation for the position of the crack tip $l(t)$ as a function of time and, as such, it is an *equation of motion* for the crack tip. It is another example of the kind discussed in Section 7.4. This equation is subject to the initial condition $l(0) = 0$ and it applies during $t_i < t < t_a$.

The features of the solution of (7.5.7) are evident. The crack tip begins to move at time $t = t_i$, it accelerates for $t_i < t < t^*$, it decelerates for $t^* < t < t_a$, and it arrests at time t_a. This response is depicted in Figure 7.13. Although these results have been obtained by study of the case of a semi-infinite crack in an otherwise unbounded body, they apply for other situations as well. For example, for a crack of finite length or for a crack near a boundary of the body, if the arrest time t_a is less than the time required for waves from the far end of the crack or from the remote boundary to significantly perturb the crack tip field, then these results apply without modification. For a finite length crack the important time is the transit time for a Rayleigh wave along the length of the crack.

7.5.2 An influence function for general loading

The results for the dynamic stress intensity factor for arbitrary motion of the crack tip in (7.3.19) and (7.5.4) are special cases due to the restrictions on the loading that were imposed. In the first case, the loading was allowed to have a general spatial variation but it was restricted to be time-independent. In the second case, the loading was allowed to have general time variation but it was restricted to have the spatial uniformity of a plane wave.

In spite of the fact that the loading was fundamentally different in the two cases, the final mathematical forms of the stress expressions were quite similar. Burridge (1976) reasoned that these special forms (7.3.19) and (7.5.4) had some common origin, and he derived a function, which he termed the influence function for the problem class, that gave the dynamic stress intensity factor directly as a linear functional over general applied crack face traction. Burridge's unifying analysis was based on the central idea developed in Section 5.7 on the weight function method, and only his final result is included here.

Suppose that the crack grows in the plane strain opening mode in the x-direction. The body is stress free and at rest for time $t \leq 0$, and the crack tip moves according to $x = l(t)$ for $t > 0$. The crack faces are subjected to normal traction $\sigma_{yy}(x, 0, t) = -p(x, t)$ for $x < l(t)$, $t > 0$. Burridge (1976) showed that the dynamic stress intensity factor for this situation is given by

$$K_I(t, l, \dot{l}) = k(\dot{l}) \, K_I(t, l, 0), \qquad (7.5.9)$$

where the stress intensity factor $K_I(t, l, 0)$ for the same crack face loading $p(x, t)$ applied to a stationary crack for which the tip is located at $x = l(t)$ is given in (5.7.37). In the present notation, this stress intensity factor expression is

$$K_I(t, l, 0) = \int_A p(x, t')\, h[x - l(t), t - t']\, dx\, dt', \qquad (7.5.10)$$

where A is the area of the x, t'-plane for which $l(t) - c_d(t - t') < x < l(t) - c_R(t - t')$ for any fixed $t > 0$. Furthermore, the function of crack speed $k(\dot{l})$ in (7.5.9) is the universal function defined in (6.4.26). While this result is complicated and difficult to apply in many situations, the main point is that the properties of the dynamic stress intensity factor results in (7.3.19) and (7.5.4) are common to all problems. That is, for any situation of mode I crack advance at an arbitrary rate under general loading the dynamic stress intensity factor is the universal function of instantaneous crack tip $k(\dot{l})$ times the stress intensity factor appropriate for a crack with a stationary tip at $x = l(t)$.

7.6 Rapid expansion of a strip yield zone

As a relatively simple application of the results on elastodynamic crack growth at nonuniform rate discussed previously in this chapter, the growth of a strip yield zone or cohesive crack tip zone is considered. The concept of a crack tip cohesive zone was introduced in Sections 5.3.2 and 6.2.4 as a device for relaxing the singular stress distribution on the crack plane ahead of the crack tip. The main idea is to think of the elastic crack as extending some distance ahead of the physical crack tip. A self-equilibrating cohesive stress acts over this distance, tending to close the crack. The size of the region over which the cohesive stress acts is determined by the condition that the net stress intensity factor at the tip of the extended crack due to both the applied loading and the cohesive stress is zero. The same idea is applied here to estimate the rate of growth of a crack tip plastic zone under stress wave loading. The cohesive stress is identified with the tensile yield stress of the material, and the length of the cohesive zone is the plastic zone size.

Consider a crack tip region under plane strain mode I conditions. The material is elastic-ideally plastic with tensile flow stress σ_0. A plane tensile pulse with wavefront parallel to the crack plane carries a jump in stress from zero to $\sigma^* < \sigma_0$, and the pulse strikes the crack at time $t = 0$. It was shown in Section 2.5 that, if the material is elastic, then the tensile stress on the crack plane is square root singular for $t > 0$ with a stress intensity factor that increases in proportion to $t^{1/2}$. If the stress carrying capacity of the material ahead of the tip is constrained to be σ_0, then a zone in which this stress level has been reached must grow out from the stationary crack tip for $t > 0$. The length of this zone will be denoted by $r_p(t)$. The way in which $r_p(t)$ varies with time will be determined from the elastodynamic crack growth solutions already in hand.

Consider a crack in the plane $y = 0$ that initially occupies the half plane $x < 0$ under plane strain conditions. The material in the crack tip region is initially stress free and at rest. At time $t = 0$ a plane stress pulse traveling at speed c_d in the y-direction strikes the crack. The stress pulse carries a jump in the stress component σ_{yy} from zero to $\sigma^* > 0$. Instantaneously, the extended crack tip begins to move in such a way that its position along the x-axis for $t > 0$ is $x = r_p(t)$. The physical crack tip remains stationary. The crack faces remain free of traction for $-\infty < x < 0$ and $t > 0$, but

$$\sigma_{yy}(x, 0^{\pm}, t) = \sigma_0 , \quad \sigma_{xy}(x, 0^{\pm}, t) = 0 \quad \text{for} \quad 0 < x < r_p(t). \quad (7.6.1)$$

Thus, whereas the incident loading pulse tends to separate the crack faces, the cohesive traction acting over $0 < x < r_p(t)$ tends to hold the crack closed.

In general, the tensile stress σ_{yy} on the crack plane ahead of the crack tip is singular as $x \to r_p^+$. The value of the stress intensity factor for arbitrary $r_p(t)$ is obtained by superimposing the results from (7.5.3) for loading by a stress pulse of magnitude σ^* and from (7.3.19) for loading by a time-independent crack face traction of magnitude σ_0. If $\dot{r}_p(t)$ here is identified with the crack growth rate l in the cited expressions, then the net stress intensity factor is

$$K_I = \sigma^* C_I k(\dot{r}_p)\sqrt{2\pi c_d t} - 2\sigma_0 k(\dot{r}_p)\sqrt{2\dot{r}_p t/\pi} \quad (7.6.2)$$

for arbitrary $r_p(t)$, where k is the universal function of crack tip speed given in (6.4.26) and $C_I = \sqrt{2(1 - 2\nu)}/\pi(1 - \nu)$.

If the strength of the material on the plane $y = 0$ ahead of the crack tip is indeed limited, then the stress distribution cannot be singular there. Thus, $K_I = 0$ for $t > 0$, which provides a condition for determining $r_p(t)$. In the present case, setting $K_I = 0$ in (7.6.2) leads to the condition that

$$\frac{\dot{r}_p(t)}{c_d} = \left(\frac{\sigma^*}{\sigma_o}\right)^2 \frac{(1 - 2\nu)}{2(1 - \nu)^2}. \tag{7.6.3}$$

Thus, the rate of growth of the plastic zone size $\dot{r}_p$ is constant for the special case of an incident step stress pulse. The stress analyses of Section 6.5 leading to the stress intensity factor results used here were based on the assumption that $l < c_R$, so the present result is subject to the restriction that $\dot{r}_p < c_R$. This limitation is of little consequence, however, because the right side of (7.6.3) is less than π^{-1} for all values of Poisson's ratio in the range $0 \le \nu \le 0.5$ and all $\sigma^* < \sigma_o$. For $\nu = 0.3$, for example, (7.6.3) gives $\dot{r}_p/c_d = 0.004, 0.102$ for $\sigma^*/\sigma_o = 0.1, 0.5$, respectively.

For an incident pulse with stress profile more complicated than a simple step, the corresponding rate of growth of the plastic zone is also more complicated. In general, this rate is not constant. For example, if the profile of the loading pulse is such that the stress at a point increases linearly in time as the wave passes, then the first term on the right side of (7.6.2) is replaced by a term proportional to $t^{3/2}$ rather than to $t^{1/2}$. Consequently, it is found that the rate of increase of the plastic zone is $\dot{r}_p \sim t^2$ rather than a constant.

Note that the validity of the result (7.6.3) does not depend in any way on the plastic zone size being small compared to $c_d t$. If it is the case that $r_p \ll c_d t$, however, then the result can be written in a somewhat different form. If $r_p \ll c_d t$, which is the case if $\sigma^* < 0.2\sigma_o$, then the plastic zone is completely enclosed within the crack tip singular field associated with the stationary physical crack tip at $x = 0$ and the incident pulse of magnitude σ^*. This singular field is characterized by the stress intensity factor

$$K_I^* = \sigma^* C_I \sqrt{2\pi c_d t} \tag{7.6.4}$$

and the plastic zone size can be written as

$$r_p = \frac{\pi}{8} \left(\frac{K_I^*}{\sigma_o}\right)^2. \tag{7.6.5}$$

This result is an elastodynamic equivalent of the equilibrium result which is commonly known as the small-scale yielding result for plastic zone size based on the strip yield model (Rice 1968). Even though there is no physical length in the present case with which the plastic zone size can be compared, it is still required that the plastic zone size must be small compared to the distance from the crack tip to the nearest wavefront. The latter distance here is on the order of $c_s t$. For a material with a shear wave speed of about $c_s = 2000 \, \text{m/s} = 2 \, \text{mm}/\mu\text{s}$ the distance from the crack tip to the closest wavefront when $t = 10 \, \mu\text{s}$ is 20 mm. Thus, the small-scale yielding approximation is valid only if r_p is less than about 2 mm at this time.

Another feature of the process of development of a crack tip plastic zone that is significant is the crack tip opening displacement. Within the framework of the strip yield model being discussed here, the crack opening displacement is the relative displacement of opposite faces of the crack at the physical crack tip. If this total displacement is denoted by $\delta_t(t)$, then

$$\delta_t(t) = u_y(0, 0^+, t) - u_y(0, 0^-, t) \qquad (7.6.6)$$

for mode I deformation. For the strip yield zone model discussed above, an expression for $\delta_t(t)$ can be worked out for the case of step stress wave loading. The expression involves rather complicated integrals that cannot be expressed in terms of elementary functions, however, and the implications for fracture initiation under dynamic loading are difficult to extract. The corresponding result for antiplane shear deformation has been obtained by Achenbach (1970c), and this result is included here. The antiplane shear problem with cohesive zone has also been studied by Glennie and Willis (1971).

Consider the mode III crack problem with a strip yield zone that corresponds to the mode I situation studied above. The geometry of the physical system is unchanged. The incident wave with wavefront parallel to the crack plane is now a horizontally polarized shear wave. It propagates at speed c_s in the y-direction, and it carries a jump in shear stress σ_{yz} from zero to τ^*. The wave arrives at the crack plane at time $t = 0$, whereupon a strip yield zone begins to grow from the crack tip in the x-direction. The cohesive shear traction in the yield zone has magnitude τ_0.

This problem can be solved for an arbitrary rate of growth of the strip plastic zone $\dot{r}_p(t)$ by means of the integral equation approach

developed in Sections 2.4 and 6.2. Indeed, a number of the steps in
the analysis can be constructed directly from the discussion in these
sections. Only certain pertinent results are included here. First of all,
it is found that the rate of growth of the plastic zone is constant for
an incident step wave, with

$$\frac{\dot{r}_p}{c_s} = \left(\frac{\tau^*}{\tau_0}\right)^2. \tag{7.6.7}$$

This result shows that $\dot{r}_p/c_s$ depends on τ^*/τ_0 for mode III deformation in the same way that it depends on σ^*/σ_0 for mode I deformation.
Furthermore, the numerical values of the coefficients in the two cases
are similar. This comparison provides some hope that the crack
opening displacement rates in the two cases will be similar as well,
but there is no basis for direct comparison.

The crack opening displacement for the case of mode III is

$$\delta_t(t) = u_z(0, 0^+, t) - u_z(0, 0^-, t). \tag{7.6.8}$$

For the problem at hand, the elastodynamic solution yields the result
that

$$\delta_t(t) = \frac{4c_s t}{\pi} \frac{\tau^*}{\mu} \tan^{-1}\left(\frac{\tau^*}{\tau_0}\right) \tag{7.6.9}$$

for $t > 0$. This result is valid for any value of τ^* in the range $0 \leq$
$\tau^* < \tau_0$ or for plastic zone size in the range $0 \leq r_p < c_s t$. The
opening displacement scales with the accumulated displacement due
to elastic strain $c_s t \tau^*/\mu$ and it depends on τ^*/τ_0 only through the
factor $\tan^{-1}(\tau^*/\tau_0)$.

A simple ductile fracture initiation condition is that the crack
will begin to grow when the crack opening displacement achieves a
critical material specific value, say δ_{cr}. In light of (7.6.9), this criterion
provides a relationship between time to fracture initiation t_f and the
intensity of applied loading in the dimensionless form

$$\frac{4}{\pi} \frac{\tau_0}{\mu} \frac{c_s t_f}{\delta_{cr}} = \left[\frac{\tau^*}{\tau_0} \tan^{-1}\left(\frac{\tau^*}{\tau_0}\right)\right]^{-1}. \tag{7.6.10}$$

For example, the dimensionless term on the right side of (7.6.10) is
about 11.5 for $\tau^*/\tau_0 = 0.3$. Thus, for a material for which $\mu/\tau_0 = 200$,

$c_s = 2\,\text{mm}/\mu\text{s}$ and $\delta_{cr} = 0.1\,\text{mm}$ the predicted fracture initiation time t_f is about $90\,\mu\text{s}$. For such a short initiation time, the magnitude of τ_o should probably be adjusted to account for the influence of material strain rate on the flow stress.

Finally, the particular form of the mode III results for small-scale yielding are examined. If $r_p \ll c_s t$, which is the case when τ^* is less than about $0.25\,\tau_o$, then the strip yield zone is completely embedded within the elastodynamic crack tip singular field. The stress intensity factor for this field is

$$K_{III}^* = 2\tau^*\sqrt{2c_s t/\pi}\,, \qquad (7.6.11)$$

from (2.3.18). Then, if τ^* is eliminated from the expression for plastic zone size in favor of K_{III}^*,

$$r_p = \frac{\pi}{8}\left(\frac{K_{III}^*}{\tau_o}\right)^2 \qquad (7.6.12)$$

which gives the plastic zone size in terms of the transient stress intensity factor. Likewise, the crack opening displacement has the simple expression

$$\delta_t = \frac{K_{III}^{*\,2}}{2\mu\tau_o}\,. \qquad (7.6.13)$$

Both of these expressions are consistent with the corresponding equilibrium results.

7.7 Uniqueness of elastodynamic crack growth solutions

A solution of the equations of linear elastodynamics with suitable boundary conditions is unique according to a classical theorem due to Neumann (Achenbach 1973). An extension of the theorem to include unbounded domains is given by Wheeler and Sternberg (1968). In each case, the statement of the theorem specifies the function classes of which the rectangular components of stress σ_{ij}, of displacement u_i, and of body force f_i are members. In particular, these functions are required to be sufficiently smooth so that the governing differential equations are valid everywhere in the body and that the divergence theorem can be applied as needed. In many cases, however, the most common smoothness conditions are violated due to concentrated loads on the boundary of the body, the existence of discrete wavefronts carrying discontinuities or singularities in certain field quantities, or geometrical features that result in stress concentrations. Some of these cases have received careful attention. For example, a version of the uniqueness theorem that takes into account concentrated loads is stated and proved by Wheeler and Sternberg (1968). At wavefronts that carry discontinuities in stress and particle velocity the governing differential equations must be replaced by kinematic and dynamic jump conditions. If the wavefronts carry algebraic singularities in stress and particle velocity, which is common in two-dimensional fields, the requisite conservation laws are satisfied if the strengths of the algebraic singularities satisfy the same conditions as jump discontinuities. The uniqueness theorem can be extended to include propagating discontinuities or algebraic singularities on wavefronts in the manner discussed for acoustic fields by Friedlander (1958). The case of discontinuities in elastodynamic fields was considered by Brockway (1972).

The purpose here is to consider modifications of the statement and proof of the elastodynamic uniqueness theorem which are required to include crack growth solutions within its range of applicability. Attention is limited to plane elastodynamic solutions for a body of homogeneous and isotropic elastic material of limited extent. Extension to unbounded regions or to three-dimensional fields can be carried out as in the discussion by Wheeler and Sternberg (1968). The crack surface may be curved, but the crack tip trajectory is supposed to have a continuously turning tangent.

Rectangular coordinates are introduced in such a way that the particle displacement is in the x_1, x_2-plane, as shown in Figure 7.14. The region of the plane occupied by the body is R and the outer boundary of R is S. The inner boundary of R is made up of the crack faces S_0 except near the ends of the crack where the boundary is augmented by vanishingly small loops S_1 and S_2 surrounding the crack tips. The shape of each loop is arbitrary, but the shape is fixed with respect to an observer moving with the crack tip. The outer boundary S is subjected to traction and/or displacement boundary conditions which result in a traction distribution T_i on S. The faces of the crack are free of traction for simplicity.

The stress and displacement fields satisfy the differential equa-

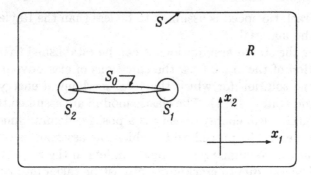

Figure 7.14. Schematic diagram of a two-dimensional body containing a rapidly growing crack. The small loops surrounding each crack tip are fixed in size and shape, and they translate with the crack tips.

tions

$$\sigma_{ij,j} + \rho f_i = \rho \ddot{u}_i, \quad \sigma_{ij} = \sigma_{ji} \quad \text{in } R,$$

$$\sigma_{ij} = C_{ijkl} u_{k,l} \quad \text{in } R, \tag{7.7.1}$$

$$\sigma_{ji} n_j = T_i \quad \text{on } S,$$

where ρ is the material mass density, C_{ijkl} is the array of elastic constants, and n_i is the unit vector normal to the boundary of R, directed away from R. The dot denotes material time differentiation and the notation is consistent with Section 1.2.1 for the rectangular components of tensors.

It was shown in Section 4.3 that the near tip fields for crack growth have universal spatial dependence in a local crack tip coordinate system for each of the three modes of deformation. These modes may coexist, in which case the local fields are the sums of the fields for the individual modes. The only feature of the crack tip field that varies from case to case is the elastic stress intensity factor, or the various stress intensity factors for mixed mode conditions. The total energy flow into the crack tip as it advances through the material is given in terms of the stress intensity factors by the generalization of the Irwin relationship in (5.3.10). This rate of energy flux into the crack tip plays a central role in the proof of uniqueness of crack growth solutions. The analysis is carried out under the assumption that the total mechanical energy of the body, kinetic energy plus potential energy, is finite, which implies that the mechanical energy density must be integrable. The motion of the crack tip is arbitrary, except that the crack tip speed is assumed to be less than the Rayleigh wave speed of the material.

Under the stated assumptions it can be established that, for any given motion of the crack tips, the equations of elastodynamics have at most one solution for which the total mechanical energy is finite (Freund and Clifton 1974). The elastic moduli are assumed to be such that the mechanical energy density is a positive definite functional of the elastic field. The procedure by which the assertion is established is based on the Neumann proof, but it differs in the respect that the change of energy due to crack growth must be taken into account.

Suppose that two solutions exist for a particular crack growth problem, each satisfying the same initial and boundary conditions. The difference between these two solutions is also a solution of the governing differential equations (7.7.1) without body force and subject to homogeneous boundary and initial conditions. If the inner product of the momentum equation $(7.7.1)_1$ with the velocity $\dot{u}_i$ is formed and the result is integrated over the region R, then

$$\int_R \left(\sigma_{ij,j} - \rho \ddot{u}_i \right) \dot{u}_i \, dR = 0 \,. \tag{7.7.2}$$

If the first term in the integrand is rewritten as $(\sigma_{ij}\dot{u}_i)_{,j} - \sigma_{ij}\dot{u}_{i,j}$ and the divergence theorem is applied in light of the homogeneous boundary conditions, then

$$\int_{S_1+S_2} \sigma_{ij} n_j \dot{u}_i \, dS - \int_R \sigma_{ij}\dot{u}_{i,j} \, dR - \int_R \rho \ddot{u}_i \dot{u}_i \, dR = 0 \,. \tag{7.7.3}$$

The order of time differentiation and integration over R can be interchanged, provided that the flux of the integrand through the time-dependent boundary of R is taken into account as in (1.2.41). Thus,

$$\int_R \rho \ddot{u}_i \dot{u}_i \, dR = \frac{d}{dt} \int_R \tfrac{1}{2}\rho \dot{u}_i \dot{u}_i \, dR - \int_{S_1+S_2} \tfrac{1}{2}\rho \dot{u}_i \dot{u}_i v_n \, dS \,,$$

$$\int_R \sigma_{ij}\dot{u}_{i,j} \, dR = \frac{d}{dt} \int_R \tfrac{1}{2}\sigma_{ij}u_{i,j} \, dR - \int_{S_1+S_2} \tfrac{1}{2}\sigma_{ij}u_{i,j} v_n \, dS \,, \tag{7.7.4}$$

where v_n is the velocity of any point on either S_1 or S_2 in the local direction of n_i. In view of these relationships, (7.7.3) becomes

$$-\int_{S_1+S_2} \left[\sigma_{ij}n_j\dot{u}_i + \tfrac{1}{2}(\sigma_{ij}u_{i,j} + \rho\dot{u}_i\dot{u}_i)v_n \right] dS$$

$$+\frac{d}{dt}\int_R \tfrac{1}{2}\rho\dot{u}_i\dot{u}_i\,dR + \frac{d}{dt}\int_R \tfrac{1}{2}\sigma_{ij}u_{i,j}\,dR = 0\,. \qquad (7.7.5)$$

In the limit as S_1 and S_2 are shrunk onto the crack tips, the first term in (7.7.5) becomes the crack tip energy flux (5.2.7) for the particular case of elastic response. The second and third terms are the time derivatives of the total kinetic and potential energy, respectively, and these energy rates are denoted here by $\dot{T}_{tot}(t)$ and $\dot{U}_{tot}(t)$. Thus, in the limit as S_1 and S_2 are shrunk onto the crack tips, the equation (7.7.5) is

$$F(t) + \dot{T}_{tot}(t) + \dot{U}_{tot}(t) = 0\,. \qquad (7.7.6)$$

For all time up until the instant that the crack begins to extend, $F = 0$ and the standard uniqueness theorem applies. In this time range, $T_{tot}+U_{tot} = 0$. The instant that the crack tip begins to extend, $F(t) > 0$ and therefore $\dot{T}_{tot} + \dot{U}_{tot} < 0$ from (7.7.6). This implies that the value of the total mechanical energy is decreasing from zero or, in other words, it is taking on negative values. But this violates the basic hypothesis on the positive definiteness of the mechanical energy. Thus, (7.7.6) is valid only if $F(t) = 0$, which implies that either the crack is not extending or that the stress intensity factor differences are zero. In either case, the standard uniqueness theorem leads directly to the desired result.

It is noted that for crack speed in the range between the Rayleigh wave speed and the shear wave speed of the material, the energy flux $F(t)$ is negative for singular mode I and mode II fields. Consequently, with reference to (7.7.6), the difference solution for crack speeds in this range may correspond to increasing mechanical energy $T_{tot}+U_{tot}$, which does not violate any of the assumptions made. Thus, uniqueness of solutions cannot be established by means of the Neumann argument for speeds in this range. It is unclear whether or not this observation has any implications of practical consequence.

8

PLASTICITY AND RATE EFFECTS
DURING CRACK GROWTH

8.1 Introduction

In this chapter, the role of material inelasticity is considered, in the form of irreversible plastic flow, dependence of the material response on the rate of deformation, or microcracking. The study of issues in dynamic fracture mechanics concerned with these effects is at an early stage. Consequently, the sections are not integrated to any significant degree. Instead, each section is intended to give an impression of the present stage of development of analytical modeling in the areas covered.

8.2 Viscoelastic crack growth

The study of crack growth in a linear viscoelastic material has been motivated primarily by interest in modeling the fracture process in relatively brittle polymeric materials, although other materials may be idealized as linear viscoelastic under some circumstances. There are numerous specific linear viscoelastic models available for stress analysis, but only some general properties are considered here for crack growth analysis. Time-dependent material response, of the kind on which the theory of viscoelasticity is based, may be of interest in the analysis of fracture phenomena at two different levels. On the one hand, the bulk properties of the body in which the crack is propagating are important in determining the way in which the

442

effect of applied loads is transferred to the crack tip region or the way in which stress is redistributed due to crack growth. If this is the case, then the entire body is assumed to exhibit time-dependent response, the crack is assumed to be sharp, and the nature of the crack tip mechanical fields is represented by a fracture characterizing parameter. Experimental data on crack propagation in Homalite-100 which suggests that this effect is significant is described by Dally and Shukla (1980). In any case, this is the domain of linear viscoelastic stress analysis in dynamic fracture. On the other hand, if the primary focus is on the influence of material rate effects in the crack tip region itself, then this region is better modeled as a cohesive zone with time-dependent cohesive stress, as in Section 7.6, or, better still, as a region of viscoplastic flow, as in Section 8.4. Linear viscoelastic boundary value problems motivated by the former objective have been solved, and some of these are described here. The solutions of these problems are very difficult to obtain, in general, and they have had a limited impact on the interpretation of experiments in dynamic fracture or on the advance of the underlying theory. Consequently, the discussion is brief.

The most common aspects of linear viscoelastic response are embodied in the convolution integral relationship between stress $\sigma_{ij}(p,t)$ and strain $\epsilon_{ij}(p,t)$ at a spatial point represented symbolically by p and at time t in an isotropic material, namely,

$$\sigma_{ij}(p,t) = \delta_{ij} \int_{-\infty}^{t} \lambda(t-\tau)\dot{\epsilon}_{kk}(p,\tau)\, d\tau + \int_{-\infty}^{t} 2\mu(t-\tau)\dot{\epsilon}_{ij}(p,\tau)\, d\tau\,,$$

$$(8.2.1)$$

where $\lambda(t)$ and $\mu(t)$ are the two *relaxation functions* that characterize the response of the material. Each function is defined for all positive values of its argument t, and each is assumed to be a positive, monotonically decreasing function of its argument. If the material is homogeneous, then these functions are independent of p, as suggested by the form of (8.2.1).

Consider the special case of homogeneous deformation, that is, when the strain $\epsilon_{ij}(p,t)$ is independent of p. Furthermore, suppose that the strain is zero up to time $t = 0$, and it is a constant ϵ_{ij}° thereafter. Then $\dot{\epsilon}_{ij}(p,t) = \epsilon_{ij}^{\circ}\delta(t)$ and $\sigma_{ij}(p,t) = \lambda(t)\epsilon_{kk}^{\circ}\delta_{ij} + 2\mu(t)\epsilon_{ij}^{\circ}$. The functions $\lambda(t)$ and $\mu(t)$ characterize the transient stress relaxation process under these conditions, thus accounting for the use of the descriptive term relaxation functions.

The instantaneous response to the suddenly imposed strain depends only on $\lambda_0 = \lambda(0^+)$ and $\mu_0 = \mu(0^+)$, the limiting values of these functions as $t \to 0^+$. These values are called the instantaneous elastic moduli, the short-time moduli, or the glassy moduli of the material. Likewise, the stress response at a long time after the strain ϵ_{ij}^o is imposed on the material depends only on $\lambda_\infty = \lambda(\infty)$ and $\mu_\infty = \mu(\infty)$. These limiting values are called the long-time moduli or the rubbery moduli. It is customary to refer to the material as a fluid if $\mu_\infty = 0$ and as a solid otherwise. A characteristic time representative of the rate of decay of a relaxation function from its instantaneous value to its long time value is commonly introduced as the *relaxation time* for the response. These basic properties are thoroughly developed by Christensen (1982).

Consider the situation of advance of a sharp crack through a viscoelastic material under conditions of essentially two-dimensional deformation. The plane of deformation is the x,y-plane. It was shown in Section 4.6 that, for a class of elastic-viscous materials, the asymptotic field near the crack tip is dominated by the instantaneous elastic response. The same result carries over to the case of a linear viscoelastic material under quite general conditions. To see that this is indeed the case, consider the integral relation in (8.2.1) under the condition of steady motion in the x-direction at speed v. The stress and strain depend on x and t only through the combination $\xi = x - vt$. If the variable of integration τ is replaced by η, where $\eta = x - v\tau$, then it is found that

$$\sigma_{ij}(\xi,y) = \delta_{ij} \int_\infty^\xi \lambda(q) \frac{\partial \epsilon_{kk}(\eta,y)}{\partial \eta}\, d\eta + \int_\infty^\xi 2\mu(q) \frac{\partial \epsilon_{ij}(\eta,y)}{\partial \eta}\, d\eta,$$
(8.2.2)

where $q = (\eta - \xi)/v$. An integration by parts for any fixed y yields

$$\sigma_{ij}(\xi,y) = \lambda_0 \epsilon_{kk}(\xi,y)\delta_{ij} + 2\mu_0 \epsilon_{ij}(\xi,y) - \lambda_\infty \epsilon_{kk}(\infty,y)\delta_{ij}$$

$$-2\mu_\infty \epsilon_{ij}(\infty,y) - v^{-1}\delta_{ij} \int_\infty^\xi \lambda'(q)\, \epsilon_{kk}(\eta,y)\, d\eta$$

$$-v^{-1} \int_\infty^\xi 2\mu'(q)\, \epsilon_{ij}(\eta,y)\, d\eta,$$
(8.2.3)

where the prime denotes the derivative of a function with respect to its argument. It is evident from this relationship that if the stress and strain are singular at $\xi = y = 0$, then the nature of this singular field is established by the relationship

$$\sigma_{ij}(\xi, y) \approx \lambda_0 \epsilon_{kk}(\xi, y)\delta_{ij} + 2\mu_0 \epsilon_{ij}(\xi, y) \qquad (8.2.4)$$

alone, provided that $\lambda'(0^+) < \infty$ and $\mu'(0^+) < \infty$. If the strain has an integrable singularity as $\xi \to 0$ then the integral term in (8.2.3) will be nonsingular if $\lambda'(0^+)$ and $\mu'(0^+)$ are finite. This condition on the limiting values of the relaxation moduli can be weakened without affecting the main result. This matter is not pursued here, but it is considered in some detail by Kostrov and Nikitin (1970).

In any case, if the nature of the crack tip field is established by (8.2.4) then the main features of this singular field are evident from the elasticity analysis of Chapter 4. The essential part of the stress–strain relationship is identical to that in linear elasticity with the elastic moduli replaced by the instantaneous or short-time viscoelastic moduli. Consequently, the near tip mode I field for crack growth at speed v in a linear isotropic viscoelastic material is given by (4.3.10) with c_s and c_d replaced by

$$c_{s0} = \sqrt{\frac{\mu_0}{\rho}}, \qquad c_{d0} = \sqrt{\frac{\lambda_0 + 2\mu_0}{\rho}}, \qquad (8.2.5)$$

respectively, where λ_0 and μ_0 are the short-time viscoelastic moduli corresponding to the elastic moduli λ and μ.

The result that the asymptotic crack tip field for a dynamically growing crack in a linear viscoelastic solid has the same form as the elastic crack tip field, provided that the limiting values of relaxation moduli as $t \to 0$ replace the elastic moduli, leads to a somewhat paradoxical conclusion (Kostrov and Nikitin 1970). Suppose that the form of the Griffith energy release rate condition or the Irwin stress intensity factor condition appropriate for dynamic crack growth is adopted. Then the only material properties appearing in the growth condition are the short-time moduli, so the growth condition does not reflect in any direct way the time-dependent nature of the material response. (For specified applied loads on the boundary of the viscoelastic body, the time-dependent bulk material properties could influence the way in which load is transferred from the boundary to the crack tip region. The relaxation properties could enter indirectly in this way.) This seems to be a consequence of the assumption that the crack tip is sharp, or that fracture occurs at a point. For any finite nonzero crack speed, the physical separation process at a sharp crack tip is infinitely fast and only the instantaneous moduli of the material are relevant. This paradox is resolved by introducing a separation zone of finite extent in the form of a cohesive zone of some length, say Λ, ahead of the moving crack tip. The idea of a crack tip cohesive zone has been developed in Sections 5.3, 5.4, 6.2, and 7.6. With a cohesive zone accompanying the crack tip, the crack opens gradually against the resistance of some cohesive stress within this zone. The time required for a crack tip to advance a distance equal to Λ at crack speed v introduces a process time Λ/v that must be compared to a characteristic relaxation time of the material to determine if the separation is "slow" or "fast." A growth criterion based on a given level of cohesive stress and a fixed level of crack tip opening displacement leads to a relationship between crack speed and crack driving force that does include the relaxation properties of the material. This has been demonstrated in detail for mode I crack growth under equilibrium conditions by Kostrov and Nikitin (1970) and Christensen (1982), and for dynamic antiplane shear crack growth in a general linear viscoelastic material by Walton (1987) and in a Maxwell material by Goleniewski (1988). Some of the important features of these models arise in the study of the growth of a crack with a viscous cohesive zone in an otherwise elastic body in Section 8.5.

A main objective of stress analysis in viscoelastic crack growth problems is to establish the dependence of the crack tip stress intensity factor on the configuration of the body in which the crack is growing and on the boundary conditions. Quasi-static crack growth in a viscoelastic material is discussed by Christensen (1982) and by Kanninen and Popelar (1985). Few exact solutions are available for the case of dynamic crack growth in a viscoelastic solid. Willis (1967b) analyzed the situation of steady propagation at speed v of a semi-infinite antiplane shear crack in an infinite body idealized as a standard linear solid. The only loading on the body was an arbitrary integrable distribution of shear traction acting on the crack faces and fixed with respect to an observer moving with the crack tip. By means of the Wiener-Hopf technique, Willis (1967b) showed that the stress intensity factor is independent of crack speed and that it depends on the applied loading in the same way as in the corresponding quasi-static elastic problem if $v < c_{s\infty}$, whereas the stress intensity factor depends on crack speed and material parameters for $c_{s\infty} < v < c_{s0}$. As before, $c_{s0} = \sqrt{\mu_0/\rho}$ and $c_{s\infty} = \sqrt{\mu_\infty/\rho}$, so that c_{s0} and $c_{s\infty}$ are the elastic wave speeds corresponding to the short-time and long-time moduli, respectively. The general results obtained by Willis have been extended by Walton (1982) to the case of general linear viscoelastic response.

Several plane strain steady state dynamic viscoelastic problems were considered by Atkinson and Coleman (1977). They assumed the material to be characterized by a single relaxation time τ, for example, $\mu(t) = (\mu_0 - \mu_\infty)e^{-t/\tau} + \mu_\infty$, and the configuration of the body to be characterized by a single length parameter h. Then, under the assumption that the parameter $v\tau/h$, where v is the steady state crack propagation speed, is small compared to unity, Atkinson and Coleman (1977) used the method of matched asymptotic expansions to obtain mode I stress intensity factors. They considered the steady growth of a semi-infinite crack along the centerline of a very long strip of width h, which is the viscoelastic equivalent of the elastic problem discussed in Section 5.4. The strip is loaded by imposing a uniform normal displacement on each edge. They found that the viscoelastic stress intensity factor is identical to the elastodynamic result (5.4.8) with the elastic moduli replaced by the long-time equivalent viscoelastic relaxation moduli. These results were subsequently refined by Popelar and Atkinson (1987), who formally solved the strip problem by means of the Wiener-Hopf technique, following the approach of Nilsson (1972) for the elastic problem. They obtained expressions for the stress intensity factor in terms of sectionally analytic functions that arise in the Wiener-Hopf factorization process by application of the appropriate Abel theorem (1.3.18). The factors were available in the form of complex integrals which had to be evaluated numerically.

Finally, two studies that represent departures from the situation of steady growth of a sharp crack in a viscoelastic body are mentioned. Both studies are based on the antiplane shear mode of crack advance. The first is that reported by Atkinson and List (1972) concerning the transient problem of the sudden onset of constant-speed crack growth at a certain instant of time in a material described by the standard linear solid material model. A uniform traction acted on the entire newly created crack surface. They obtained the transient stress intensity factor history for several cases, and they showed that the stress intensity factor increases indefinitely or approaches a limiting value, depending on the ratio of crack speed to the long-time elastic wave speed. In the second study, Walton (1987) introduced a crack tip cohesive zone with rate-independent cohesive traction within the viscoelastic body. For growth of a crack with a cohesive zone through a quite general viscoelastic material, he determined the net energy flow into the cohesive zone in terms of crack speed and material parameters. Details were presented for both a power law material and a standard linear solid.

8.3 Steady crack growth in an elastic-plastic material

The discussion in Chapters 6 and 7 of rapid crack advance through a material was based on the assumption of nominally elastic material response. This assumption led to the notion of stress intensity factor or energy release rate as a crack tip field characterizing parameter, as discussed in Chapters 4 and 5. These ideas, in turn, when considered in light of experience with rapid crack advance in real materials, led to the proposal that the actual value that the stress intensity factor or the energy release rate assumes during dynamic crack growth, say as a function of crack speed, can be interpreted as representing the resistance of the material to crack advance. These material-specific resistance measures were termed dynamic fracture toughness and dynamic fracture energy in Section 7.4. With a representation of crack driving force in hand, as either the applied level of stress intensity factor or energy release rate, along with a description of crack growth resistance as a constitutive postulate, a balance equation emerges that was termed the crack tip equation of motion in Section 7.4. These ideas collectively provide a framework for prediction of crack advance in a body of known fracture resistance subjected to specified loads, or for extraction of fracture resistance information from a laboratory experiment in which both driving force and crack motion are measured or inferred. The theoretical framework is relatively complete, in this sense.

The purpose here is to focus on a phenomenon at a finer scale of observation. In particular, it was noted above that the level of resistance to crack advance that is exhibited by a material may depend on the instantaneous crack tip speed. For materials that do not undergo a transition in fracture mode with increasing crack tip speed

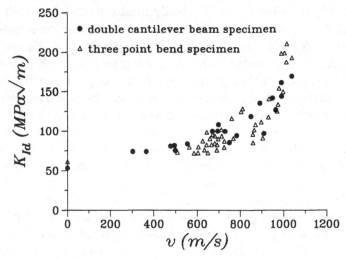

Figure 8.1. Dependence of dynamic fracture toughness on crack speed for AISI 4340 steel (Zehnder and Rosakis 1989).

and that have a relatively low rate of strain hardening in the plastic range, the dynamic fracture toughness K_d is often reported to depend on crack tip speed v in a characteristic way. An illustration of this dependence for mode I is shown in Figure 8.1 from the work of Zehnder and Rosakis (1989) for rapid crack growth in a 4340 steel. Other examples of this kind of dependence are described by Hahn et al. (1975) and Hahn, Hoagland, and Rosenfield (1976), by Kobayashi and Dally (1979), by Bilek (1978, 1980), by Rosakis, Duffy, and Freund (1984), and by Angelino (1978), all for 4340 steel in a hard condition. The detailed description of material condition for each series of experiments is given in the original articles. Still other examples are reported by Dahlberg, Nilsson, and Brickstad (1980) and Brickstad (1983a) for a high strength carbon steel, by Kanazawa et al. (1981) for a ship hull steel, and by Shukla, Agarwal, and Nigam (1988) for 7075-T6 aluminum and 4340 steel. Most of these data are summarized and compared by Rosakis and Zehnder (1985). The loading conditions and methods of measurement leading to the toughness-speed data varied widely among the experimental projects. Similar dependence of dynamic toughness on crack tip speed has also been reported for rapid crack propagation in hard plastics, but the trends in this case show features that are not yet fully understood (Ravi-Chandar and Knauss 1984c). In any case, such materials are not suitably modeled as elastic-ideally plastic materials, which is the material model upon which the discussion of this section is based.

The most significant feature of the dependence in Figure 8.1 is the increasing sensitivity of dynamic toughness to crack speed with increasing speed. Although this sensitivity might be attributed, at least in part, to strain rate dependence of the material response, it is noteworthy that the feature persists even for materials which appear to exhibit little strain rate dependence in their bulk response. Furthermore, the feature cannot be attributed to crack speed dependence of the elastic field surrounding the crack tip plastic zone. The surrounding elastic field shows little dependence on crack speed for speeds less than about 50–60 percent of the shear wave speed, whereas the sharp upturn in the variation of toughness with speed has been observed for speeds in the range of 25–30 percent of the shear wave speed. It is shown in this section that the behavior represented in Figure 8.1 can be attributed to inertial effects within the crack tip plastic zone. This conclusion is drawn from analysis of a particular antiplane shear crack problem.

Suppose that a crack grows under two-dimensional conditions through an elastic-plastic material. The potentially large stress very close to the crack edge that is predicted on the basis of an elastic model of the process must be relieved by some inelastic process. It is assumed here that the process is plastic flow, and that the material behavior can be idealized as elastic-perfectly plastic. The crack advances steadily at a constant speed v and the mechanical fields, as seen by an observer translating with the crack tip, are time-independent. The crack grows under the action of a remotely applied stress intensity factor field which is applied at a distance far from the crack tip compared to the size of the region of active plastic deformation near the crack tip. A permanently deformed but unloaded layer is left in the wake of the active plastic zone along each crack face.

The procedure to be followed within the setting described is straightforward. A solution of the problem posed is sought in the form of stress and deformation fields that satisfy the field equations in some sense, with material inertia effects included explicitly. With this solution in hand, a crack growth criterion motivated by the physics of the separation process is imposed at the level of the crack tip plastic zone. The result is a condition on the applied stress intensity factor and the crack speed that must be satisfied if the crack is to

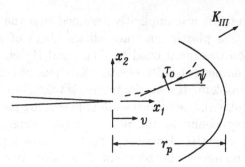

Figure 8.2. Active plastic zone region for steady dynamic growth of a mode III crack in an elastic-plastic material.

advance according to this growth criterion. This condition generates a theoretical *dynamic fracture toughness versus crack speed* relationship for the model problem.

The procedure is now followed for the case of dynamic elastic-plastic crack growth in the antiplane shear mode, which is the only case for which relatively complete results are available. Summary remarks on the less developed case of plane strain crack growth are included when corresponding results are available.

8.3.1 Plastic strain on the crack line

The active plastic zone is depicted in Figure 8.2. The crack propagates steadily at speed v through the material in the x_1-direction. The crack has traction-free faces and it occupies the half plane $x_1 < 0$, $x_2 = 0$. The active plastic zone extends a distance r_p ahead of the crack tip in the x_1-direction. The flow stress of the material in simple shear is τ_0 and the shear angle $\psi(x_1, x_2)$ at any point is shown in Figure 8.2.

The equations governing the stress and deformation fields in the active plastic zone are given as (4.4.1)–(4.4.6), and the same notation is followed here. This same problem, but with the inertial effects neglected, was studied by Rice (1968) and Chitaley and McClintock (1971). Each employed incremental plastic stress–strain relations based on the associated flow rule and a yield condition suitable for the antiplane deformation mode in a nonhardening material. They showed that the principal shear lines in the active plastic zone are straight, and Rice (1968) determined the distribution of plastic strain on the line directly ahead of the tip in terms of the unknown distance between the tip and the elastic-plastic boundary. Chitaley and McClintock (1971) went on to integrate the field equations numerically.

In the latter study, it was implicitly assumed that the principal shear lines in the active plastic zone are all members of a centered fan. This assumption was questioned by Dean and Hutchinson (1980) on the basis of a numerical finite element analysis of the steady quasi-static growth problem. If material inertia is taken into account in this steady state problem, then the system of governing partial differential equations is hyperbolic in the active plastic zone, as in the case of quasi-static growth. However, the local directions of principal shear lines are no longer characteristic directions. Instead, there exist two families of characteristic curves which coalesce to a single family corresponding to the shear lines in the limit as crack speed $v \to 0$. No clear picture of the structure of the characteristic net within the plastic zone for $v > 0$ is yet available.

An important step in this analysis is finding the shear strain ϵ_{23} on the crack line $x_2 = 0$, $0 < x_1 < r_p$ within the active plastic zone (Freund and Douglas 1982; Dunayevsky and Achenbach 1982). For this purpose, the dimensionless coordinates

$$x = x_1/r_p, \qquad y = x_2/r_p \qquad (8.3.1)$$

are introduced. Symmetry requires that $\sigma_{13} = \epsilon_{13} = 0$ on the line $y = 0$. From the definition of the shear angle ψ, a convention is adopted whereby $\psi = 0$ on the crack line. It should be noted that the plastic zone size r_p is not known beforehand, and the crack tip plastic zone analysis alone provides no means for determining it.

For steady crack growth, the governing system of partial differential equations can be reduced to (4.4.9). In terms of the dimensionless coordinates (8.3.1), these two first-order equations for the shear angle ψ and the x-component of shear strain γ are

$$\frac{\partial \psi}{\partial x} + \cos\psi \frac{\partial \gamma}{\partial x} + \sin\psi \frac{\partial \gamma}{\partial y} = 0 \,,$$

$$\cos\psi \frac{\partial \psi}{\partial x} + \sin\psi \frac{\partial \psi}{\partial y} + m^2 \frac{\partial \gamma}{\partial x} = 0 \,, \qquad (8.3.2)$$

where $m = v/c_s$. Equations (8.3.2) form a quasi-linear first-order hyperbolic system with two families of real characteristic curves in the x, y-plane. The system is equivalent to the ordinary differential equations

$$d(\psi \pm m\gamma) = 0 \qquad (8.3.3)$$

along the characteristic curves

$$\frac{dy}{dx} = \frac{\sin\psi}{\cos\psi \pm m} \qquad (8.3.4)$$

in the x, y-plane. The differential equations (8.3.4) for the characteristic curves depend on the shear angle $\psi(x,y)$, which is not known a priori, and, consequently, the characteristic network cannot be determined without determining the complete solution.

The equations in (8.3.2) are linear in the first derivatives, which is the distinguishing feature of a quasi-linear system, and the coefficients are functions of the dependent variables. A special feature is that the coefficients do not involve the independent variables. Equations such as these can be transformed into a linear first-order system by interchanging the roles of dependent and independent variables. Transformations of this type are often grouped under the heading of *hodograph transformations*, and this transformation is especially effective in cases in which a boundary with unknown location is involved (Courant and Friedrichs 1948). If the transformations

$$x = f(\psi, \gamma), \qquad y = g(\psi, \gamma) \qquad (8.3.5)$$

are introduced into (8.3.2), then the linear equations

$$\frac{\partial g}{\partial\gamma} - \cos\psi \frac{\partial g}{\partial\psi} + \sin\psi \frac{\partial f}{\partial\psi} = 0,$$

$$\cos\psi \frac{\partial g}{\partial\gamma} - \sin\psi \frac{\partial f}{\partial\gamma} - m^2 \frac{\partial g}{\partial\psi} = 0 \qquad (8.3.6)$$

are obtained. In the ψ, γ-plane, the system (8.3.6) is equivalent to the ordinary differential equations (8.3.4) along the straight characteristics curves (8.3.3). For this interpretation, dy/dx is understood to be dg/df. The ψ, γ-plane is shown in Figure 8.3.

At any point on the portion of the boundary of the active plastic zone that is advancing into the elastic region, the material is in a state of incipient flow. In particular, both the yield condition and the elastic constitutive relation are satisfied, hence

$$\gamma + \sin\psi = 0. \qquad (8.3.7)$$

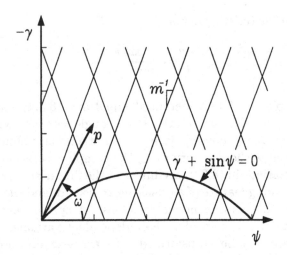

Figure 8.3.   The ψ, γ-plane showing the straight characteristics which satisfy (8.3.3) and the curve (8.3.7) which represents the leading boundary of the plastic zone.

This relation is also graphed in Figure 8.3. If the position of a point (f, g) on the elastic-plastic boundary were known in terms of ψ (or γ) on the boundary (8.3.7), then a straightforward numerical integration in the characteristic directions in the ψ, γ-plane would yield a complete description of the stress and deformation fields within the active plastic zone. Unfortunately, this information is not known a priori.

Note that the entire crack line $0 < x < 1$, $y = 0$ in the physical plane maps into the origin of the ψ, γ-plane. If the polar coordinates

$$p = \sqrt{\psi^2 + \gamma^2}\,, \qquad \omega = \tan^{-1}(-\gamma/\psi) \qquad (8.3.8)$$

are introduced, then the equations governing f and g take the form

$$-\sin(p\cos\omega) \begin{pmatrix} \cos\omega & -\sin\omega \\ \sin\omega & \cos\omega \end{pmatrix} \begin{pmatrix} f_{,p} \\ f_{,\omega}/p \end{pmatrix}$$

$$+\cos(p\cos\omega) \begin{pmatrix} \cos\omega & -\sin\omega \\ \sin\omega & \cos\omega \end{pmatrix} \begin{pmatrix} g_{,p} \\ g_{,\omega}/p \end{pmatrix}$$

$$+ \begin{pmatrix} \sin\omega & \cos\omega \\ m^2\cos\omega & -m^2\sin\omega \end{pmatrix} \begin{pmatrix} g_{,p} \\ g_{,\omega}/p \end{pmatrix} = 0\,, \qquad (8.3.9)$$

where $f_{,p} = \partial f/\partial p$ and so on. Since the entire crack line within the active plastic zone maps into the origin in the p, ω-plane, it is assumed that points near the crack line map into points near $p = 0$. For $p \ll 1$, the asymptotic approximations $\cos(p \cos \omega) \approx 1$ and $\sin(p \cos \omega) \approx p \cos \omega$ are incorporated, and the equations (8.3.9) then admit a separable solution of the form

$$f = p^n F(\omega), \qquad g = p^{n+1} G(\omega) \tag{8.3.10}$$

near $p = 0$, where n is a real number. The only value of n for which the inverse mapping will carry points near $p = 0$ into points near the line $0 < x < 1$, $y = 0$ in the physical plane is $n = 0$. With n fixed, it remains to determine the functions F and G. They are governed by a pair of first-order ordinary differential equations obtained by substituting (8.3.10) into (8.3.9) and then letting $p \to 0$. These equations take the form

$$\frac{dG}{d\omega} + \frac{1 + (1 + m^2) \sin \omega \cos \omega}{(\cos^2 \omega - m^2 \sin^2 \omega)} G = 0,$$

$$\tag{8.3.11}$$

$$\frac{dF}{d\omega} + \frac{1 - m^2}{\cos \omega (\cos^2 \omega - m^2 \sin^2 \omega)} G = 0.$$

A solution of this coupled pair of equations is sought subject to a suitable set of boundary conditions.

Points in the active plastic zone map into points above the curve $\gamma + \sin \psi = 0$ in Figure 8.3, so the differential equations (8.3.11) apply over a range of ω near $p = 0$ that increases from $\omega = \pi/4$. As the point $x = 1$, $y = 0$ is approached along the boundary of the plastic zone in the physical plane, both γ and ψ approach zero. In view of (8.3.7), however, they do so in such a way that $-\gamma/\psi \to 1$. The boundary condition

$$F(\pi/4) = 1 \tag{8.3.12}$$

follows immediately. Values of ω larger than $\pi/4$ near $p = 0$ correspond to points along $y = 0^+$ with $x < 1$. Therefore, the upper limit of the range of the equations (8.3.11) is some value of ω, say ω^*, for which x has decreased to zero, corresponding to the crack tip. Therefore, a second condition on the solution (8.3.11) is

$$F(\omega^*) = 0. \tag{8.3.13}$$

The value of ω^* is not known a priori so an additional condition is required to produce a unique solution of (8.3.11). According to Slepyan's asymptotic result (4.4.25) for this problem, ψ and γ are multiple valued at the crack tip for any value of m. For the present solution to have this feature, it is necessary that

$$G(\omega^*) = 0 . \qquad (8.3.14)$$

Upon integrating the differential equations (8.3.11) and enforcing the three conditions (8.3.12)–(8.3.14), it is found that $\omega^* = \tan^{-1}(1/m)$ and

$$F(\omega) = \frac{I(m\tan\omega)}{I(m)} ,$$

$$\qquad (8.3.15)$$

$$G(\omega) = \frac{m}{(1-m^2)I(m)} \frac{(\cos\omega - m\sin\omega)^{(1+m)/2m}}{(\cos\omega + m\sin\omega)^{(1-m)/2m}} ,$$

where I is the definite integral

$$I(z) = \int_0^{(1-z)/(1+z)} \frac{u^{(1-m)/2m}}{1+u}\, du . \qquad (8.3.16)$$

The integral can be evaluated in terms of elementary functions for any value of m for which $(1-m)/2m$ is a rational number, and it can be evaluated by numerical means for any value of m of interest.

Of primary interest for present purposes is the variation of the strain component ϵ_{23} along the crack line in the active plastic zone. The compatibility relation (4.4.2) implies that

$$\frac{\partial\epsilon_{23}}{\partial x_1} = \frac{\partial\epsilon_{13}}{\partial x_2} = \left(\frac{\tau_0}{2\mu r_p}\right)\frac{\partial\gamma}{\partial y} . \qquad (8.3.17)$$

If (8.3.17) is differentiated with respect to y, viewing ψ and γ as functions of x and y, then it is seen that

$$\frac{\partial\gamma}{\partial y} = -\frac{\sin\omega}{G(\omega)} . \qquad (8.3.18)$$

From (8.3.17) it follows that

$$\frac{2\mu\epsilon_{23}}{\tau_0} = \int_x^1 \frac{\sin\omega}{G(\omega)} \, dx' + 1 = \int_\omega^{\pi/4} \frac{\sin\omega'}{G(\omega')} \frac{dF(\omega')}{d\omega'} \, d\omega' + 1. \quad (8.3.19)$$

If the functions F and G are substituted from (8.3.15) into (8.3.19) and the integral is evaluated, then

$$2\epsilon_{23}(x_1,0) = \frac{\tau_0}{\mu} \left[-\frac{(1-m^2)}{2m^2} \ln\left(\frac{1-m^2\tan^2\omega}{1-m^2}\right) + 1 \right],$$

$$\frac{x_1}{r_p} = \frac{I(m\tan\omega)}{I(m)}. \quad (8.3.20)$$

This result is an exact, parametric representation for $\epsilon_{23}(x_1,0)$ within the active plastic zone for the problem at hand. Although this analysis of the fields within the active plastic zone was motivated by discussion of the response due to the surrounding elastic field characterized by a stress intensity factor, the result is more general. That is, the result does not depend on the surrounding elastic field, and it provides a description of the crack line plastic strain no matter what the nature of the fields surrounding the plastic zone might be. In particular, it does not depend in any way on whether or not the surrounding field meets the conditions of the small-scale yielding hypothesis.

Graphs of the strain distribution (8.3.20) are shown in Figure 8.4 in the dimensionless form of $2\mu\epsilon_{23}(x_1,0)/\tau_0$ versus x_1/r_p for several values of crack tip speed m. The main features of the variation are evident from the graphs. The total dimensionless strain at a material particle on the crack line is unity at the instant that the particle is engulfed by the active plastic zone, and the strain increases monotonically as the crack tip approaches the particle. The strain is singular at the crack tip, as already shown through the asymptotic analyses in Section 4.4. The most significant feature of the results plotted in Figure 8.4 is that the level of plastic strain at a given distance ahead of the crack tip as a fraction of plastic zone size is significantly reduced as a result of the inertial resistance of the material. For materials which fail by means of a locally ductile mechanism of separation, this suggests that an increasing driving force is required to sustain growth at increasing crack tip speed. This intuitive idea is put into a quantitative form in the next subsection.

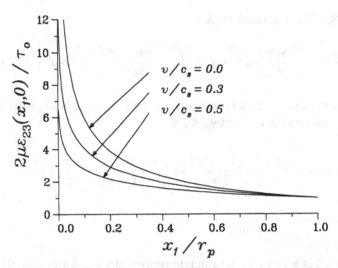

Figure 8.4. The shear strain ϵ_{23} on the crack line within the active plastic zone versus distance ahead of the crack tip for three values of crack growth speed.

Finally, recall the apparently contradictory asymptotic results for this problem obtained by Rice (1968) for $m = 0$ and any x_1 in the range $0 < x_1 < r_p$ and by Slepyan (1976) for any $m > 0$ and "small" values of x_1/r_p. These asymptotic results are given in (4.4.34) and (4.4.25), respectively. With the complete solution in hand in the form (8.3.20), this paradox can be pursued (Freund and Douglas 1982). The integral $I(z)$ in (8.3.16) has the form

$$I(z) = m \left(\frac{1-z}{1+z} \right)^{1/2m} + o(m) \qquad (8.3.21)$$

as $m \to 0$ for any z in the range $0 < z < 1$. It follows from (8.3.20) that

$$x_1/r_p \sim \exp(1 - \tan\omega) \qquad (8.3.22)$$

as $m \to 0$. If $\tan\omega$ is eliminated from (8.3.20) in favor of x_1 according to (8.3.22) and the limit as $m \to 0$ is taken, then Rice's result (4.4.34) is obtained. On the other hand, for points very near the crack tip, $m\tan\omega$ is very near unity and the integral $I(z)$ has the form

$$I(z) = \left(\frac{2m}{1+m} \right) \left(\frac{1-z}{2} \right)^{(1+m)/2m} + o(1-z) \qquad (8.3.23)$$

as $z \to 1^-$. Then, from (8.3.15),

$$\frac{x_1}{r_p} \approx L \left(\frac{1 - m \tan \omega}{2} \right)^{(1+m)/2m} , \qquad (8.3.24)$$

where L is an indeterminate scale factor. An expression for $\tan \omega$ in terms of x_1/r_p can be determined from (8.3.24) and substituted into (8.3.20). If, for any fixed value of m in the range $0 < m < 1$, the value of x_1 is sufficiently small that

$$\left(\frac{x_1}{r_p} \right)^{2m/(1+m)} \ll 1 , \qquad (8.3.25)$$

then Slepyan's result (4.4.25) is recovered. The condition (8.3.25) indicates the range of validity of the asymptotic result (4.4.25). Evidently, the range of validity vanishes as m vanishes, so the equilibrium crack growth result (4.4.34) is not recovered as $m \to 0$. Generally, for $m \leq 0.3$, the range of x_1 for which (8.3.25) can be satisfied appears to be smaller than $r_p/100$. The assumptions of a sharp crack, small strains, and homogeneous material are subject to question on this scale. The result (8.3.25) points out a difficulty in using asymptotic solutions alone to describe mechanical fields in fracture mechanics, namely, the range of validity of an asymptotic solution can be too small for that solution to be a useful approximation of crack tip fields.

8.3.2 A growth criterion

The generation of a theoretical fracture toughness versus crack speed relationship is now pursued along the lines outlined in the introduction to this section. That is, the crack grows in a body in such a way that the active plastic zone is surrounded by a region of elastic deformation and that the elastic field is characterized by the elastic stress intensity factor K_{III}. The stress intensity factor field for antiplane shear deformation is given in (4.2.15). This situation is the small-scale yielding situation of nonlinear fracture mechanics. Then, a ductile crack growth criterion is applied at the level of the crack tip plastic zone to determine the relationship that must be satisfied between the applied stress intensity factor and the crack speed for continued growth to be sustained. The particular level of applied remote stress intensity that satisfies this condition is denoted by K_{IIId}, and it is called the dynamic fracture toughness of the material. The growth criterion adopted for this calculation is the critical plastic strain criterion proposed by McClintock and Irwin (1965). According to this criterion, a crack will grow with a critical, material specific level of plastic strain at a point on the crack line at a characteristic distance ahead of the crack tip. Let the critical level of plastic strain ϵ_{23}^p be denoted by $\gamma_c^p \tau_o / 2\mu$, so that γ_c^p is the actual critical shear strain normalized by the yield strain. Likewise, let the characteristic distance be denoted by $r_c = x_c r_p$, so that x_c is the characteristic distance normalized by the plastic zone size r_p. To complete the specification of the growth criterion, the crack cannot grow for levels of plastic strain below γ_c^p at $x = x_c$, and levels of plastic strain above the critical level at $x = x_c$ are inaccessible.

Though this criterion is selected for the present calculation primarily for its simplicity, it should also be noted that the criterion captures the essence of the separation process for materials that fail by a local ductile mechanism. For example, suppose that the microstructural mechanism of crack advance in a material is the nucleation of microvoids and their subsequent ductile growth to coalescence. Assume that there is a certain distribution of void nucleation sites, say second phase particles within a certain size range, and that the flow behavior of the material is independent of the rate of deformation. Then, for a material volume in the crack tip region, a certain level of inelastic strain will be required to grow the voids from nucleation to coalescence no matter how rapidly the crack advances or how intense the applied loading might be.

For present purposes, the characteristic length x_c is eliminated in favor of the level of applied stress intensity factor that is required to initiate growth of a stationary crack in the same material under slowly applied monotonic loading. This level of applied stress intensity is denoted by K_{IIIc}. Rice (1968) showed that the plastic strain distribution on the crack line in the plastic zone in this case is

$$2\epsilon_{23}^p = \frac{\tau_o}{\mu} \left(\frac{K_{III}^2}{\pi \tau_o^2 x_1} - 1 \right) . \tag{8.3.26}$$

If the critical strain criterion is applied, then

$$\pi x_c (1 + \gamma_c^p) = \frac{K_{IIIc}^2}{r_c \tau_o^2} . \tag{8.3.27}$$

Let q be the value of $m\tan\omega$ in the parametric relationship between ϵ_{23} and x_1 defined in (8.3.20) when the growth criterion is satisfied. Then

$$\gamma_c^p = -\frac{1-m^2}{2m^2}\ln\left(\frac{1-q^2}{1-m^2}\right), \quad x_c = \frac{I(q)}{I(m)}. \tag{8.3.28}$$

The parameter x_c can be eliminated from (8.3.28) by means of (8.3.27) but the result will still depend on the unknown plastic zone size r_p. Because there is no characteristic length in the problem considered in Section 8.3.1, the plastic zone size must scale with K_{III}^2/τ_o^2. Furthermore, it must depend on crack speed in some way. Thus, assume that

$$r_p = X_p(m)\frac{K_{III}^2}{\tau_o^2}. \tag{8.3.29}$$

Then, for steady growth with the critical plastic strain criterion satisfied, (8.3.27) can be rewritten as

$$\left(\frac{K_{IIId}}{K_{IIIc}}\right)^2 = \frac{I(m)/I(q)}{\pi(\gamma_c^p+1)X_p(m)}, \tag{8.3.30}$$

where, from (8.3.28),

$$q^2 = 1 - (1-m^2)\exp[-2\gamma_c^p m^2/(1-m^2)]. \tag{8.3.31}$$

Equation (8.3.30) provides the relationship between fracture toughness and crack speed that was sought, obtained on the basis of the critical plastic strain criterion. The dependence of plastic zone size on crack speed represented by $X_p(m)$ is as yet unknown, and this dependence can only be determined with certainty from a complete solution of the problem.

8.3.3 A formulation for the complete field

The governing equations are now cast into a form appropriate for determination of the complete elastic-plastic field under small-scale yielding conditions. The dimensionless stress and strain components defined by

$$\tau_x = \sigma_{13}/\tau_o, \quad \tau_y = \sigma_{23}/\tau_o,$$
$$\gamma_x = 2\mu\epsilon_{13}/\tau_o, \quad \gamma_y = 2\mu\epsilon_{23}/\tau_o \tag{8.3.32}$$

are introduced for this purpose. Dimensionless coordinates (different from those used in Section 8.3.2) and a dimensionless displacement w are defined by

$$x = x_1 \tau_o^2 / K_{III}^2, \quad y = x_2 \alpha \tau_o^2 / K_{III}^2, \quad w = \mu \tau_o u_3 / K_{III}^2. \quad (8.3.33)$$

The parameter α is $\sqrt{1 - m^2}$.

In terms of the dimensionless variables, the equation of motion has the form

$$\frac{\partial \tau_x}{\partial x} + \alpha \frac{\partial \tau_y}{\partial y} = m^2 \frac{\partial^2 w}{\partial x^2}. \quad (8.3.34)$$

The total strain is decomposed additively into elastic and plastic parts

$$\gamma_\beta = \gamma_\beta^e + \gamma_\beta^p, \quad (8.3.35)$$

where $\beta = x$ or y, and the elastic strain is related to stress according to

$$\tau_\beta = \gamma_\beta^e = (\gamma_\beta - \gamma_\beta^p). \quad (8.3.36)$$

By making use of the strain–displacement relations

$$\gamma_x = \frac{\partial w}{\partial x}, \quad \gamma_y = \alpha \frac{\partial w}{\partial y} \quad (8.3.37)$$

and (8.3.36), the equation of motion can be rewritten in the form

$$\frac{\partial^2 w}{\partial x^2} + \frac{\partial^2 w}{\partial y^2} = \frac{1}{\alpha^2} \frac{\partial \gamma_x^p}{\partial x} + \frac{1}{\alpha} \frac{\partial \gamma_y^p}{\partial y}. \quad (8.3.38)$$

Thus, the differential equation governing the total displacement w is reduced to the Poisson equation with the right side depending on the unknown plastic strain distribution. According to the small-scale yielding hypothesis, the stress components are required to have far-field behavior that is the same as the near tip elastic solution (4.2.15). In terms of the dimensionless variables,

$$\tau_x \rightarrow -\frac{\sin \frac{1}{2}\theta}{\alpha (2\pi r)^{1/2}}, \quad \tau_y \rightarrow \frac{\cos \frac{1}{2}\theta}{(2\pi r)^{1/2}} \quad (8.3.39)$$

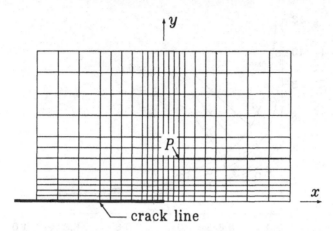

Figure 8.5. Coarse representation of the finite element mesh used to solve (8.3.38). The stress history of particle P is stored in its forward path-line to the boundary.

as $r \to \infty$ for $0 \leq \theta < \pi$, where $\theta = \arctan(y/x)$. Any plastic wake region of finite thickness is formally excluded from the asymptotic condition by enforcing (8.3.39) up to but not including $\theta = \pi$.

To provide a basis for application of a numerical finite element procedure, the governing equation (8.3.38) is recast into a variational form. A representation of the finite element mesh covering the region R over which the equation was solved is shown in Figure 8.5 (Freund and Douglas 1982). In order to solve the equations (8.3.38) for values of displacement at the mesh node points, the plastic strain must be known throughout R. This strain distribution is not known beforehand, however, so an iterative scheme is developed. The iterative scheme is based on the fact that the deformation fields are steady as seen by an observer traveling with the crack tip. This implies that the stress history of any material particle in the region is recorded on the forward path-line of that particle. For the particle P in Figure 8.5, for example, the stress history is the spatial variation of stress along the forward path-line from point P to the boundary of the region. Therefore, if the stress variation along this path-line is known, then the plastic strain at point P can be computed by integration of the incremental relation between plastic strain and stress from the boundary along the path-line to P. The cycles in the iterative procedure are then (i) finding a plastic strain distribution for an estimate of the stress distribution throughout R, and (ii) finding a total displacement field for an estimate of plastic strain throughout R.

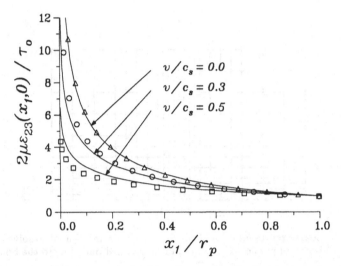

Figure 8.6. The discrete symbols show the strain distribution on the crack line within the active plastic zone obtained by the numerical procedure described in Section 8.3.3. The exact results from Figure 8.5 are included as the solid curves.

Results of calculations based on an array of 1800 rectangular elements, each comprised of four constant strain triangles, were reported by Freund and Douglas (1982). Nodal points were concentrated near the crack tip in anticipation of large field gradients there. The size of the smallest element in these calculations was about 0.3 percent of the maximum extent of the crack tip plastic zone, and the nearest boundary point (other than on the crack plane) was at a distance from the crack tip greater than ten times the maximum extent of the plastic zone. Thus, Figure 8.5 presents a very coarse representation of an actual element array.

The strain distributions on the crack line within the active plastic zone determined by means of the computational scheme are shown as discrete points in Figure 8.6 for three values of crack tip speed. The strain and distance ahead of the crack tip are normalized in the same way as in Figure 8.4, and the analytical results from the latter figure are included for comparison. The computed strain agrees quite well with the analytical result for strain up to more than ten times the plastic yield strain. The quality of the comparison is significant because the analytical result for strain on the crack line in the active plastic zone is the only exact result available to serve as a check on the numerical procedure outlined above. Once the algorithm has been tested in this way, it can be applied with greater confidence in cases for which there is no certain basis for comparison.

In order to complete the discussion of a dynamic fracture toughness versus crack speed relationship begun in Section 8.3.2, an estimate of the extent of the plastic zone in the direction of crack growth is required. More specifically, the dependence of the size of the plastic zone on the crack speed m represented by $X_p(m)$ in (8.3.29) is required. From the numerical results, this dependence is estimated to be

$$X_p(m) \approx (0.3 - 0.5 \, m^2) \qquad (8.3.40)$$

for $0 < m < 0.6$. As is evident from Figure 8.6, the variation of plastic strain with distance ahead of the crack tip near the elastic-plastic boundary is small for the higher crack speeds, which implies that the estimate (8.3.40) is less reliable for the higher crack speeds.

8.3.4 The toughness–speed relationship

With the mechanical fields in the active plastic zone related to the remote loading through (8.3.29) and (8.3.40), the theoretical dynamic fracture toughness versus crack tip speed relationship is completely specified by (8.3.30) for any values of K_{IIIc} and γ_c^p. These are the two parameters that specify the growth criterion. The normalized dynamic fracture toughness K_{IIId}/K_{IIIc} versus dimensionless crack tip speed m is shown in Figure 8.7 for three values of the critical plastic strain γ_c^p. The surface shown represents all states of the applied remote stress intensity factor and the crack speed for which steady propagation can be sustained according to the critical plastic strain criterion. The variable intercept at $m = 0$ is due to the overall increase in plastic deformation with increasing critical plastic strain. The plastically deformed material in the wake region has a stabilizing influence on the growth process, and the magnitude of this effect increases with increasing critical plastic strain. The intercept value of K_{IIId} at $m = 0$ corresponds to the so-called steady state toughness value of the theory of stable crack growth, that is, with the level of the asymptote of the resistance versus crack growth curve for a large amount of growth. The spread of intercepts on the $m = 0$ axis results from the use of the common factor K_{IIIc} to normalize all data. The intercept represents the lowest level of applied stress intensity factor that can cause crack growth; in this sense, it is the *arrest toughness* of the material.

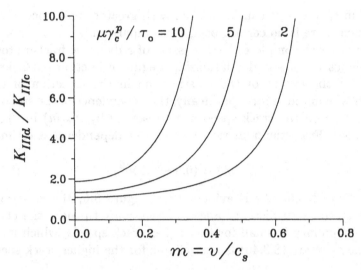

Figure 8.7. The dynamic fracture toughness versus crack speed relationship obtained by imposing the critical strain criterion described in Section 8.3.4.

The ratio of K_{IIId}/K_{IIIc} is a monotonically increasing function of crack speed for fixed critical strain, and this function takes on large values compared to unity for moderate values of m. Although there is no unambiguous way to associate a maximum attainable velocity with these results, they suggest such a terminal velocity well below the elastic wave speed of the material.

Perhaps the most significant observation on the result in Figure 8.7 is that the variation of toughness with crack speed is due to *inertial* effects alone. The material response is independent of the rate of deformation, and the crack growth criterion that is enforced involves no characteristic time. If inertial effects were neglected in this analysis, then the calculated toughness would be completely independent of speed for any given level of plastic strain. Because this theoretical result exhibits the main features observed in rapid crack growth in materials that are relatively strain rate insensitive in their bulk response and that fail in a locally ductile manner, it is concluded that material inertia on a small physical scale plays a significant role in establishing the toughness–speed relation observed in such materials. An explanation for this effect can be found in the fact that particle displacements, and therefore particle velocities, are much greater in the crack tip region for the case of elastic-plastic response than for elastic response. The greater particle velocities lead to greater inertial resistance to motion, and the model problem leads to quantitative estimates of this effect.

The equivalent plane strain problem of dynamic crack growth in an elastic-ideally plastic material has not been so fully developed. A study of a smooth punch moving along the surface of a rigid-ideally plastic solid was carried out by Spencer (1960). The field equations were referenced to a system of moving curvilinear coordinates, which were viewed as a generalization of the slip line coordinates in plane strain plasticity under equilibrium conditions. A solution of the equations of dynamic plasticity was obtained as a perturbation from known equilibrium slip line fields. The connection between equilibrium fields near the edges of a smooth punch and mode I crack tip fields was established by Rice (1968), but the matter has not been pursued for dynamic fields beyond the work of Spencer (1960). A numerical calculation leading to a fracture toughness versus crack speed relationship, analogous to Figure 8.7, has been described by Lam and Freund (1985). They adopted a critical crack tip opening angle growth criterion and derived results for dynamic mode I crack growth on the basis of the Mises yield condition and the J_2 flow theory of plasticity. Whereas in the case of antiplane shear deformation the plastic deformation is concentrated ahead of the crack tip, in this case the plastic deformation is concentrated in lobes off to the sides of the advancing crack, and the accumulated effect of this plastic strain is a net opening rate for the crack behind the tip. The region directly ahead of the crack tip is one of high hydrostatic stress but low deviatoric stress in mode I deformation. Rice and Sorensen (1978) proposed the condition that the crack must grow with the separation between the crack faces at some material specific distance, say r_m, behind the crack tip equal to a critical value, say δ_c. Results of calculations for several values of the dimensionless combination of parameters $E\delta_c/\sigma_o r_m$ are shown in Figure 8.8, where E is the Young's modulus of the material and σ_o is the uniaxial tensile flow stress. Finally, it is noted that the deformation fields obtained in this numerical simulation appeared to be qualitatively consistent with the asymptotic solution presented in Section 4.5, but the matter has not been resolved with certainty.

8.3.5 The steady state assumption

The analysis in this section has been carried out for steady state

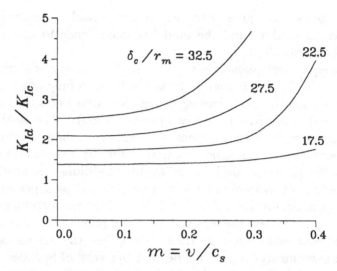

Figure 8.8. The fracture toughness versus crack speed relationship obtained by application of the numerical procedure described in Section 8.3.4 to the case of mode I crack growth in an elastic-ideally plastic material. The growth criterion imposed in this case was that the crack face opening displacement at a fixed distance behind the crack tip r_m assumed the constant value δ_c.

crack growth with constant remote stress intensity factor, but the hope is that the result can be used for nonsteady growth provided that the remote stress intensity factor does not vary too rapidly with time or position. Thus, a rough estimate of conditions under which the results of the steady state analysis can be applied for nonsteady conditions is sought. Suppose attention is focused on a particular material particle in the path of the crack tip plastic zone. For the case of a crack approaching steadily with a constant stress intensity factor, an expression for the plastic strain rate is given in (8.3.20). For purposes of comparison, consider a stationary crack at the same position in the same material and at the same level of applied stress intensity factor. If the stress intensity factor has some rate of change, then the rate of plastic strain can be calculated here as well. It might be anticipated that the steady state approximation will be useful in the nonsteady cases for which the former strain rate dominates the latter. If these rates are actually determined for the mode III problem considered in this section, using the formula (8.3.26) for the former and (8.3.20) for the latter, then it is found that the steady state approximation will be useful according to this criterion as long as

$$\tfrac{1}{2}\pi(1-m)X_p(m)\frac{c_s}{r_p} \gg \frac{\dot{K}_{III}}{K_{III}}.$$ (8.3.41)

The factor depending on m on the left side is of order one for the speed range of practical interest, so the condition simply suggests that the steady state approximation will be valid as long as the change in the stress intensity factor during the time required for an elastic wave to traverse the plastic zone is small compared to the magnitude of the stress intensity factor. This is not a severe constraint, so the steady state approximation is expected to be broadly useful.

8.4 High strain rate crack growth in a plastic solid

A particularly interesting class of dynamic fracture problems are those concerned with crack growth in materials that may or may not experience rapid growth of a sharp cleavage crack, depending on the conditions of temperature, stress state, and rate of loading. These materials can fracture by either a brittle or ductile mechanism on the microscale, and a focus of work in this area is on establishing conditions for one or the other mode to dominate. The phenomenon is most commonly observed in ferritic steels, but it has also been observed in tungsten (Liu and Shen 1984a, 1984b) and other materials. Such materials show a dependence of flow stress on strain rate, and the strain rates experienced by a material particle in the path of an advancing crack are potentially enormous. Consequently, the mechanics of rapid growth of a sharp macroscopic crack in an elastic-viscoplastic material that exhibits a fairly strong variation of flow stress with strain rate is of interest.

The general features of the process as experienced by a material particle on or near the fracture path are straightforward. As the edge of a growing crack approaches, the stress magnitude tends to increase due to the stress concentrating influence of the crack edge. The material responds by flowing at a rate related to the stress level in order to mitigate the influence of the crack edge. It appears that the essence of cleavage crack growth is the ability to elevate the stress to a critical level before plastic flow can accumulate to defeat the influence of the crack tip. In terms of the mechanical fields near the edge of an advancing crack, the rate of stress increase is determined by the elastic strain rate, whereas the rate of crack tip blunting is determined by the plastic strain rate. Thus, an equivalent observation is that the elastic strain rate near the crack edge must dominate the plastic strain rate in the crack tip region for sustained cleavage crack growth. This concept of *elastic rate dominance* is central to the model that follows. It is implicit in this approach that the material is intrinsically cleavable, and the question investigated is concerned with the way in which energy can be supplied to the immediate vicinity of the crack tip.

8.4.1 High strain rate plasticity

A material particle in the path of a rapidly advancing crack may be subjected to very high rates of plastic strain for a short period of time. For example, as a rough estimate, suppose that a crack grows in an elastic-plastic material at a speed v of about 10^3 m/s. Furthermore, suppose that a material particle in the crack path experiences a plastic strain γ^p of about 10^{-3} during the time that the active plastic zone of extent $r_p \approx 10^{-4}$ mm sweeps past the particle. An estimate of the plastic strain rate is then the accumulated plastic strain divided by the time that the particle is included within the active plastic zone,

$$\left(\dot{\gamma}^p\right)_{est.} \approx \frac{\gamma^p}{r_p/v}, \tag{8.4.1}$$

which has the value of 10^4 s^{-1} in this case. Thus, constitutive laws appropriate for the high strain rate response of materials must be adopted for a description of crack tip processes in rate-dependent materials.

At a temperature well below the melting temperature, many metals exhibit a relationship between flow stress in simple shear τ and the corresponding rate of plastic strain $\dot{\gamma}^p$ having the features illustrated schematically in Figure 8.9. The scale on the plastic strain rate axis is logarithmic. This high strain rate response apparently can be divided into two regimes. For strain rates above a certain level, hereafter called the *transition strain rate* $\dot{\gamma}_t$, the response is dominated by effects that can be represented as viscous effects, at least phenomenologically. For purposes of analysis, the response in this regime is described by

$$\dot{\gamma}^p = \dot{\gamma}_t + \dot{\gamma}_0(\tau - \tau_t)^n/\mu \quad \text{for} \quad \tau \geq \tau_t, \tag{8.4.2}$$

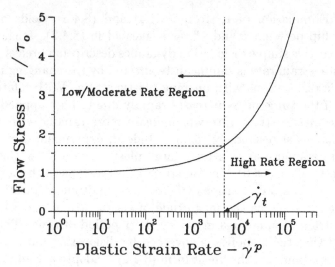

Figure 8.9. Schematic diagram of the dependence of flow stress on plastic shear strain rate for rapid deformation in simple shear.

where μ is the elastic shear modulus, τ_t is the transition flow stress defined as the flow stress when $\dot{\gamma}^p = \dot{\gamma}_t$, n is a material constant, and $\dot{\gamma}_0$ is a material constant representing the viscosity of the material. Hereafter, it will be identified as the *material viscosity*. The relation (8.4.2) is a particular case of the inelastic strain rate dependence introduced in Section 4.7. All further analysis in this section is based on the case $n = 1$ in (8.4.2).

The origin of a constitutive equation of the form (8.4.2) can be found in the "physical" theories of plasticity based on dislocation dynamics. Orowan (1934) showed that the macroscopic plastic strain rate due to N mobile dislocations per unit area with Burgers displacement b and average speed v_d is

$$\dot{\gamma}^p = bNv_d \,. \tag{8.4.3}$$

It is assumed that steady motion of a dislocation at speed v_d through the crystalline lattice requires an applied shear stress τ given by

$$\tau b = Bv_d \,, \tag{8.4.4}$$

where B is a drag coefficient. The origin of the drag on the dislocation is the background thermal motion of the lattice (Hirth and Lothe

1982). Elimination of v_d from (8.4.4) and (8.4.3) yields a linear relationship between τ and $\dot{\gamma}^p$, as suggested in (8.4.2). It should be noted that this simple dislocation dynamics description breaks down if the plastic strain rate is significantly affected by the changing number of mobile dislocations with ongoing plastic strain. With reference to (8.4.3), if the value of N increases rapidly due to high applied stress, then the plastic strain rate will increase more rapidly with applied stress than is suggested by (8.4.2). Indeed, experimental results on high rate response at relatively large plastic strain show this to be the case (Clifton 1983). In the study of crack growth, however, the main interest is on situations where the magnitude of plastic strain is about the same as the magnitude of elastic strain, and where the plastic strain accumulates in a very short period of time. Thus, it is assumed that the response is governed by dislocation motion, rather than generation, and the linear form (8.4.2) is adopted. Campbell and Ferguson (1970) carried out a series of experiments on mild steel; they give $\dot{\gamma}_t = 5 \times 10^3 \, \text{s}^{-1}$ and $\dot{\gamma}_0 = 3 \times 10^7 \, \text{s}^{-1}$ as values of the constants in (8.4.2).

For plastic strain rates below the transition strain rate $\dot{\gamma}_t$, the stress required to produce plastic flow is relatively insensitive to variations in strain rate. The plastic strain rate may drop many orders of magnitude with little reduction in flow stress. The physical mechanism of flow in this regime is believed to be the thermally activated motion of dislocations past discrete obstacles in the material lattice. For the case of iron, Frost and Ashby (1982) have proposed a relationship between flow stress and plastic strain rate of the form

$$\dot{\gamma}^p = c_p \left(\frac{\tau}{\mu}\right)^2 \exp\left\{ -\frac{\Delta F_p}{kT} \left[1 - \left(\frac{\tau}{\hat{\tau}}\right)^{3/4} \right]^{4/3} \right\} \qquad (8.4.5)$$

for $\tau \leq \tau_t$, where $\hat{\tau}$ is the flow stress at a temperature of absolute zero, T is the absolute temperature in degrees Kelvin, k is Boltzmann's constant, and ΔF_p is an activation energy. Representative values of the constants at $T = 300 \, \text{K}$ are given by Frost and Ashby (1982) as

$$\hat{\tau}/\mu_0 = 10^{-2}, \quad c_p = 10^{11} \, \text{s}^{-1}, \quad \Delta F_p/\mu_0 b^3 = 0.1, \qquad (8.4.6)$$

where μ_0 is the value of the shear modulus at $T = 300 \, \text{K}$ and $b = 2.48 \times 10^{-10} \, \text{m}$ is a lattice-spacing parameter.

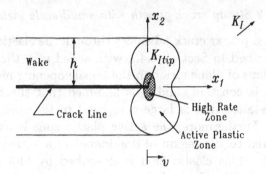

Figure 8.10. A schematic representation of steady crack growth at speed v under plane strain conditions in a rate-dependent plastic solid due to a remotely applied elastic stress intensity factor field.

To generalize the constitutive relation to a multiaxial stress state represented by σ_{ij}, let

$$\tau = \left(\tfrac{1}{2}s_{ij}s_{ij}\right)^{1/2} , \qquad (8.4.7)$$

where s_{ij} is the stress deviator. Represent the simple shear viscoplastic relation identified jointly by (8.4.2) and (8.4.5) by

$$\dot{\gamma}^p = F(\tau) . \qquad (8.4.8)$$

Then the plastic strain rate under multiaxial states of stress is

$$\dot{\epsilon}^p_{ij} = F(\tau)\frac{s_{ij}}{2\tau} , \qquad (8.4.9)$$

which reduces to (8.4.8) for simple shear. When τ drops below the slow loading yield stress τ_y, the plastic strain rate is zero. This cutoff is somewhat arbitrary, but the cutoff value has little influence on the results that are obtained.

The material is assumed to be elastically isotropic so the elastic strain rate is given by

$$\dot{\epsilon}^e_{ij} = \frac{1+\nu}{E}\dot{\sigma}_{ij} - \frac{\nu}{E}\dot{\sigma}_{kk}\delta_{ij} , \qquad (8.4.10)$$

where E is Young's modulus and ν is Poisson's ratio. The total strain rate is the sum of the elastic and plastic strain rates,

$$\dot{\epsilon}_{ij} = \dot{\epsilon}^e_{ij} + \dot{\epsilon}^p_{ij} . \qquad (8.4.11)$$

8.4.2 Steady crack growth with small-scale yielding

Suppose that a planar crack advances through the elastic-viscoplastic material described in Section 8.4.1 with speed v in the x_1-direction under conditions of plane strain in the tensile opening mode, or mode I. The speed is constant, and it is assumed that the crack tip has traveled a distance that is large compared to the size of the active plastic zone. Furthermore, the active plastic zone is assumed to be small compared to the region of dominance of a surrounding elastic singular field. This elastic field is described by (4.3.10) and it is characterized by the dynamic stress intensity factor K_I, which is assumed to be maintained at a constant level. Under these conditions the steady state, small-scale yielding situation depicted in Figure 8.10 is considered. The active plastic zone travels with the crack tip leaving behind a wake of plastic strain and, in general, an accompanying residual stress distribution. Outside of the wake region, the remote stress field in this asymptotic small-scale yielding problem is the dynamic singular field specified by K_I or, equivalently, by the dynamic energy release rate G by means of (5.3.10). This situation is analyzed within the framework of small strain theory under the assumption that the mechanical fields are constant as seen by an observer traveling with the crack tip. Thus,

$$\dot{f} = -v\frac{\partial f}{\partial x_1}, \qquad (8.4.12)$$

where the superimposed dot denotes material time differentiation and f is any scalar field or any rectangular component of an array.

An observation that is central to the analysis of this situation concerns the behavior of the stress and deformation fields near the crack tip deep within the active plastic zone. For the viscoplastic law (8.4.2), the considerations of Section 4.7 lead to the conclusion that

$$\sigma_{ij} \sim \frac{K_{Itip}}{\sqrt{2\pi r}}\Sigma_{ij}(\theta, m) \qquad (8.4.13)$$

as $r \to 0$, where r, θ are crack tip polar coordinates and $m = v/c_R$. The functions $\Sigma_{ij}(\theta, m)$ representing the angular variation of the stress distribution deep within the plastic zone are *identical* to the corresponding functions representing the angular variation of the stress distribution in the remote elastic field. The crack tip stress intensity factor K_{Itip}, however, differs from the remote stress intensity factor

K_I. The main purpose of the analysis is to establish the connection between the two stress intensities.

Because the elastic strain rate dominates the plastic strain rate near the crack tip, the local mechanical field is square root singular and there is a net energy flux out of the body at the tip which, according to (5.3.10), is

$$G_{tip} = A_I(v)\frac{1-\nu^2}{E}K_{Itip}^2. \qquad (8.4.14)$$

The rate of energy supply per unit crack advance to the crack tip region which is provided by the remote loading is G. Some of this mechanical energy is dissipated through active plastic flow, and some is locked in the wake region as stored elastic energy due to the development of incompatible plastic strain in the wake. The remainder is G_{tip}. It will be assumed subsequently that a certain level of G_{tip} must be maintained if the crack is to advance as a sharp crack. Thus, the relationship between G and G_{tip} is of central interest. The important question concerns how much of the applied energy release rate G is available to drive the crack in the form of G_{tip}. In other words, the question concerns the extent to which the intervening plastic flow is effective in *screening* the crack tip from the applied loading. It is noted in passing that $G_{tip} = 0$ for rate-independent plastic response, as well as for viscoplastic response (8.4.2) with $n > 3$.

The overall work rate balance, which is tacit in the foregoing arguments used to motivate the study of the relationship between G and G_{tip}, can be written as

$$G_{tip} = G - \frac{1}{v}\int_R \sigma_{ij}\dot{\epsilon}_{ij}^p \, dR - \int_{-h}^{h} U_e^* \, dx_2, \qquad (8.4.15)$$

where R is the area of the active plastic zone in the x_1, x_2-plane, h is the thickness of the wake on either side of the crack plane in the x_2-direction, and $U_e^*(x_2)$ is the elastic energy density locked in the wake as $x_1 \to -\infty$. This relationship states that the energy release rate at the crack tip deep within the active plastic zone equals the rate of energy supply to the crack tip region reduced by the rate of mechanical energy dissipation due to active plastic flow and further reduced by the rate at which energy is locked into the wake.

In light of the restriction to steady fields, the relationship (8.4.15) is evident from an overall work rate balance. However, as was demonstrated by Freund and Hutchinson (1985), the relationship can also be derived by direct application of the path-independent integral (5.2.9). For a path of integration which is very close to the crack tip compared to the size of the active plastic zone and which begins on one traction-free crack face, surrounds the tip, and ends on the opposite crack face, the value of the integral is G_{tip}. On the other hand, for a path of integration that is far from the crack tip compared to the size of the plastic zone, the value of the integral is

$$G - \int_{-h}^{h} U^* \, dx_2, \qquad (8.4.16)$$

where

$$U^*(x_2) = \lim_{x_1 \to -\infty} U(x_1, x_2) \qquad (8.4.17)$$

is the stress work density (5.2.3) in the wake region far behind the crack tip. In view of the path independence of (5.2.9),

$$G_{tip} = G - \int_{-h}^{h} U^* \, dx_2. \qquad (8.4.18)$$

To see that this is identical to (8.4.15), split the total stress work density U into elastic and plastic parts according to

$$U_e = \int_{-\infty}^{t} \sigma_{ij} \dot{\epsilon}_{ij}^e \, dt', \quad U_p = \int_{-\infty}^{t} \sigma_{ij} \dot{\epsilon}_{ij}^p \, dt'. \qquad (8.4.19)$$

The area integral in (8.4.15) can then be rewritten as

$$\int_R \sigma_{ij} \dot{\epsilon}_{ij}^p \, dR = \int_R \dot{U}_p \, dR = -v \int_R \frac{\partial U_p}{\partial x_1} \, dx_1 dx_2$$

$$\qquad (8.4.20)$$

$$= v \int_{-h}^{h} \lim_{x_1 \to -\infty} U_p(x_1, x_2) \, dx_2 = v \int_{-h}^{h} U_p^* \, dx_2.$$

Substitution of (8.4.20) into (8.4.15) then yields (8.4.18) once again. In any case, the expression (8.4.15) is exact.

8.4.3 An approximate analysis

A complete solution of the field equations for the asymptotic small-scale yielding problem is not available, although an approximate numerical solution of the field equations has been described by Freund, Hutchinson, and Lam (1986) and Mataga, Freund, and Hutchinson (1987). In this section, an approximate analysis that results in an explicit relationship between G and G_{tip} is outlined. The approximations that are incorporated are based on the idea of elastic rate dominance of the crack tip fields, and they are exact in the limit of vanishing plastic dissipation. The overall work rate balance (8.4.15) is the starting point for the approximate analysis.

The term in (8.4.15) representing the elastic energy stored in the wake is considered first. For steady growth of a mode I crack through a rate-independent elastic-plastic material under small-scale yielding conditions, the elastic energy locked into the wake is roughly 10 percent of the total rate of energy supply, represented here by G (Dean and Hutchinson 1980). Both inertial effects and material rate effects tend to suppress the accumulation of plastic strain in the active plastic zone which, in turn, implies smaller incompatible plastic strain in the wake. In particular, for the present case the elastic energy stored in the wake is expected to be a very small percentage of G, and this contribution to the balance in (8.4.15) will be neglected. The validity of this assumption has been supported by the numerical analyses cited above.

The plastic dissipation term in (8.4.15) is considered next. If the distribution of stress σ_{ij} throughout the plastic zone were known, then $\dot{\epsilon}_{ij}^p$ could be determined from the constitutive equation (8.4.9) and the integral over the region R in (8.4.15) could be evaluated. Of course, the stress can be determined only by solving the field equations. The stress distribution deep within the active plastic zone has the asymptotic property (8.4.13). The stress takes on its largest magnitude close to the crack tip, and the rate of plastic strain is also largest where the stress is largest. Consequently, the plastic dissipation in (8.4.15) can be estimated on the basis of the assumption that the stress everywhere in the active plastic zone is given by (8.4.13). This is the approach followed originally by Freund and Hutchinson (1985). In fact, the stress field must exhibit a transition from a remote field given by (4.3.10) to the near tip field (8.4.13) through the active plastic zone, where $K_{Itip} \leq K_I$. Consequently, the stress level in the active plastic zone given by (8.4.13) is expected to result in an estimate of the plastic dissipation that is too small.

Alternatively, it could be assumed that the stress in the plastic zone is given by (8.4.13) with K_{Itip} replaced by K_I. This approximation overestimates the stress magnitude everywhere in the plastic zone. A compromise approximation which overestimates the stress magnitude at the crack tip but which might be expected to underestimate it at points further from the tip is

$$\sigma_{ij} \approx \sqrt{\frac{K_{Itip}K_I}{2\pi r}} \, \Sigma_{ij}(\theta, m) \qquad (8.4.21)$$

everywhere within the active plastic zone. Though the assumption is motivated only by an interest in exhibiting the qualitative features of crack growth under the conditions described, it has been verified by numerical calculation that the assumption provides an accurate estimate of plastic dissipation (Mataga et al. 1987).

The integrand of the plastic dissipation term in (8.4.15) is

$$\sigma_{ij}\dot\epsilon_{ij}^p = \tau F(\tau), \qquad (8.4.22)$$

where τ is the effective shear stress defined in (8.4.7). In light of the assumption (8.4.21), τ is given by

$$\tau = \sqrt{\frac{K_{Itip}K_I}{2\pi r}} \, B(\theta, m), \qquad (8.4.23)$$

where $B(\theta, m)$ is defined by

$$B^2 = \tfrac{1}{2}\Sigma_{11}^2 + \Sigma_{12}^2 + \tfrac{1}{2}\Sigma_{22}^2 + \tfrac{1}{6}\left(2\nu^2 + 2\nu - 1\right)\left(\Sigma_{11} + \Sigma_{22}\right)^2 \quad (8.4.24)$$

everywhere in the active plastic zone. If $r_p(\theta)$ is the radial distance from the crack tip to the boundary of the active plastic zone, where $\tau = \tau_y$, then

$$r_p(\theta) = \frac{1}{2\pi}\frac{K_{Itip}K_I}{\tau_y^2}\, B(\theta, m)^2. \qquad (8.4.25)$$

The plastic dissipation term in (8.4.15) is

$$\frac{1}{v}\int_R \sigma_{ij}\dot\epsilon_{ij}^p \, dR = \frac{1}{v}\int_{-\pi}^{\pi} d\theta \int_0^{r_p(\theta)} \tau F(\tau) r \, dr. \qquad (8.4.26)$$

Next, using (8.4.23) to change the variable of integration from r to τ, it is found that

$$\frac{1}{v} \int_R \sigma_{ij} \dot{\epsilon}_{ij}^p \, dR = \frac{1}{2\pi^2 v} K_{Itip}^2 K_I^2 C(m) \int_{\tau_y}^{\infty} \tau^{-4} F(\tau) \, d\tau , \qquad (8.4.27)$$

where

$$C(m) = \int_{-\pi}^{\pi} B(\theta, m)^4 \, d\theta . \qquad (8.4.28)$$

The integral with respect to τ in (8.4.27) is independent of crack speed m. It is convenient to separate this integral into contributions from the high strain rate regime with $\tau_t \leq \tau < \infty$ and from the low or moderate regime with $\tau_y \leq \tau < \tau_t$. The former contribution is easily evaluated because $F(\tau)$ is a linear function in this regime, whereas the latter contribution is small by comparison for most realistic values of material parameters and it is therefore neglected. Thus,

$$\int_{\tau_y}^{\infty} \tau^{-4} F(\tau) \, d\tau \approx \frac{\dot{\gamma}_0}{6\mu\tau_t^2} + \frac{\dot{\gamma}_t}{3\tau_t^3} . \qquad (8.4.29)$$

With this estimate of plastic dissipation, the work rate balance takes the form

$$G_{tip} = G - \frac{\dot{\gamma}_0 \mu}{3c_s \tau_t^2} \left(1 + \frac{2\dot{\gamma}_t \mu}{\dot{\gamma}_0 \tau_t} \right) D(m) G G_{tip} , \qquad (8.4.30)$$

where the stress intensity factors K_I and K_{Itip} have been replaced by equivalent expressions in terms of energy rates G and G_{tip} by means of (5.3.10), and

$$D(m) = \frac{C(m) c_s / c_R}{\pi^2 m \left[(1 - \nu) A_I(m) \right]^2} . \qquad (8.4.31)$$

Values of the function $D(m)$ have been determined numerically for $\nu = 0.3$ and the result shown in Figure 8.11. The function has a minimum at $m = 0.55$, which is an observation that has some significance for the details to follow.

Suppose that the criterion for the crack to propagate steadily is that the crack tip energy release rate G_{tip} must take on a material specific value,

$$G_{tip} = G_{tip}^c . \qquad (8.4.32)$$

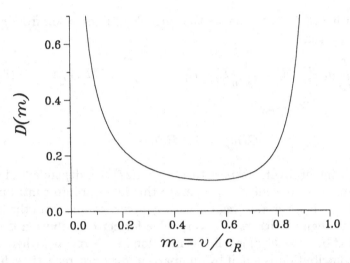

Figure 8.11. The function $D(m)$ defined in (8.4.31) versus dimensionless crack speed $m = v/c_R$ for $\nu = 0.3$.

For present purposes G_{tip}^c is assumed to be a material constant, independent of temperature and crack speed. The quantity G_{tip} represents the energy per unit crack advance that flows out of the body at the crack tip and that is not otherwise taken into account in the continuum analysis. The growth criterion (8.4.32) implies that this is the energy per unit crack advance that is absorbed by the material separation process. The quantity G_{tip}^c should perhaps be assumed to be a function of v and T, but not enough is known about this process to provide a basis for assigning dependence. With (8.4.32) imposed, relation (8.4.30) yields the value of G required to drive the crack at speed v as

$$\frac{G}{G_{tip}^c} = \frac{1}{[1 - D(m)P_c]},$$ (8.4.33)

where P_c is the dimensionless collection of material parameters

$$P_c = \frac{\dot{\gamma}_0 \mu G_{tip}^c}{3\tau_t^2 c_s} \left[1 + \frac{2\dot{\gamma}_t \mu}{\dot{\gamma}_0 \tau_t} \right].$$ (8.4.34)

The term in the expression for P_c that is perhaps the most strongly temperature dependent is τ_t, which decreases with increasing temperature for iron, for example. In this sense, P_c is a temperature-like parameter in this model.

The ratio G/G_{tip}^c is plotted against crack speed m for three values of P_c in Figure 8.12. For the corresponding value of P_c, each curve indicates all states represented by G and m for which steady crack growth can be sustained under the conditions assumed. The general shape of the curve for any fixed value of P_c could have been anticipated from the basic features of the model. At low crack speed, there is sufficient time for plastic strain to accumulate, so a larger remote energy release rate is required to maintain a critical crack tip energy release rate. This effect diminishes with increasing crack speed, which accounts for the falling branch of the curve with increasing m. At high crack speed, on the other hand, inertial resistance of the material is called into play. Again, a high driving force is required to sustain steady growth, and this effect accounts for the rising branch of the curve with increasing m. The combination of these effects leads to the variation of driving force versus speed with a minimum as shown. The fact that the curves become asymptotically infinite as $v \rightarrow 0$ is a shortcoming of the model. As the crack speed becomes small, there is time for plastic strain to accumulate. However, there is no provision in the model for this case (it is based on the assumption of elastic rate dominance), and the limiting response as the rate sensitivity vanishes is not rate-independent plastic response, in the sense considered in the preceding section. Modeling of the transition between the crack propagation situations in which elastic strain dominates, as in this section, and in which plastic strain dominates, as in the preceding section, is incomplete, although some ideas are summarized in Section 8.5.

8.4.4 Rate effects and crack arrest

Although the analysis is based on the assumption that the process is steady state, some implications for nonsteady crack propagation can be inferred. For example, for any state above the curve for the appropriate value of P_c, the crack is overdriven and it will tend to accelerate toward the rising branch of the curve. In this sense, the rising branch of each curve represents states that are stable under perturbations in driving force, whereas the falling branch represents unstable states. As a second example, consider the situation in which a crack is initiated in a material at values of G and m that correspond to a point above the steady state curve for the prevailing value of P_c, and the crack then propagates under loading that produces a dynamic energy release rate that decreases as the crack becomes

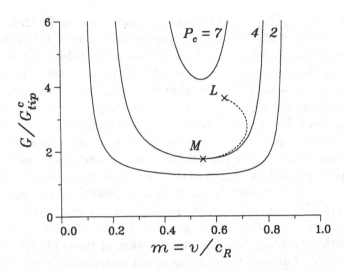

Figure 8.12. The ratio of remote crack tip driving force G to the crack tip driving force G_{tip}^c versus dimensionless crack speed m as given in (8.4.33) for three values of the dimensionless parameter P_c in (8.3.34). The dashed line is a possible representation of the G-v history for fracture initiation at state L if the crack grows into a region of decreasing driving force.

longer. Point L in Figure 8.12 indicates such a starting point. The crack tip apparently accelerates toward the stable branch of the steady state curve with G decreasing, as suggested by the dashed curve in the figure. Once the stable branch is reached, the crack can continue to grow but with diminishing speed by following this branch in the direction of decreasing G toward the minimum in the steady state curve. The existence of this minimum is perhaps the most significant feature of the result shown in Figure 8.12. It implies that, for a given value of P_c, there is some minimum level of applied energy release rate below which rapid growth of a sharp crack cannot be sustained.

Although the model being developed here does not include events after the driving force falls below this minimum, it is possible to speculate on the matter. For G below this minimum, the growth criterion (8.4.32) can no longer be satisfied, which implies that growth must be arrested. The graphs in Figure 8.12 suggest that this occurs from a relatively high crack propagation speed, hence arrest is abrupt rather than gradual. The abruptness of crack arrest in rate sensitive materials was observed by Clark and Irwin (1966) in hard plastics, and by Ishikawa and Tsuya (1976) and deWit and Fields (1987) in

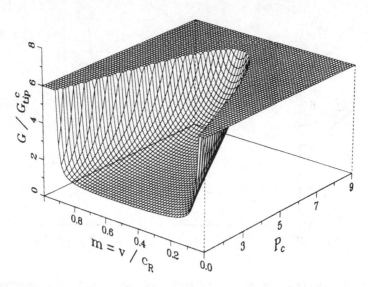

Figure 8.13. Surface of G/G^c_{tip} over the plane of dimensionless crack speed m and dimensionless material parameter P_c according to (8.4.33). The surface is truncated arbitrarily at $G/G^c_{tip} = 6$ to permit better visualization of its features.

structural steels. Once the crack tip motion is stopped, it is no longer possible for elastic deformation rates to dominate plastic rates, so plastic deformation can develop freely. The crack tip is expected to remain stationary until some ductile growth criterion is satisfied or until the stress is elevated within the plastic zone, presumably by a combination of strain hardening and rate effects, to a level sufficiently great to reinitiate cleavage fracture. It is emphasized that this sequence of events is speculative; the model itself provides no direct information on events that might occur when the propagation condition can no longer be satisfied.

The graph in Figure 8.13 gives the locus of state combinations of G, v, and P_c for which steady state propagation of a sharp crack can be sustained. As was discussed in connection with Figure 8.12, the implication is that if a cleavage crack can be initiated for a state combination G, v, P_c that is above the surface, then the crack will accelerate to a state on the stable branch of the surface (that is, the side with increasing G at fixed P_c). If the driving force diminishes as the crack advances, or if the local value of P_c increases as the crack advances due to a temperature increase, say, then the state combination will move toward the minimum point on the surface at

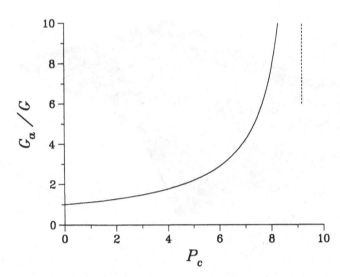

Figure 8.14. The crack arrest fracture energy G_a versus material parameter P_c implied by (8.4.33). The arrest toughness K_{Ia} can be determined by means of (5.3.10).

the local value of P_c. If the driving force is further decreased, or if P_c is further increased, then growth of a sharp cleavage crack cannot be sustained according to the model. The implication is that the crack will arrest abruptly from a fairly large speed, and a plastic zone will then grow from the arrested crack.

Of special significance is the observation that, at any given temperature, the variation of required driving force with crack speed has an absolute minimum, say G_a/G^c_{tip}. This implies that it is impossible to sustain cleavage crack growth at that value of P_c with a driving force below this minimum. Thus, this minimum as a function of P_c can be interpreted as the variation of the *crack arrest toughness* for the material with the local value of P_c or, for most practical purposes, with material temperature. This minimum is plotted against P_c in Figure 8.14. The dependence of P_c on temperature depends on the material properties, as seen from (8.4.34). Perhaps the most significant variation arises from the temperature dependence of the transition flow stress τ_t. Based on high strain rate data for mild steel reported by Campbell and Ferguson (1970), the parameter P_c increases monotonically from a value of about one to a value of about nine as the temperature increases from 0 K to 300 K for this material. Crack arrest data for A533 grade B pressure vessel steel is reported and discussed by Pugh et al. (1988). For this material, arrest of rapid fracture coincides with the end of cleavage crack growth.

Finally, it should be noted that the foregoing argument on crack arrest due to the combined roles of rate and temperature effects is based on global energy considerations, and is not concerned with the details of the microstructural mechanism by which rapid cleavage crack growth gives way to slower ductile growth. The accumulation of plastic strain ahead of the advancing crack to mitigate the stress concentrating effect of the crack edge, with the ductile growth of voids leading to coalescence as a consequence, is one possibility. Another possibility for polycrystalline materials is that, as plastic strain accumulation ahead of the cleavage crack tip increases for whatever reason, not all grains in the crack path can be cleaved. Some grains are left behind the main crack front as ligaments connecting the crack faces (Hoagland et al. 1972). With the relief of stress triaxiality, these ligaments then must be extended plastically to open the crack. As the distance of a ligament behind the main crack increases, so does its restraining effect on further crack advance, and a distribution of these ligaments may be responsible for arrest of a running cleavage crack.

8.5 Fracture mode transition due to rate effects

In Section 8.3 a model of the process of dynamic crack growth through an elastic-plastic material is developed for the case when some crack tip plastic deformation measure is required to take on a material specific value by the "ductile" growth criterion. Similarly, in Section 8.4, a model is developed that provides a description of rapid crack growth in a rate-dependent elastic-plastic material under conditions that permit crack advance in a cleavage mode. The notion of elastic rate dominance within the crack tip field is identified as a necessary condition for advance of a sharp crack in a "brittle" cleavage mode through an elastic-viscoplastic material. The descriptive terms "brittle" and "ductile" are used here to suggest material separation mechanisms that are stress-controlled and deformation controlled, respectively. These connotations, which embody gross idealizations of actual physical processes of material separation, are not universal but they are adopted for the discussion in this section.

Both of these lines of study of nonlinear fracture phenomena have led to conceptual pictures of rapid crack advance by means of stress-controlled and deformation-controlled mechanisms that are quite complete. However, the two lines of study have few intersections, in the sense that models have not been developed for which *either* type of criterion can be applied, depending on the prevailing conditions. The availability of such models would make it possible to analyze the process of fracture mode conversion as it is observed to occur, most commonly in fracture initiation and crack arrest. The purpose of this section is to outline a modest step in the direction of developing such a model. The analysis is based on the simple one-dimensional strip yield model of crack tip plasticity. In spite of the limitations, the form of the results establishes a framework for pursuing the matter through computational or other less direct methods in the future. The deficiencies of the strip yield model of the crack tip plastic zone for describing fracture processes that occur under essentially plane strain conditions have been identified elsewhere. As was noted in Section 8.3, the influence of the wake of the active plastic zone is very small in the high strain rate crack growth regime, so the absence of a wake effect with the strip yield model is not viewed as a serious drawback in considering processes in this regime.

The presentation is organized in the following way. A boundary value problem is formulated that models the process of steady dynamic growth of a crack under plane strain, small-scale yielding conditions. The influence of material strain rate sensitivity is included by assuming that the resistance to opening within the strip yield zone depends on the opening rate. The analysis follows the work of Glennie (1971b), who considered the case of general yielding. Then, it is assumed that the crack grows steadily according to a critical crack tip opening displacement criterion or a critical stress at a characteristic distance criterion to infer the relationship that must be satisfied between the remotely applied stress intensity factor (or dynamic fracture toughness) and the crack speed in order to sustain growth. Finally, the criteria are imposed simultaneously and, under the assumption that the criterion that is easiest to satisfy will be the one that is operative, the complete toughness–speed relationship is generated.

8.5.1 Formulation

Consider a planar crack extending in an otherwise unbounded elastic solid under two-dimensional plane strain conditions. The crack tip moves at constant speed v and the traction distribution on the crack

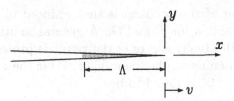

Figure 8.15. A representation of steady growth of a crack under plane strain conditions with a cohesive zone of length Λ.

faces is time invariant in a reference frame translating with the crack tip. Under these conditions, a steady state solution of the governing equations is sought. A rectangular x, y-coordinate system is established in the plane of deformation with its origin fixed at the moving crack tip. The crack grows in the plane $y = 0$ and the crack occupies the half plane $x < 0$; see Figure 8.15. The general solution procedure is outlined in Section 6.2.1 and the notation of that section is adopted here.

The boundary conditions imposed on the crack faces that provide the essential features of the strip yield model are

$$\sigma_{yy}(x, 0^{\pm}) = \begin{cases} \sigma(x) & \text{for } -\Lambda < x < 0, \\ 0 & \text{for } -\infty < x < -\Lambda, \end{cases}$$

$$\sigma_{xy}(x, 0^{\pm}) = 0 \qquad \text{for} \quad -\infty < x < 0.$$

(8.5.1)

Thus, $\sigma(x)$ represents the cohesive stress in the strip yield zone of length Λ. In terms of the unknown functions F and G introduced in Section 6.2 for analysis of steady elastodynamic crack growth, the boundary conditions take the form

$$(1 + \alpha_s^2)\left[F_+''(x) + F_-''(x)\right] + 2\alpha_s\left[G_+''(x) + G_-''(x)\right]$$
$$= -2\sigma(x)H(\Lambda + x)/\mu, \quad (8.5.2)$$

$$2\alpha_d\left[F_+''(x) - F_-''(x)\right] + (1 + \alpha_s^2)\left[G_+''(x) - G_-''(x)\right] = 0$$

for $-\infty < x < 0$, where $H(\cdot)$ is the unit step function and the symmetries consistent with mode I crack growth have been incorporated. The functions are further restricted by the requirement that their far-field behavior must match the square root singular stress intensity factor field for steady dynamic growth (4.3.10).

The solution of the problem is thus reduced to the solution of a Hilbert arc problem for F and G. A general solution is presented by Muskhelishvili (1953) and only the result is presented here. The gradient of the opening displacement $u_y(x, 0^+)$ in the x-direction along the crack face $y = 0^+$ is found to be

$$\frac{\partial u_y}{\partial x} = \frac{v^2}{c_s^2} \frac{\alpha_d}{D} \frac{1}{\mu\sqrt{-x}} \left[\frac{1}{\pi} \int_0^{-\Lambda} \frac{\sqrt{-x'}\,\sigma(x')}{x - x'}\, dx' - \frac{K_I}{\sqrt{2\pi}} \right], \qquad (8.5.3)$$

where K_I is the remotely applied dynamic stress intensity factor and $D = 4\alpha_d\alpha_s - (1 + \alpha_s^2)^2$ as in (4.3.8). For the strip yield zone model, it is required that the displacement gradient must be nonsingular at $x = 0$, the leading edge of the zone. Therefore, the cohesive stress $\sigma(x)$ is related to the applied stress intensity factor by

$$K_I = \sqrt{\frac{2}{\pi}} \int_{-\Lambda}^0 \frac{\sigma(x)}{\sqrt{-x}}\, dx. \qquad (8.5.4)$$

It is convenient to normalize the spatial coordinate in the direction of crack growth by the length of the yield zone. In terms of the dimensionless coordinates

$$\xi = -x'/\Lambda \quad \text{and} \quad \eta = -x/\Lambda \qquad (8.5.5)$$

the displacement gradient (8.5.3) becomes

$$\frac{\partial u_y}{\partial x}(x, 0^+) = \omega(\eta) = \frac{v^2}{c_s^2} \frac{\alpha_d}{D} \frac{\sqrt{\eta}}{\mu\pi} \int_0^1 \frac{\sigma(\xi)}{\sqrt{\xi}(\xi - \eta)}\, d\xi. \qquad (8.5.6)$$

This result provides a single relationship between the opening in the cohesive zone and the cohesive traction resisting the opening; this result follows from the elastodynamic field equations outside of the yield zone. This relationship must be augmented by a constitutive assumption on the response within the yield zone.

8.5.2 A rate-dependent cohesive zone

A fundamental objective in selecting a constitutive relation for the response in the yield zone is to incorporate the salient features of viscoplastic material behavior while still retaining a mathematical structure that is amenable to analysis. This objective is met by choosing the cohesive traction at any point $\sigma(\eta)$ such that it depends on the local rate of opening in the cohesive zone $\dot{u}_y$ according to

$$\sigma(\eta) = \sigma_o \left(1 + \dot{u}_y / \dot{u}_y^o\right) \qquad (8.5.7)$$

for $\dot{u}_y \geq 0$, where σ_o and $\dot{u}_y^o$ are material constants. For very slow deformations, the response is ideally plastic with flow stress σ_o. For rapid deformation, the local magnitude of the cohesive stress is elevated by an amount determined by the reference opening rate $\dot{u}_y^o$. The arguments for selecting a value for $\dot{u}_y^o$ that will have relevance to plane strain crack growth are oblique at best. In any case, this model permits a qualitative illustration of the phenomenon of interest.

An argument is proposed that is based on the idea that the "average cohesive stress" in the yield zone should be increased above σ_o by a factor of two or three under the most extreme conditions of crack growth. The actual numerical value of such a factor is of little significance, but existence of such a factor is essential to the argument. As an estimate of "average opening rate" within the yield zone, consider the crack tip opening displacement $\delta = 2u_y(-\Lambda, 0^+)$ divided by the time required for the crack to advance one plastic zone length Λ at speed v, that is,

$$(\dot{u}_y)_{est} = v\delta/2\Lambda. \qquad (8.5.8)$$

Under quasi-static conditions of small-scale yielding

$$\Lambda = \frac{\pi}{8}\left(\frac{K_I}{\sigma_o}\right)^2, \qquad \delta = \frac{(1-\nu)}{2}\frac{K_I^2}{\mu\sigma_o}, \qquad (8.5.9)$$

where ν is Poisson's ratio, so that in this case the opening rate estimate is

$$(\dot{u}_y)_{est} = \frac{2(1-\nu)}{\pi}\frac{v\sigma_o}{\mu}. \qquad (8.5.10)$$

This combination of terms with a numerical coefficient of order unity is used to set the scale for opening rate, and the reference opening rate $\dot{u}_y^o$ is chosen to be

$$\dot{u}_y^o = \frac{c_s}{\beta}\frac{\sigma_o}{\mu}, \qquad (8.5.11)$$

where β is a dimensionless viscosity parameter. In terms of β, the relation (8.5.7) becomes

$$\sigma(\eta) = \sigma_0 \left[1 + \beta \frac{\mu}{\sigma_0} \frac{\dot{u}_y}{c_s}\right], \qquad (8.5.12)$$

or, for steady crack growth at speed v,

$$\sigma(\eta) = \sigma_0 \left[1 - \beta \frac{\mu}{\sigma_0} \frac{v}{c_s} \omega(\eta)\right]. \qquad (8.5.13)$$

Thus, for a crack speed that is a modest fraction of c_s and an "average crack opening" over the length of the yield zone that is on the order of σ_0/μ (in the range from 0.01 to 0.001, say) the material rate effect on the flow stress will be significant if β has a value greater than about two or three. The limiting value of $\beta = 0$ corresponds to the inviscid ideally plastic case. The relationship (8.5.13) is used in the analysis to follow, and numerical results are presented for β in the range $0 \le \beta \le 10$.

If $\omega(\eta)$ is eliminated from (8.5.6) and (8.5.13), the result is the singular integral equation

$$f(\eta) + \frac{\tan(\pi\gamma)}{\pi} \int_0^1 \frac{f(\lambda)}{\lambda - \eta} d\lambda = \frac{1}{\sqrt{\eta}}, \qquad 0 < \eta < 1, \qquad (8.5.14)$$

where

$$f(\eta) = \frac{\sigma(\eta)}{\sigma_0 \sqrt{\eta}}, \qquad \tan(\pi\gamma) = \beta \frac{v^3}{c_s^3} \frac{\alpha_d}{D}. \qquad (8.5.15)$$

The parameter γ is in the range $0 \le \gamma \le \frac{1}{2}$. For $v/c_s \ll 1$, the parameter has the asymptotic form $\tan(\pi\gamma) = (1-\nu)\beta v/c_s + O(v^3/c_s^3)$, and $\gamma \to \frac{1}{2}$ as $v/c_R \to 1$. A solution of (8.5.14) is sought which meets the conditions that $\sigma(0) = \sigma_0$ and $\sigma(\eta)\sqrt{1-\eta}/\sigma_0 \to 0$ as $\eta \to 1^-$.

A general solution of the integral equation (8.5.14) subject to the end conditions of interest here is given by Muskhelishvili (1953). One form of this solution is

$$\frac{\sigma(\eta)}{\sigma_0} = 1 + \frac{\sin(\pi\gamma)}{\pi} \frac{\eta^{\gamma+\frac{1}{2}}}{(1-\eta)^\gamma} \int_0^1 \frac{(1-s)^\gamma}{s^{\frac{1}{2}}(1-s\eta)} ds \qquad (8.5.16)$$

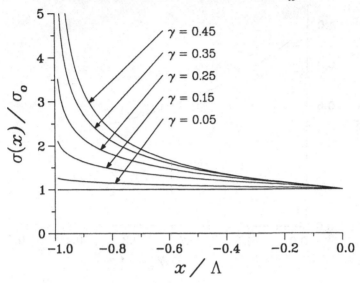

Figure 8.16. Distribution of the traction $\sigma(x)$ within the cohesive zone from (8.5.17) for five values of parameter γ.

for $0 \leq \eta < 1$. The integral in (8.5.16) is identified as a hypergeometric function $_2F_1$ which, in turn, has a representation in terms of the Gamma function $\Gamma(\cdot)$. Consequently, alternate representations of this result are

$$\frac{\sigma(\eta)}{\sigma_o} = 1 + \frac{\sin(\pi\gamma)}{\sqrt{\pi}} \frac{\eta^{\gamma+\frac{1}{2}}}{(1-\eta)^\gamma} \frac{_2F_1(1,\frac{1}{2};\gamma+\frac{3}{2};\eta)}{\Gamma(\gamma+\frac{3}{2})}$$

$$= 1 + \frac{\sin(\pi\gamma)}{\pi} \frac{\eta^{\gamma+\frac{1}{2}}}{(1-\eta)^\gamma} \sum_{k=0}^{\infty} \frac{\Gamma(k+\frac{1}{2})}{\Gamma(\gamma+k+\frac{3}{2})} \eta^k .$$

(8.5.17)

The infinite series converges for $0 \leq \eta < 1$. For $\gamma \rightarrow 0$, the right side of (8.5.16) is one. The dependence of $\sigma(\eta)/\sigma_o$ on η for five nonzero values of γ is shown in Figure 8.16. It is evident from Figure 8.16 that, as the influence of rate sensitivity becomes greater with other factors held fixed, the strength of the crack edge singularity increases and the fraction of the yield zone over which the singular solution dominates becomes greater. These same features were suggested by the finite element numerical results reported by Freund and Douglas (1983) for dynamic mode III crack growth in a viscoplastic material.

With the cohesive stress determined, the relationship between the yield zone size Λ and the applied stress intensity factor K_I can be

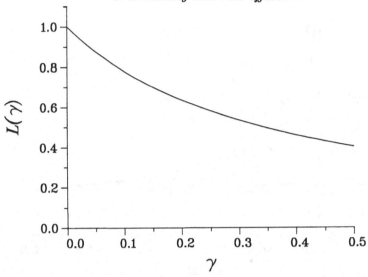

Figure 8.17. Normalized length of the cohesive zone $L(\gamma)$ versus γ according to (8.5.18).

found by means of (8.5.4). The power series representation of $\sigma(\eta)$ can be integrated term-by-term with the result that

$$\Lambda\frac{8}{\pi}\left(\frac{\sigma_o}{K_I}\right)^2 = \frac{1}{\pi}\left[\frac{\Gamma(\gamma+1/2)}{\Gamma(\gamma+1)}\right]^2 = L(\gamma). \qquad (8.5.18)$$

The quantity $L(\gamma)$ is the ratio of the yield zone size for any K_I, β and v/c_s to the zone size for any K_I but with $\beta \to 0$ and $v/c_s \to 0$. The ratio $L(\gamma)$ is plotted for the full range of γ in Figure 8.17 where it is evident that the function decreases monotonically from $L(0) = 1$ and the limiting value is $L(1/2) = 4/\pi^2$. This is the distance from the crack tip at which the stress distribution σ_{yy} for the case of elastic material response, that is, for $\beta \to \infty$, equals the yield stress σ_o.

According to (5.3.9), the rate of energy flux into the crack tip region per unit crack advance is

$$G = (1 - \nu)A_I(v)K_I^2/2\mu. \qquad (8.5.19)$$

For $\gamma < \frac{1}{2}$, all of this energy is dissipated within the yield zone, but it is interesting to note the part of the yield zone in which most energy

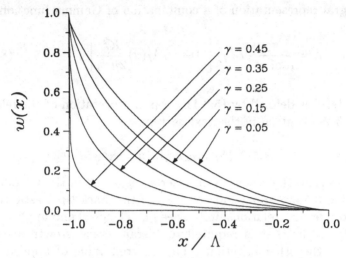

Figure 8.18. The fraction of total energy flux into the crack tip region G that is dissipated in the leading $-x/\Lambda$ fractional part of the cohesive zone according to (8.5.20).

is dissipated. To illustrate this distribution, values of the ratio

$$w(\eta) = -\frac{2}{G} \int_x^0 \sigma(x') u_{y,x}\,(x',0^+)\,dx'$$

$$= \frac{\pi L(\gamma)}{2\tan(\pi\gamma)} \int_0^\eta \frac{\sigma(\lambda)}{\sigma_o}\left[1 - \frac{\sigma(\lambda)}{\sigma_o}\right]d\lambda$$

(8.5.20)

have been computed for several values of γ. In short, $w(\eta)$ is the fraction of the energy G that is dissipated between the leading edge of the yield zone at $x = 0$ and the intermediate position $x = -\eta\Lambda$. Thus, $w(\eta)$ must increase monotonically from $w(0) = 0$ to $w(1) = 1$. The function has been evaluated numerically for several values of γ, and the results are shown in Figure 8.18. From the figure, it is evident that, as material rate effects become more significant, the energy flowing into the crack tip region is dissipated in a smaller fractional interval near the physical crack tip.

The crack tip opening displacement $\delta = 2u_y(-\Lambda,0^+)$ is obtained by integrating $\omega(\eta)$ from $\eta = 0$ to $\eta = 1$. The resulting expression is an integral representation of a combination of Gamma functions, so

$$\delta = \frac{\pi\gamma}{\tan(\pi\gamma)}L(\gamma)\left[(1-\nu)A_I(v)\frac{K_I^2}{2\mu\sigma_o}\right],\qquad(8.5.21)$$

where $A_I(v)$ is defined by (5.3.11). An interpretation of the relationship (8.5.21) is given in the next section.

8.5.3 The crack growth criteria

In this section, the two growth criteria suggested in the introduction to Section 8.5 are applied at the level of the crack tip plastic zone in order to infer the relationship between crack speed and applied stress intensity factor that is required to sustain crack growth according to one or the other criterion. The inferred value of applied stress intensity factor is once again the theoretical dynamic fracture toughness. The main focus is on the theoretical fracture toughness versus crack speed relationship because the results of experiments are most commonly reported in terms of the toughness–speed relationship, and it is the input required in order to do fracture mechanics calculations to predict crack growth.

The two criteria adopted are the critical crack tip opening displacement criterion and the critical stress at a characteristic distance criterion. The two criteria are applied separately at first, and then later they are evaluated with respect to each other. For there to be a basis for comparison, the criteria must be expressed in terms of the same set of material parameters. The important parameters here are σ_o, the slow loading flow stress, and δ_c, the critical crack tip opening displacement for very slow crack growth. A critical stress intensity factor K_{Ic} is related to these parameters through

$$\sigma_o\delta_c = \frac{1-\nu}{2\mu}K_{Ic}^2.\qquad(8.5.22)$$

This connection follows from overall energy balance under the condition when all of the energy flowing into the crack tip region is dissipated in the cohesive zone with uniform cohesive stress σ_o and crack tip opening displacement δ_c.

The crack tip opening displacement δ is given in terms of the applied stress intensity factor in (8.5.21). A crack growth criterion usually associated with a "ductile" mode of crack advance is the requirement that $\delta = \delta_c$ at all times. If the relation (8.5.21) is incorporated into this condition and if the result is viewed as an equation for K_I, then

$$\frac{K_I}{K_{Ic}} = \left[A_I(v) L(\gamma) \frac{\pi\gamma}{\tan(\pi\gamma)} \right]^{-\frac{1}{2}}. \qquad (8.5.23)$$

This is the principal result for the critical crack tip opening displacement criterion. The right side depends only on the crack speed v/c_s and the viscosity parameter β. For $\beta = 0$, the well-known result that $K_I/K_{Ic} = [A_I(v)]^{-1/2}$ is recovered. This result follows directly from the conservation of energy statement that $G = \sigma_o \delta_c$. For some fixed value of $\beta > 0$, the dynamic toughness is an increasing function of crack speed. As the crack moves more rapidly, the material is deformed more rapidly and a larger cohesive stress is required in order to achieve the requisite crack tip opening displacement. This increase of stress in the crack tip region translates into an increased likelihood of cleavage initiation in those materials that are intrinsically cleavable. This point of view leads to the second growth criterion.

As a simple cleavage crack growth criterion, it is required that the stress at a point in the yield zone at some fixed distance ahead of the crack tip must always have a material specific value. In order to have a basis for comparison of the two criteria, the distance is expressed as a multiple, say q, of δ_c and the critical stress is expressed as a multiple, say p, of σ_o. The distance ahead of the tip $q\delta_c$ is expected to be a small fraction of Λ and the critical stress level is expected to be significantly larger than σ_o, so only the singular part of the crack tip stress distribution is used, that is,

$$\frac{\sigma(\eta)}{\sigma_o} \sim \frac{\Sigma(\gamma)}{(1-\eta)^\gamma}, \qquad \Sigma(\gamma) = \frac{\sin(\pi\gamma)}{\sqrt{\pi}\,\Gamma(\gamma + \frac{1}{2})\gamma}. \qquad (8.5.24)$$

The mathematical statement of the growth criterion is then

$$\sigma(\eta) = p\sigma_o \quad \text{when} \quad \eta = 1 - q\delta_c/\Lambda \qquad (8.5.25)$$

or

$$\frac{K_I}{K_{Ic}} = \sqrt{\frac{4q(1-\nu)}{\pi L(\gamma)} \frac{\sigma_o}{\mu} \left[\frac{p\sqrt{\pi}\,\Gamma(\gamma + \frac{1}{2})\gamma}{\sin(\pi\gamma)} \right]^{1/\gamma}}. \qquad (8.5.26)$$

Numerical results are presented below for $q = 2$ and $p = 3$. As is evident from (8.5.26), a value of σ_o/μ is also required for the calculation, and 0.005 was chosen as a representative value.

The interpretation of the result (8.5.26) is quite straightforward. If either the crack speed v/c_s or the viscosity β is small, then there is little elevation of the stress in the yield zone due to rate effects. Consequently, a large applied stress intensity factor is required in order to satisfy the stress-controlled growth criterion. As the crack tip speed increases for some fixed level of viscosity, the local stress is elevated due to rate sensitivity and a smaller applied stress intensity factor is required. This effect becomes stronger as the viscosity increases.

It should be noted that the definition of $\Sigma(\gamma)$ in (8.5.24) is not completely unambiguous. The form shown was obtained by evaluating the coefficient of $(1 - \eta)^{-\gamma}$ at $\eta = 1$ for any value of γ, and then by using the identity

$$2F_1(a_1, a_2; b_1; 1) = \frac{\Gamma(b_1)\Gamma(b_1 - a_1 - a_2)}{\Gamma(b_1 - a_1)\Gamma(b_1 - a_2)} \qquad (8.5.27)$$

which applies for $b_1 > a_1 + a_2$, which is the case here. This asymptotic expansion about $\eta = 1$ is not uniform in γ. The fact that $\Sigma(\gamma)(1-\eta)^{-\gamma}$ takes on the correct value of $\sigma(\eta)/\sigma_o$ as $\gamma \to 0$ for any $\eta < 1$ appears to be fortuitous.

It is clear that the relationships (8.5.23) and (8.5.26) have opposite trends. For the deformation-controlled criterion, K_I/K_{Ic} takes on *minimum* values when either the crack speed or the viscosity (or both) is small, and this ratio *increases* as either v/c_s or β increases. On the other hand, for the stress-controlled criterion, K_I/K_{Ic} takes on its largest values when either the crack speed or the viscosity is small, and this ratio *decreases* as either v/c_s or β increases. Thus, for small viscosity and/or for low crack speed, the level of K_I required to satisfy the deformation-controlled criterion is less than the level required to satisfy the stress-controlled criterion. For larger crack speed or viscosity, however, the relative magnitudes of the requisite K_I levels for the two criteria are reversed.

If conditions of the body are such that crack growth can occur in either mode, and that the actual mode is established as the one which becomes operative at the lower level of applied K_I, then a composite surface of theoretical toughness versus crack speed and viscosity can be constructed. This surface is obtained by selecting the lower K_I/K_{Ic}

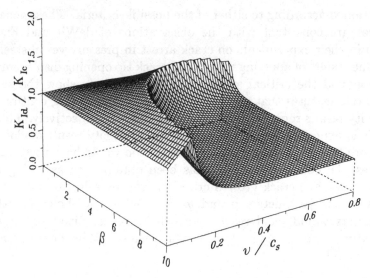

Figure 8.19. Surface of dynamic fracture toughness $K_I = K_{Id}$ versus crack speed v/c_s and viscosity parameter β. The ridge corresponds to a change in growth criterion.

value of the two relationships for any given values of v/c_s and β. The result is shown in Figure 8.19. With reference to Figure 8.19, the following crack growth behavior is represented. Suppose that a cracked body characterized by a particular value of β is loaded so that the crack begins to advance from speed $v = 0$, and that the applied stress intensity factor increases as the crack advances. The result implies that the crack will accelerate with the separation on a local scale occurring according to a ductile mechanism. In terms of the surface in Figure 8.19, the state of K_I and v follows the surface along a path for which β = constant, starting from $K_I = K_{Ic}$ and $v = 0$. The crack accelerates until a speed corresponding to the position of the ridge in the surface is attained. At this point, separation converts to a stress controlled mechanism due to the rate-induced elevation of flow stress. The only way for the applied stress intensity factor to further increase is for the crack speed to suddenly become very large, so the state ends up on the rapidly rising portion of the surface associated with inertial effects (not shown in Figure 8.19). Thereafter, if the applied stress intensity factor decreases, then the state falls to the minimum in the path for fixed β. Further decrease of the applied stress intensity factor implies crack arrest, or no further growth can be sustained according to either of the possible criteria. These general features are consistent with the observations of deWit and Fields (1987) in their experiments on crack arrest in pressure vessel steels.

The results of applying the critical crack tip opening displacement criterion and the critical stress criterion separately show the same qualitative features that were found in the corresponding continuum plasticity results reported in Sections 8.3 and 8.4, respectively. Therefore, it is expected that if the continuum plasticity result analogous to Figure 8.19 could be generated, then it too would have these same features. Of course, this result has been obtained by assuming the simplest possible crack growth criterion, and by selecting particular values of the parameters p and q in (8.5.25). A choice of other parameters would yield results with differing magnitudes, but the qualitative features appear to arise nonetheless within the framework of the model.

8.6 Ductile void growth

The examination of fracture surfaces of a wide range of metals and other materials following tensile crack growth under plane strain conditions leads to the conclusion that the process of crack advance is essentially the nucleation of voids or cavities at material inhomogeneities, and the subsequent ductile growth of these voids to coalescence. The voids nucleate at second phase particles in the ductile matrix, due to either decohesion at particle-matrix interfaces or cracking of the particles. The process of material failure by void growth to coalescence has been reviewed by Tvergaard (1989). Certain features of the elastic-plastic crack tip field, which provides the environment in which the mechanism operates, are important for this process. Among these are the high triaxial stress condition within the small strain region ahead of the crack tip that serves to nucleate voids and the zone of large plastic straining directly ahead of the tip that accommodates the ductile expansion of voids necessary for coalescence with the main crack. The mechanical process of the ductile growth of cylindrical and spherical voids in plastic materials has been described by McClintock (1968) and Rice and Tracey (1969), respectively, who showed the strong influence of the mean normal stress component on the rate of growth of an isolated void.

McClintock (1968) and Rice and Johnson (1970) developed models of the void growth process within the crack tip field with a view toward relating microstructural features to fracture mechanics parameters for fracture initiation and stable crack growth. Though these models can be viewed only as rough approximations, they do appear to capture the essence of the hole growth mechanism when it is not terminated prior to coalescence by strain localization. It should be emphasized that the assumption of a fully ductile separation mechanism on the microscale does not necessarily imply extensive plastic flow in the body containing the crack. Indeed, it is quite possible to observe fully ductile separation at a crack tip in a material for which the plastic zone size is small and conditions of small-scale yielding are satisfied. Likewise, extensive plastic flow in the body does not imply that the separation mechanism is a ductile mechanism. It could just as well be cleavage induced by material rate effects or strain hardening as a result of the plastic flow.

Consider the situation deep within the active plastic zone near the edge of a crack advancing rapidly in the plane strain tensile mode through an elastic-plastic material. Based on the asymptotic analysis described in Section 4.5 and on the calculations of Lam and Freund (1985), it appears that there is a region of high stress triaxiality ahead of the advancing crack which serves as a driving force for nucleation of voids in the material. However, the concentration of plastic strain at the crack tip is not nearly as great for crack growth, even quasi-static growth, as it is for a stationary crack tip subject to monotonic loading. For example, the crack tip opening displacement obtained from a small strain continuum plasticity analysis is zero for steady growth, whereas it has a finite value proportional to $(K_I/\sigma_o)^2$ for a stationary crack under small-scale yielding conditions, where K_I is the remote elastic stress intensity factor and σ_o is the tensile yield stress.

This difference is significant, for it implies that the macroscopic plastic field cannot provide the large strain zone ahead of the crack tip that is necessary to accommodate void growth. Thus, if a large strain region is to exist, and it seems that it must, then an alternative is that the void growth process itself must lead to such a region. This requires the sympathetic growth of voids in the crack tip region to form a small region of cavitated material adjacent to the tip. Suppose that voids grow by ductile expansion to coalescence without the intervention of some other mode of joining. Then, for a given distribution of void nucleation sites, the weaker plastic strain concentration for growth implies that a greater remote driving force is required to sustain growth than to initiate growth. If the crack is growing rapidly, then

the velocity of material particles on the surfaces of the expanding voids
are potentially enormous, although occurring over a small volume
of material. If the inertial resistance to this motion is manifested
macroscopically as an effect on the applied driving force required to
sustain fracture, then the apparent dynamic fracture toughness should
be an increasing function of crack speed, even without the influence
of material rate sensitivity. This was indeed suggested by the analysis
presented in Section 8.3 for the case of antiplane shear crack growth,
and the purpose of this section is to examine the influence of inertia
of individual voids undergoing rapid ductile expansion. Some observations on the role of material strain rate effects are also included.

8.6.1 Spherical expansion of a void

In order to examine the influence of material inertia on a small-scale
on the ductile void growth process, a simple spherically symmetric
model introduced by Carroll and Holt (1972) is adopted. This model
was further developed within the framework of dynamic spall fracture
by Johnson (1981). Consider a thick spherical shell of incompressible
material with inner and outer radii $r = a(t)$ and $r = b(t)$, respectively,
where r is the eulerian coordinate representing distance from the fixed
center of the shell. The material is initially at rest with void radius
$a_0 = a(0)$. At time $t = 0$, a uniform normal traction of magnitude σ_b
begins to act on the outer surface. The inner surface is traction free.
It is assumed that elastic effects are negligible and that the magnitude
of σ_b is sufficient to produce ductile expansion of the shell.

   With reference to the discussion of ductile void growth and coalescence as a mechanism of crack advance, the parameter a_0 is identified
with a representative physical dimension of the inhomogeneity that
causes nucleation of the void, and $2b_0 = 2b(0)$ is identified with the
spacing of void nucleation sites. If a body with a periodic array of such
voids is subjected to a mean normal stress σ_b resulting in spherical
growth of the voids, then the body "fractures" at some later time, say
$t = t^*$, for which $a(t^*) = b_0$. The objective of this model calculation
is to determine the relationship between σ_b and t^*. With reference
to a crack growth process, $2b_0/t^*$ and σ_b provide crude estimates of
crack speed and crack tip field intensity, respectively.

   The velocity in the radial direction of a particle with instantaneous radial coordinate r is $v_r(r,t)$. Material incompressibility

requires that

$$\frac{\partial v_r}{\partial r} + 2\frac{v_r}{r} = 0, \tag{8.6.1}$$

or that

$$v_r(r,t) = \dot{a}a^2/r^2. \tag{8.6.2}$$

Furthermore, the inner and outer radii are related by

$$b^3 - b_0^3 = a^3 - a_0^3, \tag{8.6.3}$$

identically in time.

The momentum balance equation in terms of true stress components in the eulerian coordinate r is

$$\frac{\partial \sigma_r}{\partial r} + \frac{2(\sigma_r - \sigma_t)}{r} = \rho\left(\frac{\partial v_r}{\partial t} + v_r\frac{\partial v_r}{\partial r}\right), \tag{8.6.4}$$

where ρ is the material mass density, σ_r is the radial component of normal stress, and σ_t is the component of normal stress in any direction transverse to the radial direction. Substitution of the expression (8.6.2) for v_r into (8.6.4) and integration of the result with respect to r from a to b yields

$$\sigma_b = \rho\left[(a\ddot{a} + 2\dot{a}^2)\left(1 - \frac{a}{b}\right)\right] - \tfrac{1}{2}\dot{a}^2\left(1 - \frac{a^4}{b^4}\right) - 2\int_a^b \frac{\sigma_r - \sigma_t}{r}\,dr. \tag{8.6.5}$$

This result is independent of the constitutive description of the material, except for the assumption of incompressibility. It is tacitly assumed in writing (8.6.5), however, that the normal stress difference $\sigma_r - \sigma_t$ is completely determined by the velocity field by means of a constitutive description of flow for either rate-independent or rate-dependent material response.

To examine the influence of inertia on void growth, suppose that the material is perfectly plastic with tensile flow stress of magnitude σ_o. Note that material elements in the radial direction shorten as the spherical shell expands, so $\sigma_r - \sigma_t = -\sigma_o$ in this case and

$$\int_a^b \frac{\sigma_r - \sigma_t}{r}\,dr = -\sigma_o \ln\frac{b}{a}. \tag{8.6.6}$$

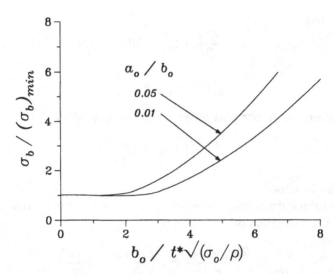

Figure 8.20. Relationship between the tensile traction applied to the surface $r = b(t)$ and the time t^* defined by $a(t^*) = b_0$, obtained by numerical integration of (8.6.5).

Thus, the minimum applied stress σ_b required to produce flow in the sphere is $(\sigma_b)_{min} = 2\sigma_o \ln(b_0/a_0)$. For any given value of σ_b, the relation (8.6.5) provides a second-order ordinary differential equation for $a(t)$ subject to the initial conditions that

$$a(0) = a_0, \qquad \dot{a}(0) = 0. \qquad (8.6.7)$$

This equation can be solved numerically for any values of $b_0/a_0 > 1$ and $\sigma_b/(\sigma_b)_{min} > 1$. A particular result is shown in Figure 8.20 for the case of $b_0/a_0 = 20$ in the form of a graph of $\sigma_b/(\sigma_b)_{min}$ versus $b_0\sqrt{\rho/\sigma_o}/t^*$. Recall that t^* is defined by the condition that $a(t^*) = b_0$. It is evident from the result that any applied stress only slightly larger than $(\sigma_b)_{min}$ will result in growth of the void to critical size in a time $t^* > \frac{1}{2}b_0\sqrt{\rho/\sigma_o}$. However, if the void must be grown to critical size in a shorter time, then a larger stress σ_b is required to overcome the inertial resistance.

Based on this simple model calculation, the time $\frac{1}{2}b_0\sqrt{\rho/\sigma_o}$ appears to have particular significance for assessing the influence of material inertia on the physical scale of the ductile hole growth mechanism. This time can be estimated for high strength steel or aluminum alloys for which the crack growth mechanism is predominantly ductile void growth. Of greater interest, perhaps, is the "speed" $2b_0/t^*$ which, as noted above, is a crude estimate of crack propagation speed if the initial void nucleation site spacing is $2b_0$ and the time required to grow the voids to the critical size is t^*. Thus, microscale inertia is significant for $b_0/a_0 = 20$ if

$$2b_0/t^* \geq 4\sqrt{\sigma_0/\rho} \,. \tag{8.6.8}$$

If the parameter $\sqrt{\sigma_0/\rho}$ has a value of 300 m/s, which is typical for high strength alloys, then (8.6.8) implies that the crack speed at which local inertial effects result in an observable influence on macroscopic dynamic fracture toughness is about 1200 m/s. This speed is somewhat greater than the observed crack speed at which the measured or inferred dynamic fracture toughness for such materials begins to show a dramatic increase with speed, which is usually in the range of 500–1000 m/s. The model can provide only a very rough approximation, and a systematic study of the influence of microstructural features on its implications is not yet available.

Up to this point, the spherical expansion model has not included an influence due to material rate effects. Some information on the magnitude of material viscosity effects can be obtained in the following way. Suppose that the rectangular components of the plastic stretching rate tensor d_{ij}^p depend on the instantaneous deviatoric stress s_{ij} components according to

$$d_{ij}^p = \tfrac{1}{2}\dot{\gamma}_0 \left(\frac{\tau - \tau_0}{\tau_0}\right)^n \frac{s_{ij}}{\tau}, \quad \tau \geq \tau_0 = \sigma_0/\sqrt{3} \,, \tag{8.6.9}$$

where $\dot{\gamma}_0$ is a material parameter with dimensions of strain rate, n is a material constant, and τ is given in terms of s_{ij} by (8.4.7). The parameter $\dot{\gamma}_0$ is essentially the strain rate at which rate effects become significant for homogeneous straining.

For the case of spherical expansion, $d_r^p = \partial v_r/\partial r = -2d_t^p$ and

$$\sigma_t - \sigma_r = \sigma_0 + \sigma_0 \left(\frac{2\sqrt{3}}{\dot{\gamma}_0} \frac{\dot{a}a^2}{r^3}\right)^{1/n} \,. \tag{8.6.10}$$

In this case, the last term of (8.6.5) becomes

$$\int_a^b \frac{\sigma_t - \sigma_r}{r}\,dr = \sigma_0 \ln\frac{b}{a} + \frac{\sigma_0 n}{3}\left(\frac{2\sqrt{3}}{\dot{\gamma}_0}\frac{\dot{a}}{a}\right)^{1/n}\left[1 - (a/b)^{3/n}\right].$$
$$\tag{8.6.11}$$

It is evident that (8.6.11) reduces to the rate-independent limit
when the viscosity parameter $\dot{\gamma}_0 \to \infty$. The question of the relative
magnitudes of the two terms on the right side of (8.6.11) has not
been considered in detail. A rough estimate of the magnitude of the
viscosity term relative to the strength term can be obtained in the
following way. Suppose that $\dot{b}/b$ is identified as a representative overall
reference plastic strain rate, say $\dot{\gamma}^p_{ref}$, for the voiding material. From
the incompressibility condition (8.6.3) it follows that $\dot{a} \approx \dot{\gamma}^p_{ref} b^3/a^2$.
For the early stages of voiding when $a \ll b$, the ratio of the viscosity
term to the strength term is roughly equal to

$$\frac{n}{3\ln(b/a)} \left(2\sqrt{3} \frac{\dot{\gamma}_{ref}}{\dot{\gamma}_0} \frac{b^3}{a^3} \right)^{1/n} . \tag{8.6.12}$$

For $n = 5$, $\dot{\gamma}^p_{ref} = \dot{\gamma}_0/10$, and $b/a = 25$ this ratio has a value of about
2.9, implying that viscosity effects are relatively large compared to
strength effects in the early stages of void expansion at relatively
modest values of overall strain rate. The factor $(b/a)^{3/n}$ has a strong
influence on the ratio (8.6.12). For example, if the viscosity is linear,
that is, if $n = 1$, but the values of other parameters are unchanged
then the ratio has a value of about 560. For ductile materials, the
strains around expanding cavities are quite large and an exponent n
in the range from 3 to 10 might be reasonable (Clifton 1983).

A similar conclusion on the magnitude of viscosity effects compared to the inertial effects was reached by Curran et al. (1987). They
considered the spherical expansion model under the assumption of
linear viscous material resistance to flow and no residual strength,
that is, they assumed $n = 1$ and $\sigma_0 = 0$ in (8.6.11). In this case,
the process is insensitive to the value of b as long as $b \gg a$, and the
differential equation governing void expansion (8.6.5) becomes

$$\ddot{a} + \frac{3}{2}\frac{\dot{a}^2}{a} + \frac{16}{3}\frac{\mu}{\rho\dot{\gamma}_0}\frac{\dot{a}}{a^2} = \frac{\sigma_b}{\rho a} . \tag{8.6.13}$$

They included an effect of material strength indirectly by assuming
that σ_b is actually the excess of the applied stress over a threshold
stress given by $\frac{2}{3}\sigma_0 [1 + \ln(2\mu/\sigma_0)]$, where μ is the elastic shear modulus. This threshold stress, which was obtained by Hill (1950), is the
remote isotropic tension required to produce yielding at the traction-
free surface of a spherical cavity in an incompressible elastic-ideally

plastic material. Through numerical integration of this equation for material parameters corresponding to a particular aluminum alloy, they concluded that in dynamic ductile fracture the threshold stress and the material viscosity are dominant; inertial effects become significant only for quite large voids, greater than about $10\,\mu$m or so for the particular material considered.

Although this area requires extensive quantitative modeling to sort out the complicated features of the interconnected phenomena involved, some qualitative observations on the process of rapid growth of a macroscopic crack through a material that separates by means of a void growth mechanism can be made. As the speed of the crack increases, voids must be grown at an ever increasing rate from very small size to coalescence in order to accommodate crack advance. The spherical expansion model suggests that an ever increasing applied stress (σ_b in the model) is required to drive this process, and this stress is identified with the crack driving force. The increase in crack driving force with crack speed for materials that separate by the mechanism is surely evident in observations.

There are other ramifications of this stress increase that are connected with the prevailing crack tip stress fields. For one thing, the presence of higher stress levels in the crack tip region implies that void nucleation sites that might have remained dormant during slow crack growth can be activated during rapid crack growth. A second observation is based on the spatial structure of the crack tip fields. To illustrate this point, consider the plane strain elastic crack tip field for which the stress intensity factor is the driving force. In this case, if the crack driving force is doubled, then the linear dimension of the region around the crack tip over which the stress magnitude exceeds some given value is increased by a factor of four and the area of this region is increased by a factor of sixteen. The implication of the observation is that the increased resistance of the material near the crack tip to high rate voiding causes the region of high stress, and consequently the region over which damage accumulates, to spread over a significantly larger volume of the material around the crack tip. This suggests that the size of the region near a crack tip over which the physical process of voiding progresses is larger and more diffuse during rapid crack growth than during slow growth. This increased size results in a more textured fracture surface. It also implies a connection between crack tip stress fields and the size of the region over which the physical separation mechanism is operative. A tacit

assumption in the fracture mechanics approach is that this scale is
set by the material microstructure and that it is independent of the
surrounding stress field. This point has been discussed qualitatively
by Broberg (1977).

8.6.2 A more general model

The process of dynamic growth of a void in a plastic material subjected to a far-field strain rate and mean normal stress was studied
by Glennie (1972), who applied the results to a model of dynamic
fracture by the void growth mechanism. The features of the model
and the principal results are summarized here. Consider the flow of
an isotropic rigid-ideally plastic material that deforms according to
an associated flow rule. Suppose that an unbounded body of this
material experiences a spatially uniform stretching rate, or rate of
strain with respect to the instantaneous configuration, say d_{ij}. The
deviatoric stress is determined from the strain rate as

$$s_{ij} = \frac{\sigma_o d_{ij}}{\sqrt{3 d_{pq} d_{pq}/2}} = \lambda^{-1} d_{ij}, \qquad (8.6.14)$$

where σ_o is the flow stress in simple tension. If v_i is the velocity field
with respect to spatial or eulerian coordinates x_i, then the momentum
equation

$$\frac{\partial \sigma_{ij}}{\partial x_j} = \rho \left(\frac{\partial v_i}{\partial t} + v_j \frac{\partial v_i}{\partial x_j} \right) \qquad (8.6.15)$$

yields the mean normal stress field, where $\sigma_{ij} = s_{ij} + \frac{1}{3}\sigma_{kk}\delta_{ij}$ is the
true stress and ρ is the material mass density. The particle velocity
field is

$$v_i = d_{ij} x_j, \qquad (8.6.16)$$

where one material point has been fixed arbitrarily at the origin.
The material spin rate is zero everywhere. The momentum balance
equation is satisfied when the mean normal stress is

$$\tfrac{1}{3}\sigma_{kk} = \bar{\sigma} + \tfrac{1}{2}\rho \left(\dot{d}_{ik} + d_{ij}d_{jk} \right) x_i x_k, \qquad (8.6.17)$$

where $\bar{\sigma}$ is an arbitrary function of time.

With the uniform deformation rate field as a reference, suppose
that a spherical void with traction-free surface is introduced into

the body, centered at the origin of the coordinate system, without changing the remote conditions on deformation rate or mean normal stress. Two assumptions are then made concerning the deformation field in the presence of the expanding void. In a somewhat simplified form, it is assumed that the void is small enough that the background strain it experiences is spatially uniform and that the velocity field in the presence of the void rapidly approaches the velocity field for a uniform rate of deformation as distance from the origin becomes large compared to the void radius.

On the basis of these assumptions, a variational principle was applied in order to determine the rate of growth of the void in terms of the applied mean normal stress $\bar{\sigma}$ and the background uniform deformation rate, say d_{ij}^{∞}. The procedure is similar to that developed by Rice and Tracey (1969) for quasi-static void expansion.

Under these conditions, which do not correspond to spherical symmetry, the void does not grow in a self-similar fashion. However, at this level of approximation, the size of the void is described by a single length parameter, say $a(t)$, which can be interpreted as the mean void radius, as the radius of some equivalent spherical void, or as the cube root of the void volume. With this understanding, Glennie (1972) found that the expansion of the void is governed by an ordinary differential equation of the form

$$\dot{a} = 0.372\lambda \sinh\left[\frac{3}{2\sigma_{\mathrm{o}}}\left\{\bar{\sigma} - \rho\left(a\ddot{a} + \tfrac{3}{2}\dot{a}^2 - \tfrac{1}{9}\lambda^2 a^2\right)\right\}\right], \qquad (8.6.18)$$

where $\lambda = \sqrt{3d_{ij}^{\infty}d_{ij}^{\infty}/2}$ represents the influence of the background deformation rate and $\bar{\sigma}$ is the remotely applied mean normal stress.

Implications of the result (8.6.18) for fracture dynamics are examined by Glennie (1972) through numerical solution of the differential equation and the development of a model of crack tip processes. Conditions under which material inertia is important are evident from (8.6.18). It is known from the results of McClintock (1968) and Rice and Tracey (1969) that the presence of a mean normal stress has a strong influence on void growth rate, as can be seen from the exponential dependence of $\dot{a}$ on $\bar{\sigma}/\sigma_{\mathrm{o}}$ in (8.6.18) when $\rho = 0$. It is evident from this equation that the effect of inertia is to counteract the mean normal stress. Consequently, inertial effects become significant when the variation of $a(t)$ is such that the inertial term in the argument of the hyperbolic sine function in (8.6.18) becomes comparable to the mean normal stress.

Consider the particular case when the void grows at essentially constant rate from a relatively small initial size to a final size b_0 in a time t^*, that is, $a(t) \approx b_0 t/t^*$. The notation of the previous subsection is adopted here. Furthermore, suppose that the mean normal stress $\bar{\sigma}$ is about $3\sigma_0$, as implied by the small strain crack tip solution for an ideally plastic material. This, in turn, implies that inertial effects are important when $3\sigma_0 \approx \frac{3}{2}\rho(b_0/t^*)^2$ or

$$2b_0/t^* \approx 4.83\sqrt{\sigma_0/\rho}\,, \tag{8.6.19}$$

which is quite similar to the result obtained in (8.6.8).

While these ductile void growth models provide only rough approximations to the physically important mechanism of crack advance, they do suggest that inertial resistance to material motion on the small scale may be important in the interpretation of the apparent dynamic fracture resistance of real materials under realizable conditions. Other influences, such as void interaction, strain hardening, and material rate sensitivity, will require thorough study before these effects can be understood. Another complicating factor in examining the process is that voids can join by mechanisms other than simple expansion up to coalescence. For example, fracture surfaces reveal some evidence of plastic strain localization in ligaments between voids, resulting in coalescence by shearing of the ligament. The mechanism appears to require movement of less material than continued void expansion, so it may be promoted by inertial effects.

8.7 Microcracking and fragmentation

If a macroscopically homogeneous body of brittle material fractures due to quasi-static loading or stress wave loading of low intensity, the process normally consists of the formation of a single dominant crack that grows across the load carrying section of the body, thereby unloading adjacent material. Apparently, this crack growth arises when a particular flaw in the body is stressed to a critical condition, and the growth process becomes catastrophic before growth of any other potential fracture site is activated. On the other hand, if a body of brittle material is subjected to intense impulsive loading or to radiant deposition of a nonmechanical form of energy that is rapidly converted to mechanical energy, then a higher density of fractures can be nucleated and they can grow to a significant size without arresting each other. That is, the loading is sufficiently intense to cause more than one flaw, and possibly very many flaws, to precipitate crack growth before these potential fracture sites are unloaded by stress waves from a single growing crack. This seems to require an initial energy density, in the form of kinetic energy or elastic energy with predominantly compressive stresses, that is much greater than the body could sustain as elastic energy with predominantly tensile stresses. The result of many cracks growing is the development of distributed damage in the material. If many cracks grow to intersections with other cracks or if they grow to the body surface, then the body is split into many fragments. The fracture resistance of rocks, ceramics, and other brittle materials under impulsive fracture conditions is commonly observed to be significantly greater than the resistance under slow loading conditions. Consequently, the process is often viewed as rate dependent on the macroscopic scale. The origin of the rate dependence appears to be inertial and frictional effects on the microscale.

The fragmentation process involves a number of physical mechanisms. It most cases, the process is too complex to be viewed as being completely deterministic, and the evolution of the process must be considered in terms of statistical characteristics. Pioneering work incorporating the statistical aspects for ductile fragmentation was reported by Mott (1947). The statistical and geometrical aspects of brittle fragmentation have been considered by Grady and Kipp (1985). The process is extremely complex and no theoretical models of broad applicability are available. The phenomenon is important in mining, materials processing, and other applications, and some steps toward a rational basis for observed phenomena have been provided. One conceptual approach based on overall energy balance is discussed here on the basis of a particular model and followed by a brief discussion of a more mechanistic approach.

8.7.1 Overall energy considerations

A specific example is considered in order to obtain an estimate of the size of a fragment due to impulsive fracture of a brittle material. Consider the process of fragmentation of a thin circular ring or cylinder due to sudden radial expansion. Suppose that the ring has mass density ρ and a uniform cross-sectional area A. At a certain instant, the ring is impulsively loaded so that it begins to expand symmetrically from rest with radial speed v_0. If the mean radius of the ring is r, then the tensile strain rate in the circumferential direction in the ring is $\dot{\epsilon}_0 = v_0/r$. Suppose that the ring expands and then splits into n equal-sized fragments, each of circumferential length d such that $nd = 2\pi r$. If the initial loading is very intense, then the fragment size can be estimated in terms of material parameters and initial conditions by means of an approach introduced by Grady (1982). The loading is intense, for example, if the initial kinetic energy in the ring is very large compared to the elastic energy in the ring when cracks begin to form.

Consider one fragment of length d, and view the velocity of any particle in the fragment as the sum of the velocity of the center of mass of the fragment and the circumferential velocity relative to the translation of the center of mass in the radial direction. The kinetic energy associated with the translation of the center of mass is initially very large. The work of the forces producing the fractures is small by comparison, so the contribution to the kinetic energy due to translation of the center of mass does not change significantly. Thus, the final radial speed of the center of mass of each fragment is only slightly less than the initial radial speed.

The principal source of energy available to drive the fractures is the kinetic energy of the material arising from motion relative to the center of mass. If a fragment is viewed as a straight rod for purposes of estimating this kinetic energy, then the relative velocity is approximately $v_{rel}(x) = \dot{\epsilon}_0 x$, where x is the coordinate along the rod measured from the center of mass. The associated kinetic energy is

$$T_{rel} = \tfrac{1}{2} A\rho \int_{-d/2}^{d/2} \dot{\epsilon}_0^2 x^2 \, dx = \tfrac{1}{24} A\rho \dot{\epsilon}_0^2 d^3 \,. \tag{8.7.1}$$

For purposes of comparison, the kinetic energy associated with radial motion is about $\tfrac{1}{2}\rho A d \dot{\epsilon}_0^2 r^2 = 12 T_{rel} r^2/d^2$. Suppose that Γ_c is the average specific fracture energy for the material and that the area of each fracture surface is the cross-sectional area A. Then a fracture energy of $\tfrac{1}{2}\Gamma_c A$ is consumed at each end of the fragment from the energy supply within the fragment. If this is set equal to the kinetic energy supply in (8.7.1) then

$$\Gamma_c = \tfrac{1}{24}\rho \dot{\epsilon}_0^2 d^3 \,. \tag{8.7.2}$$

Thus, an estimate of fragment size d in terms of the initial state represented by $\dot{\epsilon}_0$ and the material parameters Γ_c and ρ is

$$d = \left\{ \frac{24\Gamma_c}{\rho\dot{\epsilon}_0^2} \right\}^{1/3} . \tag{8.7.3}$$

If Γ_c is eliminated in favor of a fracture toughness through the relationship $\Gamma_c \approx K_{cr}^2/\rho c^2$, where K_{cr} is a representative fracture toughness for high speed crack growth which is several times larger than K_{Ic} for the material, then

$$d = 2.9 \left\{ \frac{K_{cr}}{\rho c_0 \dot{\epsilon}_0} \right\}^{2/3} , \tag{8.7.4}$$

where $c_0 = \sqrt{E/\rho}$ is the bar wave speed for the ring material. To consider a specific case, consider a material for which the relevant crack growth toughness K_{cr} is about $50\,\text{MPa}\sqrt{\text{m}}$, the mass density ρ is about $7000\,\text{kg/m}^3$, and the pertinent elastic wave speed c is about $5000\,\text{m/s}$. If the ring is subjected to an axially symmetric, radial impulse that induces a circumferential strain rate of $5 \times 10^4\,\text{s}^{-1}$ then the estimate of fragment size on the basis of (8.7.4) is $d \approx 3\,\text{mm}$. If the initial radius of the cylinder r is about $3\,\text{cm}$ then the kinetic energy associated with radial motion is about 100 times larger than that due to motion of the material in each fragment relative to the fragment center of mass. These parameters are within the range of parameters reported by Weimer and Rogers (1979) for a series of experiments on fragmentation of a high strength tool steel. The specimen configuration was a long thin-walled cylinder, and the specimens were loaded by internal explosive charges. Grady (1982) examined the data in terms of the energy balance model, with the conclusions that the formula (8.7.4) overestimates the fragment size by about 30 percent and that the observed variation of fragment size with toughness measure seemed to follow the 2/3 power dependence suggested by (8.7.4).

A feature of the process that was exploited above, namely, that the energy consumed in fracturing is extremely small compared to the overall energy of the fragmenting solid, often emerges as a principal difficulty in analyzing fragmentation by means of overall energy methods. For example, Tilly and Sage (1970) studied the process of impact of glass spheres against various surfaces. For values of the parameters in a representative impact, suppose that a glass particle with diameter of 3 mm is projected against a steel surface at a speed of 275 m/s. The kinetic energy of the particle is then about 1.3 J. Upon impact, the glass sphere shatters into about 8000 fragments of mean diameter 150 μm. The total surface area of the fragments is about $6 \times 10^{-4}\,\mathrm{m}^2$. If the specific fracture energy is taken to be $10\,\mathrm{J/m}^2$, then the net energy dissipated in fragmentation is only about 0.006 J. Even if the area of fracture surface created is underestimated by an order of magnitude, the energy dissipated in fracturing is still small compared to other energies involved in the process. Relatively large amounts of energy must be dissipated through frictional interaction of the fragments or it must be radiated into the impacted body by means of stress waves.

8.7.2 Time-dependent strength under pulse loading

Brittle materials such as rocks, glasses, and ceramics exhibit virtually no macroscopic plastic deformation when subjected to tensile loading. Furthermore, for loading that is applied relatively slowly, the response up to and including fracture is insensitive to the rate of loading. The deformation appears to be elastic for the early stages of loading. Because these materials are incapable of plastic deformation except at elevated temperature, their response is very sensitive to the existence of structural flaws in their microstructure. Examples of such flaws in rocks and ceramics are poorly bonded grain boundaries, inclusion particles, and discontinuities in elastic stiffness at grain boundary junctions. In glass and brittle plastics, the flaw structure, aside from surface cracks, has not yet been fully characterized.

Once the level of applied loading is increased to the point where one flaw can be extended in a cracklike manner, it grows catastrophically into a macroscopic crack completely splitting the sample. The size and severity of these "critical flaws" in nominally identical specimens can vary, accounting for the scatter of experimental data on the strength of brittle materials. Similarly, the likelihood of a material sample including a flaw that is critical at any given stress level increases with the absolute physical size of the sample, accounting for the "size effect" of experimental data on the strength of brittle materials. Simply stated, if the tensile strengths of two samples of nominally the same material are measured, the smaller sample typically will have greater strength because the larger sample typically will contain a more severe strength limiting flaw than the smaller sample. Such observations are meaningless when actually applied to only two samples, and they must be understood as statistical inferences based on measurements on many samples.

The statistical approach to brittle fracture was pioneered by Weibull (1939). He imagined that all of the flawed elements in a material sample were arranged in series, and that the sample would fracture when the common load was increased to the strength level of the critically flawed "weakest link". Furthermore, he proposed a statistical distribution of flaw sizes throughout the material to account for the size effect of strength mentioned above. The distribution, a sample of which will be used in the development to follow, is expressed in terms of parameters that can be determined for a given material only through an appropriate series of experiments. The Weibull statistical approach is discussed within a general framework by Freudenthal (1968).

The situation is quite different for intense pulse loading of a brittle material. In this case, the weakest link concept is not applicable. Suppose that a sample of material is deformed very rapidly so that a stress level significantly greater than that required to precipitate growth of the most severe flaw is achieved in a short time. Once the stress level necessary to produce fracture at the most severe flaw is reached, a finite additional time may elapse before fracture can actually begin. This delay time was called the incubation time in Section 7.5, and an illustration of the idea is given in (7.5.6). During this time, the macroscopic stress on the sample can continue to increase as the macroscopic deformation proceeds. As it does so, the critical level of stress to produce crack growth can be reached at many other defects in the material. Thus, crack growth can be initiated at very many flaws if the loading is applied rapidly.

As the flaws begin to grow as microcracks and as the overall deformation proceeds, the stress at first continues to rise. As the microcracks become longer, however, the effective stiffness of the sample diminishes and, correspondingly, the rate of increase of stress corresponding to a fixed rate of deformation diminishes. At some instant of time, the rate of change of stress on the sample becomes zero even though the deformation rate is maintained at a fixed level. Thereafter, the stress level falls as the deformation proceeds. Eventually, the growing microcracks coalesce into one or more macroscopic cracks. In an average sense, this occurs when each growing microcrack has extended to a length equal to the initial spacing of activated flaws. The maximum stress acting on the sample during this process is called the impact strength σ_s at constant strain rate, and the time at which the maximum occurs after stressing begins is identified as the minimum duration of the high rate deformation that is required to produce a macroscopic failure. These qualitative ideas are next given a simple mathematical structure for a particular plane strain model.

Consider a sample of brittle material initially at rest and stress free. At a certain time instant, the sample begins to undergo homogeneous tensile straining at a constant strain rate $\dot{\epsilon}_0$. The macroscopic tensile stress required to sustain this rate of deformation is $\sigma(t)$, and the elastic modulus of the material in the absence of microcracks is E. For purposes of this discussion, the instantaneous elastic modulus of the material with many microcracks is denoted by $\overline{E}$. This modulus is defined as the elastic modulus that would be perceived for the sample for a given microcrack distribution if the sample were deformed homogeneously under equilibrium conditions with no extension of the microcracks. This definition probably results in an estimate of the apparent stiffness that is too low because it neglects the finite time required for dynamically growing microcracks to interact by means of stress waves.

If the sample is large enough that its macroscopic response is representative of the material, then the size of the sample cannot enter the relationship between macroscopic stress and strain. But if the effective stiffness is assumed to depend on an average microcrack length, say a, then an additional length scale must be identified on dimensional grounds. This length is assumed to be the average initial spacing between flaws that are activated or, equivalently, the initial density of activated flaws per unit volume (per unit area in two dimensions).

In his original statistical approach, Weibull assumed that there was some stress level, say σ_w, below which no microcracks would grow for a given material. For stress levels $\sigma \geq \sigma_w$, he assumed that the number of flaws per unit volume (or per unit area in two dimensions) having a strength equal to or less than σ is Poisson distributed. On the basis of probability arguments he concluded that the density of flaws with strength less than or equal to σ has the form

$$n(\sigma) = \frac{1}{b_w^\alpha} \left(\frac{\sigma - \sigma_w}{\sigma_w} \right)^{m\alpha}, \qquad (8.7.5)$$

where b_w is a material length scale, m is a positive material constant, and $\alpha =1$, 2, or 3 for flaw distributions in one, two, or three dimensions, respectively. The parameters σ_w, b_w and m can only be determined through experimental measurements. Although the weakest link idea was used heuristically by Weibull (1939) in arriving at (8.7.5), the idea has been shown to lead more directly to the same result on the basis of the asymptotic theory of extreme values in large samples of a statistical population. In any case, the characteristic length parameter for purposes of comparison to the average microcrack size is $[n(\sigma)]^{-1/\alpha}$.

On the basis of features identified up to this point, the time-dependent stress acting on the sample for constant strain rate is

$$\sigma(t) = \overline{E}\left(a(t)\sqrt{n(\sigma)}\right)\dot{\epsilon}_0 t + \sigma_w \qquad (8.7.6)$$

for $\alpha = 2$, where the dependence of $\overline{E}$ on the microcrack distribution has been made explicit. It is tacitly assumed in writing (8.7.6) that $\sigma(0) = \sigma_w$. This equation can be solved for σ as a function of time if $a(t)$ is known. A general framework for writing equations of motion for crack tips is presented in Section 7.4. For example, with reference to (7.4.2), if the equilibrium stress intensity factor $K_I(t,a,0)$ for a microcrack of length $2a(t)$ acted upon by stress $\sigma(t)$ is assumed to be $\sigma\sqrt{\pi a}$, then an evolution equation for $a(t)$ is

$$\frac{\dot{a}}{c_R} = 1 - \frac{\overline{E}\,\Gamma}{(1-\bar{\nu}^2)\pi a^2 \sigma}, \qquad (8.7.7)$$

where Γ is the specific fracture energy. The initial conditions for this differential equation are established by the initial flaw size. This equation of motion implies that if a crack grows to twice the initial flaw size under constant stress then the speed has increased to $0.75\,c_R$ if Γ is a constant. Because microcracks are expected to grow to many times their initial sizes, the initial length is relatively unimportant (especially at this crude level of modeling) and the growth speed is large for most of the growth time interval. Consequently, a large constant speed is assumed here, that is, $a(t) \approx vt$ is assumed for the time dependence of the average crack length, where v is a constant in the range $0 < v < c_R$.

A specific form of the dependence of $\overline{E}$ on its argument is required for a calculation. Budiansky and O'Connell (1976) presented results for the elastic moduli of materials with a dilute distribution of randomly oriented elliptic microcracks. The analysis, which was based on the self-consistent theory of material inhomogeneities, was extended to the case of dilute distributions of microcracks with non-random orientations by Hoenig (1979). For present purposes involving one-dimensional deformation based on plane strain crack analysis, the effective modulus result of Delameter, Herrmann, and Barnett (1975) is adopted. They determined the effective modulus of a two-dimensional elastic solid containing a rectangular array of identical cracks, each perpendicular to the tensile stress axis. For the case of cracks of length $2a$ with centers spaced at a constant interval b from each other on planes separated by a distance b, the effective modulus given by Delameter et al. (1975) is

$$\overline{E} = \frac{E}{1 - \frac{4}{\pi} \ln \left(\cos \frac{\pi a}{b} \right)}, \quad a < \tfrac{1}{2}b. \qquad (8.7.8)$$

This expression does not presume that the microcrack concentration is dilute, but it is based on a microcrack distribution with an unrealistically high degree of geometric regularity.

For this estimate of the elastic modulus of the microcracked material, the expression (8.7.6) for stress becomes

$$\frac{\sigma(t)}{\sigma_w} = \frac{\dot{\epsilon}_o t E / \sigma_w + 1}{1 - \frac{4}{\pi} \ln \left(\cos \left[(\pi v t / b_w) \left(\sigma(t) / \sigma_w - 1 \right)^m \right] \right)}. \qquad (8.7.9)$$

To illustrate the time dependence of stress, values of the various parameters must be selected. For several rock types (granodiorite, basalt, and limestone) Vardar and Finnie (1977) concluded that $m\alpha$ in (8.7.5) has a value in the range from 3 to 4.5. Here, $\alpha = 2$, so $m = 2$ is selected. They also concluded that σ_w has a value in the range 2 to 20 MPa, and b_w has a value that ranges from 10 μm to 2 cm, with the lower portion of the range being more representative. Values of the strain rate parameter

$$r = \frac{\dot{\epsilon}_o b_w}{v} \frac{E}{\sigma_w} \qquad (8.7.10)$$

are chosen to be 1, 2, 5, 25, and 125. Thus, if $v = 10^3$ m/s, $b_w = 100$ μm and $E/\sigma_w = 2 \times 10^4$, the value $r = 1$ corresponds to a strain

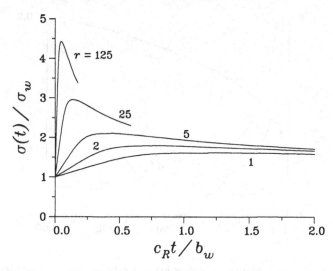

Figure 8.21. Stress $\sigma(t)$ acting on a sample of microcracking material deforming at constant strain rate $\dot\epsilon_0$ versus time t according to (8.7.9). The rate parameter r is defined in (8.7.10).

rate of $500\,\mathrm{s}^{-1}$. Evaluation of (8.7.9) leads to the stress histories shown in Figure 8.21, where c_R is the Rayleigh wave speed. The stress variation has the general features anticipated in the preliminary discussion.

The maximum stress for a given value of $\dot\epsilon_0$, which was termed the impact strength σ_s at constant strain rate, is readily found from (8.7.9) by the condition that

$$\sigma(t_s) = \sigma_s \quad \text{for} \quad \frac{d\sigma}{dt}(t_s) = 0 . \qquad (8.7.11)$$

Differentiation of (8.7.9) with respect to time with $d\sigma/dt = 0$ yields the equation

$$1 - \frac{4}{\pi}\ln\left(\cos\pi\xi\right) - 4\xi\tan\pi\xi = 0 , \qquad (8.7.12)$$

where $\xi = (\sigma/\sigma_w - 1)^m vt/b_w$. The only root of this equation in the range $0 < \xi < 0.5$ is $\xi = 0.302$. Therefore, the relationship

$$\frac{\sigma_s}{\sigma_w} = 1 + \left(0.302\frac{b_w}{vt_s}\right)^{1/m} \qquad (8.7.13)$$

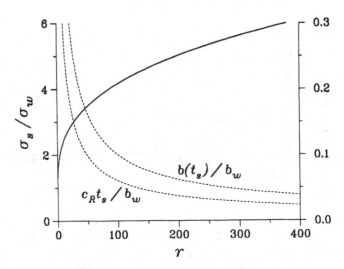

Figure 8.22. Impact strength σ_s versus loading rate parameter r implied by the constant strain rate model. Also shown by the dashed curves are t_s, the time required for the stress to reach the maximum level, and $b(t_s)$, a measure of the amount of crack growth at that time.

defines the locus of stress maxima in the graph of $\sigma(t)$ versus t; see Figure 8.21 for example. The dependence of σ_s and t_s on strain rate $\dot{\epsilon}_0$ can be established by solving (8.7.9) and (8.7.11) for these quantities. No simple expressions are evident, but numerically solving for the values of the parameters used to generate Figure 8.21 yields the result shown in Figure 8.22. Also shown in the figure is $b(t_s)$ versus strain rate. This quantity can be interpreted as an estimate of the spacing of activated flaw sites at the time that the stress reaches its maximum value σ_s.

It is interesting to note that Tuler and Butcher (1968) proposed a criterion for dynamic fracture due to damage accumulation that, in effect, anticipated the form deduced from mechanistic modeling. They suggested that if $\sigma(t)$ is a tensile stress pulse amplitude and t_s is the pulse duration, then a necessary condition for fracture due to damage is

$$\int_0^{t_s} \left[\frac{\sigma(t)}{\sigma_w} - 1\right]^{\beta} dt = \text{constant}, \qquad (8.7.14)$$

where σ_w is the stress threshold for damage accumulation and β is a positive constant. If it is assumed that stress increases linearly from

σ_w at $t = 0$ to σ_s at $t = t_s$ in the present model, then the integral

$$\int_0^{t_s} \left[\left(\frac{\sigma_s}{\sigma_w} - 1 \right) \frac{t}{t_s} \right]^{\beta} dt \qquad (8.7.15)$$

does indeed have a constant value if $\beta = m$. The observation follows immediately from integration of (8.7.15) and substitution of (8.7.13).

The analysis of this simple model has been presented to illustrate the qualitative features of transient microcrack damage accumulation in a brittle material under intense stress pulse loading. The physical processes involved are very complex, and a number of important observations have been made on the basis of detailed modeling and experimental studies. Shockey et al. (1974) carried out impact experiments on a polycrystalline quartzite rock. They characterized the microcrack structure of the damaged material, and developed a model of fragmentation on the basis of these observations. The model was embedded in a computational wave propagation code so that the impact experiment could be simulated numerically. Vardar and Finnie (1977) considered experiments on several rocks in which the material near the surface of the samples was heated rapidly by means of short bursts of energetic electrons. The resulting expansion of the material near the traction-free surface produced severe microcrack damage. They analyzed their results in the same spirit as the model described above to infer values of dynamic strength. Impact methods were also used by Grady and Kipp (1979) to create controlled dynamic fracture in polycrystalline quartzite rock samples. They showed that their observations were consistent with a micromechanical fracture model based on dynamic elastic fracture mechanics. Detailed computations on microcrack-induced damage evolution in brittle rock under pulse loading were reported by Taylor, Chen, and Kuszmaul (1986). They have proposed detailed constitutive equations for large-scale numerical simulation of rock fracture behavior.

Many aspects of the process of microcrack growth and coalescence have been neglected up to this point in the study of the phenomenon. For example, the dynamic interaction of growing microcracks by means of stress waves is not well understood, nor is the process of crack coalescence. On the latter point, the experiments described by Melin (1983) on the coalescence of initially coplanar tensile cracks suggest that the process may consume more energy than is commonly assumed. Another issue that is not well understood is the influence of the inertial resistance to opening of microcracks on the apparent effective stiffness of a crack elastic body. Crack branching and bifurcation are other phenomena that are of potential importance in the fragmentation process but for which a mechanistic understanding is not yet available. The prospect for more complete theories of dynamic microcracking and fragmentation of rocks, ceramics, concrete, and glasses hinges on progress on these and related issues.

Finally, it is noted that some one-dimensional wave propagation processes may be analyzed on the basis of the simple model outlined above. As an example, consider the experimental procedure described by Gran, Gupta, and Florence (1987). Suppose that a long cylindrical rod of a brittle material is loaded in static triaxial compression. The axial pressure is then released suddenly and simultaneously at the ends of the rod. A stress relaxation wave propagates away from each end and, barring any dissipation in the material, an axial tensile stress equal in magnitude to the initial compression is induced at the center of the rod according to one-dimensional wave theory. If the level of tensile stress is sufficiently high, then microcracking is induced near the center of the rod as the zone of tensile stress expands. If the known rise time of the stress pulse is assumed to determine the strain rate at the central cross section of the rod, then the result in Figure 8.21 provides an indication of the stress level at which fracture will occur at that section and the time at which it will occur. One-dimensional computations of the kind described by Taylor et al. (1986) provide a means of studying the evolution of damage as a function of time and position along the rod.

BIBLIOGRAPHY

ABDEL-LATIF, A. I. A., BRADT, R. C., AND TRESSLER, R. E. (1977), Dynamics of fracture mirror boundary formation in glass, *International Journal of Fracture* 13, pp. 349–59.

ABOUDI, J., AND ACHENBACH, J. D. (1981), Rapid mode III crack propagation in a strip of viscoplastic work-hardening material, *International Journal of Solids and Structures* 17, pp. 879–90.

ABOUDI, J., AND ACHENBACH, J. D. (1983a), Arrest of fast mode I fracture in an elastic-viscoplastic transition zone, *Engineering Fracture Mechanics* 18, pp. 109–19.

ABOUDI, J., AND ACHENBACH, J. D. (1983b), Numerical analysis of fast mode I fracture of a strip of viscoplastic work-hardening material, *International Journal of Fracture* 21, pp. 133–47.

ACHENBACH, J. D. (1970a), Crack propagation generated by a horizontally polarized shear wave, *Journal of the Mechanics and Physics of Solids* 18, pp. 245–59.

ACHENBACH, J. D. (1970b), Extension of a crack by a shear wave, *Zeitschrigt für angewandte Mathematik und Physik* 21, pp. 887–900.

ACHENBACH, J. D. (1970c), Brittle and ductile extension of a finite crack by a horizontally polarized shear wave, *International Journal of Engineering Science* 8, pp. 947–66.

ACHENBACH, J. D., AND NUISMER, R. (1971), Fracture generated by a dilatational wave, *International Journal of Fracture* 7, pp. 77–88.

ACHENBACH, J. D. (1973), *Wave Propagation in Elastic Solids*. Amsterdam: North-Holland.

ACHENBACH, J. D. (1974), Dynamic effects in brittle fracture, in *Mechanics Today, Vol. 1*, ed. S. Nemat-Nasser. Elmsford, NY: Pergamon, pp. 1–57.

ACHENBACH, J. D., AND VARATHARAJULU, V. K. (1974), Skew crack propagation at the diffraction of a transient stress wave, *Quarterly of Applied Mathematics* 32, pp. 123–35.

ACHENBACH, J. D. (1975), Bifurcation of a running crack in antiplane shear, *International Journal of Solids and Structures* 11, pp. 130–1314.

ACHENBACH, J. D., AND BAZANT, Z. P. (1975), Elastodynamic near-tip stress and displacement fields for rapidly propagating cracks in orthotropic media, *Journal of Applied Mechanics* 42, pp. 183–9.

ACHENBACH, J. D., AND GAUTESEN, A. K. (1977), Elastodynamic stress intensity factors for a semi-infinite crack under 3-D loading, *Journal of Applied Mechanics* 44, pp. 243–9.

ACHENBACH, J. D. (1979), Dynamic crack tip fields according to deformation theory, *Journal of Applied Mechanics* 46, pp. 707–8.

ACHENBACH, J. D., AND KHETAN, R. P. (1979), Kinking of a crack under dynamic loading conditions, *Journal of Elasticity* 9, pp. 113–29.

ACHENBACH, J. D., KEER, L. M., AND MENDELSOHN, D. A. (1980), Elastodynamic analysis of an edge crack, *Journal of Applied Mechanics* 47, pp. 551–6.

ACHENBACH, J. D., AND DUNAYEVSKY, V. (1981), Fields near a rapidly propagating crack tip in an elastic-perfectly plastic material, *Journal of the Mechanics and Physics of Solids* 29, pp. 283–303.

ACHENBACH, J. D., KANNINEN, M. F., AND POPELAR, C. H. (1981), Crack tip fields for fast fracture of an elastic-plastic material, *Journal of the Mechanics and Physics of Solids* 29, pp. 211–25.

ACHENBACH, J. D., AND NEIMITZ, A. (1981), Fast fracture and crack arrest according to the Dugdale model, *Engineering Fracture Mechanics* 14, pp. 385–95.

ACHENBACH, J. D., KUO, M. K., AND DEMPSEY, J. P. (1984), Mode III and mixed mode I–II crack kinking under stress wave loading, *International Journal of Solids and Structures* 20, pp. 395–410.

ACHENBACH, J. D., AND KUO, M. K. (1985), Conditions for crack kinking under stress wave loading, *Engineering Fracture Mechanics* 22, pp. 165–80.

ACHENBACH, J. D., LI, Z. L., AND NISHIMURA, N. (1985), Dynamic fields generated by rapid crack growth, *International Journal of Fracture* 27, pp. 215–27.

ACHENBACH, J. D., AND NISHIMURA, N. (1985), Effect of inertia on finite near tip deformation for fast mode III crack growth, *Journal of Applied Mechanics* 52, pp. 281–6.

ACHENBACH, J. D., AND NISHIMURA, N. (1986), Large deformations near a propagating crack tip, *Engineering Fracture Mechanics* 23, pp. 183–99.

ADLER, W. F., AND HOOKER, S. V. (1978a), Water drop impact damage in zinc sulfide, *Wear* 48, pp. 103–19.

ADLER, W. F., AND HOOKER, S. V. (1978b), Rain erosion mechanisms in brittle materials, *Wear* 50, pp. 11–38.

ADLER, W. F., AND HOOKER, S. V. (1978c), Rain erosion behavior of polymethylmethacrylate, *Journal of Materials Science* 13, pp. 1015–25.

AHMAD, J., JUNG, J., BARNES, C. R., AND KANNINEN, M. F. (1983), Elastic-plastic finite element analysis of dynamic fracture, *Engineering Fracture Mechanics* 17, pp. 235–46.

AKI, K. (1966), Generation and propagation of G-waves from the Niigata earthquake of June 16, 1964. Estimation of earthquake movement, released energy and stress–strain drop from G-wave spectrum, *Bulletin of the Earthquake Research Institute* 44, pp. 23–88.

AKI, K., AND RICHARDS, P. G. (1980), *Quantitative Seismology*. New York: Freeman.

AKITA, Y., AND IKEDA, K. (1961), Theory of brittle crack initiation and propagation – A theoretical analysis of ESSO test, *Welding Journal Research Supplement* 40, pp. 138–44s.

ANDREWS, D. J. (1976a), Rupture propagation with finite stress in antiplane strain, *Journal of Geophysical Research* 81, pp. 3575–82.

ANDREWS, D. J. (1976b), Rupture velocity of plane strain shear cracks, *Journal of Geophysical Research* 81, pp. 5679–87.

ANDREWS, D. J. (1985), Dynamic plane strain shear rupture with a slip weakening friction law calculated by a boundary integral method, *Bulletin of the Seismological Society of America* 75, pp. 1–21.

ANG, D. D. (1960), Elastic waves generated by a force moving along a crack, *Journal of Mathematics and Physics* 38, pp. 246–56.

ANG, D. D., AND KNOPOFF, L. (1964), Diffraction of scalar elastic waves by a finite crack, *Proceedings of the National Academy of Science* 51, pp. 593–8.

ANGELINO, G. C. (1978), Influence of the geometry on unstable crack extension and determination of dynamic fracture mechanics parameters, in *Fast Fracture and Crack Arrest*, Special Technical Publication 627, eds. G. T. Hahn and M. F. Kanninen. Philadelphia: American Society for Testing and Materials, pp. 392–407.

ANTHONY, S. R., CHUBB, J. P., AND CONGLETON, J. (1970), The crack-branching velocity, *Philosophical Magazine* 22, pp. 1201–16.

AOKI, S., KISHIMOTO, K., KONDO, H., AND SAKATA, M. (1978), Elastodynamic analysis of crack by finite element method using singular element, *International Journal of Fracture* 14, pp. 59–68.

AOKI, S., KISHIMOTO, K., IZUMIHARA, Y., AND SAKATA, M. (1980), Dynamic analysis of cracked linear viscoelastic solids by finite element method using singular element, *International Journal of Fracture* 16, pp. 97–109.

AOKI, S., KISHMOTO, K., AND SAKATA, M. (1987), Finite element computation of dynamic stress intensity factor for a rapidly propagating crack using $\hat{J}$– integral, *Computational Mechanics* 2, pp. 54–62.

AOKI, S. (1988), On the mechanics of dynamic fracture, *JSME International Journal, Series I* 31, pp. 487–99.

ARCHULETA, R. J., AND BRUNE, J. N. (1975), Surface strong motion associated with a stick-slip even in a foam rubber model of earthquakes, *Bulletin of the Seismological Society of America* 65, pp. 1059–71.

ARCHULETA, R. J., AND FRAZIER, G. A. (1978), Three-dimensional numerical simulations of dynamic faulting in a half space, *Bulletin of the Seismological Society of America* 68, pp. 541–72.

ARCHULETA, R. J., AND DAY, S. M. (1980), Dynamic rupture in layered medium: The 1966 Parkfield earthquake, *Bulletin of the Seismological Society of America* 70, pp. 671–89.

ARCHULETA, R. J. (1982), Analysis of near source static and dynamic measurements from the 1979 Imperial Valley earthquake, *Bulletin of the Seismological Society of America* 72, pp. 1927–56.

ASHURST, W. T., AND HOOVER, W. G. (1976), Microscopic fracture studies in the two-dimensional triangular lattice, *Physical Review* B14, pp. 1465–73.

ATKINSON, C. (1965), The propagation of fracture in aeolotropic materials, *International Journal of Fracture Mechancs* 1, pp. 47–55.

ATKINSON, C., AND HEAD, A. K. (1966), The influence of elastic anisotropy on the propagation of fracture, *International Journal of Fracture* 2, pp. 489–505.

ATKINSON, C. (1967), A simple model of a relaxed expanding crack, *Arkiv fur Fysik* 35, pp. 469–76.

ATKINSON, C. (1968), On axially symmetric expanding boundary value problems in classical elasticity, *International Journal of Engineering Science* 6, pp. 27–35.

ATKINSON, C., AND ESHELBY, J. D. (1968), The flow of energy into the tip of a moving crack, *International Journal of Fracture* 4, pp. 3–8.

ATKINSON, C., AND LIST, R. D. (1972), A moving crack problem in a viscoelastic solid, *International Journal of Engineering Science* 10, pp. 309–22.

ATKINSON, C. (1974), On the dynamic stress and displacement field associated with a crack propagating across the interface between two media, *International Journal of Engineering Science* 12, pp. 491–506.

ATKINSON, C., AND COLEMAN, C. J. (1977), On some steady-state moving boundary problems in the linear theory of viscoelasticity, *Journal of the Institute for Mathematics and Applications* 2, pp. 85–106.

ATKINSON, C. (1979), A note on some dynamic crack problems in linear viscoelasticity, *Archiwum Mechanik Stosowanej* 31, pp. 829–49.

ATKINSON, C., AND POPELAR, C. H. (1979), Antiplane dynamic crack propagation in a viscoelastic strip, *Journal of the Mechanics and Physics of Solids* 27, pp. 431–9.

ATKINSON, C., AND SMELSER, R. E. (1982), Invariant integrals for thermo-viscoelasticity and applications, *International Journal of Solids and Structures* 18, pp. 533–49.

ATKINSON, C., BASTERO, J. M., AND MIRANDA, I. (1986), Path-independent integrals in fracture dynamics using auxiliary fields, *Engineering Fracture Mechanics* 25, pp. 53–62.

ATLURI, S. N., AND NISHIOKA, T. (1983), Path independent integrals, energy release rates and general solutions of near-tip fields in mixed mode dynamic fracture mechanics, *Engineering Fracture Mechanics* 18, pp. 1–22.

ATLURI, S. N., AND NISHIOKA, T. (1985), Numerical studies in dynamic fracture mechanics, *International Journal of Fracture* 27, pp. 245–61.

AYERS, D. J. (1976), Dynamic plastic analysis of ductile fracture – the Charpy specimen, *International Journal of Fracture* 12, pp. 567–78.

BAKER, B. R. (1962), Dynamic stresses created by a moving crack, *Journal of Applied Mechanics* 29, pp. 449–58.

BARBEE, T. W., JR., SEAMAN, L., CREWDSON, R., AND CURRAN, D. (1972), Dynamic fracture criteria for ductile and brittle metals, *Journal of Materials* 7, pp. 2812–25.

BARENBLATT, G. I. (1959a), The formation of equilibrium cracks during brittle fracture: General ideas and hypotheses, axially symmetric cracks, *Applied Mathematics and Mechanics (PMM)* 23, pp. 622–36.

BARENBLATT, G. I. (1959b), Concerning equilibrium cracks forming during brittle fracture: The stability of isolated cracks, relationship with energetic theories, *Applied Mathematics and Mechanics (PMM)* 23, pp. 1273–82.

BARENBLATT, G. I., AND CHEREPANOV, G. P. (1960), On the wedging of brittle bodies, *Applied Mathematics and Mechanics (PMM)* 24, pp. 993–1014.

BARENBLATT, G. I., SALGANIK, R. L., AND CHEREPANOV, G. P. (1962), On the nonsteady motion of cracks, *Applied Mathematics and Mechanics (PMM)* 26, pp. 469–77.

BARENBLATT, G. I., AND GOLDSTEIN, G. V. (1972), Wedging of an elastic body by a slender wedge moving with a constant super-Rayleigh subsonic velocity, *International Journal of Fracture Mechancs* 8, pp. 427–34.

BARSOM, J. M., AND ROLFE, S. T. (1987), *Fracture and Fatigue Control of Structures*, 2nd edition. Englewood Cliffs, NJ: Prentice-Hall.

BARSTOW, F. E., AND EDGERTON, H. E. (1939), Glass fracture velocity, *Journal of the American Ceramic Society* 22, pp. 302–7.

BASS, B. R., PUGH, C. E., AND WALKER, J. K. (1987), Elastodynamic fracture analysis of large crack arrest experiments, *Nuclear Engineering and Design* 98, pp. 157–69.

BATEMAN, H. (1955), *Electrical and Optical Wave Motion*. Mineola, NY: Dover.

BAZANT, Z. P., GLAZIK, J. L., AND ACHENBACH, J. D. (1976), Finite element analysis of wave diffraction by a crack, *Journal of the Engineering Mechanics Division, ASCE* 102, pp. 479–96.

BENBOW, J. J., AND ROESSLER, F. C. (1957), Experiments on controlled fractures, *Proceedings of the Physical Society (London)* 70, p. 201.

BERGKVIST, H. (1973), The motion of a brittle crack, *Journal of the Mechanics and Physics of Solids* 21, pp. 229–39.

BERGKVIST, H. (1974), Crack arrest in elastic sheets, *Journal of the Mechanics and Physics of Solids* 22, pp. 491–502.

BERRY, J. P. (1960a), Some kinetic considerations of the Griffith criterion for fracture – I. Equations of motion at constant force, *Journal of the Mechanics and Physics of Solids* 8, pp. 194–206.

BERRY, J. P. (1960b), Some kinetic considerations of the Griffith criterion for fracture – II. Equations of motion at constant displacement, *Journal of the Mechanics and Physics of Solids* 8, pp. 207–16.

BERTOLOTTI, R. L. (1973), Fracture toughness of polycrystalline Al_2O_3, *Journal of the American Ceramic Society* 56, p. 107.

BIENIAWSKI, Z. (1967), Stability concept of brittle fracture propagation in rock, *Engineering Geology* 2, pp. 149–62.

BILBY, B. A. (1980), Tewksbury Lecture: Putting fracture to work, *Journal of Materials Science* 15, pp. 535–56.

BILEK, Z. J., AND BURNS, S. J. (1974), Crack propagation in wedged double cantilever beam specimens, *Journal of the Mechanics and Physics of Solids* 22, pp. 85–95.

BILEK, Z. J. (1978), The dependence of the fracture toughness of high strength steel on crack velocity, *Scripta Metallurgica* 12, pp. 1101–6.

BILEK, Z. J. (1980), Some comments on dynamic crack propagation in a high strength steel, in *Crack Arrest Methodology and Applications*, Special Technical Publication 711, eds. G. T. Hahn and M. F. Kanninen. Philadelphia: American Society for Testing and Materials, pp. 240–7.

BIRCH, M. W., AND WILLIAMS, J. G. (1978), The effect of rate on the impact fracture toughness of polymers, *International Journal of Fracture* 14, pp. 69–84.

BLEICH, H. H., AND MATTHEWS, A. T. (1967), Step load moving with superseismic velocity on the surface of an elastic-plastic half space, *International Journal of Solids and Structures* 3, pp. 819–52.

BLUHM, J. I., AND MARDIROSIAN, M. M. (1963), Fracture arrest capabilities of annularly reinforced cylindrical pressure vessels, *Experimental Mechanics* 3, pp. 57–66.

BLUHM, J. I. (1969), Fracture arrest, in *Fracture*, Vol. 5, ed. H. Liebowitz. New York: Academic, pp. 1–63.

BOATWRIGHT, J., AND FLETCHER, J. B. (1984), The partition of radiated energy between P and S waves, *Bulletin of the Seismological Society of America* 74, pp. 361–76.

BODNER, S. R. (1973), Stress waves due to fracture of glass in bending, *Journal of the Mechanics and Physics of Solids* 21, pp. 1–8.

BOHME, W., AND KALTHOFF, J. F. (1982), The behavior of notched bend specimens in impact testing, *International Journal of Fracture* 20, pp. 139–43.

BORZYKH, A. A., AND CHEREPANOV, G. P. (1980), On the theory of fracture of solids submitted to powerful pulsed electron beams, *Applied Mathematics and Mechanics (PMM)* 44, pp. 801–7.

BOWDEN, F. P., AND BRUNTON, J. H. (1961), The deformation of solids by liquid impact at supersonic speeds, *Proceedings of the Royal Society (London)* A26, pp. 433–50.

BOWDEN, F. P., AND FIELD, J. E. (1964), The brittle fracture of solids by liquid impact, by solid impact, and by shock, *Proceedings of the Royal Society (London)* A28, pp. 331–52.

BOWDEN, F. P., BRUNTON, J. H., FIELD, J. E., AND HAYES, A. D. (1967), Controlled fracture of brittle solids and interruption of electric current, *Nature* 216, pp. 38–42.

BRADLEY, W. B., AND KOBAYASHI, A. S. (1970), An investigation of propagating cracks by dynamic photoelasticity, *Experimental Mechanics* 10, pp. 106–13.

BRICKSTAD, B., AND NILSSON, F. (1980), Numerical evaluation by FEM of crack propagation experiments, *International Journal of Fracture* 16, pp. 71–84.

BRICKSTAD, B. (1983a), A FEM analysis of crack arrest experiments, *International Journal of Fracture* 21, pp. 177–94.

BRICKSTAD, B. (1983b), A viscoplastic analysis of rapid crack propagation experiments in steel, *Journal of the Mechanics and Physics of Solids* 31, pp. 307–27.

BROBERG, K. B. (1960), The propagation of a brittle crack, *Archiv fur Fysik* 18, pp. 159–92.

BROBERG, K. B. (1964), On the speed of a brittle crack, *Journal of Applied Mechanics* 31, pp. 546–7.

BROBERG, K. B. (1967), Discussion of fracture from the energy point of view, in *Recent Progress in Applied Mechanics*, eds. K. B. Broberg, J. Hult, and F. Niordson. New York: Wiley, pp. 125–51.

BROBERG, K. B. (1977), On the effects of plastic flow at fast crack growth, in *Fast Fracture and Crack Arrest*, Special Technical Publication 627, eds. G. T. Hahn and M. F. Kanninen. Philadelphia: American Society for Testing and Materials, pp. 243–56.

BROBERG, K. B. (1978), On transient sliding motion, *Geophysical Journal of the Royal Astronomical Society* 52, pp. 397–432.

BROCK, L. M. (1983), The dynamic stress intensity factor due to arbitrary screw dislocation motion, *Journal of Applied Mechanics* 50, pp. 383–9.

BROCK, L. M., JOLLES, M., AND SCHROEDL, M. (1985), Dynamic impact over a subsurface crack: Applications to the dynamic tear test, *Journal of Applied Mechanics* 52, pp. 287–90.

BROCK, L. M., AND ROSSMANITH, H. P. (1985), Analysis of the reflection of point force induced crack surface waves by a crack edge, *Journal of Applied Mechanics* 52, pp. 57–61.

BROCK, L. M. (1986), Transient dynamic Green's functions for a cracked plane, *Quarterly of Applied Mathematics* 44, pp. 265–75.

BROCK, L. M., AND JOLLES, M. (1987), Dislocation–crack edge interaction in dynamic brittle fracture and crack propagation, *International Journal of Solids and Structures* 23, pp. 607–19.

BROCKENBROUGH, J. R., SURESH, S., AND DUFFY, J. (1988), An analysis of dynamic fracture in microcracking brittle solids, *Philosophical Magazine* A58, pp. 619–34.

BROCKWAY, G. S. (1972), On the uniqueness of singular solutions to boundary-initial value problems in linear elastodynamics, *Archive for Rational Mechanics and Analysis* 48, pp. 213–44.

BRUNE, J. N. (1973), Earthquake modelling by stable slip along pre-cut surfaces in stressed foam rubber, *Bulletin of the Seismological Society of America* 63, pp. 2105–29.

BUDIANSKY, B., AND O'CONNELL, R. J. (1976), Elastic moduli of a cracked solid, *International Journal of Solids and Structures* 12, pp. 81–97.

BUECKNER, H. F. (1970), A novel principle for the computation of stress intensity factors, *Zeitschrigt für angewandte Mathematik und Mechanik* 50, pp. 529–46.

BUI, H. D., EHRLACHER, A., AND NGUYEN, Q. S. (1980), Propagation de fissure en thermoelasticite dynamique, *Journal de Mecanique* 19, pp. 698–723.

BULL, T. H. (1956), The tensile strengths of liquids under dynamic loading, *Philosophical Magazine* 1, pp. 153–65.

BURGERS, P. (1980), An analysis of dynamic linear elastic crack propagation in antiplane shear by finite differences, *International Journal of Fracture* 16, pp. 261–74.

BURGERS, P., AND FREUND, L. B. (1980), Dynamic growth of an edge crack in a half-space, *International Journal of Solids and Structures* 16, pp. 265–74.

BURGERS, P. (1982), Dynamic propagation of a kinked or bifurcated crack in antiplane strain, *Journal of Applied Mechanics* 49, pp. 371–6.

BURGERS, P., AND DEMPSEY, J. P. (1982), Two analytical solutions for dynamic crack bifurcation in antiplane strain, *Journal of Applied Mechanics* 49, pp. 366–70.

BURGERS, P. (1983), Dynamic kinking of a crack in plane strain, *International Journal of Solids and Structures* 19, pp. 735–52.

BURGERS, P., AND DEMPSEY, J. P. (1984), Plane strain dynamic crack bifurcation, *International Journal of Solids and Structures* 20, pp. 609–18.

BURNS, S. J., AND WEBB, W. W. (1966), Plastic deformation during cleavage of LiF, *Transactions of the Metallurgical Society of AIME* 236, pp. 1165–74.

BURNS, S. J. (1968), Transverse fracture markings generated by unsteady cleavage velocities, *Philosophical Magazine* 18, pp. 625–35.

BURNS, S. J., AND WEBB, W. W. (1970a), Fracture surface energies and dislocation processes during dynamical cleavage of LiF. I. Theory, *Journal of Applied Physics* 41, pp. 2078–85.

BURNS, S. J., AND WEBB, W. W. (1970b), Fracture surface energies and dislocation processes during dynamical cleavage of LiF. II. Experiments, *Journal of Applied Physics* 41, pp. 2086–95.

BURNS, S. J. (1972), Fracture surface energies from dynamical cleavage analysis, *Philosophical Magazine* 25, pp. 131–8.

BURRIDGE, R. (1965), The effect of source rupture velocity on the ratio of S to P corner frequencies, *Bulletin of the Seismological Society of America* 65, pp. 667–75.

BURRIDGE, R., AND WILLIS, J. R. (1969), The self-similar problem of the expanding elliptical crack in an anisotropic solid, *Proceedings of the Cambridge Philosophical Society* 66, pp. 443–68.

BURRIDGE, R., AND HALIDAY, G. S. (1971), Dynamic shear cracks with friction as models for shallow focus earthquakes, *Geophysical Journal of the Royal Astronomical Society* 25, pp. 261–83.

BURRIDGE, R. (1973), Admissible speeds for plane-strain self-similar shear cracks with friction but lacking cohesion, *Geophysical Journal of the Royal Astronomical Society* 35, pp. 439–55.

BURRIDGE, R., AND LEVY, C. (1974), Self similar circular shear cracks lacking cohesion, *Bulletin of the Seismological Society of America* 64, pp. 1789–808.

BURRIDGE, R. (1976), An influence function for the intensity factor in tensile fracture, *International Journal of Engineering Science* 14, pp. 725–34.

BURRIDGE, R., AND KELLER, J. B. (1978), Peeling, slipping and cracking – some one-dimensional free boundary problems in mechanics, *SIAM Review* 20, pp. 31–61.

BURRIDGE, R., CONN, G., AND FREUND, L. B. (1979), The stability of a plane strain shear crack with finite cohesive force running at intersonic speeds, *Journal of Geophysical Research* 84, pp. 2210–22.

BURRIDGE, R., AND MOON, R. (1981), Slipping on a frictional fault plane in three dimensions: A numerical simulation of a scalar analogue, *Geophysical Journal of the Royal Astronomical Society* 67, pp. 325–42.

BYKOVTSEV, A. S., AND CHEREPANOV, G. P. (1980a), On one model of source of tectonic earthquake, *Doklady Akademii Nauk SSSR* 251, pp. 1353–6.

BYKOVTSEV, A. S., AND CHEREPANOV, G. P. (1980b), On modelling the earthquake source, *Applied Mathematics and Mechanics (PMM)* 44, pp. 557–64.

BYKOVTSEV, A. S. (1983), On conditions for starting two collinear dislocation ruptures, *Applied Mathematics and Mechanics (PMM)* 47, pp. 700–3.

BYKOVTSEV, A. S., AND TAVBAEV, J. S. (1984), On star-like system of propagating dislocation ruptures, *Applied Mathematics and Mechanics (PMM)* 48, pp. 148–51.

BYKOVTSEV, A. S., AND KRAMAROVSKII, D. B. (1987), The propagation of a complex fracture area – the exact three-dimensional solution, *Applied Mathematics and Mechanics (PMM)* 51, pp. 89–98.

CAGNIARD, L. (1962), *Reflection and Refraction of Progressive Seismic Waves*, translated by E. A. Flinn and C. H. Dix. New York: McGraw-Hill.

CAMPBELL, J. D., AND FERGUSON, W. G. (1970), The temperature and strain rate dependence of the shear strength of mild steel, *Philosophical Magazine* 21, pp. 63–82.

CARDENAS-GARCIA, J. F., AND HOLLOWAY, D. C. (1988), On the Lamb solution and Rayleigh wave induced cracking, *Experimental Mechanics* 28, pp. 105–9.

CARLSSON, A. J. (1962), Experimental studies of brittle fracture propagation, *Transactions of the Royal Institute of Technology, Stockholm*, Number 189.

CARLSSON, A. J. (1963), Method for continuous measurement of instantaneous crack propagation velocities, *Transactions of the Royal Institute of Technology, Stockholm*, Number 207.

CARRIER, G. F., KROOK, M., AND PEARSON, C. E. (1966), *Functions of a Complex Variable*. New York: McGraw-Hill.

CARROLL, M. M., AND HOLT, A. C. (1972), Static and dynamic pore-collapse relations for ductile porous materials, *Journal of Applied Physics* 43, pp. 1626–36.

CHAMPION, C. R. (1988), The stress intensity factor history for an advancing crack under three-dimensional loading, *International Journal of Solids and Structures* 24, pp. 285–300.

CHANG, S. J. (1971), Diffraction of plane dilatational waves by a finite crack, *Quarterly Journal of Mechanics and Applied Mathematics* 24, pp. 423–43.

CHATTERJEE, A. K., AND KNOPOFF, L. (1983), Bilateral propagation of a spontaneous two-dimensional antiplane shear crack under the influence of cohesion, *Geophysical Journal of the Royal Astronomical Society* 73, pp. 449–73.

CHATTERJEE, A. K., AND KNOPOFF, L. (1984), Spontaneous growth of an in-plane shear crack, *International Journal of Solids and Structures* 20, pp. 963–78.

CHAUDHRI, M. M., WELLS, J. K., AND STEPHENS, A. (1981), Dynamic hardness, deformation and fracture of simple ionic crystals at very high rates of strain, *Philosophical Magazine* 43, pp. 643–64.

CHEN, E. P., AND SIH, G. C. (1973), Running crack in an incident wave field, *International Journal of Solids and Structures* 9, pp. 897–919.

CHEN, E. P., AND SIH, G. C. (1975), Scattering of plane waves by a propagating crack, *Journal of Applied Mechanics* 42, pp. 705–11.

CHEN, E. P., AND SIH, G. C. (1977), Transient response of cracks to impact, in *Elastodynamic Crack Problems*, ed. G. C. Sih. Leyden: Noordhoff Publishing, pp. 1–58.

CHEN, Y. M., AND WILKINS, M. L. (1976), Stress analysis of crack problems with a three-dimensional, time dependent computer program, *International Journal of Fracture* 12, pp. 607–17.

CHEREPANOV, G. P., AND SOKOLINSKY, V. B. (1972), On fracturing of brittle bodies by impact, *Engineering Fracture Mechanics* 4, pp. 205–14.

CHEREPANOV, G. P., AND AFANASEV, E. F. (1973), Some dynamic problems of the theory of elasticity – a review, *Applied Mathematics and Mechanics (PMM)* 37, pp. 584–606.

CHEREPANOV, G. P., AND AFANASEV, E. F. (1974), Some dynamic problems of the theory of elasticity – a review, *International Journal of Engineering Science* 12, pp. 665–90.

CHEREPANOV, G. P. (1979), *Mechanics of Brittle Fracture*, translated by R. de Wit and W. C. Cooley. New York: McGraw-Hill.

CHEREPANOV, G. P. (1980), On the theory of fracture of brittle bodies by explosion, *Mechanics of Solids (Mechanika Tverdogo Tela)* 15, pp. 155–68.

CHI, Y. C., LEE, S., CHO, K., AND DUFFY, J. (1989), The effects of tempering and test temperature on the dynamic fracture initiation behavior of an AISI 4340 VAR steel, *Materials Science and Engineering* A114, pp. 105–26.

CHITALEY, A. D., AND McCLINTOCK, F. A. (1971), Elastic-plastic mechanics of steady crack growth under antiplane conditions, *Journal of the Mechanics and Physics of Solids* 19, pp. 147–63.

CHRISTENSEN, R. M. (1982), *Theory of Viscoelasticity: An Introduction*. New York: Academic Press.

CHRISTIE, D. G. (1952), An investigation of cracks and stress waves in glass and plastics by high-speed photography, *Transactions of the Society of Glass Technology* 36, pp. 74–89.

CHRISTIE, D. G., AND KOLSKY, H. (1952), The fractures produced in glass and plastics by the passage of stress waves, *Transactions of the Society of Glass Technology* 36, pp. 65–73.

CHRISTIE, D. G. (1955), A multiple spark camera for dynamic stress analysis, *The Journal of Photographic Science* 3, pp. 153–9.

CLARK, A. B. J., AND IRWIN, G. R. (1966), Crack propagation behaviors, *Experimental Mechanics*, pp. 321–30.

CLAYTON, J. Q. (1987), Modes of cleavage initiation and propagation in spheriodized steels, *Scripta Metallurgica* 21, pp. 993–6.

CLIFTON, R. J. (1983), Dynamic plasticity, *Journal of Applied Mechanics* 50, pp. 941–52.

COHEN, L. J., AND BERKOWITZ, H. M. (1971), Time dependent fracture criteria for 6061-T6 aluminum under stress wave loading in uniaxial strain, *International Journal of Fracture Mechancs* 7, pp. 183–96.

COLE, J. D., AND HUTH, J. H. (1958), Stresses produced in a half plane by moving loads, *Journal of Applied Mechanics* 25, pp. 433–6.

CONGLETON, J., AND PETCH, N. J. (1965), The surface energy of a running crack in Al_2O_3, MgO and glass from crack branching measurements, *International Journal of Fracture* 1, pp. 14–9.

CONGLETON, J., AND PETCH, N. J. (1967), Crack branching, *Philosophical Magazine* 16, pp. 749–60.

COOK, R. F., Crack propagation thresholds: A measure of surface energy, *Journal of Materials Research* 1, pp. 852–60.

COSTIN, L. S., DUFFY, J., AND FREUND, L. B. (1977), Fracture initiation in metals under stress wave loading conditions, in *Fast Fracture and Crack Arrest*, Special Technical Publication 627, eds. G. T. Hahn and M. F. Kanninen. Philadelphia: American Society for Testing and Materials, pp. 301–18.

COSTIN, L. S., AND DUFFY, J. (1979), The effect of loading rate and temperature on the initiation of fracture in a mild, rate sensitive steel, *Journal of Engineering Materials and Technology* 101, pp. 258–64.

COSTIN, L. S., SERVER, W. L., AND DUFFY, J. (1979), Dynamic fracture initiation: A comparison of two experimental methods, *Journal of Engineering Materials and Technology* 101, pp. 168–72.

COTTERELL, B. (1964), On the nature of moving cracks, *Journal of Applied Mechanics* 31, pp. 12–6.

COTTERELL, B. (1965), On brittle fracture paths, *International Journal of Fracture Mechancs* 1, pp. 96–103.

COUQUE, H., ASARO, R. J., DUFFY, J., AND LEE, S. H. (1988), Correlations of microstructure with dynamic and quasi-static fracture in plain carbon steel, *Metallurgical Transactions* 19A, pp. 2119–206.

COURANT, R., AND FRIEDRICHS, K. O. (1948), *Supersonic Flow and Shock Waves*. New York: Springer-Verlag.

COURANT, R., AND HILBERT, D. (1962), *Methods of Mathematical Physics*, Vol. II. New York: Interscience.

CRAGGS, J. W. (1960), On the propagation of a crack in an elastic-brittle solid, *Journal of the Mechanics and Physics of Solids* 8, pp. 66–75.

CRAGGS, J. W. (1966), The growth of a disk shaped crack, *International Journal of Engineering Science* 4, pp. 113–24.

CROSLEY, P. B., AND RIPLING, E. J. (1969), Dynamic fracture toughness of A5333 steel, *Journal of Basic Engineering* 91, pp. 525–34.

CROSLEY, P. B., AND RIPLING, E. J. (1977), Characteristics of a run-arrest segment of crack extension, in *Fast Fracture and Crack Arrest*, Special Technical Publication 627, eds. G. T. Hahn and M. F. Kanninen. Philadelphia: American Society for Testing and Materials, pp. 203–27.

CROSLEY, P. B., AND RIPLING, E. J. (1980), Significance of crack arrest toughness (K_{Ia}) testing, in *Crack Arrest Methodology and Applications*, Special Technical Publication 711, eds. G. T. Hahn and M. F. Kanninen. Philadelphia: American Society for Testing and Materials, pp. 321–37.

CROSLEY, P. B., AND RIPLING, E. J. (1980), Comparison of crack arrest methodologies, in *Crack Arrest Methodology and Applications*, Special Technical Publication 711, eds. by G. T. Hahn and M. F. Kanninen. Philadelphia: American Society for Testing and Materials, pp. 211–27.

CROUCH, B. A., AND WILLIAMS, J. G. (1987a), Application of a dynamic numerical solution to high speed fracture experiments – I. Analysis of experimental geometries, *Engineering Fracture Mechanics* 26, pp. 541–51.

CROUCH, B. A., AND WILLIAMS, J. G. (1987b), Application of a dynamic numerical solution to high speed fracture experiments – II. Results and a thermal blunting model, *Engineering Fracture Mechanics* 26, pp. 553–66.

CURRAN, D. R., SHOCKEY, D. A., AND SEAMAN, L. (1973), Dynamic fracture criteria for a polycarbonate, *Journal of Applied Physics* 44, pp. 4025–38.

CURRAN, D. R., SEAMAN, L., AND SHOCKEY, D. A. (1987), Dynamic failure of solids, *Physics Reports* 147, pp. 253–388.

DAHLBERG, L., AND NILSSON, F. (1977), Some aspects of testing crack propagation toughness, in *International Conference on Dynamic Fracture Toughness*. Cambridge: The Welding Institute, pp. 281–91.

DAHLBERG, L., NILSSON, F., AND BRICKSTAD, B. (1980), Influence of specimen geometry on crack propagation and arrest toughness, in *Crack Arrest Methodology and Applications*, Special Technical Publication 711, eds. G. T. Hahn and M. F. Kanninen. Philadelphia: American Society for Testing and Materials, pp. 89–108.

DALLY, J. W., AND KOBAYASHI, T. (1978), Crack arrest in duplex specimens, *International Journal of Solids and Structures* 14, pp. 121–9.

DALLY, J. W. (1979), Dynamic photoelastic studies of fracture, *Experimental Mechanics* 19, pp. 349–61.

DALLY, J. W., AND SHUKLA, A. (1980), Energy loss in Homalite-100 during crack propagation and arrest, *Engineering Fracture Mechanics* 13, pp. 807–17.

DALLY, J. W., FOURNEY, W. L., AND IRWIN, G. R. (1985), On the uniqueness of the stress intensity factor–crack velocity relationship, *International Journal of Fracture* 27, pp. 159–68.

DALLY, J. W. (1987), Dynamic photoelasticity and its application to stress wave propagation, fracture mechanics and fracture control, in *Static and Dynamic Photoelasticity and Caustics*, ed. A. Lagarde. New York: Springer-Verlag, pp. 247–406.

DALLY, J. W., AND BARKER, D. B. (1988), Dynamic measurements of initiation toughness at high loading rates, *Experimental Mechanics*, pp. 298–303.

DAS, S., AND AKI, K. (1977a), A numerical study of two-dimensional spontaneous rupture propagation, *Geophysical Journal of the Royal Astronomical Society* 50, pp. 643–68.

DAS, S., AND AKI, K. (1977b), Fault plane with barriers: a versatile earthquake model, *Journal of Geophysical Research* 82, pp. 565–70.

DAS, S. (1980), A numerical method for determination of source time functions for general three-dimensional rupture propagation, *Geophysical Journal of the Royal Astronomical Society* 62, pp. 591–604.

DAS, S. (1981), Three-dimensional spontaneous rupture propagation and implications for the earthquake source mechanisms, *Geophysical Journal of the Royal Astronomical Society* 67, pp. 375–93.

DAS, S. (1985), Application of dynamic shear crack models to the study of the earthquake faulting process, *International Journal of Fracture* 26, pp. 263–76.

DAS, S., AND KOSTROV, B. V. (1987), On the numerical boundary integral equation method for three-dimensional dynamic shear crack problems, *Journal of Applied Mechanics* 54, pp. 99–104.

DATTA, S. K. (1979), Diffraction of SH-waves by edge cracks, *Journal of Applied Mechanics* 46, pp. 101–6.

DATTA, S. K., SHAH, A. H., AND FORTUNKO, C. M. (1982), Diffraction of medium and long wavelength horizontally polarized shear waves by edge cracks, *Journal of Applied Physics* 53, pp. 2895–903.

DAVISON, L., AND STEVENS, A. L. (1973), Thermomechanical constitution of spalling elastic bodies, *Journal of Applied Physics* 44, pp. 668–74.

DAVISON, L., STEVENS, A. L., AND KIPP, M. E. (1977), Theory of spall damage accumulation in ductile materials, *Journal of the Mechanics and Physics of Solids* 25, pp. 11–28.

DAVISON, L., AND GRAHAM, R. A. (1979), Shock compression of solids, *Physics Reports* 55, pp. 255–379.

DAY, S. M. (1982a), Three-dimensional finite difference simulation of fault dynamics: Rectangular faults with fixed rupture velocity, *Bulletin of the Seismological Society of America* 72, pp. 705–27.

DAY, S. M. (1982b), Three-dimensional simulation of spontaneous rupture: The effect of nonuniform pre-stress, *Bulletin of the Seismological Society of America* 72, pp. 1881–902.

DE HOOP, A. T. (1958), *Representation Theorems for the Displacement in an Elastic Solid and Their Application to Elastodynamic Diffraction Theory*, Doctoral Dissertation, Technical University of Delft.

DE HOOP, A. T. (1961), A modification of Cagniard's method for solving seismic pulse problems, *Applied Scientific Research* B8, pp. 349–56.

DE WIT, R., AND FIELDS, R. J. (1987), Wide plate crack arrest testing, *Nuclear Engineering and Design* 98, pp. 149–55.

DE WIT, R., LOW, S. R., AND FIELDS, R. J. (1988), Wide-plate crack arrest testing: Evolution of experimental procedures, in *Fracture Mechanics: Nineteenth Symposium*, Special Technical Publication 969. Philadelphia: American Society for Testing and Materials, pp. 679–90.

DEAN, R. H., AND HUTCHINSON, J. W. (1980), Quasi-static steady crack growth in small scale yielding, in *Fracture Mechanics: Twelfth Symposium*, Special Technical Publication 700. Philadelphia: American Society for Testing and Materials, pp. 383–405.

DELAMETER, W. R., HERRMANN, G., AND BARNETT, D. M. (1975), Weakening of an elastic solid by a rectangular array of cracks, *Journal of Applied Mechanics* 42, pp. 74–80.

DEMPSEY, J. P., KUO, M. K., AND ACHENBACH, J. D. (1982), Mode III crack kinking under stress wave loading, *Wave Motion* 4, pp. 181–90.

DEMPSEY, J. P., KUO, M. K., AND BENTLEY, D. L. (1986), Dynamic effects in mode III crack bifurcation, *International Journal of Solids and Structures* 22, pp. 333–53.

DIAZ, M., AND LUND, F. (1989), The inertia of a crack near a dislocation, *Philosophical Magazine* A60, pp. 139–45.

DMOWSKA, R., AND RICE, J. R. (1983), Fracture theory and its seismological applications, in *Continuum Theories in Solid Earth Geophysics*, ed. R. Teisseyre. Warsaw: PWN-Polish Scientific Publishers, pp. 187–255.

DOLL, W. (1973), An experimental study of the heat generated in the plastic region of a running crack in different polymeric materials, *Engineering Fracture Mechanics* 5, pp. 259–68.

DOLL, W. (1975a), Investigations of the crack branching energy, *International Journal of Fracture* 11, pp. 184–6.

DOLL, W. (1975b), A molecular weight dependent fracture transition in polymethylmethacrylate, *Journal of Materials Science* 10, pp. 935–42.

DOLL, W. (1976), Application of an energy balance and an energy method to dynamic crack propagation, *International Journal of Fracture* 12, pp. 595–605.

DOLL, W., WEIDMANN, G. W., AND SCHINKER, M. (1977), Isothermal and adiabatic effects associated with crack propagation in polymethylmethacrylate and polystyrene, *Journal of Non-Crystalline Solids*, pp. 612–7.

DOORNBOS, D. J. (1984), On the determination of radiated seismic energy and related source parameters, *Bulletin of the Seismological Society of America* 74, pp. 395–416.

DRUGAN, W. J., AND RICE, J. R. (1984), Restrictions on quasi-statically moving surfaces of strong discontinuity in elastic-plastic solids, in *Mechanics of Material Behavior*, eds. G. J. Dvorak and R. T. Shield. New York: Elsevier, pp. 59–73.

DRUGAN, W. J., AND SHEN, Y. (1987), Restrictions on dynamically propagating surfaces of strong discontinuity in elastic-plastic solids, *Journal of the Mechanics and Physics of Solids* 35, pp. 771–87.

DUFFY, A. R., MCCLURE, G. M., EIBNER, R. J., AND MAXEY, W. A. (1969), Fracture design practices for pressure piping, in *Fracture*, Vol. 5, ed. H. Liebowitz. New York: Academic, pp. 159–232.

DUGDALE, D. C. (1960), Yielding of steel sheets containing slits, *Journal of the Mechanics and Physics of Solids* 8, pp. 100–4.

DULANEY, E. N., AND BRACE, W. F. (1960), Velocity behavior of a growing crack, *Journal of Applied Physics* 31, pp. 2233-6.

DUNAYEVSKY, V., AND ACHENBACH, J. D. (1982), Boundary layer phenomenon in the plastic zone near a rapidly propagating crack tip, *International Journal of Solids and Structures* 18, pp. 1-12.

DVORAK, G. J. (1971a), Formation of plastic enclaves at running brittle cracks, *International Journal of Fracture* 7, pp. 251-67.

DVORAK, G. J. (1971b), A model of brittle fracture propagation: Part I.- Continuum aspects, *Engineering Fracture Mechanics* 3, pp. 351-79.

DVORAK, G. J., AND TANG, H. C. (1973), Influence of material properties on dynamic fracture toughness of steels, *Engineering Fracture Mechanics* 5, pp. 91-106.

EDGERTON, H. E., AND BARSTOW, F. E. (1941), Further studies of glass fracture with high-speed photography, *Journal of the American Ceramic Society* 24, pp. 131-7.

EFTIS, J., AND KRAFFT, J. M. (1965), A comparison of the initiation with the rapid propagation of a crack in a mild steel plate, *Journal of Basic Engineering* 87, pp. 257-63.

EMBLEY, G. T., AND SIH, G. C. (1973), Sudden appearance of a crack in a bent plate, *International Journal of Solids and Structures* 9, pp. 1349-59.

ENGEL, P. A. (1978), *Impact Wear of Materials*. New York: Elsevier.

ERDOGAN, F. (1968), Crack propagation theories, in *Fracture*, Vol. 2, ed. H. Liebowitz. New York: Academic, pp. 498-586.

ERMAK, A. A. (1978), The temperature field in the vicinity of a moving crack, *Soviet Mining Science*, pp. 28-32.

ESHELBY, J. D. (1953), The equation of motion of a dislocation, *Physical Review* 90, pp. 248-55.

ESHELBY, J. D. (1956), The continuum theory of lattice defects, in *Progress in Solid State Physics*, Vol. 3, eds. F. Seitz and D. Turnbull. New York: Academic, pp. 79-144.

ESHELBY, J. D. (1969a), The elastic field of a crack extending nonuniformly under general anti-plane loading, *Journal of the Mechanics and Physics of Solids* 17, pp. 177-99.

ESHELBY, J. D. (1969b), The starting of a crack, in *Physics of Strength and Plasticity*, ed. A. S. Argon. Cambridge, MA: MIT Press, pp. 263-75.

ESHELBY, J. D. (1970), Energy relations and the energy-momentum tensor in continuum mechanics, in *Inelastic Behavior of Solids*, eds. M. F. Kanninen et al. New York: McGraw-Hill, pp. 77-115.

ETHERIDGE, J. M., DALLY, J. W., AND KOBAYASHI, T. (1978), A new method of determining the stress intensity factor K from isochromatic fringe loops, *Engineering Fracture Mechanics* 10, pp. 81-93.

EVANS, A. G. (1974), Slow crack growth in brittle materials under dynamic loading conditions, *International Journal of Fracture* 10, pp. 251-9.

EVANS, A. G., AND WILSHAW, T. R. (1977), Dynamic solid particle damage in brittle materials: an appraisal, *Journal of Materials Science* 12, pp. 97-116.

FIELD, F. A., AND BAKER, B. R. (1962), Crack propagation under shear displacements, *Journal of Applied Mechanics* 29, pp. 436-7.

FINKEL', V. M., MURAVIN, G. B., LEZVINSKAYA, L. M. (1981), Energy flux density when a longitudinal shear crack propagates, *Soviet Journal of Nondestructive Testing*, pp. 485-9.

FINKEL', V. M., KOROLEV, A. P., AND SAVEL'EV, A. M. (1979), Self-arrest potential of rapid cracks in ferrosilicon at low temperatures, *Strength of Materials (Problemy Prochnosti)*, pp. 1138-43.

FISHER, B. (1971), The product of distributions, *Quarterly Journal of Mathematics* 22, pp. 291–8.

FISHER, J. C., AND HOLLOMON, J. H. (1947), A statistical theory of fracture, *Transactions of AIME* 171, pp. 546–61.

FLITMAN, L. M. (1963), Waves generated by sudden crack in a continuous elastic medium, *Applied Mathematics and Mechanics (PMM)* 27, pp. 938–53.

FONSECA, J. G., ESHELBY, J. D., AND ATKINSON, C. (1971), The fracture mechanics of flint-knapping and allied processes, *International Journal of Fracture Mechancs* 7, pp. 421–33.

FORWOOD, C. T., AND FORTY, A. J. (1965), The interaction of cleavage cracks with inhomogeneities in sodium chloride crystals, *Philosophical Magazine* 11, pp. 1067–82.

FORWOOD, C. T. (1968), The work of fracture in crystals of sodium chloride containing cavities, *Philosophical Magazine* 17, pp. 657–67.

FOSSUM, A. F., AND FREUND, L. B. (1975), Nonuniformly moving shear crack model of a shallow focus earthquake mechanism, *Journal of Geophysical Research* 80, pp. 3343–7.

FOSSUM, A. F. (1978), Influence of rotary inertia on steady-state crack propagation in a finite plate subjected to out-of-plane bending, *Journal of Applied Mechanics* 45, pp. 130–4.

FOURNEY, W. L., HOLLOWAY, D. C., AND DALLY, J. W. (1975), Fracture initiation and propagation from a center of dilatation, *International Journal of Fracture* 11, pp. 1011–29.

FREDRICKS, R. W., AND KNOPOFF, L. (1960), The reflection of Rayleigh waves by a high impedance obstacle on a half space, *Geophysics* 25, pp. 1195–202.

FREDRICKS, R. W. (1961), Diffraction of an elastic pulse in a loaded half space, *Journal of the Acoustical Society of America* 33, pp. 17–22.

FREUDENTHAL, A. M. (1968), Statistical approach to brittle fracture, in *Fracture*, Vol. 2, ed. H. Liebowitz, New York: Academic, pp. 592–619.

FREUND, L. B., AND PHILLIPS, J. W. (1970), Near front stress singularity for impact on an elastic quarter space, *Journal of the Acoustical Society of America* 47, pp. 942–3.

FREUND, L. B. (1971), The oblique reflection of a Rayleigh wave from a crack tip, *International Journal of Solids and Structures* 7, pp. 1199–210.

FREUND, L. B. (1972a), Crack propagation in an elastic solid subjected to general loading. I. Constant rate of extension, *Journal of the Mechanics and Physics of Solids* 20, pp. 129–40.

FREUND, L. B. (1972b), Crack propagation in an elastic solid subjected to general loading. II. Nonuniform rate of extension, *Journal of the Mechanics and Physics of Solids* 20, pp. 141–52.

FREUND, L. B. (1972c), Energy flux into the tip of an extending crack in an elastic solid, *Journal of Elasticity* 2, pp. 341–9.

FREUND, L. B. (1972d), Surface waves guided by a slit in an elastic solid, *Journal of Applied Mechanics* 39, pp. 1027–32.

FREUND, L. B. (1972e), The initial wave emitted by a suddenly extending crack in an elastic solid, *Journal of Applied Mechanics* 39, pp. 601–2.

FREUND, L. B. (1973a), Crack propagation in an elastic solid subjected to general loading. III. Stress wave loading, *Journal of the Mechanics and Physics of Solids* 21, pp. 47–61.

FREUND, L. B. (1973b), The response of an elastic solid to nonuniformly moving surface loads, *Journal of Applied Mechanics* 40, pp. 699–704.

FREUND, L. B., AND CLIFTON, R. J. (1974), On the uniqueness of elastodynamic solutions for running cracks, *Journal of Elasticity* 4, pp. 293–9.

FREUND, L. B., AND RICE, J. R. (1974), On the determination of elastodynamic crack tip stress fields, *International Journal of Solids and Structures* 10, pp. 411–7.

FREUND, L. B. (1974a), Crack propagation in an elastic solid subjected to general loading. IV. Obliquely incident stress pulse, *Journal of the Mechanics and Physics of Solids* 22, pp. 137–46.

FREUND, L. B. (1974b), The stress intensity factor due to normal impact loading of the faces of a crack, *International Journal of Engineering Science* 12, pp. 179–89.

FREUND, L. B., PARKS, D. M., AND RICE, J. R. (1975), Running ductile fracture in a pressurized line pipe, in *Mechanics of Crack Growth*, Special Technical Publication 590. Philadelphia: American Society for Testing and Materials, pp. 243–62.

FREUND, L. B., AND HERRMANN, G. (1976), Dynamic fracture of a beam or plate in plane bending, *Journal of Applied Mechanics* 43, pp. 112–6.

FREUND, L. B. (1976a), The analysis of elastodynamic crack tip fields, in *Mechanics Today*, Vol. III, ed. S. Nemat-Nasser, Elmsford, NY: Pergamon, pp. 55–91.

FREUND, L. B. (1976b), Dynamic crack propagation, in *The Mechanics of Fracture*, ed. F. Erdogan. New York: American Society of Mechanical Engineers, pp. 105–34.

FREUND, L. B. (1977), A simple model of the double cantilever beam crack propagation specimen, *Journal of the Mechanics and Physics of Solids* 25, pp. 69–79.

FREUND, L. B., LI, V. C. F., AND PARKS, D. M. (1979), An analysis of a wire-wrapped mechanical crack arrester for pressurized pipelines, *Journal of Pressure Vessel Technology* 101, pp. 51–8.

FREUND, L. B. (1979a), The mechanics of dynamic shear crack propagation, *Journal of Geophysical Research* 84, pp. 2199–209.

FREUND, L. B. (1979b), A one-dimensional dynamic crack propagation model, in *Mathematical Problems in Fracture*, ed. R. Burridge. Providence, RI: American Mathematical Society, pp. 21–37.

FREUND, L. B., AND PARKS, D. M. (1980), Analytical interpretation of running ductile fracture experiments in gas pressurized linepipe, in *Crack Arrest Methodology and Applications*, Special Technical Publication 711, eds. G. T. Hahn and M. F. Kanninen. Philadelphia: American Society for Testing and Materials, pp. 359–78.

FREUND, L. B. (1981), Influence of reflected Rayleigh waves on a propagating edge crack, *International Journal of Fracture* 7, pp. R83–6.

FREUND, L. B., AND DOUGLAS, A. S. (1982), The influence of inertia on elastic-plastic antiplane shear crack growth, *Journal of the Mechanics and Physics of Solids* 30, pp. 59–74.

FREUND, L. B., AND DOUGLAS, A. S. (1983), Dynamic growth of an antiplane shear crack in a rate-sensitive elastic-plastic material, in *Elastic-Plastic Fracture Mechanics*, Special Technical Publication 803, eds. C. F. Shih and J. Gudas. Philadelphia: American Society for Testing and Materials, pp. 5–20.

FREUND, L. B., AND HUTCHINSON, J. W. (1985), High strain-rate crack growth in rate-dependent plastic solids, *Journal of the Mechanics and Physics of Solids* 33, pp. 169–91.

FREUND, L. B., HUTCHINSON, J. W., AND LAM, P. S. (1986), Analysis of high strain rate elastic-plastic crack growth, *Engineering Fracture Mechanics* 23, pp. 119–29.

FREUND, L. B. (1987a), The stress intensity factor history due to three-dimensional loading of the faces of a crack, *Journal of the Mechanics and Physics of Solids* 35, pp. 61–72.

FREUND, L. B. (1987b), The apparent fracture energy for dynamic crack growth with fine scale periodic fracture resistance, *Journal of Applied Mechanics* 54, pp. 970–3.

FRIEDLANDER, F. G. (1958), *Sound Pulses*. Cambridge: Cambridge University Press.

FROST, H. J., AND ASHBY, M. J. (1982), *Deformation Mechanism Maps*. Elmsford, NY: Pergamon.

FULLER, K. N. G., FOX, P. G., AND FIELD, J. E. (1975), The temperature rise at the tip of fast-moving cracks in glassy polymers, *Proceedings of the Royal Society (London)* A341, pp. 537–57.

FYFE, I. M., AND RAJENDRAN, A. M. (1980), Dynamic pre-strain and inertia effects on the fracture of metals, *Journal of the Mechanics and Physics of Solids* 28, pp. 17–26.

GAO, Y. C., AND NEMAT-NASSER, S. (1983a), Dynamic fields near a crack tip growing in an elastic-perfectly plastic solid, *Mechanics of Materials* 2, pp. 47–60.

GAO, Y. C., AND NEMAT-NASSER, S. (1983b), Near-tip dynamic fields for a crack advancing in a power-law elastic-plastic material: modes I, II, and III, *Mechanics of Materials* 2, pp. 305–17.

GAO, Y. C., AND NEMAT-NASSER, S. (1984), Mode II dynamic fields near a crack tip growing in an elastic-perfectly plastic solid, *Journal of the Mechanics and Physics of Solids* 32, pp. 1–19.

GAO, Y. C. (1985), Asymptotic dynamic solution to the mode I propagating crack tip field, *International Journal of Fracture* 29, pp. 171–80.

GELFAND, I. M., AND FOMIN, S. V. (1963), *Calculus of Variations*. Englewood Cliffs, NJ: Prentice-Hall.

GILATH, I., ELIEZER, S., DARIEL, M. P., AND KORNBLIT, L. (1988), Brittle-to-ductile transition in laser induced spall at ultrahigh strain rate in 6061-T6 aluminum, *Applied Physics Letters* 52, pp. 1207–9.

GILATH, I., SALZMANN, D., GIVON, M., DARIEL, M., KORNBLIT, L., AND BARNOY, T. (1988), Spallation as an effect of laser-induced shock waves, *Journal of Materials Science* 23, pp. 1825–8.

GILLIS, P. P., AND GILMAN, J. J. (1964), Double-cantilever cleavage mode of crack propagation, *Journal of Applied Physics* 35, pp. 647–58.

GILMAN, J. J., KNUDSEN, C., AND WALSH, W. P. (1958), Cleavage cracks and dislocations in LiF crystals, *Journal of Applied Physics* 29, pp. 601–7.

GILMAN, J. J., AND TULER, F. R. (1970), Dynamic fracture by spallation in metals, *International Journal of Fracture Mechancs* 6, pp. 169–82.

GLENN, L. A. (1976), The fracture of a glass half space by projectile impact, *Journal of the Mechanics and Physics of Solids* 24, pp. 93–106.

GLENN, L. A., AND JANACH, W. (1977), Failure of granite cylinders under impact loading, *International Journal of Fracture* 13, pp. 301–17.

GLENN, L. A., GOMMERSTADT, B. Y., AND CHUDNOVSKY, A. (1986), A fracture mechanics model of fragmentation, *Journal of Applied Physics* 60, pp. 1224–6.

GLENNIE, E. B., AND WILLIS, J. R. (1971), An examination of the effects of some idealized models of fracture on accelerating cracks, *Journal of the Mechanics and Physics of Solids* 19, pp. 11–30.

GLENNIE, E. B. (1971a), A rate-independent crack model, *Journal of the Mechanics and Physics of Solids* 19, pp. 255–72.

GLENNIE, E. B. (1971b), The unsteady motion of a rate-dependent crack model, *Journal of the Mechanics and Physics of Solids* 19, pp. 329–38.

GLENNIE, E. B. (1972), The dynamic growth of a void in a plastic material and an application to fracture, *Journal of the Mechanics and Physics of Solids* 20, pp. 415–29.

GODSE, R., RAVICHANDRAN, G., AND CLIFTON, R. J. (1989), Micromechanisms of dynamic crack propagation in an AISI 4340 steel, *Materials Science and Engineering* A112, pp. 79–88.

GOLDSMITH, W., AND KATSAMANIS, R. (1979), Fracture of notched polymeric beams due to central impact, *Experimental Mechanics* 19, pp. 235–43.

GOL'DSHTEIN, R. V., AND KAPTSOV, A. V. (1983), Integrodifferential equations of the three-dimensional nonsteady dynamic problem of a crack in an elastic medium, *Mechanics of Solids (Mechanika Tverdogo Tela)* 18, pp. 74–9.

GOLENIEWSKI, G. (1988), Dynamic crack growth in a viscoelastic material, *International Journal of Fracture* 37, pp. R39–44.

GORITSKII, V. M. (1981), Ductile-brittle transition and nonmetallic inclusions in materials with a bcc lattice, *Strength of Materials (Problemy Prochnosti)*, pp. 96–105.

GRADY, D. E., AND HOLLENBACH, R. E. (1979), Dynamic fracture strength of rock, *Geophysics Research Letters* 6, pp. 73–6.

GRADY, D. E., AND KIPP, M. E. (1979), The micromechanics of impact fracture of rock, *International Journal of Rock Mechanics and Mining Science* 16, pp. 293–302.

GRADY, D. E., AND LIPKIN, J. (1980), Criteria for impulsive rock fracture, *Geophysics Research Letters* 7, pp. 255–8.

GRADY, D. E. (1981), Fragmentation of solids under impulsive stress loading, *Journal of Geophysical Research* 86, pp. 1047–54.

GRADY, D. E. (1982), Local inertial effects in dynamic fracture, *Journal of Applied Physics* 53, pp. 322–5.

GRADY, D. E., AND BENSON, D. A. (1983), Fragmentation of metal rings by electromagnetic loading, *Experimental Mechanics* 23, pp. 393–400.

GRADY, D. E., AND KIPP, M. E. (1985), Geometric statistics and dynamic fragmentation, *Journal of Applied Physics* 58, pp. 1210–22.

GRADY, D. E. (1988), The spall strength of condensed matter, *Journal of the Mechanics and Physics of Solids* 36, pp. 353–84.

GRAN, J. K., GUPTA, Y. M., AND FLORENCE, A. L. (1987), An experimental method to study the dynamic tensile failure of brittle geologic materials, *Mechanics of Materials* 6, pp. 113–25.

GREENWOOD, J. H. (1972), The measurement of fracture speeds in silicon and germanium, *International Journal of Fracture* 8, pp. 183–93.

GREGORY, R. D. (1966), The attenuation of a Rayleigh wave in a half space by a surface impedance, *Proceedings of the Cambridge Philosophical Society* 62, pp. 811–27.

GRIFFITH, A. A. (1920), The phenomenon of rupture and flow in solids, *Philosophical Transactions of the Royal Society (London)* A221, pp. 163–98.

GÜNTHER, W. (1962), Uber einige randintegrale der elastomechanik, *Abhandlungen der Braunschweigischen Wissenschaftlichen Gesellschaft* 14, pp. 53–72.

GUO, Q., LI, Z., AND LI, K. (1988), Dynamic effects on the near crack line fields for crack growth in an elastic perfectly-plastic solid, *International Journal of Fracture* 36, pp. 71–81.

GUR, Y., JAEGER, Z., AND ENGLMAN, R. (1984), Fragmentation of rock by geometrical simulation of crack motion – I. Independent planar cracks, *Engineering Fracture Mechanics* 20, pp. 783–800.

GURTIN, M. E. (1972), The linear theory of elasticity, in *Handbuch der Physik*, Vol. VIa/2, ed. S. Flügge. Berlin: Springer-Verlag, pp. 1–295.

GURTIN, M. E. (1976), On a path-independent integral for elastodynamics, *International Journal of Fracture* 12, pp. 643–4.

GURTIN, M. E., AND YATOMI, C. (1980), On the energy release rate in elastodynamic crack propagation, *Archive for Rational Mechanics and Analysis* 74, pp. 231–47.

GUZ', I. S., AND KATANCHIK, V. N. (1981), Interaction of a transverse wave with a stationary macrocrack, *Strength of Materials (Problemy Prochnosti)*, pp. 1046–53.

HAHN, G. T., SARRATE, M., KANNINEN, M. F., AND ROSENFIELD, A. R. (1973), A model for unstable shear crack propagation in pipes containing gas pressure, *International Journal of Fracture* 9, pp. 209–22.

HAHN, G. T., HOAGLAND, R. G., ROSENFIELD, A. R., AND SEJNOHA, R. (1974), Rapid crack propagation in a high strength steel, *Metallurgical Transactions* 5, pp. 475–82.

HAHN, G. T., HOAGLAND, R. G., KANNINEN, M. F., AND ROSENFIELD, A. R. (1975), Crack arrest in steels, *Engineering Fracture Mechanics* 7, pp. 583–91.

HAHN, G. T., HOAGLAND, R. G., AND ROSENFIELD, A. R. (1976), Influence of metallurgical factors on the fast fracture energy absorption rates, *Metallurgical Transactions* 7, pp. 49–54.

HAHN, G. T., AND KANNINEN, M. F. (1981), Dynamic fracture toughness parameters for HY80 and HY130 steels and their weldments, *Engineering Fracture Mechanics* 14, pp. 725–40.

HAHN, G. T. (1984), The influence of microstructure on brittle fracture toughness, *Metallurgical Transactions* 15A, pp. 947–59.

HALICIOGLU, T., AND COOPER, D. M. (1986), A computer simulation of the crack propagation process, *Materials Science and Engineering* 79, pp. 157–63.

HALL, E. O. (1953), The brittle fracture of metals, *Journal of the Mechanics and Physics of Solids* 1, pp. 227–33.

HALL, W. J., KIHARA, H., SOETE, W., AND WELLS, A. A. (1976), *Brittle Fracture of Welded Plate*. Englewood Cliffs, NJ: Prentice-Hall.

HANSON, M. E., AND SANFORD, A. R. (1971), Some characteristics of a propagating brittle tensile crack, *Geophysical Journal of the Royal Astronomical Society* 24, pp. 231–44.

HANSON, M. E. (1975), Some effects of macroscopic flaws on dynamic fracture patterns near a pressure-driven fracture, *International Journal of Rock Mechanics and Mining Science* 12, pp. 311–23.

HASSON, D. F., AND JOYCE, J. A. (1981), The effect of a higher loading rate on the J_{Ic} fracture toughness transition temperature of HY steels, *Journal of Engineering Materials and Technology* 103, pp. 133–41.

HAWONG, J. S., KOBAYASHI, A. S., DADKHAH, M. S., KANG, B. S.-J., AND RAMULU, M. (1987), Dynamic crack curving and branching under biaxial loading, *Experimental Mechanics* 27, pp. 146–53.

HELLAN, K. (1978a), Debond dynamics of an elastic strip – I. Timoshenko beam properties and steady motion, *International Journal of Fracture* 14, pp. 91–100.

HELLAN, K. (1978b), Debond dynamics of an elastic strip – II. Simple transient motion, *International Journal of Fracture* 14, pp. 173–84.

HELLAN, K. (1981), An alternative one-dimensional study of dynamic crack growth in dcb test specimens, *International Journal of Fracture* 17, pp. 311–9.

HELLAN, K. (1984), *Introduction to Fracture Mechanics*. New York: McGraw-Hill.

HILL, R. (1950), *The Mathematical Theory of Plasticity.* Oxford: Oxford University Press.

HILLE, E. (1959), *Analytic Function Theory.* Vol. I, New York: Blaisdell.

HIRTH, J. P., AND LOTHE, J. (1982), *Theory of Dislocations,* 2nd edition. New York: McGraw-Hill.

HOAGLAND, R. G., ROSENFIELD, A. R., AND HAHN, G. T. (1972), Mechanisms of fast fracture and arrest in steels, *Metallurgical Transactions* 3, pp. 123–36.

HOENIG, A. (1979), Elastic moduli of a non-randomly cracked body, *International Journal of Solids and Structures* 15, pp. 137–54.

HOFF, R., RUBIN, C. A., AND HAHN, G. T. (1985), Strain-rate dependence of the deformation at the tip of a stationary crack, in *Fracture Mechanics: Sixteenth Symposium,* Special Technical Publication 868, eds. M. F. Kanninen and A. T. Hooper. Philadelphia: American Society for Testing and Materials, pp. 409–30.

HOFF, R., RUBIN, C. A., AND HAHN, G. T. (1987), Viscoplastic finite element analysis of rapid fracture, *Engineering Fracture Mechanics* 26, pp. 445–61.

HOMMA, H., USHIRO, T., AND NAKAZAWA, H, (1979), Dynamic crack growth under stress wave loading, *Journal of the Mechanics and Physics of Solids* 27, pp. 151–62.

HOMMA, H., SHOCKEY, D. A., AND MURAYAMA, Y. (1983), Response of cracks in structural materials to short pulse loads, *Journal of the Mechanics and Physics of Solids* 31, pp. 261–79.

HORI, M., AND NEMAT-NASSER, S. (1988), Dynamic response of crystalline solids with microcavities, *Journal of Applied Physics* 64, pp. 856–63.

HUANG, N. C. (1983), The mechanics of splitting a strip through penetration with a sharp wedge, *International Journal of Fracture* 23, pp. 23–5.

HUDAK, S. J., DEXTER, R. J., FITZGERALD, J. H., AND KANNINEN, M. F. (1986), The influence of specimen boundary conditions on the fracture toughness of running cracks, *Engineering Fracture Mechanics* 23, pp. 201–13.

HUDSON, G., AND GREENFIELD, M. (1947), The speed of propagation of brittle cracks in steel, *Journal of Applied Physics* 18, pp. 405–7.

HUI, C. Y., AND RIEDEL, H. (1981), The asymptotic stress and strain field near the tip of a growing crack under creep conditions, *International Journal of Fracture* 17, pp. 409–25.

HULL, D., AND BEARDMORE, P. (1966), Velocity of propagation of cleavage cracks in tungsten, *International Journal of Fracture Mechancs* 2, pp. 468–87.

HULT, J. A., AND McCLINTOCK, F. A. (1956), Elastic-plastic stress and strain distributions around sharp notches under repeated shear, *Ninth International Congress for Applied Mechanics,* Brussels 8, pp. 51–8.

HUSSEINI, M. I., JOVANOVICH, D. B., RANDALL, M. J., AND FREUND, L. B. (1975), The fracture energy of earthquakes, *Geophysical Journal of the Royal Astronomical Society* 43, pp. 367–85.

HUSSEINI, M. I., AND RANDALL, M. J. (1976), Rupture velocity and radiation efficiency, *Bulletin of the Seismological Society of America* 66, pp. 1173–87.

HUSSEINI, M. I. (1977), Energy balance for motion along a fault, *Geophysical Journal of the Royal Astronomical Society* 49, pp. 699–714.

HUTCHINSON, J. W. (1968), Plastic stress and strain fields at a crack tip, *Journal of the Mechanics and Physics of Solids* 16, pp. 337–47.

IDA, Y. (1972), Cohesive force across the tip of a longitudinal shear crack and Griffith's specific surface energy, *Journal of Geophysical Research* 77, pp. 3796–805.

IDA, Y., AND AKI, K. (1972), Seismic source time function of propagating longitudinal shear cracks, *Journal of Geophysical Research* 77, pp. 2034–44.

540 Bibliography

IDA, Y. (1973), Stress concentration and unsteady propagation of longitudinal shear cracks, *Journal of Geophysical Research* 78, pp. 3418–29.

IRWIN, G. R. (1948), Fracture dynamics, in *Fracturing of Metals*. Cleveland: American Society of Metals, pp. 147–66.

IRWIN, G. R., AND KIES, J. A. (1952), Fracturing and fracture dynamics 38, *Welding Journal Research Supplement*, pp. 95–6s.

IRWIN, G. R., AND KIES, J. A. (1954), Critical energy rate analysis of fracture strength, *Welding Journal Research Supplement* 33, pp. 193–8s.

IRWIN, G. R. (1957), Analysis of stresses and strains near the end of a crack traversing a plate, *Journal of Applied Mechanics* 24, pp. 361–4.

IRWIN, G. R. (1958), Fracture, in *Encyclopedia of Physics*, Vol. VI, ed. S. Flügge. New York: Springer-Verlag, pp. 551–90.

IRWIN, G. R. (1960), Fracture mechanics, in *Structural Mechanics*, eds. J. N. Goodier and N. J. Hoff. Elmsford, NY: Pergamon, pp. 557–91.

IRWIN, G. R. (1964), Crack-toughness testing of strain-rate sensitive materials, *Journal of Engineering for Power* 86, pp. 444–50.

IRWIN, G. R. (1969), Basic concepts for dynamic fracture testing, *Journal of Basic Engineering* 91, pp. 519–24.

IRWIN, G. R., AND PARIS, P. C. (1971), Fundamental aspects of crack growth and fracture, in *Fracture*, Vol. 3, ed. H. Liebowitz, New York: Academic, pp. 1–46.

IRWIN, G. R., DALLY, J. W., KOBAYASHI, T., FOURNEY, W. L., ETHERIDGE, M. J., AND ROSSMANITH, H. P. (1979), On the determination of the à–K relationship for birefringent polymers, *Experimental Mechanics* 9, pp. 121–8.

ISHIKAWA, K., AND TSUYA, K. (1976), A note on brittle crack deceleration in structural steel, *International Journal of Fracture* 12, pp. 95–100.

IVES, K. D., SHOEMAKER, A. K., AND MCCARTNEY, R. F. (1974), Pipe deformation during a running shear fracture in a line pipe, *Journal of Engineering Materials and Technology* 96, pp. 309–17.

JAHANSHAHI, A. (1967), A diffraction problem and crack propagation, *Journal of Applied Mechanics* 34, pp. 100–3.

JAIN, D. L., AND CONWAL, R. P. (1972), Diffraction of a plane shear elastic wave by a circular rigid disk and a penny shaped crack, *Quarterly of Applied Mathematics* 30, pp. 283–97.

JOHNSON, J. N. (1981), Dynamic fracture and spallation in ductile solids, *Journal of Applied Physics* 52, pp. 2812–25.

JOHNSON, J. N. (1983), Ductile fracture of rapidly expanding rings, *Journal of Applied Mechanics* 50, pp. 593–600.

JOHNSON, J. N., AND ADDESSIO, F. L. (1988), Tensile plasticity and ductile fracture, *Journal of Applied Physics* 64, pp. 6699–712.

JOHNSON, J. W., AND HOLLOWAY, D. G. (1966), On the shape and size of the fracture zones on glass fracture surfaces, *Philosophical Magazine* 14, pp. 731–43.

JOHNSON, J. W., AND HOLLOWAY, D. G. (1968), Microstructure of the mist zone on glass fracture surfaces, *Philosophical Magazine* 18, pp. 899–910.

JONES, D. S. (1952), A simplifying technique in the solution of a class of diffraction problems, *Quarterly Journal of Mathematics* 3, pp. 189–96.

KALTHOFF, J. F. (1971), On the characteristic angle for crack branching in brittle materials, *International Journal of Fracture Mechancs* 7, pp. 478–80.

KALTHOFF, J. F., WINKLER, S., AND BEINERT, J. (1976), Dynamic stress intensity factors for arresting cracks in dcb specimens, *International Journal of Fracture* 12, pp. R317-9.

KALTHOFF, J. F., BEINERT, J., AND WINKLER, S. (1977), Measurements of dynamic stress intensity factors for fast running and arresting cracks in dcb specimens, in *Fast Fracture and Crack Arrest*, Special Technical Publication 627, eds. G. T. Hahn and M. F. Kanninen. Philadelphia: American Society for Testing and Materials, pp. 161–76.

KALTHOFF, J. F., AND SHOCKEY, D. A. (1977), Instability of cracks under impulse loading, *Journal of Applied Physics* 48, pp. 986–93.

KALTHOFF, J. F., WINKLER, S., AND BEINERT, J. (1977), The influence of dynamic effects in impact testing, *International Journal of Fracture* 13, pp. 528–31.

KALTHOFF, J. F., BEINERT, J., WINKLER, S., AND KLEMM, W. (1980), Experimental analysis of dynamic effects in different crack arrest test specimens, in *Crack Arrest Methodology and Applications*, Special Technical Publication 711, eds. G. T. Hahn and M. F. Kanninen. Philadelphia: American Society for Testing and Materials, pp. 109–27.

KALTHOFF, J. F. (1985), On the measurement of dynamic fracture toughness – A review of recent work, *International Journal of Fracture* 27, pp. 277–98.

KALTHOFF, J. F. (1986), Fracture behavior under high rates of loading, *Engineering Fracture Mechanics* 23, pp. 289–98.

KALTHOFF, J. F. (1987), The shadow optical method of caustics, in *Static and Dynamic Photoelasticity and Caustics*, ed. by A. Lagarde. New York: Springer-Verlag, pp. 407–522.

KAMEDA, J. (1986), A kinetic model for ductile-brittle fracture mode transition behavior, *Acta Metallurgica* 12, pp. 2391–8.

KANAZAWA, T., MACHIDA, S., TERAMOTO, T., AND YOSHINARI, H. (1981), Study on fast fracture and crack arrest, *Experimental Mechanics* 21, pp. 78–88.

KANEMITSU, Y., AND TANAKA, Y. (1987), Mechanism of crack formation in glass after high-power laser pulse radiation, *Journal of Applied Physics* 62, pp. 1208–11.

KANEMITSU, Y., TANAKA, Y., AND HARADA, Y. (1987), Surface transformations in glass initiated by laser-driven shock, *Japanese Journal of Applied Physics* 26, pp. 212–4.

KANNINEN, M. F. (1968), An estimate of the limiting speed of a propagating ductile crack, *Journal of the Mechanics and Physics of Solids* 16, pp. 215–28.

KANNINEN, M. F. (1973), An augmented double cantilever beam model for studying crack propagation and arrest, *International Journal of Fracture* 9, pp. 83–91.

KANNINEN, M. F. (1974), A dynamic analysis of unstable crack propagation and arrest in the dcb test specimen, *International Journal of Fracture* 10, pp. 415–30.

KANNINEN, M. F., SAMPATH, S. G., AND POPELAR, C. H. (1976), Steady-state crack propagation in pressurized pipelines without backfill, *Journal of Pressure Vessel Technology* 98, pp. 56–64.

KANNINEN, M. F. (1985), Application of dynamic fracture mechanics for the prediction of crack arrest in engineering structures, *International Journal of Fracture* 27, pp. 299–312.

KANNINEN, M. F., AND POPELAR, C. H. (1985), *Advanced Fracture Mechanics*. Oxford: Oxford University Press.

KATSAMANIS, F. G., AND DELIDES, C. (1988), Fracture surface energy measurements of PMMA: A new experimental approach, *Journal of Physics D: Applied Physics* 21, pp. 79–86.

KEEGSTRA, P. N. R. (1976), A transient finite element crack propagation model for nuclear pressure vessel steels, *Journal of the Institution of Nuclear Engineers* 17, pp. 89–96.

KELLEY, J. M., AND SUN, C. T. (1979), A singular finite element for computing time dependent stress intensity factors, *Engineering Fracture Mechanics* 12, pp. 13–22.

KERKHOF, F. (1970), *Bruchvorgange in Gläsern*. Frankfurt: Verlag der Deutschen Glastechnischen Gesellschaft.

KERKHOF, F. (1973), Wave fractographic investigations of brittle fracture dynamics, in *Dynamic Crack Propagation*, ed. G. C. Sih. Leyden: Noordhoff, pp. 3–35.

KIES, J. A., SULLIVAN, A. M., AND IRWIN, G. R. (1950), Interpretation of fracture markings, *Journal of Applied Physics* 21, pp. 716–20.

KIKUCHI, M. (1975), Inelastic effects on crack propagation, *Journal of Physics of the Earth* 23, pp. 161–72.

KIM, K. S. (1985a), A stress intensity factor tracer, *Journal of Applied Mechanics* 52, pp. 291–7.

KIM, K. S. (1985b), Dynamic fracture under normal impact loading of the crack faces, *Journal of Applied Mechanics* 52, pp. 585–92.

KING, W. W., MALLUCK, J. F., ABERSON, J. A., AND ANDERSON, J. M. (1976), Application of running-crack eigenfunctions to finite-element simulation of crack propagation, *Mechanics Research Communications* 3, pp. 197–202.

KING, W. W. (1978), Toward a singular element for propagating cracks, *International Journal of Fracture* 14, pp. R7–R10.

KINRA, V. K. (1976), Stress pulses emitted during fracture in tension, *International Journal of Solids and Structures* 12, pp. 803–8.

KINRA, V. K., AND KOLSKY, H. (1977), The interaction between bending fractures and the emitted stress waves, *Engineering Fracture Mechanics* 10, pp. 423–32.

KINRA, V. K., AND VU, B. Q. (1980), Brittle fracture of plates in tension: Virgin waves and boundary reflections, *Journal of Applied Mechanics* 47, pp. 45–50.

KIPP, M. E., GRADY, D. E., AND CHEN, E. P. (1980), Strain rate dependent fracture initiation, *International Journal of Fracture* 16, pp. 471–8.

KIPP, M. E., AND GRADY, D. E. (1985), Dynamic fracture growth and interaction in one dimension, *Journal of the Mechanics and Physics of Solids* 33, pp. 399–415.

KIRKALDY, D. (1863), *Experiments on Wrought Iron and Steel*, 2nd edition. New York: Scribner.

KISHIMOTO, K., AOKI, S., AND SAKATA, M. (1980a), Computer simulation of fast crack propagation in brittle material, *International Journal of Fracture* 16, pp. 3–13.

KISHIMOTO, K., AOKI, S., AND SAKATA, M. (1980b), Dynamic stress intensity factors using J-integral and finite element method, *Engineering Fracture Mechanics* 13, pp. 387–94.

KISHIMOTO, K., AOKI, S., AND SAKATA, M. (1980c), On the path independent $\hat{J}$-integral, *Engineering Fracture Mechanics* 13, pp. 841–50.

KLEPACZKO, J. R. (1984), Loading rate spectra for fracture initiation in metals, *Theoretical and Applied Fracture Mechanics* 1, pp. 181–91.

KLEPACZKO, J. R. (1985), Fracture initiation under impact, *International Journal of Impact Engineering* 3, pp. 191–210.

KNAUSS, W. G., AND RAVI-CHANDAR, K. (1985), Some basic problems in stress wave dominated fracture, *International Journal of Fracture* 27, pp. 127–43.

KNIGHT, C. G., SWAIN, M. V., AND CHAUDHRI, M. M. (1977), Impact of small steel spheres on glass surfaces, *Journal of Materials Science* 12, pp. 1573–86.

KNOPOFF, L., AND GILBERT, F. (1959), First motion methods in theoretical seismology, *Journal of the Acoustical Society of America* 31, pp. 1161–8.

KNOPOFF, L., MOUTON, J. O., AND BURRIDGE, R. (1973), The dynamics of a one-dimensional fault in the presence of friction, *Geophysical Journal of the Royal Astronomical Society* 35, pp. 169–84.

KNOPOFF, L., AND CHATTERJEE, A. K. (1982), Unilateral extension of a two-dimensional shear crack under the influence of cohesive forces, *Geophysical Journal of the Royal Astronomical Society* 68, pp. 7–25.

KNOTT, J. F. (1979), *Fundamentals of Fracture Mechanics.* London: Butterworth.

KNOWLES, J. K., AND STERNBERG, E. (1972), On a class of conservation laws in linearized and finite elastostatics, *Archive for Rational Mechanics and Analysis* 44, pp. 187–211.

KNOWLES, J. K. (1977), The finite anti-plane shear field near the tip of a crack for a class of incompressible elastic solids, *International Journal of Fracture* 13, pp. 611–39.

KOBAYASHI, A. S., WADE, B. G., BRADLEY, W. B., AND CHIU, S. T. (1974), Crack branching in Homalite-100 plates, *Engineering Fracture Mechanics* 6, pp. 81–92.

KOBAYASHI, A. S., AND CHAN, C. F. (1976), A dynamic photoelastic analysis of dynamic tear test specimen, *Experimental Mechanics* 16, pp. 176–81.

KOBAYASHI, A. S., EMERY, A. F., AND MALL, S. (1976), Dynamic finite element and dynamic photoelastic analyses of two fracturing Homalite-100 plates, *Experimental Mechanics* 16, pp. 321–8.

KOBAYASHI, A. S., AND MALL, S. (1978), Dynamic fracture toughness of Homalite-100, *Experimental Mechanics* 18, pp. 11–8.

KOBAYASHI, A. S., EMERY, A. F., LOVE, W. J., AND CHAO, Y.-H. (1988), Subsize experiments and numerical modeling of axial rupture of gas transmission lines, *Journal of Pressure Vessel Technology* 110, pp. 155–60.

KOBAYASHI, T., AND DALLY, J. W. (1979), Dynamic photoelastic determination of the *à–K* relation for 4340 steel, in *Crack Arrest Methodology and Applications*, Special Technical Publication 711, eds. G. T. Hahn and M. F. Kanninen. Philadelphia: American Society for Testing and Materials, pp. 189–210.

KOBAYASHI, T., YAMAMOTO, H., AND MATSUO, K. (1988), Evaluation of dynamic fracture toughness of heavy wall ductile cast iron for container, *Engineering Fracture Mechanics* 30, pp. 397–407.

KOLSKY, H., AND CHRISTIE, D. G. (1952), The fractures produced in glass and plastics by the passage of stress waves, *Transactions of the Society of Glass Technology* 36, pp. 65–73.

KOLSKY, H. (1953), *Stress Waves in Solids.* Mineola, NY: Dover.

KOLSKY, H., AND RADER, D. (1969), Stress waves and fracture, *Fracture*, Vol. 1, ed. H. Liebowitz. New York: Academic, pp. 533–69.

KOSTROV, B. V. (1964a), The axisymmetric problem of propagation of a tensile crack, *Applied Mathematics and Mechanics (PMM)* 28, pp. 793–803.

KOSTROV, B. V. (1964b), Self-similar problems of propagation of shear cracks, *Applied Mathematics and Mechanics (PMM)* 28, pp. 1077–87.

KOSTROV, B. V. (1966), Unsteady propagation of longitudinal shear cracks, *Applied Mathematics and Mechanics (PMM)* 30, pp. 1241–8.

KOSTROV, B. V., AND NIKITIN, L. V. (1970), Some general problems of mechanics of brittle fracture, *Archiwum Mechaniki Stosowanej* 22, pp. 749–75.

KOSTROV, B. V. (1973), Seismic moment, earthquake energy and seismic flow of rocks, *Publications of the Institute of Geophysics, Polish Academy of Science* 62, pp. 25–47.

KOSTROV, B. V. (1974), Seismic moment and energy of earthquakes and seismic flow of rocks, *Izvestia: Earth Physics* 1, pp. 23–40.

KOSTROV, B. V. (1975), On the crack propagation with variable velocity, *International Journal of Fracture* 11, pp. 47–56.

KOSTROV, B. V., AND OSAULENKO, V. I. (1976), Crack propagation at arbitrary variable rate under the action of static loads, *Mechanics of Solids (Mechanika Tverdogo Tela)* 11, pp. 76–92.

KOSTROV, B. V., AND DAS, S. (1989), *Principles of Earthquake Source Mechanics*. Cambridge: Cambridge University Press.

KOTOUL, M., AND BILEK, Z. (1987), Waves of fracture in brittle bodies, *Engineering Fracture Mechanics* 27, pp. 517–38.

KOVALEVA, I. N., AND KOSMODEM'YANSKI, A. A. (1980), Plane fracture waves in brittle solids, *Mechanics of Solids (Mechanika Tverdogo Tela)* 15, pp. 118–21.

KRAFFT, J. M., SULLIVAN, A. M., AND IRWIN, G. R. (1957), Relationship between the fracture ductility transition and strain hardening characteristics of a low carbon steel, *Journal of Applied Physics* 28, pp. 379–80.

KRAFFT, J. M., AND SULLIVAN, A. M. (1963), Effects of speed and temperature on crack toughness and yield strength in mild steel, *Transactions of the American Society of Metals* 56, pp. 160–75.

KRAFFT, J. M., AND IRWIN, G. R. (1965), Crack velocity considerations, in *Fracture Toughness Testing and Its Applications*, Special Technical Publication 381. Philadelphia: American Society for Testing and Materials, pp. 114–32.

KRASOWKSY, A. J., KASHTALYAN, YU. A., AND KRASIKO, V. N. (1983), Brittle-to-ductile transition in steels and the critical transition temperature, *International Journal of Fracture* 23, pp. 297–315.

KUHN, G., AND MATCZYNSKI, M. (1974), A dynamic problem of a crack in a plate strip, *Engineering Transactions, Polska Akademia Nauk* 22, pp. 469–85.

KULAKHMETOVA, S. A., SARAIKIN, V. A., AND SLEPYAN, L. I. (1984), Plane problem of a crack in a lattice, *Mechanics of Solids (Mechanika Tverdogo Tela)* 19, pp. 102–8.

KULAKHMETOVA, S. A. (1985), Dynamics of a crack in an anisotropic lattice, *Soviet Physics* 30, pp. 254–5.

KUMAR, A. M., HAHN, G. T., RUBIN, C. A., AND XU, N. (1988), A crack arrest toughness test for tough materials, *Engineering Fracture Mechanics* 31, pp. 793–805.

KUNDU, T., AND MAL, A. K. (1981), Diffraction of elastic waves by a surface crack on a plate, *Journal of Applied Mechanics* 48, pp. 570–6.

KUPPERS, H. (1967), The initial course of crack velocity in glass plates, *International Journal of Fracture Mechancs* 3, pp. 13–7.

KUPRADZE, V. D. (1963), Dynamical problems in elasticity, in *Progress in Solid Mechanics*, Vol. III, eds. I. N. Sneddon and R. Hill. New York: North-Holland, pp. 1–258.

LAM, P. S., AND FREUND, L. B. (1985), Analysis of dynamic growth of a tensile crack in an elastic-plastic material, *Journal of the Mechanics and Physics of Solids* 33, pp. 153–67.

LAMB, H. (1904), On the propagation of tremors over the surface of an elastic solid, *Philosophical Transactions of the Royal Society (London)* A203, pp. 1–12.

LANDONI, J. A., AND KNOPOFF, L. (1981), Dynamics of one-dimensional crack with variable friction, *Geophysical Journal of the Royal Astronomical Society* 64, pp. 151–61.

LARDNER, R. W. (1974), *Mathematical Theory of Dislocations and Fracture*. Toronto: University of Toronto Press.

LAWN, B. R., AND WILSHAW, T. R. (1975), *Fracture of Brittle Solids*. Cambridge: Cambridge University Press.

LEE, C. S., LIVINE, T., AND GERBERICH, W. W. (1986), The acoustic emission measurement of cleavage initiation near the ductile brittle transition temperature in steel, *Scripta Metallurgica* 20, pp. 1137–40.

LEIGHTON, J. T., CHAMPION, C. R., AND FREUND, L. B. (1987), Asymptotic analysis of steady dynamic crack growth in an elastic-plastic material, *Journal of the Mechanics and Physics of Solids* 35, pp. 541–63.

LI, Z. L., YANG, J. L., AND LEE, H. (1988), Temperature fields near a running crack tip, *Engineering Fracture Mechanics* 30, pp. 791–9.

LIN, B.-S. (1985), Elastic perfectly-plastic fields at a rapidly propagating crack tip, *Applied Mathematics and Mechanics* (trans. from Chinese) 6, pp. 1017–25.

LIN, I. H., AND THOMSON, R. M. (1986), Dynamic cleavage in ductile materials, *Journal of Materials Research* 1, pp. 73–80.

LINGER, K. R., AND HOLLOWAY, D. G. (1968), The fracture energy of glass, *Philosophical Magazine* 18, pp. 1269–80.

LIU, J. M., AND SHEN, B. W. (1984a), Dependence of dynamic fracture resistance on crack velocity in tungsten: Part I. Single crystals, *Metallurgical Transactions* 15A, pp. 1247–51.

LIU, J. M., AND SHEN, B. W. (1984b), Dependence of dynamic fracture resistance on crack velocity in tungsten: Part II. Bicrystals and polycrystals, *Metallurgical Transactions* 15A, pp. 1253–8.

LO, K. K. (1983), Dynamic crack-tip fields in rate sensitive solids, *Journal of the Mechanics and Physics of Solids* 31, pp. 287–305.

LOEBER, J. F., AND SIH, G. C. (1975), Diffraction of antiplane shear waves by a finite crack, *Journal of the Acoustical Society of America* 44, pp. 90–8.

MA, C. C., AND BURGERS, P. (1986), Mode III crack kinking with delay time: An analytical approximation, *International Journal of Solids and Structures* 22, pp. 883–99.

MA, C. C., AND FREUND, L. B. (1986), The extent of the stress intensity factor field during crack growth under dynamic loading conditions, *Journal of Applied Mechanics* 53, pp. 303–10.

MA, C. C., AND BURGERS, P. (1987), Dynamic mode I and mode II crack kinking including delay time effects, *International Journal of Solids and Structures* 23, pp. 897–918.

MA, C. C. (1988), Initiation, propagation, and kinking of an in-plane crack, *Journal of Applied Mechanics* 55, pp. 587–95.

MA, C. C., AND BURGERS, P. (1988), Initiation, propagation and kinking of an antiplane crack, *Journal of Applied Mechanics* 55, pp. 111–9.

MACHIDA, S., YOSHINARI, H., AND KANAZAWA, T. (1986), Some recent experimental work in Japan on fast fracture and crack arrest, *Engineering Fracture Mechanics* 23, pp. 251–64.

MADARIAGA, R. (1976), Dynamics of an expanding circular fault, *Bulletin of the Seismological Society of America* 66, pp. 639–66.

MAKOVEI, V. A. (1984), Determination of the dynamic characteristics of fracture toughness on ring specimens in loading using a vertical impact tester, *Strength of Materials (Problemy Prochnosti)*, pp. 887–90.

MAL, A. K. (1968a), Diffraction of elastic waves by a penny shaped crack, *Quarterly of Applied Mathematics* 26, pp. 231–8.

MAL, A. K. (1968b), Dynamic stress intensity factors for a non-symmetric loading of a penny shaped crack, *International Journal of Engineering Science* 6, pp. 725–33.

MAL, A. K. (1970a), Interaction of elastic waves with a Griffith crack, *International Journal of Engineering Science* 8, pp. 763–76.

MAL, A. K. (1970b), Interaction of elastic waves with a penny shaped crack, *International Journal of Engineering Science* 8, pp. 381–8.

MALL, S., KOBAYASHI, A. S., AND URABE, Y. (1978), Dynamic photoelastic and dynamic finite element analyses of dynamic tear test specimens, *Experimental Mechanics* 18, pp. 449–56.

MALLUCK, J. F., AND KING, W. W. (1977), Simulations of fast fracture in the dcb specimen using Kanninen's model, *International Journal of Fracture* 13, pp. 655–65.

MALLUCK, J. F., AND KING, W. W. (1980), Fast fracture simulated by conventional finite elements: a comparison of two energy release algorithms, in *Crack Arrest Methodology and Applications*, Special Technical Publication 711, ed. by G. T. Hahn and M. F. Kanninen. Philadelphia: American Society for Testing and Materials, pp. 38–53.

MALVERN, L. E. (1951), Plastic wave propagation in a bar of material exhibiting a strain rate effect, *Quarterly of Applied Mathematics* 8, pp. 405–11.

MALVERN, L. E. (1969), *Introduction to the Mechanics of a Continuous Medium.* Englewood Cliffs, NJ: Prentice-Hall.

MANNION, L. F., AND PIPKIN, A. C. (1983), Dynamic fracture of idealized fiber reinforced materials, *Journal of Elasticity* 13, pp. 395–409.

MANOGG, P. (1966), Investigation of the rupture of a plexiglass plate by means of an optical method involving high speed filming of the shadows originating around holes drilled in the plate, *International Journal of Fracture* 2, pp. 604–13.

MANSINHA, L. (1964), The propagating fracture of constant shape, *Journal of the Mechanics and Physics of Solids* 12, pp. 353–60.

MASLOV, L. A. (1980), Motion of a crack in a discrete medium, *Mechanics of Solids (Mechanika Tverdogo Tela)* 15, pp. 106–9.

MATAGA, P. A., FREUND, L. B., AND HUTCHINSON, J. W. (1987), Crack tip plasticity in dynamic fracture, *Journal of the Physics and Chemistry of Solids* 48, pp. 985–1005.

MAUE, A. W. (1959), Die beugung elastischer wellen an der halbebene, *Zeitschrigt für angewandte Mathematik und Mechanik* 12, pp. 1–10.

McCLINTOCK, F. A., AND SUKHATME, S. P. (1960), Travelling cracks in elastic materials under longitudinal shear, *Journal of the Mechanics and Physics of Solids* 8, pp. 187–93.

McCLINTOCK, F. A., AND IRWIN, G. R. (1965), Plasticity aspects of fracture mechanics, in *Fracture Toughness Testing and Its Applications*, Special Technical Publication 381. Philadelphia: American Society for Testing and Materials, pp. 84–113.

McCLINTOCK, F. A. (1968), A criterion for ductile fracture by the growth of holes, *Journal of Applied Mechanics* 35, pp. 363–71.

McLACHLAN, N. W. (1963), *Complex Variable Theory and Transform Calculus.* Cambridge: Cambridge University Press.

McMAHON, C. J., JR., AND COHEN, M. (1965), Initiation of cleavage in polycrystalline iron, *Acta Metallurgica* 13, pp. 591–604.

MECHOLSKY, J. J., RICE, R. W., AND FREIMAN, S. W. (1974), Prediction of fracture energy and flaw size in glasses from measurements of mirror size, *Journal of the American Ceramic Society* 57, pp. 440–3.

MECHOLSKY, J. J., FREIMAN, S. W., AND RICE, R. W. (1976), Fracture surface analysis of ceramics, *Journal of Materials Science* 11, pp. 1310–9.

MELIN, S. (1983), Why do cracks avoid each other?, *International Journal of Fracture* 23, pp. 37–45.

MEYERS, M. A., AND AIMONE, C. T. (1983), Dynamic fracture (spalling) of metals, *Progress in Materials Science* 28, pp. 1–96.

MIKUMO, T., AND MIYATAKE, T. (1979), Earthquake sequences on a frictional fault model with a non-uniform strength and relaxation time, *Geophysical Journal of the Royal Astronomical Society* 59, pp. 497–522.

MILES, J. W. (1960), Homogeneous solutions in elastic wave propagation, *Quarterly of Applied Mathematics* 18, pp. 37–59.

MIYATAKE, T. (1980), Numerical simulations of earthquake source process by a three-dimensional crack model. Part I. Rupture process, *Journal of Physics of the Earth* 28, pp. 565–98.

MOCK, W., AND HOLT, W. H. (1983), Fragmentation behavior of Armco iron and HF-1 steel explosive-filled cylinders, *Journal of Applied Physics* 54, pp. 2344–51.

MOTT, N. F. (1947), Fragmentation of shell cases, *Proceedings of the Royal Society (London)* A300, pp. 300–8.

MOTT, N. F. (1948), Brittle fracture in mild steel plates, *Engineering* 165, pp. 16–8.

MUSKHELISHVILI, N. I. (1953), *Singular Integral Equations*. Leyden: Noordhoff.

NAIT ABDELAZIZ, M., NEVIERE, R., AND PLUVINAGE, G. (1987), Experimental method for J_{Ic} computation on fracture of solid propellants under dynamic loading conditions, *Engineering Fracture Mechanics* 28, pp. 425–34.

NAKAMURA, T., SHIH, C. F., AND FREUND, L. B. (1985a), Computational methods based on an energy integral in dynamic fracture, *International Journal of Fracture* 27, pp. 229–43.

NAKAMURA, T., SHIH, C. F., AND FREUND, L. B. (1985b), Elastic-plastic analysis of a dynamically loaded circumferentially notched round bar, *Engineering Fracture Mechanics* 22, pp. 437–52.

NAKAMURA, T., SHIH, C. F., AND FREUND, L. B. (1986), Analysis of a dynamically loaded three-point-bend ductile fracture specimen, *Engineering Fracture Mechanics* 25, pp. 323–39.

NAKAMURA, T., SHIH, C. F., AND FREUND, L. B. (1988), Three-dimensional transient analysis of a dynamically loaded three-point-bend ductile fracture specimen, in *Nonlinear Fracture Mechanics*, Special Technical Publication 995. Philadelphia: American Society for Testing and Materials, pp. 217–41.

NIGAM, H., AND SHUKLA, A. (1988), Comparison of the techniques of transmitted caustics and photoelasticity as applied to fracture, *Experimental Mechanics* 28, pp. 123–33.

NIKOLAEVSKII, V. N. (1980), Dynamics of fracture fronts in brittle solids, *Mechanics of Solids (Mechanika Tverdogo Tela)* 15, pp. 93–102.

NIKOLAEVSKII, V. N. (1981), Limit velocity of fracture front and dynamic strength of brittle solids, *International Journal of Engineering Science* 19, pp. 41–56.

NIKOLIC, R. R., AND RICE, J. R. (1988), Dynamic growth of antiplane shear cracks in ideally plastic crystals, *Mechanics of Materials* 7, pp. 163–73.

NILSSON, F. (1972), Dynamic stress-intensity factors for finite strip problems, *International Journal of Fracture* 8, pp. 403–11.

NILSSON, F. (1973), A path-independent integral for transient crack problems, *International Journal of Solids and Structures* 9, pp. 1107–15.

NILSSON, F. (1974a), Crack propagation experiments on strip specimens, *Engineering Fracture Mechanics* 6, pp. 397–403.

NILSSON, F. (1974b), A note on the stress singularity at a nonuniformly moving crack tip, *Journal of Elasticity* 4, pp. 73–5.

NILSSON, F. (1977a), Steady mode III crack propagation followed by non-steady growth, *International Journal of Solids and Structures* 13, pp. 543–8.

NILSSON, F. (1977b), Sudden arrest of steadily moving cracks, in *Dynamic Fracture Toughness*. Cambridge: The Welding Institute, pp. 249–57.

NILSSON, F., AND BRICKSTAD, B. (1985), Dynamic fracture mechanics – rapid crack growth in linear and nonlinear materials, in *Elastic-Plastic Fracture Mechanics*, Special Technical Publication 803, eds. C. F. Shih and J. P. Gudas. Philadelphia: American Society for Testing and Materials, pp. 427–78.

NILSSON, F., AND STÅHLE, P. (1988), Crack growth criteria and crack tip models, *Solid Mechanics Archives* 13, pp. 193–238.

NISHIOKA, T., AND ATLURI, S. N. (1980a), Numerical modeling of dynamic crack propagation in finite bodies, by moving singular elements – Part I. Formulation, *Journal of Applied Mechanics* 47, pp. 570–6.

NISHIOKA, T., AND ATLURI, S. N. (1980b), Numerical modeling of dynamic crack propagation in finite bodies, by moving singular elements – Part II. Results, *Journal of Applied Mechanics* 47, pp. 577–82.

NISHIOKA, T., AND ATLURI, S. N. (1982), Finite element simulation of fast fracture in steel dcb specimen, *Engineering Fracture Mechanics* 16, pp. 157–75.

NISHIOKA, T., AND ATLURI, S. N. (1983), Path-independent integrals, energy release rates, and general solutions of near tip fields in mixed mode dynamic fracture mechanics, *Engineering Fracture Mechanics* 18, pp. 1–22.

NOBLE, B. (1958), *Methods Based on the Wiener-Hopf Technique*. Elmsford, NY: Pergamon.

NUISMER, R. J., AND ACHENBACH, J. D. (1972), Dynamically induced fracture, *Journal of the Mechanics and Physics of Solids* 20, pp. 203–22.

O'DONOGHUE, P. E., PREDEBON, W. W., AND ANDERSON, C. E. (1988), Dynamic launch process for preformed fragments, *Journal of Applied Physics* 63, pp. 337–48.

OBREIMOFF, J. W. (1930), The splitting strength of mica, *Proceedings of the Royal Society (London)* A127, pp. 290–7.

OGDEN, R. W. (1984), *Non-Linear Elastic Deformations*. Chichester: Horwood.

OH, K. P. L., AND FINNIE, I. (1970), On the location of fracture in brittle solids – II. Due to wave propagation in a slender rod, *International Journal of Fracture Mechancs* 6, pp. 333–9.

OROWAN, E. (1934), Crystal plasticity – III. On the mechanism of the glide process, *Zeitschrift für Physik* 89, pp. 634–44.

OROWAN, E. (1949), Fracture and strength of solids, *Reports of Progress in Physics* 12, pp. 185–232.

OROWAN, E. (1955), Energy criteria of fracture, *Welding Journal Research Supplement* 34, pp. 157–60s.

OSTLUND, S., AND GUDMUNDSON, P. (1988), Asymptotic crack tip fields for dynamic fracture of linear strain-hardening solids, *International Journal of Solids and Structures* 24, pp. 1141–58.

OWEN, D. R. J., AND SHANTARAM, S. (1977), Numerical study of dynamic crack growth by the finite element method, *International Journal of Fracture* 13, pp. 821–37.

PALMER, A. C., AND RICE, J. R. (1973), The growth of slip surfaces in the progressive failure of overconsolidated clay, *Proceedings of the Royal Society (London)* A332, pp. 527–48.

PAPADOPOULOS, M. (1963), Diffraction of plane elastic waves by a crack, with application to a problem of brittle fracture, *Journal of the Australian Mathematical Society* 3, pp. 325–39.

PARKS, D. M., AND FREUND, L. B. (1978), On the gas dynamics of running ductile fracture in a pressurized linepipe, *Journal of Pressure Vessel Technology* 100, pp. 13–7.

PASKIN, A., SOM, D. K., AND DIENES, G. J. (1983), The dynamic properties of moving cracks, *Acta Metallurgica* 31, pp. 1841–8.

PAXSON, T. L., AND LUCAS, R. A. (1973), An experimental investigation of the velocity characteristics of a fixed boundary fracture model, in *Dynamic Crack Propagation*, ed. G. C. Sih. Leyden: Noordhoff, pp. 415–26.

PERZYNA, P. (1963), The constitutive equations for rate sensitive plastic materials, *Quarterly of Applied Mathematics* 20, pp. 321–32.

PIPKIN, A. C. (1984), Crack speeds in an ideal fiber reinforced material, *Quarterly of Applied Mathematics* 42, pp. 267–74.

PONCELET, E. F. (1946), Fracture and comminution of brittle solids, *American Institute of Mining, Metallurgical and Petroleum Engineers* 169, p. 37.

POPELAR, C. H., AND ATKINSON, C. (1981), Dynamic crack propagation in a viscoelastic strip, *Journal of the Mechanics and Physics of Solids* 28, pp. 79–93.

POYNTON, W. A., SHANNON, R. W. E., AND FEARNEHOUGH, G. D. (1974), The design and application of shear fracture propagation studies, *Journal of Engineering Materials and Technology* 96, pp. 323–9.

PRATT, P. L., AND STOCK, T. (1965), The distribution of strain about a running crack, *Proceedings of the Royal Society (London)* A28, pp. 73–82.

PUGH, C. E., AND WHITMAN, G. D. (1987), HSST crack arrest studies overview, *Nuclear Engineering and Design* 98, pp. 141–7.

PUGH, C. E., NAUS, D. J., BASS, B. R., NANSTAD, R. K., DE WIT, R., FIELDS, R. J., AND LOW, S.R. (1988), Wide-plate crack arrest tests utilizing a prototypical pressure vessel steel, *International Journal of Pressure Vessels and Piping* 31, pp. 165–85.

RADER, D. (1967), On the dynamics of crack growth in glass, *Proceedings of the SESA* 24, pp. 160–7.

RADOK, J. R. M. (1956), On the solution of problems of dynamic plane elasticity, *Quarterly of Applied Mathematics* 14, pp. 289–98.

RAJENDRAN, A. M., AND FYFE, I. M. (1982), Inertia effects on the ductile failure of thin rings, *Journal of Applied Mechanics* 49, pp. 31–6.

RAMULU, M., KOBAYASHI, A. S., AND KANG, B. S.-J. (1982), Dynamic crack curving and branching in line-pipe, *Journal of Pressure Vessel Technology* 104, pp. 317–22.

RAMULU, M., AND KOBAYASHI, A. S. (1983), Dynamic crack curving – a photoelastic evaluation, *Experimental Mechanics* 23, pp. 1–9.

RAMULU, M., KOBAYASHI, A. S., KANG, B. S.-J., AND BARKER, D. (1983), Further studies on dynamic crack branching, *Experimental Mechanics* 23, pp. 431–7.

RAVENHALL, F. W. (1977), The application of stress wave emission to crack propagation in metals – a crack propagation model, *Acustica* 37, pp. 307–16.

RAVI-CHANDAR, K., AND KNAUSS, W. G. (1982), Dynamic crack-tip stresses under stress wave loading – a comparison of theory and experiment, *International Journal of Fracture* 20, pp. 209–22.

RAVI-CHANDAR, K., AND KNAUSS, W. G. (1984a), An experimental investigation into dynamic fracture: I. Crack initiation and arrest, *International Journal of Fracture* 25, pp. 247–62.

RAVI-CHANDAR, K., AND KNAUSS, W. G. (1984b), An experimental investigation into dynamic fracture: II. Microstructural aspects, *International Journal of Fracture* 26, pp. 65–80.

RAVI-CHANDAR, K., AND KNAUSS, W. G. (1984c), An experimental investigation into dynamic fracture: III. On steady state crack propagation and crack branching, *International Journal of Fracture* 26, pp. 141–54.

RAVI-CHANDAR, K., AND KNAUSS, W. G. (1984d), An experimental investigation into dynamic fracture: IV. On the interaction of stress waves with propagating cracks, *International Journal of Fracture* 26, pp. 189–200.

RAVICHANDRAN, G., AND CLIFTON, R. J. (1989), Dynamic fracture under plane wave loading, *International Journal of Fracture* 40, pp. 157–201.

REGAZZONI, G., JOHNSON, J. N., AND FOLLANSBEE, P. S. (1986), Theoretical study of the dynamic tensile test, *Journal of Applied Mechanics* 53, pp. 519–28.

RICE, J. R. (1968), Mathematical analysis in the mechanics of fracture, in *Fracture*, Vol. 2, ed. H. Liebowitz. New York: Academic, pp. 191–311.

RICE, J. R., AND ROSENGREN, G. F. (1968), Plane strain deformation near a crack tip in a power law hardening material, *Journal of the Mechanics and Physics of Solids* 16, pp. 1–12.

RICE, J. R., AND TRACEY, D. M. (1969), On the ductile enlargement of voids in triaxial stress fields, *Journal of the Mechanics and Physics of Solids* 17, pp. 201–17.

RICE, J. R. (1970), On the structure of stress–strain relations for time dependent plastic deformation in metals, *Journal of Applied Mechanics* 37, pp. 728–37.

RICE, J. R., AND JOHNSON, M. A. (1970), The role of large crack tip geometry changes in plane strain fracture, in *Inelastic Behavior of Solids*, eds. M. F. Kanninen et al. New York: McGraw-Hill, pp. 641–72.

RICE, J. R. (1972), Some remarks on elastic crack tip stress fields, *International Journal of Solids and Structures* 8, pp. 751–8.

RICE, J. R., AND SORENSEN, E. P. (1978), Continuing crack tip deformation and fracture for plane strain crack growth in elastic-plastic materials, *Journal of the Mechanics and Physics of Solids* 26, pp. 163–86.

RICE, J. R., MCMEEKING, R. M., PARKS, D. M., AND SORENSEN, E. P. (1979), Recent finite element studies in plasticity and fracture mechanics, *Computer Methods in Applied Mechanics and Engineering* 17, pp. 411–42.

RICE, J. R. (1980), The mechanics of earthquake rupture, in *Physics of the Earth's Interior*, ed. E. Boschi. Amsterdam: North Holland, pp. 555–649.

RICE, J. R., DRUGAN, W. J., AND SHAM, T. L. (1980), Elastic-plastic analysis of growing cracks, in *Fracture Mechanics: Twelfth Symposium*, Special Technical Publication 700. Philadelphia: American Society for Testing and Materials, pp. 189–219.

RICHARDS, P. G. (1973), The dynamic field of a growing plane elliptical shear crack, *International Journal of Solids and Structures* 9, pp. 843–61.

RICHARDS, P. G. (1976), Dynamic motions near an earthquake fault: A three-dimensional solution, *Bulletin of the Seismological Society of America* 66, pp. 1–32.

RIPLING, E. J., CROSLEY, P. B., AND WIERSMA, S. J. (1986), A review of static crack arrest concepts, *Engineering Fracture Mechanics* 23, pp. 21–33.

ROBERTS, D. K., AND WELLS, A. A. (1954), The velocity of brittle fracture, *Engineering* 178, pp. 820–1.

ROBERTSON, I. A. (1967), Diffraction of a plane longitudinal wave by a penny shaped crack, *Proceedings of the Cambridge Philosophical Society* 63, pp. 229–38.

ROBERTSON, T. S. (1953), Propagation of brittle fracture in steel, *Journal of the Iron and Steel Institute* 175, pp. 361–74.

ROSAKIS, A. J., AND FREUND, L. B. (1981), The effect of crack tip plasticity on the determination of dynamic stress intensity factors by the optical method of caustics, *Journal of Applied Mechanics* 48, pp. 302–8.

ROSAKIS, A. J., DUFFY, J., AND FREUND, L. B. (1984), The determination of dynamic fracture toughness of AISI 4340 steel by the shadow spot method, *Journal of the Mechanics and Physics of Solids* 32, pp. 443–60.

ROSAKIS, A. J., AND ZEHNDER, A. T. (1985), On the dynamic fracture of structural metals, *International Journal of Fracture* 27, pp. 169–86.

ROSE, L. R. F. (1976a), An approximate (Wiener-Hopf) kernel for dynamic crack problems in linear elasticity and viscoelasticity, *Proceedings of the Royal Society (London)* 349, pp. 497–521.

ROSE, L. R. F. (1976b), Recent theoretical and experimental results on fast brittle fracture, *International Journal of Fracture* 12, pp. 799–813.

ROSE, L. R. F. (1976c), On the initial motion of a Griffith crack, *International Journal of Fracture* 12, pp. 829–41.

ROSE, L. R. F. (1981), The stress wave radiation from growing cracks, *International Journal of Fracture* 17, pp. 45–60.

ROSENFIELD, A. R., AND MAJUMDAR, B. S. (1987), A micromechanical model for cleavage-crack reinitiation, *Metallurgical Transactions* 18A, pp. 1053–9.

ROSSMANITH, H. P., AND SHUKLA, A. (1981), Dynamic photoelastic investigation of interaction of stress waves with running cracks, *Experimental Mechanics*, pp. 415–22.

ROSSMANITH, H. P. (1983), How mixed is dynamic mixed-mode crack propagation? – A dynamic photoelasticity study, *Journal of the Mechanics and Physics of Solids* 31, pp. 251–60.

RUDNICKI, J. W., AND FREUND, L. B. (1981), On energy radiation from seismic sources, *Bulletin of the Seismological Society of America* 71, pp. 583–95.

RUDNICKI, J. W. (1980), Fracture mechanics applied to the Earth's crust, *Annual Reviews in Earth and Planetary Science* 8, pp. 489–525.

RUSSO, R. (1986a), Theorems in linear elastodynamics for plane cracked bodies, *Theoretical and Applied Fracture Mechanics* 6, pp. 187–95.

RUSSO, R. (1986b), On the dynamics of an elastic body containing a moving crack, *Journal of Elasticity* 16, pp. 367–74.

RYDHOLM, G., FREDRIKSSON, B., AND NILSSON, F. (1978), Numerical investigations of rapid crack propagation, in *Numerical Methods in Fracture Mechanics*, eds. A. R. Luxmore and D. J. R. Owen. Swansea: University of Swansea, pp. 660–72.

SANDERS, W. T. (1972), On the possibility of a supersonic crack in a crystal lattice, *Engineering Fracture Mechanics* 4, pp. 145–53.

SCHARDIN, H. (1959), Velocity effects in fracture, in *Fracture*, eds. B. L. Averbach et al. Cambridge, MA: MIT Press, pp. 297–330.

SCHINDLER, H. J. (1981), Path-stability of a crack crossing a beam in bending, *Journal of Applied Mathematics and Physics (ZAMP)* 32, pp. 570–81.

SCHINDLER, H. J., AND KOLSKY, H. (1983), Multiple fractures produced by the bending of brittle beams, *Journal of the Mechanics and Physics of Solids* 31, pp. 427–37.

SCHINDLER, H. J., AND SAYIR, M. (1984), Path of a crack in a beam due to dynamic flexural fracture, *International Journal of Fracture* 25, pp. 95–107.

SCHIRRER, R. (1978), The effects of a strain rate dependent Young's modulus upon the stress and strain fields around a running crack tip, *International Journal of Fracture* 14, pp. 265–79.

SEAMAN, L., CURRAN, D. R., AND SHOCKEY, D. A. (1976), Computational models for ductile and brittle fracture, *Journal of Applied Physics* 47, pp. 4814–26.

SEAMAN, L., CURRAN, D. R., AND MURRI, W. J. (1985), A continuum model for dynamic microfracture and fragmentation, *Journal of Applied Mechanics* 52, pp. 593–600.

552 Bibliography

SECOR, G. (1972), Dynamic fracture of simulated solid propellant, *International Journal of Fracture Mechancs* 8, pp. 299–309.

SEEBASS, A. R. (1983), Functions of a complex variable, in *Handbook of Applied Mathematics*, ed. C. E. Pearson. New York: Van Nostrand Reinhold, pp. 226–70.

SHAH, A. H., WONG, K. C., AND DATTA, S. K. (1986), Dynamic stress intensity factors for buried planar and non-planar cracks, *International Journal of Solids and Structures* 22, pp. 845–57.

SHAH, A. H., CHIN, Y. F., AND DATTA, S. K. (1987), Elastic wave scattering by surface-breaking planar and nonplanar cracks, *Journal of Applied Mechanics* 54, pp. 761–7.

SHAND, E. B. (1954a), Experimental study of fracture of glass: I. The fracture process, *Journal of the American Ceramic Society* 37, pp. 52–60.

SHAND, E. B. (1954b), Experimental study of fracture of glass: II. Experimental data, *Journal of the American Ceramic Society* 37, pp. 559–72.

SHAND, E. B. (1959), Breaking stress of glass determined from dimensions of fracture mirrors, *Journal of the American Ceramic Society* 42, pp. 474–7.

SHANNON, R. W. E., AND WELLS, A. A. (1974), Brittle crack propagation in gas filled pipelines – A model study using thin walled unplasticised PVC pipe, *International Journal of Fracture* 10, pp. 471–86.

SHAR, E. N. (1980), Stress state of a straight isolated cut, loaded from without by concentrated forces and growing at a constant rate, *Journal of Applied Mechanics and Technical Physics (USSR)*, pp. 144–152.

SHMUELY, M., AND ALTERMAN, Z. S. (1973), Crack propagation analysis by finite differences, *Journal of Applied Mechanics* 40, pp. 902–8.

SHMUELY, M., AND PERETZ, D. (1976), Static and dynamic analysis of the dcb problem in fracture mechanics, *International Journal of Solids and Structures* 12, pp. 67–79.

SHMUELY, M. (1977), Analysis of fast fracture and crack arrest by finite differences, *International Journal of Fracture* 13, pp. 443–54.

SHOCKEY, D. A., CURRAN, D. R., SEAMAN, L., ROSENBERG, J., AND PETERSEN, C. F. (1974), Fragmentation of rock under dynamic loads, *International Journal of Rock Mechanics and Mining Science* 11, pp. 303–17.

SHOCKEY, D. A., SEAMAN, L., DAO, K. C., AND CURRAN, D. R. (1980), Kinetics of void growth in fracturing A533B tensile bars, *Journal of Pressure Vessel Technology* 102, pp. 14–21.

SHOCKEY, D. A., KALTHOFF, J. F., AND ERLICH, D. C. (1983), Evaluation of dynamic crack instability, *International Journal of Fracture* 22, pp. 217–29.

SHOCKEY, D. A., KALTHOFF, J. F., KLEMM, W., AND WINKLER, S. (1983), Simultaneous measurements of stress intensity and toughness for fast-running cracks in steel, *Experimental Mechanics* 23, pp. 140–5.

SHOEMAKER, A. K., AND ROLFE, S. T. (1969), Static and dynamic low temperature *K* behavior of steels, *Journal of Basic Engineering* 91, pp. 512–8.

SHOEMAKER, A. K., AND ROLFE, S. T. (1971), The static and dynamic low temperature crack toughness performance of seven structural steels, *Engineering Fracture Mechanics* 2, pp. 319–39.

SHOEMAKER, A. K., AND MCCARTNEY, R. F. (1974), Displacement considerations for a ductile propagating fracture in line pipe, *Journal of Engineering Materials and Technology* 96, pp. 318–22.

SHOEMAKER, A. K., AND SEELEY, R. S. (1983), Summary report of round robin testing by the ASTM Group E24.01.06 on rapid loading plane strain fracture toughness testing, *Journal of Testing and Evaluation* 11, pp. 261–72.

SHUKLA, A., AGARWAL, R. K., AND NIGAM, H. (1988), Dynamic fracture studies of 7075-T6 aluminum and 4340 steel using strain gages and photoelastic coatings, *Engineering Fracture Mechanics* 31, pp. 501–15.

SHUKLA, A., AND CHONA, R. (1988), The stress field surrounding a rapidly propagating curving crack, in *Fracture Mechanics: Eighteenth Symposium*, Special Technical Publication 945, eds. D. T. Read and R. P. Reed. Philadelphia: American Society for Testing and Materials, pp. 86–99.

SIH, G. C. (1968), Some elastodynamic problems of cracks, *International Journal of Fracture* 4, pp. 51–68.

SIH, G. C., AND LOEBER, J. F. (1968), Flexural wave scattering at a through crack in an elastic plate, *Engineering Fracture Mechanics* 1, pp. 369–78.

SIH, G. C., AND LOEBER, J. F. (1969), Normal compression and radial shear waves scattering at a penny shaped crack in an elastic solid, *Journal of the Acoustical Society of America* 46, pp. 711–21.

SIH, G. C. (1970), Dynamic aspects of crack propagation, in *Inelastic Behavior of Solids*, eds. M. F. Kanninen et al. New York: McGraw-Hill, pp. 607–39.

SIH, G. C., AND CHEN, E. P. (1972), Crack propagating in a strip of material under plane extension, *International Journal of Engineering Science* 10, pp. 537–51.

SIH, G. C., EMBLEY, G. T., AND RAVERA, R. S. (1972), Impact response of a finite crack in plane extension, *International Journal of Solids and Structures* 8, pp. 977–93.

SIH, G. C., AND EMBLEY, G. T. (1973), Sudden twisting of a penny shaped crack, *International Journal of Solids and Structures* 9, pp. 1349–59.

SIMONOV, I. V. (1983), Behavior of solutions of dynamic problems in the neighborhood of the edge of a cut moving at transonic speed in an elastic medium, *Mechanics of Solids (Mechanika Tverdogo Tela)* 18, pp. 100–6.

SLEPYAN, L. I. (1976), Crack dynamics in an elastic-plastic body, *Mechanics of Solids (Mechanika Tverdogo Tela)* 11, pp. 126–34.

SLEPYAN, L. I. (1978), Calculation of the size of the crater formed by a high-speed impact, *Soviet Mining Journal* 14, pp. 465–71.

SLEPYAN, L. I. (1981a), Dynamics of brittle fracture in lattice, *Soviet Physics Doklady* 26, pp. 538–40.

SLEPYAN, L. I. (1981b), Crack propagation in high-frequency lattice vibrations, *Soviet Physics Doklady* 26, pp. 900–2.

SLEPYAN, L. I. (1982), Antiplane problem of a crack in a lattice, *Mechanics of Solids (Mechanika Tverdogo Tela)* 16, pp. 101–14.

SLEPYAN, L. I. (1984), Dynamics of brittle fracture in media with a structure, *Mechanics of Solids (Mechanika Tverdogo Tela)* 19, pp. 114–22.

SMITH, E. (1968), Crack bifurcation in brittle solids, *Journal of the Mechanics and Physics of Solids* 16, pp. 329–36.

SMITH, E. (1981), The origin of dynamic effects during the arrest of a propagating crack, *Archive for Mechanics* 33, pp. 313–8.

SMITH, E. (1987), The effective fracture energy associated with cleavage crack growth in b.c.c. iron, *Materials Science and Engineering* 85, pp. 53–8.

SMITH, G. C., AND KNAUSS, W. G. (1975), Experiments on critical stress intensity factors resulting from stress wave loading, *Mechanics Research Communications* 2, pp. 187–92.

SNEDDON, I. N. (1946), The distribution of stress in the neighborhood of a crack in an elastic solid, *Proceedings of the Royal Society (London)* A187, pp. 229–60.

SNEDDON, I. N. (1952), The stress produced by a pulse of pressure moving along the surface of a semi-infinite solid, *Rendiconti del Circolo Matematico di Palermo* 1, pp. 57–62.

SNEDDON, I. N. (1958), Note on a paper by J. R. M. Radok, *Quarterly of Applied Mathematics* 16, p. 197.

SOKOLINSKY, V. B. (1981), On the nature of dynamic strength of brittle solids, *Engineering Fracture Mechanics* 14, pp. 753–8.

SOKOLOWSKI, M. (1977), On a one-dimensional model of the fracture process, *Engineering Transactions (Rozprawy Inzynierskie)* 25, pp. 369–93.

SPENCER, A. J. M. (1960), The dynamic plane deformation of an ideal plastic-rigid solid, *Journal of the Mechanics and Physics of Solids* 8, pp. 262–79.

STAHL, B., AND KEER, L. M. (1972), Vibration and stability of cracked rectangular plates, *International Journal of Solids and Structures* 8, pp. 69–91.

STAKGOLD, I. (1968), *Boundary Value Problems of Mathematical Physics*. Vol. II. New York: MacMillan.

STEPANOV, G. V., AND MAKOVEI, V. A. (1982), Determination of fracture toughness characteristics under a pulsed loading, *Strength of Materials (Problemy Prochnosti)*, pp. 1252–6.

STEPANOV, G. V. (1984), Unsteady crack propagation under dynamic loading, *Strength of Materials (Problemy Prochnosti)*, pp. 1432–6.

STEPANOV, G. V., AND MAKOVEI, V. A. (1984a), Effect of loading rate on the fracture toughness of quenched steel, *Strength of Materials (Problemy Prochnosti)*, pp. 802–6.

STEPANOV, G. V., AND MAKOVEI, V. A. (1984b), Propagation of a rapid crack in annular specimens of brittle steel, *Strength of Materials (Problemy Prochnosti)*, pp. 1255–60.

STERNBERG, E. (1960), On the integration of the equations of motion in the classical theory of elasticity, *Archive for Rational Mechanics and Analysis* 6, pp. 34–50.

STEVENS, A. L., DAVISON, L., AND WARREN, W. E. (1972), Spall fracture in aluminum monocrystals: a dislocation dynamics approach, *Journal of Applied Physics* 43, pp. 4922–7.

STEVERDING, B. (1969), Brittleness and impact resistance, *Journal of the American Ceramic Society* 52, pp. 133–6.

STEVERDING, B., AND LEHNIGK, S. H. (1970), Response of cracks to impact, *Journal of Applied Physics* 41, pp. 2096–9.

STEVERDING, B., AND LEHNIGK, S. H. (1971), Collision of stress pulses with obstacles and dynamics of fracture, *Journal of Applied Physics* 42, pp. 3231–8.

STOCK, T. A. C. (1967), Stress field intensity factors for propagating brittle cracks, *International Journal of Fracture Mechancs* 3, pp. 121–9.

STOCKL, H., AND AUER, F. (1976), Dynamic behavior of a tensile crack: finite difference simulation of fracture experiments, *International Journal of Fracture* 12, pp. 345–57.

STROH, A. N. (1954), The formation of cracks as a result of plastic flow, *Proceedings of the Royal Society (London)* A223, pp. 404–14.

STROH, A. N. (1955), The formation of cracks in plastic flow II, *Proceedings of the Royal Society (London)* A232, pp. 548–60.

STROH, A. N. (1957), A theory of the fracture of metals, *Advances in Physics, Philosophical Magazine Supplement* 6, pp. 418–65.

STROH, A. N. (1960), A simple model of a propagating crack, *Journal of the Mechanics and Physics of Solids* 8, pp. 119–22.

STROH, A. N. (1962), Steady state problems in anisotropic elasticity, *Journal of Mathematics and Physics* 41, pp. 77–103.

SUNG, J. C., AND ACHENBACH, J. D. (1987), Temperature at a propagating crack tip in a viscoplastic material, *Journal of Thermal Stresses* 10, pp. 243–62.

SUZUKI, S., HOMMA, H., AND KUSAKA, R. (1988), Pulsed holographic microscopy as a measurement method of dynamic fracture toughness for fast propagating cracks, *Journal of the Mechanics and Physics of Solids* 36, pp. 631–53.

TADA, H., PARIS, P. C., AND IRWIN, G. R. (1985), *The Stress Analysis of Cracks Handbook*. St. Louis: Del Research Corporation.

TAKAHASHI, K. (1987), Dynamic fracture instability in glassy polymers as studied by ultrasonic fractography, *Polymer Engineering and Science* 27, pp. 25–32.

TAKAHASHI, K., AND ARAKAWA, K. (1987), Dependence of crack acceleration on the dynamic stress intensity factor of polymers, *Experimental Mechanics* 27, pp. 195–200.

TAKEUCHI, H., AND KIKUCHI, M. (1973), A dynamical model of crack propagation, *Journal of Physics of the Earth* 21, pp. 27–37.

TANG, H. C., AND DVORAK, G. J. (1973), The configuration of moving crack tip zones, *Engineering Fracture Mechanics* 5, pp. 79–90.

TAYLOR, J. W., HARLOW, F. H., AND AMSDEN, A. A. (1978), Dynamic plastic instabilities in stretching plates and shells, *Journal of Applied Mechanics* 45, pp. 105–10.

TAYLOR, L. M., CHEN, E. P., AND KUSZMAUL, J. S. (1986), Microcrack induced damage accumulation in brittle rock under dynamic loading, *Computer Methods in Applied Mechanics and Engineering* 55, pp. 301–20.

THAU, S. A., AND LU, T. H. (1971), Transient stress intensity factors for a finite crack in an elastic solid caused by a dilatational wave, *International Journal of Solids and Structures* 7, pp. 731–50.

THOMSON, R. (1986), Physics of fracture, *Solid State Physics* 39, pp. 1–129.

TILLY, G. P., AND SAGE, W. (1970), The interaction of particles and material behavior in erosion processes, *Wear* 16, pp. 447–65.

TIPPER, C. F. (1962), *The Brittle Fracture Story*. Cambridge: Cambridge University Press.

TRUESDELL, C., AND TOUPIN, R. A. (1960), The classical field theories, in *Handbuch der Physik*, Vol. III/1, ed. S. Flügge. Berlin: Springer-Verlag, pp. 226–793.

TSAI, Y. M., AND KOLSKY, H. (1967), A study of the fractures produced in glass blocks by impact, *Journal of the Mechanics and Physics of Solids* 15, pp. 263–78.

TSAI, Y. M. (1973a), Exact stress distribution, crack shape and energy for a running penny shaped crack in an infinite elastic solid, *International Journal of Fracture* 9, pp. 157–69.

TSAI, Y. M., AND CHEN, Y. T. (1978), Propagation of a flexure crack: experiment and analysis, *International Journal of Fracture* 14, pp. 281–91.

TSAI, Y. M. (1981), Initiation and propagation of a penny shaped crack in a finitely deformed incompressible elastic medium, *Engineering Fracture Mechanics* 14, pp. 627–36.

TULER, F. R., AND BUTCHER, B. M. (1968), A criterion for the time dependence of dynamic fracture, *International Journal of Fracture* 4, pp. 431–7.

TURNER, R. D. (1956), The diffraction of a cylindrical pulse by a half plane, *Quarterly of Applied Mathematics* 14, pp. 63–73.

TVERGAARD, V., AND NEEDLEMAN, A. (1988), An analysis of the temperature and rate dependence of Charpy v-notch energies for a high nitrogen steel, *International Journal of Fracture* 37, pp. 197–216.

TVERGAARD, V. (1989), Material failure by void growth to coalescence, in *Advances in Applied Mechanics*, eds. J. W. Hutchinson and T. Y. Wu. New York: Academic, in press.

VAN DER POL, B., AND BREMMER, H. (1955), *Operational Calculus*, 2nd edition. Cambridge: Cambridge University Press.

VAN DER ZWAAG, S., AND FIELD, J. E. (1982a), Liquid jet impact damage on zinc sulphide, *Journal of Materials Science* 17, pp. 2625–36.

VAN DER ZWAAG, S., AND FIELD, J. E. (1983b), Rain erosion damage in brittle materials, *Engineering Fracture Mechanics* 17, pp. 367–79.

VAN DER ZWAAG, S., AND FIELD, J. E. (1983c), Indentation and liquid impact studies on coated germanium, *Philosophical Magazine* A48, pp. 767–77.

VARDAR, O., AND FINNIE, I. (1977), The prediction of fracture in brittle solids subjected to very short duration tensile stresses, *International Journal of Fracture* 13, pp. 115–31.

VASUDEVAN, N., AND KNAUSS, W. G. (1988), An approximate analysis of the effect of micro fractures on the caustic of a dynamically moving crack tip, *International Journal of Fracture* 36, pp. 121–35.

VELIKOTNYI, A. V., AND SMETANIN, B. I. (1975), On the problem of steady vibrations of a plane with a slit, *Applied Mathematics and Mechanics (PMM)* 39, pp. 179–82.

VICKERS, G. W., AND JOHNSON, W. (1972), The development of an impression and the threshold velocity for erosion damage in α-brass and perspex due to repeated jet impact, *International Journal of Mechanical Sciences* 14, pp. 765–77.

VU, B. Q., AND KINRA, V. K. (1981), Brittle fracture of plates in tension: Static field radiated by a suddenly stopping crack, *Engineering Fracture Mechanics* 15, pp. 107–14.

WALTON, J. R., AND NACHMAN, A. (1979), The propagation of a crack by a rigid wedge in an infinite power law viscoelastic body, *Journal of Applied Mechanics* 46, pp. 605–10.

WALTON, J. R. (1982), On the steady-state propagation of an antiplane shear crack in an infinite general linearly viscoelastic body, *Quarterly of Applied Mathematics* 40, pp. 37–52.

WALTON, J. R. (1985), The dynamic steady-state propagation of an antiplane shear crack in a general linearly viscoelastic layer, *Journal of Applied Mechanics* 52, pp. 853–6.

WALTON, J. R. (1987), The dynamic energy release rate for a steadily propagating antiplane shear crack in a linearly viscoelastic body, *Journal of Applied Mechanics* 54, pp. 635–41.

WARD, G. N. (1955), *Linearized Theory of Steady High-Speed Flow*. New York: Cambridge University Press.

WATSON, G. N. (1966), *A Treatise on the Theory of Bessel Functions*. Cambridge: Cambridge University Press.

WEBB, D. (1969), A note on a penny shaped crack expanding under nonuniform internal pressure, *International Journal of Engineering Science* 7, pp. 525–30.

WEIBULL, W. (1939), A statistical theory of the strength of materials, *Ingeniors Vetenskaps Akademien Handlingar*, Number 151.

WEICHERT, R., AND SCHONERT, K. (1974), Heat generation at the tip of a moving crack, *Journal of the Mechanics and Physics of Solids* 22, pp. 127–33.

WEICHERT, R., AND SCHONERT, K. (1978), On the temperature rise at the tip of a fast running crack, *Journal of the Mechanics and Physics of Solids* 26, pp. 151–62.

WEIMER, R. J., AND ROGERS, H. C. (1979), Dynamic fracture phenomena in high-strength steels, *Journal of Applied Physics* 50, pp. 8025–30.

WEINER, J. H., AND PEAR, M. (1975), Crack and dislocation propagation in an idealized crystal model, *Journal of Applied Physics* 46, pp. 2398–405.

WELLS, A. A., AND POST, D. (1958), The dynamic stress distribution surrounding a running crack – a photoelastic analysis, *Proceedings of the SESA* 16, pp. 69–92.

WELLS, A. A. (1961), Influence of residual stresses and metallurgical changes on low stress brittle fracture in welded steel plates, *Welding Journal Research Supplement* 40, pp. 182–92s.

WESENBERG, D. L., AND SAGARTZ, M. J. (1977), Dynamic fracture of 6061-T6 aluminum cylinders, *Journal of Applied Mechanics* 44, pp. 643–6.

WHEELER, L. T., AND STERNBERG, E. (1968), Some theorems in classical elastodynamics, *Archive for Rational Mechanics and Analysis* 31, pp. 51–90.

WHITTAKER, E. T., AND WATSON, G. N. (1927), *Modern Analysis*. Cambridge: Cambridge University Press.

WIEDERHORN, S. M. (1969), Fracture surface energy of glass, *Journal of the American Ceramic Society* 52, pp. 99–105.

WIEDERHORN, S. M., AND HOCKEY, B. J. (1983), Effect of material parameters on the erosion resistance of brittle materials, *Journal of Materials Science* 18, pp. 766–80.

WILLIAMS, D. P. (1973), Relation between hysteresis and the dynamic crack growth resistance of rubber, *International Journal of Fracture* 9, pp. 449–62.

WILLIAMS, J. G. (1987), The analysis of dynamic fracture using lumped mass-spring models, *International Journal of Fracture* 33, pp. 47–59.

WILLIAMS, J. G., AND ADAMS, G. C. (1987), The analysis of instrumented impact tests using a mass-spring model, *International Journal of Fracture* 33, pp. 209–22.

WILLIAMS, M. L. (1957), On the stress distribution at the base of a stationary crack, *Journal of Applied Mechanics* 24, pp. 109–14.

WILLIS, J. R. (1967a), A comparison of the fracture criteria of Griffith and Barenblatt, *Journal of the Mechanics and Physics of Solids* 15, pp. 151–62.

WILLIS, J. R. (1967b), Crack propagation in viscoelastic media, *Journal of the Mechanics and Physics of Solids* 15, pp. 229–40.

WILLIS, J. R. (1973), Self-similar problems in elastodynamics, *Philosophical Transactions of the Royal Society (London)* 274, pp. 435–91.

WILLIS, J. R. (1975), Equations of motion for propagating cracks, *The Mechanics and Physics of Fracture, The Metals Society*, pp. 57–67.

WILSON, M. L., HAWLEY, R. H., AND DUFFY, J. (1980), The effect of loading rate and temperature on fracture initiation in 1020 hot-rolled steel, *Engineering Fracture Mechanics* 13, pp. 371–85.

WINKLER, S., CURRAN, D. R., AND SHOCKEY, D. A. (1970a), Crack propagation at supersonic velocities, *International Journal of Fracture* 6, pp. 151–8.

WINKLER, S., CURRAN, D. R., AND SHOCKEY, D. A. (1970b), Crack propagation at supersonic velocities, *International Journal of Fracture* 6, pp. 271–8.

WRIGHT, T. W. (1969), Impact on an elastic quarter space, *Journal of the Acoustical Society of America* 45, pp. 935–43.

WU, K. C. (1988), Stress intensity factors due to non-elastic strains and body forces for steady dynamic crack extension in an anisotropic elastic material, *International Journal of Solids and Structures* 24, pp. 805–15.

WYSS, M., AND BRUNE, J. N. (1968), Seismic moment, stress, and source dynamics for earthquakes in the California-Nevada region, *Journal of Geophysical Research* 73, pp. 4681–94.

YATOMI, C. (1981), The effect of surface energy on temperature rise around a fast running crack, *Engineering Fracture Mechanics* 14, pp. 759–62.

YATOMI, C. (1982), On Irwin's and Achenbach's expressions for the energy release rate, *International Journal of Fracture* 82, pp. 233–6.

YOFFE, E. H. (1951), The moving Griffith crack, *Philosophical Magazine* 42, pp. 739–50.

ZEHNDER, A. T., AND ROSAKIS, A. J. (1989), Dynamic fracture initiation and propagation in 4340 steel under impact loading, *International Journal of Fracture* 30, in press.

ZENER, C., AND HOLLOMON, J. H. (1944), Effect of strain rate upon plastic flow of steel, *Journal of Applied Physics* 15, pp. 22–32.

ZIMMERMANN, C., KLEMM, W., AND SCHONERT, K. (1984), Dynamic energy release rate and fracture heat in polymethylmethacrylate (pmma) and a high strength steel, *Engineering Fracture Mechanics* 20, pp. 777–82.

ZLATIN, N. A., AND IOFFE, B. S. (1973), Time dependence of the resistance to spalling, *Soviet Physics: Technical Physics* 17, pp. 1390–3.

INDEX

fracture toughness, dynamic, 141
  versus crack speed, 448–50, 465–7
fragmentation, 508–12, 518
fundamental solution, 342, 358
  and superposition, 341, 350–2,
    362–3

gradient
  of a scalar, 14
  of a vector, 15
grazing incidence, 129
Green's function, 65
Green's method of solution, 65–7,
    370
Griffith energy criterion, 4–8

Hadamaard's lemma, 70
Hamilton's principle, 422–4
heat flux vector, 230
Helmholtz decomposition, 26
higher order terms, 58, 94–7, 169–70
hodograph transformation, 453
homogeneous function, 63
homogeneous solutions, method of,
    313
Hooke's law, 22
hoop stress, *see* circumferential ten-
    sile stress
hyperbolic system of partial differ-
    ential equations, 178, 251, 411,
    452

ideally plastic material, 177, 186
identity theorem of analytic func-
    tions, 32
impact strength of brittle material,
    514
incompressibility condition, 188
inertia of crack tip, 398
inertial effects, 2–3, 153–4, 235
influence function, 431–2; *see also*
    weight function
inner product, 13
Irwin fracture criterion, 10, 60, 140
Irwin relationship, 10, 222–3
Irwin-Williams asymptotic stress
    field, 58–9

J-integral, 223, 264
Jones's method, 78
Jordan's lemma, 34
jump conditions, 18, 198, 411

kinetic energy, 8, 18, 225

Lamb's problem, 109
Lamé displacement potentials, 26
  displacement in terms of, 28
  stress in terms of, 28
Lamé elastic constants, 22–3
Laplace transform, 33–6
  Abel asymptotic theorem, 36–7
  convolution formula, 33
  inversion formula, 33, 35
  one-sided transform, 33
  two-sided transform, 35
Laplace's equation, 30
  general solution, 30
linear elastic fracture mechanics, 55
linear elastodynamics, 22–9
linear momentum, balance of, 17
  propagating discontinuity, 18, 199,
    411
Liouville's theorem, 32, 91

mass of crack tip, apparent, 398
maximum plastic work inequality,
    200
microcracking, 508–9
  fragment size, 511
  time-dependent strength, 512
modes of deformation, 38, 57
modulus of cohesion, 308
moving loads on crack faces, 105

Navier's equation, 23
Neumann's uniqueness theorem, *see*
    uniqueness theorem
node release method, 241–2
nonuniform crack tip motion
  mode I stress intensity factor, 390,
    428
  mode II stress intensity factor,
    393
  mode III stress intensity factor,
    374, 375
numerical techniques, *see* computa-
    tional methods

optical measurement methods, 12
overshoot, dynamic, 57, 123

path-independent integral, 223, 229,
    264–9
penny-shaped crack, *see* circular
    crack

Printed in the United States
By Bookmasters